Peter Kramer
Mario Dal Cin (Eds.)

Groups, Systems, and Many-Body Physics

VIEWEG TRACTS IN PURE AND APPLIED PHYSICS

Volume 4

Peter Kramer
Mario Dal Cin (Eds.)

Groups, Systems and Many-Body Physics

with contributions by
A. O. Barut, R. Block, M. Dal Cin, G. G. Emch, M. Hazewinkel, L. Jansen, G. John, P. Kramer, A. Rieckers, R. W. J. Roël, K. Scheerer, and H. Stumpf

Springer Fachmedien Wiesbaden GmbH

CIP-Kurztitelaufnahme der Deutschen Bibliothek

Groups, systems and many-body physics
Peter Kramer; Mario Dal Cin (Eds.). With contributions
by A. O. Barut ... – Braunschweig, Wiesbaden:
Vieweg, 1980.
(Vieweg tracts in pure and applied physics; Vol. 4)
ISBN 978-3-528-08444-8 ISBN 978-3-663-06825-9 (eBook)
DOI 10.1007/978-3-663-06825-9

NE: Kramer, Peter [Hrsg.]; Barut, Asim O. [Mitarb.]

Originally published by Friedr. Vieweg & Sohn Verlagsgesellschaft mbH, Braunschweig in 1980
Softcover reprint of the hardcover 1st edition 1980

Set by Vieweg, Braunschweig

Bookbinder: W. Langelüddecke, Braunschweig
Cover design: Peter Morys, Salzhemmendorf

ISBN 978-3-528-08444-8

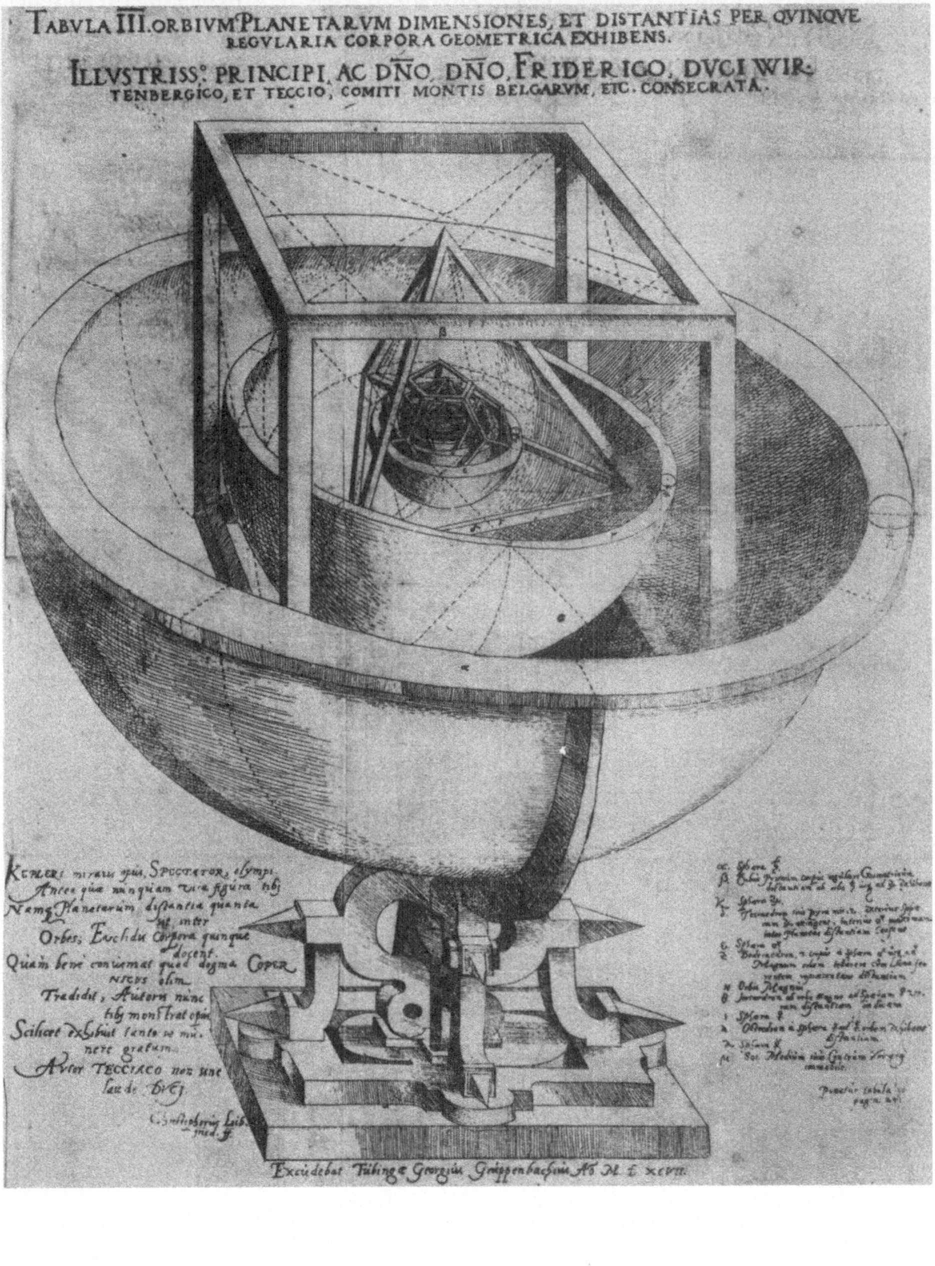
TABVLA III. ORBIVM PLANETARVM DIMENSIONES, ET DISTANTIAS PER QVINQVE REGVLARIA CORPORA GEOMETRICA EXHIBENS.
ILLVSTRISS.° PRINCIPI, AC DÑO, DÑO, FRIDERICO, DVCI WIRTENBERGICO, ET TECCIO, COMITI MONTIS BELGARVM, ETC. CONSECRATA.

The engraving illustrates Johannes Kepler's attempt to relate the distances between spheres of planets to the five platonic bodies. He noted that the cube, tetrahedron, dodecahedron, icosahedron and octahedron could be fitted in between the spheres of Copernicus. This coincidence he considered as a striking proof of the heliocentric hypothesis. Moreover, he proposed to examine the kinematics assumed by Copernicus with respect to shortcomings responsible for deviations from the geometric pattern yielded by the platonic bodies. When Kepler, with the help of his teacher Michael Maestlin, published his construction in 1597 at Tübingen under the title "Mysterium Cosmographicum", he was already working at Graz. At that time, he did not know that he would devote the whole of his life to the examination proposed, which led him to the study of planetary motion in general. The result was a new form of celestial kinematics with consequences perhaps more profound than those of the theory of Copernicus.

Although Kepler's present-day fame rests on his laws of planetary motion, these laws give us only an inkling of the world-view from which they evolved. The directions of his work, as exemplified by his Harmonice Mundi, published in Linz in 1619 and containing the third law, were not well appreciated for a long time to come. He wanted to anchor his results on the ground of symmetries underlying the whole structure of the world. For symmetry he looked in regular and semiregular bodies, in tesselations of plane and space, and in the study of harmonies. He lived in a century when for the first time attention was paid to the geometrical shape of crystals. Facing the problem of hexagonal flakes of snow he tried to solve it by the hypothesis of closest packing of small particles.

The Mysterium Cosmographicum always ranged high in his own opionion. He failed in his hope that this work would open him the way back to Tübingen. After his great discoveries he published a revised edition in 1621 at Frankfurt. Today, Kepler's ideas are alive and give shape to wide areas of present-day physics.

Matthias Schramm, Professor of the History of Science, Tübingen

Preface

The authors of the present book share the view that groups and semigroups play a fundamental role in the structure of the complex systems which they are studying. A serious effort was made to implement this point of view by presenting the fundamental concepts pertaining to groups and semigroups before going into the various fields of application. The first two chapters are written in this spirit. The following seven chapters deal with groups in relation to specific systems and lead from basic notions to high-level applications. The systems under study are in all cases characterized by a high degree of complexity as found in the physics of many degrees of freedom and in the theory of automata and systems.

In 1977 the authors from the University of Tübingen (M. Dal Cin, G. John, P. Kramer, A. Rieckers, K. Scheerer and H. Stumpf) organized an International Summer School on Groups and Many-Body Physics. The lectures presented at this School dealt specifically with this interplay of groups and complex systems. The contributions of this book cover the fields which were treated in a condensed form at the Summer School.

We now give a brief description of the various chapters and their interrelation. In chapter I, M. Dal Cin, G. John, P. Kramer, and K. Scheerer discuss the basic concepts of group and semigroup theory. The exposition is centered around the notions of group action, subgroup structures, topological properties, and representation theory. Various examples of groups are treated for use in other chapters. A. Rieckers presents in chapter II the fundamentals of algebraic quantum theory. This theory has been developed for the study of quantum systems with infinitely many degrees of freedom as encountered in quantum field theory and in non-relativistic many-body systems. Both the physical motivation and the mathematical notions of the theory are introduced. Special attention is given to symmetry transformations. The chapter is also intended to prepare the ground for the analysis of symmetry breaking given by G. G. Emch in chapter V.

In chapter III, L. Jansen, R. W. J. Roël and R. Block expose indirect exchange phenomena in molecules and solids. These authors explain the concept of exchange and its origins in quantum chemistry and solid state physics, develop in full detail the theory of the symmetric groups involved, and discuss specific applications of the theory. Chapter IV by P. Kramer treats composite nucleon systems in a non-relativistic theory. The analysis of exchange phenomena reveals common ground with the methods used by L. Jansen and R. W. J. Roël. The full linear group and groups and semigroups of linear canonical transformations are implemented to describe the many-nucleon system through the interaction of composite particles.

In chapter V, G. G. Emch applies algebraic quantum theory to the mathematically advanced theory of symmetry breaking. The most important concepts, the achievements and the open problems of the theory are presented. The treatment comprises simple and historical examples as well as most recent discoveries.

A. O. Barut describes in chapter VI the structure of composite relativistic systems. The emphasis is on global properties of these systems before an examination of the internal structure. Reducible representations of the Poincaré group and dynamical groups are shown to describe the global properties of composite particles. Representative examples, corresponding wave equations and composite particle interactions are treated in detail.

In chapter VII, H. Stumpf discusses new representation spaces of the Poincaré group. The representations arise from the well-known connection between non-unitary representations of the Poincaré group and an indefinite metric in quantum field theory. The new representation spaces arise by considering these representations in the framework of functional quantum field theory. The motivation for functional quantum field theory and the basic concept of this theory are outlined. Particular attention is given to the role of relativistic symmetry in the problems of unitarization, regularization and for observable quantities in general.

The last two chapters on automata and linear dynamical systems pursue a combined algebraic and system theoretic approach, which might have its relevance also to physics. After all, system theory and physics deal with the same physical and artificial objects and, in fact, dynamical system theory has its origin in Newtonian mechanics. In system theory, however, the holistic point of view dominates. It is germane to system thinking.

Chapter VIII by M. Dal Cin outlines the algebraic theory of automata. This chapter has the double task of introducing the reader to the theory of finite time-discrete information processing systems and of exemplifying the application of group and semigroup theoretic results to system theory. The diagram technique introduced in chapter I proves very helpful in this task. In contrast to physics, emphasis is placed exclusively on finite groups and their structures. No use of representation theory is made. Even so the embarrassing richness of the group structures gives rise to many interesting results.

In chapter IX, M. Hazewinkel exploits the fact that the external (input-output) description of a linear dynamical system is degenerate, much as in atomic physics energy levels may be degenerate. It is shown that there is an internal symmetry group associated with this degeneracy. The chapter is concerned with those aspects of system theory which relate to the existence of this symmetry.

All these contributions deal with diverse systems from a common point of view and are written at a level that makes the book suitable for use as a textbook and as a reference for research. The editors hope that this book with its interdisciplinary spirit is an invitation to new approaches.

Finally, the editors would like to express their dept to Mrs. El-Sheikh and Mrs. Adler for preparing the manuscript. They also thank the Vieweg Verlag and the advisors of this series, Prof. Dr. H. Stumpf and Prof. K. Wildermuth for their cooperation.

The Editors: Peter Kramer
Mario Dal Cin

List of contributing authors

Asim O. Barut, Department of Physics, University of Colorado,
Boulder, Colorado 80309, USA

Ruud Block, Institute of Theoretical Chemistry,
University of Amsterdam, Amsterdam, The Netherlands

Mario Dal Cin, Institut für Informationsverarbeitung,
Universität Tübingen, 7400 Tübingen, Germany

Gérard G. Emch, Departments of Mathematics and Physics,
The University of Rochester, NY 14627, USA

Michiel Hazewinkel, Faculteit der Economische Wetenschappen,
Erasmus University Rotterdam, Rotterdam, The Netherlands

Laurens Jansen, Institute of Theoretical Chemistry,
University of Amsterdam, Amsterdam, The Netherlands

Gero John, Institut für Theoretische Physik,
Universität Tübingen, 7400 Tübingen, Germany

Peter Kramer, Institut für Theoretische Physik,
Universität Tübingen, 7400 Tübingen, Germany

Alfred Rieckers, Institut für Theoretische Physik,
Universität Tübingen, 7400 Tübingen, Germany

Ruud W. J. Roël, Institute of Theoretical Chemistry,
University of Amsterdam, Amsterdam, The Netherlands

Klaus Scheerer, Institut für Theoretische Physik,
Universität Tübingen, 7400 Tübingen, Germany

Harald Stumpf, Institut für Theoretische Physik,
Universität Tübingen, 7400 Tübingen, Germany

Table of Contents

Chapter I

Fundamentals on Semigroups, Groups and Representations

Chapter II

Fundamentals of Algebraic Quantum Theory

(*A. Rieckers*)

Chapter III

Pauli Principle and Indirect Exchange Phenomena in Molecules and Solids

Chapter IV

Groups and Semigroups for Composite Nucleon Systems

(*P. Kramer*)

Chapter V

An Algebraic Approach for Spontaneous Symmetry Breaking in Quantum Statistical Mechanics

(*Gérard G. Emch*)

Chapter VI

Dynamical Groups for the Motion of Relativistic Composite Systems

(*A. O. Barut*)

Chapter VII

New Representation Spaces of the Poincaré Group and Functional Quantum Theory

(H. Stumpf)

Chapter VIII

The Algebraic Theory of Automata

(M. Dal Cin)

Chapter IX

On the (Internal) Symmetry Groups of Linear Dynamical Systems

(*Michiel Hazewinkel*)

Chapter I

Fundamentals on Semigroups, Groups and Representations

P. Kramer, M. Dal Cin, G. John, K. Scheerer

1 Groups and group action

1.1 Preliminaries

We shall use the notion of a set without giving a definition. When we speak in what follows of a set X with elements $x \in X$ we tacitly assume that X is a subset of some bigger set. The elements of X are specified by a property P. Hence a set X with elements x will be specified in the form

$$X = \{x \mid x \text{ has property } P\}$$

Examples: (1) $\mathbb{R} = \{x \mid x \text{ is a real number}\}$
(2) $\mathbb{R}^+ = \{x \mid x \in \mathbb{R} \text{ and } x > 0\}$
(3) $\mathbb{N} = \{0, 1, 2, \ldots\}$
(4) $\emptyset$ = empty set

The number of elements of a set X will be denoted by $|X|$. We recall for given sets A, B the concepts of their union $A \cup B$, intersection $A \cap B$ and difference $B - A$ which are illustrated in Fig. 1.1.

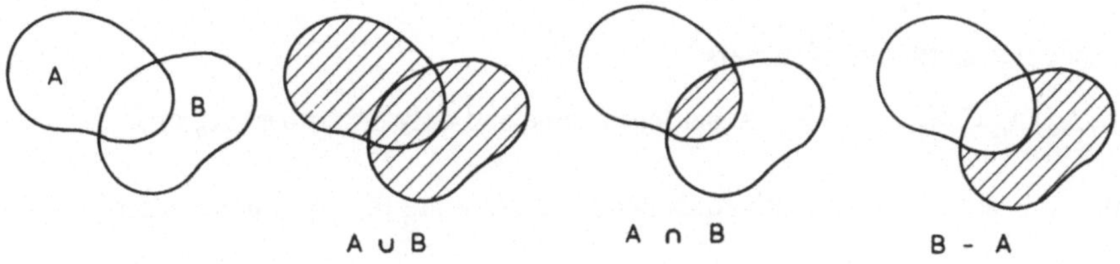

Fig. 1.1 Union, intersection and difference of two sets A and B

The cartesian product $A \times B$ of two sets A, B is the set of ordered pairs (a, b),

$$A \times B = \{(a, b) \mid a \in A \text{ and } b \in B\}$$

Examples: (1) $\mathbb{R} \times \mathbb{R}$: ordered pairs of real numbers
(2) $A = \{x \mid x \in \mathbb{R} \text{ and } c \leqslant x \leqslant d\}$
$B = \{y \mid y \in \mathbb{R} \text{ and } e \leqslant y \leqslant f\}$

The cartesian product $A \times B$ is illustrated in Fig. 1.2.

For B = A the set Diag (A × A) is defined as

$$\text{Diag}(A \times A) = \{(a, a) \mid a \in A\}$$

A relation R on a set A is a subset of A × A.

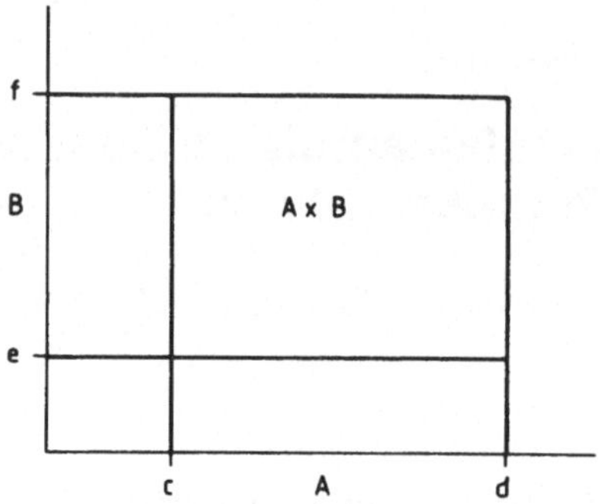

Fig. 1.2
A cartesian product A × B of two sets A, B.

Example: $A = \{x \mid x \in \mathbb{R}\}$
$R = \{(a, b) \mid a \in A, b \in A \text{ and } a - b < 0\}$

For $(a, b) \in R$ we write aRb and call the relation

reflexive iff aRa for all $a \in A$
symmetric iff aRb implies bRa
transitive if aRb, bRc together imply aRc.

Example: The relation $R = \{(a, b) \mid a - b \leqslant 0\}$ on $\mathbb{R}$ is reflexive and transitive, but not symmetric.

1.1 Definition: An equivalence relation E on the set X is a reflexive, symmetric and transitive relation. It is usually written as $a \sim b$ instead of aRb.

1.2 Definition: A partition Π of a set X is a collection of subsets $A_\alpha \subset X$ such that

$$\bigcup_\alpha A_\alpha = X, \quad A_\alpha \cap A_\beta = \emptyset \text{ if } \alpha \neq \beta. \quad A_\alpha \neq \emptyset$$

It may be specified as the set of subsets

$$\Pi = \{A_\alpha \mid A_\alpha \subset X, \quad \text{any } x \in X \text{ belongs to one and only one subset } A_\beta \text{ of } X\}$$

Some important sets associated with equivalence relations are the equivalence class [x] of an element,

$$[x] = \{z \mid z \in X \text{ and } z \sim x\}$$

and the quotient set X/E,

$$X/E = \{C \mid C \subset X \text{ and } C = [x] \text{ for some } x \in X\}$$

The fundamental correspondence between equivalence relations and partitions is expressed by

1.3 Proposition: Any equivalence relation E on a set X yields precisely one partition of X into equivalence classes. Any partition Π of a set X defines an equivalence relation E on X.

We pass now to elementary concepts associated with a function.

1.4 Definition: A function

$$f : A \to B$$

is a subset f of $A \times B$ such that for any $a \in A$ there exists one and only one $b \in B$ such that $(a, b) \in f$. The domain of f is

$$D_f = A,$$

the range R_f of f is

$$R_f = \{b \mid (a, b) \in f \text{ for some } a \in A\}.$$

We write $b = f(a)$ for $(a, b) \in f$ and $R_f = f(A)$.
A function $f: A \to B$ is said to be

injective iff $a \neq a'$ implies $f(a) \neq f(a')$
surjective iff $R_f = f(A) = B$
bijective iff it is injective and surjective.

In the definition of a function, the sets A and B may coincide. The identity function or map is defined as

$$i_A : A \to A, \quad i_A(a) = a \quad \text{for all } a \in A.$$

1.5 Definition: A composition of two functions

$$g: A \to B \qquad f: B \to C$$

is a new function

$$f \circ g : A \to C$$

such that

$$(f \circ g)(c) = f(g(c)).$$

The composition $f \circ g$ is represented by the commutative diagram

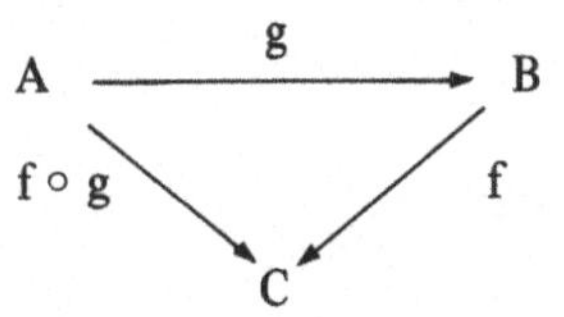

The ranges are indicated in Fig. 1.3.

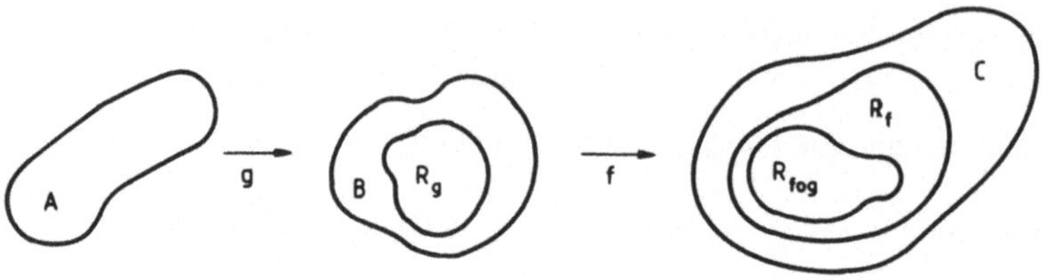

Fig. 1.3 The composition $f \circ g: A \to C$ of two functions $f: B \to C$ and $g: A \to B$

Clearly for the identity function we have

$$i_B \circ f = f \circ i_A = f .$$

1.6 Proposition: The composition of functions is associative, that is, for three functions f, g, h whose domains and ranges allow for the construction of $(f \circ g) \circ h$ and $f \circ (g \circ h)$ we have

$$(f \circ g) \circ h = f \circ (g \circ h) .$$

The commutative diagram for associativity of composition is given by

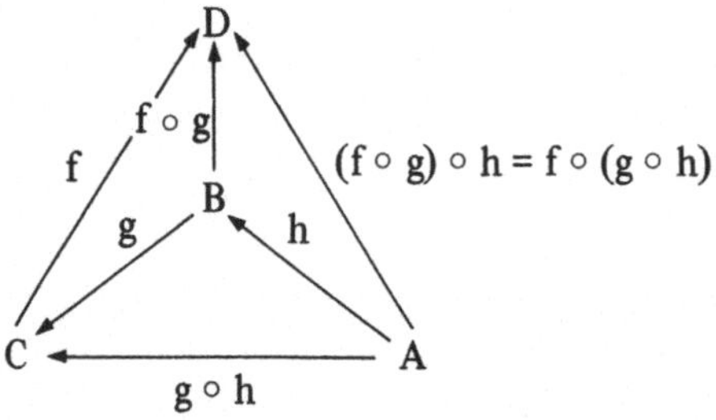

1.2 Algebraic composition, semigroups and groups

We proceed now to the concepts associated with groups and semigroups.

1.7 Definition: An internal binary composition is a function

$$h: A \times A \to A$$

with domain $A \times A$.

1.8 Definition: An external binary composition is a function

$$j: B \times A \to A$$

with domain $B \times A$.

We write a general internal binary composition with $h(a', a'') = a$ in the form

$$a' \nabla a'' = a$$

and denote by (A, ∇) a set with a composition ∇.

1.9 Definition: A unit element e in a set (A, ∇) is an element $e \in A$ such that

$$e \nabla a = a \nabla e = a \quad \text{for all } a \in A .$$

If a unit element e exists in a set (A, ∇) it is easily shown to be uniquely defined.

1.10 Definition: For a set (A, ∇) with unit element and given $a \in A$, any element b fulfilling

$$b \nabla a = a \nabla b = e$$

is called the inverse of a and is denoted as $b = a^{-1}$.

1.11 Definition: The composition ∇ on A is called

commutative iff $a' \nabla a'' = a'' \nabla a'$ for all $a', a'' \in A$
associative iff $(a' \nabla a'') \nabla a''' = a' \nabla (a'' \nabla a''')$ for all $a', a'', a''' \in A$.

The concepts appearing in the last three definitions serve to introduce the following important sets with internal composition.

1.12 Definition: A semigroup is a set with an associative composition. A monoid is a set with an associative composition and a unit element. A group is a set with associative composition, unit element and an inverse for every element.
Given sets with composition laws, we may consider those functions which are compatible with an internal composition.

1.13 Definition: A morphism of two sets with algebraic compositions (A, ∇) and (A', ∇') is a function

$$f: (A, \nabla) \rightarrow (A', \nabla')$$

such that

$$f(a \nabla b) = f(a) \nabla' f(b) .$$

A morphism is called

monomorphic if it is injective,
epimorphic if it is surjective,
isomorphic if it is bijective.

If

$$f: (A, \nabla) \rightarrow (A', \nabla')$$

is an epimorphism, we call (A', ∇') the homomorphic image of (A, ∇).

In particular, our interest is in groups and we briefly define the following special terms and concepts for groups. For a group G with elements $g', g'' \in G$ we write the composition as $g'g''$.

1.14 Definition: Two elements $g', g'' \in G$ are conjugate if

$$g' = g\, g''\, g^{-1} \quad \text{for some } g \in G\,.$$

1.15 Definition: A subgroup H of G is a subset of G which is a group under the same composition. We write the subgroup property as $H \leqslant G$ or $H < G$ (if $H \neq G$).

1.16 Definition: For $H \leqslant G$, the left cosets of H in G are the sets gH and the right cosets are the sets Hg.

1.17 Proposition: The left cosets of a subgroup $H \leqslant G$ are a partition of G.

Proof: The identity of two cosets implies

$$gH = g'H \quad \text{or} \quad g'^{-1} g \in H\,.$$

This relation for g', g is easily shown to be an equivalence relation and hence from proposition 1.3 yields a partition of G. ■

An important generalization of cosets are the double cosets to be discussed in section 1.3.

1.14′ Definition: Two subgroups $H \leqslant G$ and $H' \leqslant G$ are conjugate if $H' = gHg^{-1}$ for some $g \in G$.

1.15′ Definition: A subgroup $N \leqslant G$ is normal or invariant if $gNg^{-1} = N$ for all $g \in G$. We write this property as $N \trianglelefteq G$.

For a finite group G we denote its order as $|G|$.

1.3 Transformation groups

In this section we consider the action of a group on a set. The concepts associated with the action provide a unified viewpoint for the structure of groups.

1.18 Definition: Let X be a set and T be a subset of bijections of X, $T \times X \to X$, such that

(a) $i_X \in T$
(b) if $t \in T$ then $t^{-1} \in T$
(c) if $t_\alpha \in T$ and $t_\beta \in T$ then $t_\alpha \circ t_\beta \in T$.

1.19 Proposition: The subset T of bijections of X is a group called a transformation group acting on the set X.

1.20 Definition: For a set X with transformation group T, the orbit Tx of T at the element x is the set

$$Tx = \{x' \mid x' = tx \text{ for some } t \in T\}.$$

1.21 Proposition: On a set X with transformation group T, the relation

$$R: x' R x \text{ iff } x' = tx \text{ for some } t \in T$$

is an equivalence relation and leads to a partition of X into orbits under T.

1.22 Definition: A set X is a homogeneous space under T, or T acts transitively on X, iff X consists of a single orbit under T.

1.23 Definition: For (T, X), the isotropy or stability group $T_0 \leqslant T$ at $x_0 \in X$ is the subgroup of T given by

$$T_0 = \{t \mid t x_0 = x_0,\ t \in T\}.$$

We now show that various properties of a group may be recognized in terms of group actions.

Example 1.1: Let T = G act on X = G according to

$$\begin{aligned} &T \times X \to X \\ &(g, g') \to g g'. \end{aligned}$$

This action is easily shown to yield any group G as a transformation group which is known as Cayley's theorem. The orbit of T at g' is Gg' = G, hence X = G is a homogeneous space for T = G.

Example 1.2: Let T = G act on X = G according to

$$\begin{aligned} &T \times X \to X \\ &(g, g') \to g g' g^{-1}. \end{aligned}$$

Again this action yields a transformation group on G. The orbit of T = G at $g' \in X = G$ is the conjugacy class of g', the collection of these orbits yields the partition of G into conjugacy classes. The isotropy group at g_0' is called the normalizer $N(g_0')$ of g_0' in G.

Example 1.3: Let $T = H \leqslant G$ act on X = G according to

$$\begin{aligned} &T \times X \to X \\ &(h, g') \to h g'. \end{aligned}$$

Under this action, the orbit of T = H at $g' \in X = G$ is the right coset H g'. As mentioned in proposition 1.17, the right cosets partition G. The isotropy group at $g_0' \in X = G$ consists of the identity element e.

Example 1.4: Let $H \leqslant G, K \leqslant G$ be subgroups of G and let $T = H \times K$ act on $X = G$ according to

$$(H \times K) \times X \rightarrow X$$
$$((h, k), g') \rightarrow h\, g'\, k^{-1}.$$

Under this transformation group, the orbit at $g' \in X = G$ is the set $H\, g'\, K$ called the double coset of $g' \in G$. To compute the isotropy group at g' we proceed as follows: For $(h, k) \in T_0$ we must have $h\, g'\, k^{-1} = g'$ or

$$h\, g'\, k^{-1}\, (g')^{-1} = e.$$

This requires that k be given in terms of h as

$$g' k^{-1} (g')^{-1} = h^{-1}.$$

Moreover for this equation to hold, h must be an element of the subgroup $H \cap g'K(g')^{-1} \leqslant G$. Hence

$$T_0 = \{(h, k) \mid h \in H \cap g'K(g')^{-1},\ k = (g')^{-1}\, h\, g'\}.$$

In other words, T_0 is isomorphic to $H \cap g'K(g')^{-1}$.

The equivalence relation between elements in the same double coset yields a partition of the group G into double cosets.

Example 1.5: Let $T = G$ act on the set X of all left cosets of a subgroup $H \leqslant G$ according to

$$T \times X \rightarrow X$$
$$(g, g'H) \rightarrow g\, g'H.$$

This action is easily seen to yield $T = G$ acting as a transformation group on the coset space X. The orbit of $T = G$ at $x = g'H$ is $G\, g'H = G\, H = G$, hence X is a homogeneous space under G. The isotropy group T_0 at $x_0 = g_0'\, H$ is defined by

$$T_0 = \{g \mid g\, g_0'\, H = g_0'\, H\}$$

and is found to be $g_0'\, H\, (g_0')^{-1}$.

Example 1.5 provides for each pair (G, H) with $H \leqslant G$ a transitive action of G on the left coset space which we shall denote as G/H. The converse of this result is

1.24 Proposition: Any homogeneous space X under a transformation group $T = G$ is in one-to-one correspondence to a coset space of G with an appropriate subgroup $H \leqslant G$.

Proof: For the given homogeneous space we choose a reference element x_0 with isotropy group T_0. Since X is homogeneous there is an element $g' \in G$ such that $g'x = x_0$ for any $x \in X$. Define the function

$$f : X \rightarrow G/T_0$$
$$f(x) = g'T_0.$$

The function f is found to be injective and surjective and moreover commutes with the action of $T = G$ on the set X according to the commutative diagram

$$\begin{array}{ccc} x & \xrightarrow{\ f\ } & g'T_0 \\ {\scriptstyle g}\downarrow & & \downarrow{\scriptstyle g} \\ gx & \xrightarrow[\ f\ ]{} & gg'T_0 \end{array}$$

Finally, any change of the reference element x_0 yields an isotropy group conjugate and hence isomorphic to T_0. ■

2 Examples of groups

2.1 The symmetric group S (n)

2.1 Definition: The symmetric group $S(n)$ is the set of all bijections of a set X with n elements.

The bijections of X we call permutations p. A specific permutation will be indicated by

$$p = \begin{bmatrix} p(s) \\ s \end{bmatrix} = \begin{bmatrix} p(1) & p(2) & & p(n) \\ 1 & 2 & \ldots & n \end{bmatrix}$$

$$= \begin{bmatrix} 1 & 2 & \ldots & n \\ p^{-1}(1) & p^{-1}(2) & & p^{-1}(n) \end{bmatrix}$$

and the product will be taken by the composition of functions,

$$(p_1 p_2)(i) = (p_1 \circ p_2)(i) = p_1(p_2(i)) \qquad i = 1, 2, \ldots, n$$

2.2 Proposition: The group $S(n)$ acts transitively on the set X, X is in one-to-one correspondence to the coset space $S(n) / S(n-1)$.

Proof: Choose in X the element x_n labelled by the number n. Its isotropy group according to definition 1.23 is the group $S(n-1) < S(n)$. Since X is easily seen to be a homogeneous space, it is by proposition in one-to-one correspondence to the coset space $S(n) / S(n-1)$. ■

Example: The left cosets of $S(2) < S(3)$ are

$$x_1 = \begin{bmatrix} 2 & 3 & 1 \\ 1 & 2 & 3 \end{bmatrix}, \begin{bmatrix} 3 & 2 & 1 \\ 1 & 2 & 3 \end{bmatrix}$$

$$x_2 = \begin{bmatrix} 1 & 3 & 2 \\ 1 & 2 & 3 \end{bmatrix}, \begin{bmatrix} 3 & 1 & 2 \\ 1 & 2 & 3 \end{bmatrix}$$

$$x_3 = \begin{bmatrix} 1 & 2 & 3 \\ 1 & 2 & 3 \end{bmatrix}, \begin{bmatrix} 2 & 1 & 3 \\ 1 & 2 & 3 \end{bmatrix}$$

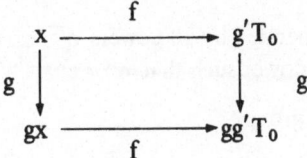

Computation of the action of S(3) according to example 1.5 yields

$$p\,x_i = x_{p(i)} \qquad i = 1, 2, 3\,.$$

2.3 Definition: The cyclic group C(n) is the group generated by all powers q^m of a single element q with $m \leqslant n$ and n being the smallest power such that $q^n = e$.

Clearly any element of a finite group generates a cyclic group.

2.4 Definition: The cycle of a permutation $p \in S(n)$ is the orbit on X under the cyclic group C_p generated by p.

2.5 Proposition: The classes of conjugate elements of S(n) are in one-to-one correspondence to the partitions $\lambda = [\lambda_1 \lambda_2 \ldots \lambda_j]$ of the integer n with the restriction $\lambda_1 \geqslant \lambda_2 \geqslant \ldots \geqslant \lambda_j > 0$.

Proof: For $p' \in S(n)$, we define the numbers of the elements of X sitting in the orbits under $C_{p'}$ as $\lambda_1 \lambda_2 \ldots \lambda_j$. Under conjugation of p' with $p \in S(n)$ we have $C_{pp'p^{-1}} = p\, C_{p'} p^{-1}$, hence the numbers λ_i are class functions.
Conversely assume that for two permutations $p', p'' \in S(n)$ the orbits on X under $C_{p'}$ and $C_{p''}$ respectively may be paired into orbits of equal length. If a bijection $f: X \to X$ is defined by $f(i) = j$ for i and j sitting in the same position of paired orbits, the bijection f connects p' and p'' according to $p' = f \circ p \circ f^{-1}$. Hence p' and p'' are in the same class of conjugate elements. ∎

Consider subgroups of S(n) of the type

$$H = S(\tilde{w}_1) \times S(\tilde{w}_2) \times \ldots \times S(\tilde{w}_{\tilde{j}}) \qquad \tilde{w}_1 + \tilde{w}_2 + \ldots + \tilde{w}_{\tilde{j}} = n$$
$$K = S(w_1) \times S(w_2) \times \ldots \times S(w_j) \qquad w_1 + w_2 + \ldots + w_j = n$$

2.6 Definition: A double coset symbol k for the weights $\tilde{w} = (\tilde{w}_1 \tilde{w}_2 \ldots \tilde{w}_{\tilde{j}})$, $w = (w_1 w_2 \ldots w_j)$ is a $\tilde{j} \times j$ matrix $k = (k_{i\,l})$ with integer non-negative elements fulfilling

$$\sum_{l=1}^{j} k_{i\,l} = \tilde{w}_i, \quad i = 1\,2 \ldots \tilde{j} \qquad \sum_{i=1}^{\tilde{j}} k_{i\,l} = w_l, \quad l = 1\,2 \ldots j$$

2.7 Proposition: The double cosets of (S(n), H, K) are in one-to-one correspondence to the double coset symbols k.
For the proof and more detailed analysis we refer to Kramer and Seligman [4].

Example: For (S(6), S(4) × S(2), S(3) × S(3)) the double coset symbols are

$$\begin{Bmatrix} 3 & 1 \\ 0 & 2 \end{Bmatrix}, \quad \begin{Bmatrix} 2 & 2 \\ 1 & 1 \end{Bmatrix}, \quad \begin{Bmatrix} 1 & 3 \\ 2 & 0 \end{Bmatrix}$$

These three double cosets admit the construction of generators z_k which are given in the same order by

$$z_1 = \begin{bmatrix} 1 & 2 & 3 & | & 4 & | & 5 & 6 \\ 1 & 2 & 3 & | & 4 & | & 5 & 6 \end{bmatrix}$$

$$z_2 = \begin{bmatrix} 1 & 2 & | & 3 & | & 4 & 5 & | & 6 \\ 1 & 2 & | & 4 & & 5 & | & 3 & | & 6 \end{bmatrix}$$

$$z_3 = \begin{bmatrix} 1 & | & 2 & 3 & | & 4 & 5 & 6 \\ 1 & | & 4 & 5 & 6 & | & 2 & 3 \end{bmatrix}$$

The construction of these permutations runs as follows: Reading the columns of the second symbol from left to right yields the numbers 2, 1, 2, 1. These numbers are used to separate the top line of z_2 by bars and to identify four subsets of numbers. Reading now the rows of the second symbol from top to bottom yields the numbers 2, 2, 1, 1. These numbers are used to separate the bottom line of z_2 and to interchange those subsets of the top line which were associated with the permuted numbers 1, 2. For the proof of this assertion compare [7].

2.2 Matrix (semi-)groups and inner products [5]

2.8 Definition: Consider the set of all linear invertible operators from $\mathbb{R}^m$ to $\mathbb{R}^m$ or from $\mathbb{C}^m$ to $\mathbb{C}^m$. When the operators are written as matrices with respect to some basis, these sets are called the general linear (real or complex) group $GL(m, \mathbb{R})$ or $GL(m, \mathbb{C})$ respectively. The restriction of these groups to subgroups of determinant equal to one are called unimodular groups $SL(m, \mathbb{R})$ and $SL(m, \mathbb{C})$.

More structure is introduced into these linear spaces by the concept of an inner product.

2.9 Definition: A non-degenerate inner product on $\mathbb{C}^m$ is a map $\langle , \rangle$

$$\langle , \rangle : \mathbb{C}^m \times \mathbb{C}^m \to \mathbb{C}$$
$$(x, y) \to \langle x, y \rangle$$

such that for all $x, y, z \in \mathbb{C}^m$, $\lambda \in \mathbb{C}$

(1) $\langle x, \lambda y \rangle = \lambda \langle x, y \rangle$

(2) $\langle x, y + z \rangle = \langle x, y \rangle + \langle x, z \rangle$

(3) $\langle x, y \rangle = 0$ for all $x \in \mathbb{C}^m$ implies $y = 0$

2.10 Definition: An inner product is called

sesquilinear if (1) $\langle \mu y, x \rangle = \bar{\mu} \langle y, x \rangle$

(2) $\langle y + z, x \rangle = \langle y, x \rangle + \langle z, x \rangle$

bilinear if (1)′ $\langle \mu y, x \rangle = \mu \langle y, x \rangle$

(2)′ $\langle y + z, x \rangle = \langle y, x \rangle + \langle z, x \rangle$

sesquilinear hermitian if sesquilinear and

$$(3) \quad \langle y, x\rangle = \overline{\langle x, y\rangle}$$

bilinear symmetric if bilinear and

$$(3)' \quad \langle y, x\rangle = \langle x, y\rangle$$

bilinear antisymmetric if bilinear and

$$(3)'' \quad \langle y, x\rangle = -\langle x, y\rangle$$

2.11 Proposition: The set G of linear operators on $\mathbb{C}^m$ fulfilling for all $x, y \in \mathbb{C}^m$ and an inner product $\langle , \rangle$

$$\langle gx, gy\rangle = \langle x, y\rangle$$

is a group called the group of linear isometries of $\langle , \rangle$.
By introduction of standard bases for the various inner products one arrives at a definition of important groups of isometries. These are displayed in the following table.

Standard inner product $\langle , \rangle$	Group of isometries G	
$\sum_1^r x_i y_i - \sum_{r+1}^{r+s} x_i y_i$	Orthogonal group	$O(r, s, \mathbb{R})$
$\sum_1^q x_i y_{i+q} - \sum_1^q x_{i+q} y_i$	Symplectic group	$Sp(2q, \mathbb{C}), Sp(2q, \mathbb{R})$
$\sum_1^r \bar{x}_i y_i - \sum_{r+1}^{r+s} \bar{x}_i y_i$	Unitary group	$U(r, s)$

Among these matrix groups there are groups of great physical interest like the rotation group $O(3, \mathbb{R})$, the Lorentz group $O(3, 1, \mathbb{R})$, the proper unitary group $U(m, 0)$ and the group of linear canonical transformations in phase space $Sp(2q, \mathbb{R})$.
An inner product may also be associated with a semigroup as shown in the following example.

2.12 Proposition: For $\langle , \rangle$ sesquilinear and hermitian on $\mathbb{C}^m$, the sets

$$U^{>}(r, s) = \{h \mid \langle hx, hx\rangle - \langle x, x\rangle > 0 \text{ for all } x \in \mathbb{C}^m\}$$
$$U^{\geqslant}(r, s) = \{h \mid \langle hx, hx\rangle - \langle x, x\rangle \geqslant 0 \text{ for all } x \in \mathbb{C}^m\}$$

are semigroups which we shall call strictly length-increasing or length-increasing semigroups.

Proof: To show the semigroup properties it suffices to show that the product of two length-increasing transformations is length-increasing. ■

Note that $U^{>}(r, s)$ has no unit element while $U^{\geqslant}(r, s)$ contains the group $U(r, s)$.

2.3 Inhomogeneous matrix groups

The matrix groups considered in section 2.2 may be augmented by the inclusion of the translation group A on $\mathbb{C}^m$ or $\mathbb{R}^m$ to yield inhomogeneous groups of transformation.

2.13 Definition: For G being a matrix group acting on $\mathbb{C}^m$ and A being the translation group, the inhomogeneous group I G is defined as

$$A \times G \times \mathbb{C}^m \to \mathbb{C}^m$$
$$(a, g, x) \to g x + a\,.$$

In this extended notation, $I0(3, \mathbb{R})$ is the Euclidean group acting on $\mathbb{R}^3$, $I0(3, 1, \mathbb{R})$ is the Poincaré-group acting on Minkowski space.
The composition law of the group I G is obtained from its action as

$$\{a', g'\}\,\{a'', g''\} = \{a' + g' a'', g' g''\}\,.$$

The unit element is

$$\{0, e\}$$

and the inverse is given by

$$\{a, g\}^{-1} = \{-g^{-1} a, g^{-1}\}\,.$$

From the multiplication rule one finds

$$\{a', g''\}\,\{a, e\}\,\{a', g'\}^{-1} = \{g' a, e\}$$

which proves that the translation subgroup is an invariant subgroup of I G.

2.14 Proposition: Any inhomogeneous matrix group I G acts transitively on $\mathbb{C}^m$, the corresponding coset space is I G/G and equivalent of the translation group A.

Proof: The transitive property is obvious from the presence of the translation group. The isotropy group of the point 0 of $\mathbb{C}^m$ is the group G and hence from proposition 1.24, the coset space is I G/G. ∎

2.15 Proposition: The inhomogeneous matrix group I G is a semidirect product of its normal subgroup A of translations and the subgroup of homogeneous transformations G.

Proof: Any element $\{a, g\}$ of I G may be written as a product

$$\{a, g\} = \{a, e\}\,\{0, g\}$$

where the first factor belongs to the subgroup A and the second one to the subgroup G.

These two subgroups have only the identity element in common. Writing the product of two elements as

$$\begin{aligned}
&\{a', g'\}\,\{a'', g''\} \\
&= \{a', e\}\,\{0, g'\}\,\{a'', e\}\,\{0, g''\} \\
&= \{a', e\}\,\{0, g'\}\,\{a'', e\}\,\{0, (g')^{-1}\}\,\{0, g'\}\,\{0, g''\} \\
&= \{a', e\}\,\{g'a'', e\}\,\{0, g'g''\}
\end{aligned}$$

shows that G induces an automorphism $A \to A$ as required in the definition of semidirect products given in section 3.4. ∎

3 Subgroup structures of groups and semigroups

This section provides an introduction to the structure theory of finite and infinite groups. The basic concept is that of a subgroup diagram. A condensed but still elementary theory of subgroup diagrams will be given. A few remarks on the structure of finite semigroups are made at the end of this section.

3.1 Preliminaries: complexes and lattices

Let $(G, \cdot)$ be a group. In the following we denote $(G, \cdot)$ by G and omit the multiplication point; e is always the neutral element of G. A set with only one element and this element will be identified, e.g., e denotes the neutral element of G, its subgroup $(\{e\}, \cdot)$ and the subset $\{e\}$. Subsets of G are called *complexes.* For any two complexes A and B of G we define a product $AB = \{ab \mid a \in A, b \in B\}$. The inverse of a complex A is $A^{-1} = \{a^{-1} \mid a \in A\}$. The subgroup of G generated by a complex A (Erzeugnis) is $\langle A \rangle = \cap \{U \mid A \subseteq U \leqslant G\}$ and $\langle A, B \rangle := \langle A \cup B \rangle$. If N is a normal subgroup of G (i.e., $N \trianglelefteq G$) then the set of cosets G/N is a group with respect to the multiplication of complexes; G/N is the *factor group of* G *modulo* N.

3.1 *Exercise:* Show, that $A \leqslant G$ is normal in G iff $(gA)(g'A) = (gg')A$ for all $g, g' \in G$.

3.2 Some simple facts: Let A, B be subgroups of G. Then

(1) $A \cap B \leqslant G$,
(2) $\langle A, B \rangle \leqslant G$,
(3) $AB \subseteq \langle A, B \rangle$,
(4) $|AB| = \dfrac{|A|\,|B|}{|A \cap B|}$, if A and B are finite,
(5) $AB = \langle A, B \rangle$ iff $AB = BA$,
(6) $\langle A, N \rangle = AN$ if $N \trianglelefteq A$,
(7) $\langle U, N \rangle = UN \trianglelefteq G$ if $U, N \trianglelefteq G$.

Only (5) needs a proof.

Proof (5): A complex $U \subset G$ is a subgroup of G iff $UU = U$ and $U^{-1} = U$. Hence,

(a) if $AB \leqslant G$ then $AB = (AB)^{-1} = B^{-1} A^{-1} = BA$,

(b) if $AB = BA$ and $A \leqslant G$, $B \leqslant G$, then $(AB)(AB) = A(BA)B = A(AB)B = (AA)(BB) = AB$ and $(AB)^{-1} = B^{-1}A^{-1} = BA = AB$.

Thus, $\langle A, B\rangle \subseteq AB$, since $A \cup B \subseteq AB \leqslant G$.
Hence (3): $\langle A, B\rangle = AB$. ■

The kind of mathematical structure to be studied involves the notion of a lattice. A *lattice* is a set V with two commutative and associative (binary) operations $\sqcap$ and $\sqcup$ such that for all $u, v \in V$ the relations (1) $u \sqcap (u \sqcup v) = u$ and (2) $u \sqcup (u \sqcap v) = u$ hold. Power sets and Boolean algebras are familiar examples of lattices. In this chapter we shall study the properties of subgroup lattices.

3.3 Theorem/Definition: Let G be a group. The set of all subgroups (A, B ect.) of G together with the two operations $\langle A, B\rangle$ and $A \cap B$ is a lattice. This lattice is denoted by $\mathcal{U}(G) = \{A \mid A \leqslant G\}$.

Proof: Let $A, B \in \mathcal{U}(G)$.

(1) $A \cap \langle A, B\rangle = A \cap \langle A \cup B\rangle = A$ since $A \leqslant \langle A \cup B\rangle$.

(2) $\langle A, (A \cap B)\rangle = \langle A \cup (A \cap B)\rangle = \langle A\rangle = A$.

Both operations are commutative and associative ($A \cap B$ corresponds to $A \sqcap B$ and $\langle A, B\rangle$ to $A \sqcup B$). ■

The physical motivation for studying subgroup lattices is the interest in symmetry breaking mechanisms. Many physical interactions lower the symmetry of a physical system from the original symmetry group G to a subgroup and, furtherdown, to subgroups of subgroups of G. Therefore, it is advantageous to have some understanding of the subgroup lattice $\mathcal{U}(G)$ before tackling the complexities of these interactions. Furthermore, as we have seen in Sec. 1, the homogeneous spaces of a symmetry G can be distinguished by an examination of $\mathcal{U}(G)$. In system theory, subgroup lattices are closely related to the physical structures of certain input-output systems. In the following we shall investigate the most important substructures of subgroup lattices.

3.2 Subgroup diagrams

The subgroup diagram of a group G consists of three types of building elements and two conventions [1]. The building elements are points, edges, and double edges.

(1) Points represent groups (A, B, C, etc.).

(2) Edges represent the subgroup relation ($A \leqslant B$).

(3) Double edges represent the normal subgroup relation ($A \trianglelefteq B$).

3.4 *Example:* The subgroup diagram of Fig. 3.1 has the following interpretation: $B \trianglelefteq A$, $C \leqslant A$, $D \leqslant B$ and $D \leqslant C$. It follows that $\langle B, C\rangle \leqslant A \leqslant G$ and $B \cap C \geqslant D \geqslant e$.
Of course: $e \leqslant D \leqslant A$ and $e \leqslant C$ (not shown).

First convention: Trivial relations are (sometimes) not shown in the diagrams.

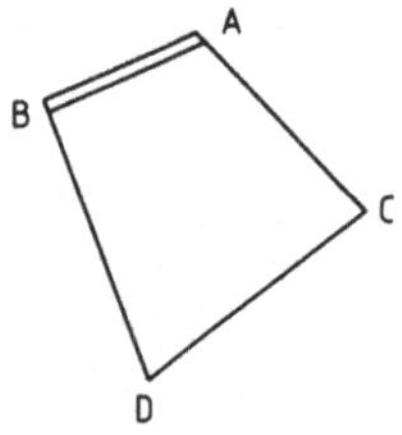

Fig. 3.1

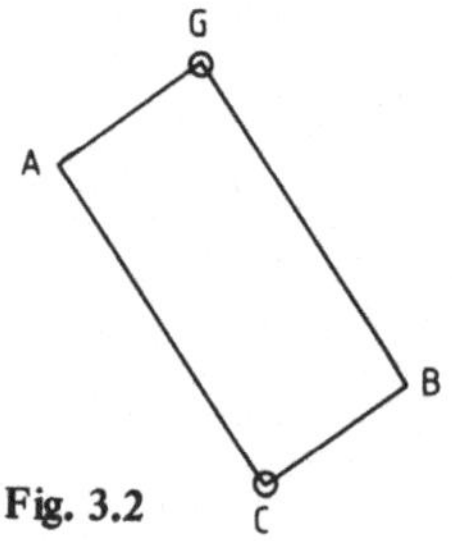

Fig. 3.2

Second convention: A parallelogram as in Fig. 3.2 means that $G = AB = BA$ (hence, $G = \langle A, B\rangle$) and $C = A \cap B$. (The length of an edge $_B{-}A$ can be interpreted as the number of cosets of A modulo B, if G is finite. This justifies the notion of a parallelogram).

Two subgroups A, B are called *complementary* in G if $G = AB$ and $A \cap B = e$. That is, $C = e$ in Fig. 3.2. If $A \trianglelefteq G$ in Fig. 3.2 then $C \trianglelefteq B$. (Exercise).

The following lemmas are very useful when subgroup lattices are constructed.

3.5 Lemma: The diagrams (1) and (2) of Fig. 3.3 imply the diagram (3).

Proof: $DC = (DL)C = D(LC) = DB = A$;
$D \cap C = D \cap (B \cap C) = (D \cap B) \cap C = L \cap C = M$. ■

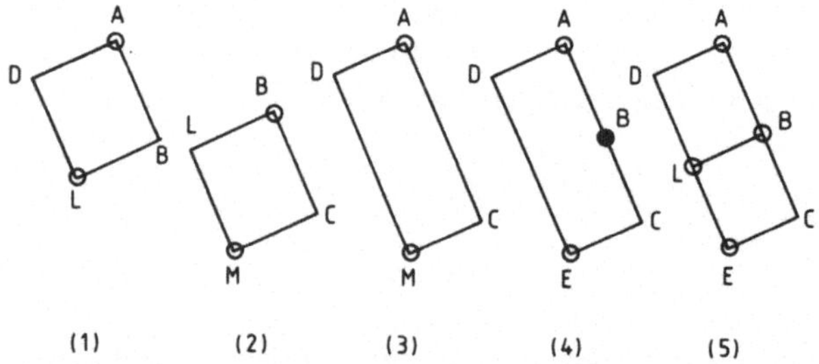

Fig. 3.3

3.6 Lemma: From diagram (4) follows the existence of a unique $L \in \mathcal{U}(G)$ such that diagram (5) holds. Vice versa: The existence of L implies the existence of B, iff $LC = CL$.

Proof: (first part only): First we prove, that $(D \cap B)C = DC \cap BC$. $(D \cap B)C \subseteq DC \cap BC$ is obvious. Let $a \in DC \cap BC$ then there are $d \in D, b \in B$ and $c, c' \in C$ such that $dc = bc' = a$. Hence, $d = bc'c^{-1} \in BC = B$ and $d \in B \cap D$.

Thus, $a \in (B \cap D)C$ and $DC \cap BC \subsetneq (D \cap B)C$.
Now, let $L = D \cap B \in \mathcal{U}(G)$; L is unique.

(i) $A \supseteq DB \supseteq DC = A$, hence, $DB = A$ and ABLD is a parallelogram.

(ii) $LC = (D \cap B)C = DC \cap BC = DC \cap B = A \cap B = B$
and $L \cap C = (D \cap B) \cap C = D \cap C = E$. Hence, BCEL is a parallelogram. ■

3.7 Lemma: A subgroup A of G with $|G/A| = 2$ is normal in G.

Proof: $G = A \cup gA = A \cup Ag$ and, hence, $gA = Ag$ for $g \in G - A$. ■

3.8 *Exercise:* The commutator of two complexes A and B of G is
$[A, B] := \langle \{a^{-1}b^{-1}ab \mid a \in A, b \in B\} \rangle$.
The commutator group of G is $[G, G]$.
Let $A, B \trianglelefteq G$. Show that $[A, B] \trianglelefteq \langle A, B \rangle$.

3.9 *Example:* Fig. 3.4 shows the lattice of the symmetric group S(4) acting on $X = \{1, 2, 3, 4\}$. A(4) is the alternating group on X. It is the commutator group of S(4).

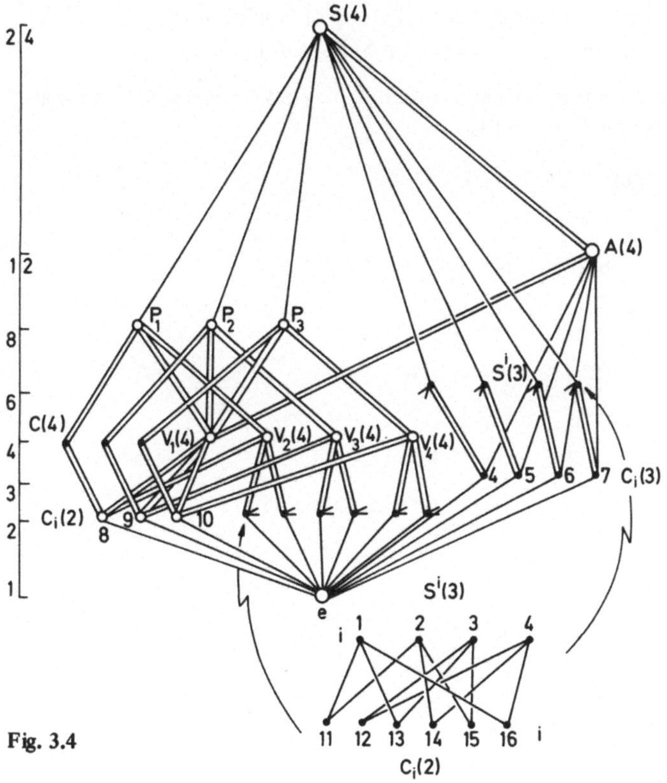

Fig. 3.4

Tab. 3.1: The group elements of S(4)

i	p_i	i	p_i	i	p_i
1	(1324)	9	(13) (24)	17	(1423)
2	(1234)	10	(14) (23)	18	(1432)
3	(1243)	11	(12)	19	(1342)
4	(123)	12	(34)	20	(132)
5	(124)	13	(13)	21	(142)
6	(134)	14	(24)	22	(143)
7	(234)	15	(14)	23	(243)
8	(12) (34)	16	(23)	24	e

A(4) describes the rotational symmetry of a tetrahedron and S(4) that of a cube; S(4) = (Sym X, X) where Sym X is the group of all bijections of X. Similarly, $S^i(3) = (\mathrm{Sym}(X-i), X-i)$. The symmetric groups have been introduced in Sec. 2. $V_i(4)$ are Klein's Four Groups. Their group table is given by Tab. 3.2; P_j are the 2-Sylow Groups of S(4).

3.10 Definition: A p-subgroup of a group G is a subgroup of G with order $p^n (n \in \mathbb{N})$ and p prime. A maximal p-subgroup of G is called a p-Sylow Group of G.

The p-Sylow Groups of G are conjugate to each other. A p-Sylow Group of G is normal in G iff there is no other p-Sylow Group of G.

Examples of subgroups of S(4):

$\langle p_2 \rangle = \{e, p_2, p_9, p_{18}\} \simeq C(4)$,

$P_2 = \langle p_2, p_{13} \rangle = \{e, p_2, p_8, p_9, p_{10}, p_{13}, p_{14}, p_{18}\} \simeq D(4)$,

the dihedral group of order 8,

$V_1(4) = \{e, p_8, p_9, p_{10}\}$ is abelian.

Tab. 3.2: $V(4) \simeq D(2)$

•	e	a	b	c
e	e	a	b	c
a	a	e	c	b
b	b	c	e	a
c	c	b	a	e

3.3 Morphisms and subgroups

Morphisms are important for studying subgroup lattices because they are exactly the maps from one group to another one that preserve the group structure. The special value of morphisms to physics is that they relate the subgroup structure of concrete symmetry groups of physical objects to the subgroup structure of linear transformation groups (see Sec. 5). At the roots of this fact lies the Parallelogram Lemma.

3.11 Let G and H be two groups. If $R \trianglelefteq S \leqslant G$ and if $\phi : G \to H$ is an homomorphism such that $\mathrm{Ker}\,\phi \subseteq R$, then the so called *Parallelogram Lemma* holds. That is, $R\phi \trianglelefteq S\phi$ and S/R is isomorphic to $S\phi/R\phi$. Moreover, it can be shown that $V := \{S \mid \mathrm{Ker}\,\phi \leqslant S \leqslant G\}$ and $\bar{V} := \{U \mid U \leqslant G\phi\}$ are isomorphic lattices.

This lemma is given (through illustration) by Fig. 3.5. Its proof is given in any textbook on group theory. We use for morphisms the (calculator) notation
$\phi : x \mapsto x\phi := \phi(x)$. If ϕ is an isomorphism we write $G \underset{\phi}{\simeq} H$ or $G \simeq H$. Let $U \subset G$, $\phi|_U$ is the restriction of ϕ on U.
Interpretation of Fig. 3.5: $\bar{e}$ is the neutral element of H. The kernel of ϕ, $\mathrm{Ker}\,\phi := \bar{e}\phi^{-1}$, is a normal subgroup of G; R is normal in S by assumption and ψ is an isomorphism between the factor groups S/R and $S\phi/R\phi$.

3.12 *Remark:* $G\phi$ is not necessarily normal in H. For example, take $S < G$ not normal in G. The map $i_S : S \to G : g \mapsto g$ is an homomorphism and $i_S(S) = S$ is not normal in G.

A consequence of the Parallelogram Lemma is the following theorem which gives us some more information about $U(G)$.

3.13 Theorem: From the diagram (6) of Fig. 3.6 follows the diagram (7).

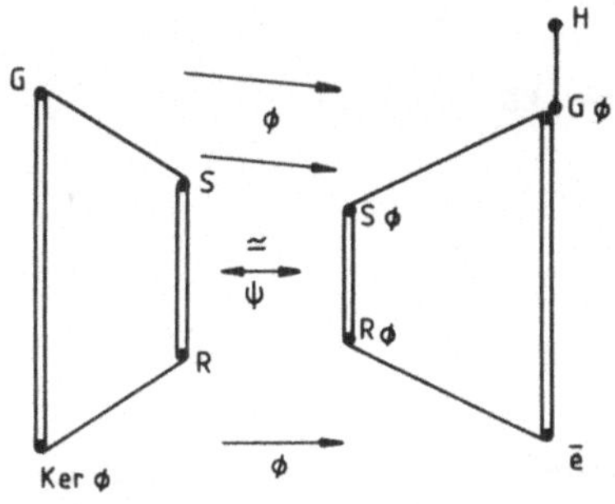

Fig. 3.5

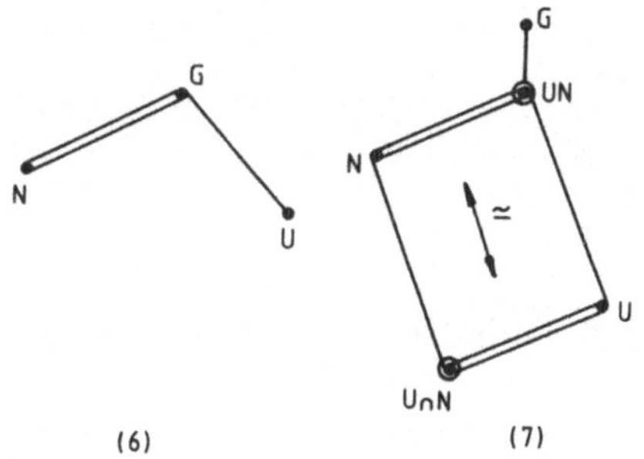

Fig. 3.6

That is, from $N \trianglelefteq G$ and $U \leqslant G$ follow:

(1) $\langle U, N\rangle = UN = NU$

(2) $N \trianglelefteq UN$, $U \cap N \trianglelefteq U$ (3.1)

(3) $\frac{UN}{N}$ is isomorphic to $\frac{U}{U \cap N}$.

Proof: Only (3) needs a proof.

Consider the (canonical) homomorphism $\phi : UN \to \frac{UN}{N} : a \mapsto aN$.

(i) $\frac{UN}{N} = \{(un)N \mid u \in U, n \in N\} = \{uN \mid u \in U\} = \{(ue)\phi \mid u \in U\} = U\phi$.

(ii) $\mathrm{Ker}\,(\phi|_U) = \mathrm{Ker}\,\phi \cap U = N \cap U$.

That is Fig. 3.7 through illustration.

The remaining steps follow from the Parallelogram Lemma. ∎

3.14 An automorphism of a group G is an isomorphism $\phi : G \to G$. For example, let $h \in G$. The maps $\phi_h : g \mapsto hgh^{-1}$ are(inner)automorphisms of G. The set of all automorphisms of G itself is a group Aut(G), the automorphism group of G. The set Inn(G) of inner automorphisms is a subgroup of Aut(G). Moreover, $\mathrm{Inn}(G) \trianglelefteq \mathrm{Aut}(G)$.

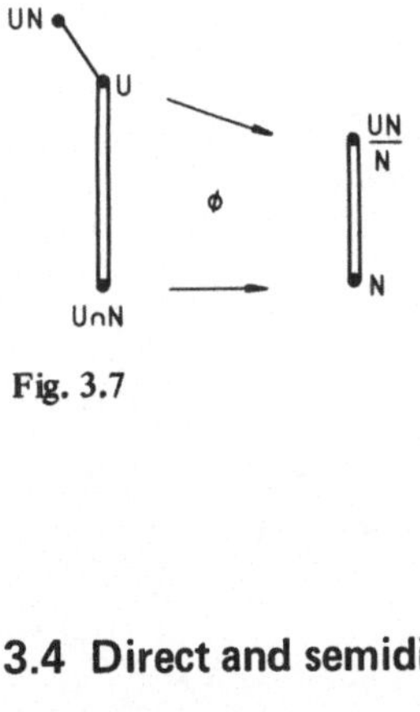

Fig. 3.7

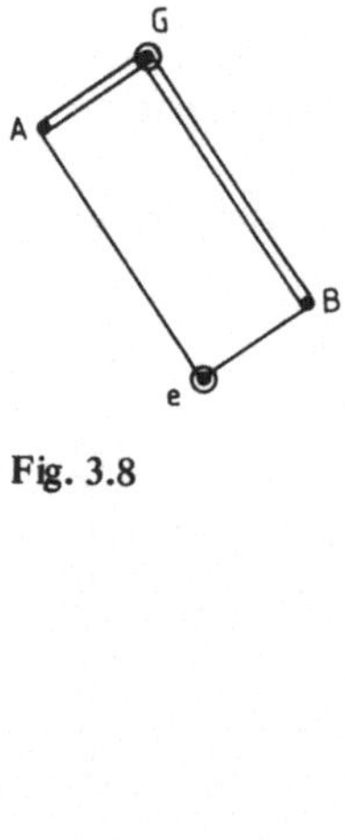

Fig. 3.8

3.4 Direct and semidirect products

We now introduce two special subgroup diagrams to describe direct and semidirect products. Many groups in physics are direct or semidirect products [2]. Sec. 6 illustrates their crucial role in representation theory.

3.15 Definition: A group G is called the (interior) *direct product* of two subgroups A and B, if A and B are complementary, normal subgroups of G, that is, if Fig. 3.8 holds. The normal subgroups A and B are said to be direct factors of G. We write $G = A \times B$.

Now, any two groups A and B can be combined into a third group C – the (exterior) dire product of A and B – in the following way:
$C = (A \times B, \cdot)$ where $A \times B$ is the cartesian product of the sets of group elements. The multiplication law is given by

$$(a, b) \cdot (\hat{a}, \hat{b}) = (a\hat{a}, b\hat{b}) \text{ and } (a, b)^{-1} = (a^{-1}, b^{-1}) \qquad (3.2)$$

with $a, \hat{a} \in A$, $b, \hat{b} \in B$.

3.16 Theorem: If G is the direct product of two subgroups A and B, then G is isomorphic to $C = (A \times B, \cdot)$. The direct product of two groups A and B is the (interior) direct product of its subgroups $(A, \bar{e})$ and (e, B) where $e(\bar{e})$ is the neutral element of A(B).

Proof: Easy. ∎

3.17 *Examples:* V(4) is the direct product of two cyclic subgroups of order 2 (c.f. Fig. 3.4).

A trigger flip-flop is an electronic memory device with only two stable states. The state of this device is changed by an external (trigger) signal. Typically, fewer than four transistors are needed for such a bistable device. The state transitions of a trigger flip-flop form a permutation group, viz C(2). The direct product V(4) is the group of a parallel connection of two such flip-flops whereas C(4) is the group of a series of two flip-flops (see Chap. 8). Shift registers and random access computer memories are made up of linked bistable devices. Their (state transformation) groups provide many examples of finite groups and semigroups with regularly composed subgroup lattices.

3.18 Definition: A group G is the (interior) *semidirect product* of its subgroups A and B if $A \trianglelefteq G = AB$ and $A \cap B = e$, that is, if Fig. 3.9 holds.
We denote G by $A \rtimes B$. It follows, that $B \simeq G/A$ (see Sec. 3.2). G is the direct product of A and B iff $G = A \rtimes B$ and $G = B \rtimes A$. Furthermore, we have for $g = ab$, $\bar{g} = \bar{a}\bar{b}$ with $a, \bar{a} \in A$ and $b, \bar{b} \in B$: $g\bar{g} = (ab)(\bar{a}\bar{b}) = a(b\bar{a}b^{-1})(b\bar{b})$. That is, the multiplication law in G is determined by the multiplication laws in A and B and the action of B on A given by $a \to bab^{-1}$ (b induces an automorphism of A since $A \trianglelefteq G$). This observation leads us to the following construction of an (exterior) semidirect product.

3.19 Definition: Let A, B be groups and $\sigma : B \to \mathrm{Aut}(A)$ an homomorphism. The semidirect product of A by B is: $\hat{G} := (A \times B, \cdot)$ where $A \times B$ is again the cartesian product. The multiplication law is given by:

$$(a, b) \cdot (\bar{a}, \bar{b}) = (a(\bar{a}(b\sigma)), b\bar{b}) . \qquad (3.3)$$

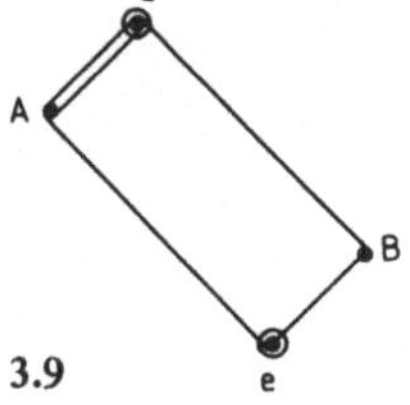

Fig. 3.9

3.20 Theorem:

(i) $\hat{G}$ is a group.

(ii) $A \simeq \hat{A} := (A, \bar{e}) \trianglelefteq \hat{G}$, $B \simeq \hat{B} := (e, B) \leqslant \hat{G}$ and

(iii) $\hat{G} = \hat{A} \curlyvee \hat{B}$.

We identify $\hat{A}$, $\hat{B}$ with A, B and write $\hat{G} = A \curlyvee_\sigma B$ or $A \curlyvee B$. If $b\sigma = i_A$ is the trivial automorphism (for all b), then $\hat{G} = A \times B$. Hence, there are, in general, more than one semidirect products of A by B. However, observe that, the only semidirect product of C(4) by C(3) is the direct product, since the only homomorphism $C(3) \to \mathrm{Aut}(C(4)) = C(2)$ is $g \mapsto i_{C(4)}$. On the other hand, there does exist a nontrivial semidirect product $C(3) \curlyvee C(4)$ corresponding to the canonical homomorphism $C(4) \to \mathrm{Aut}(C(3)) = C(2)$.

Proof:

(i) Check the group axioms; the neutral element is $(e, \bar{e}) =: 1$ and the inverse of (a, b) is $(a, b)^{-1} = (a^{-1}(b\sigma)^{-1}, b^{-1})$.

(ii) $\hat{G} = \hat{A}\hat{B} = \hat{B}\hat{A}$, $\hat{A} \cap \hat{B} = 1$.

(iii) Let $N_G(U)$ be the normalizer of $U \subseteq G$ in G, i.e., $N_G(U) = \{g \in G \mid gUg^{-1} = U\}$. In order to show that $\hat{A} \trianglelefteq \hat{G}$, observe that $\hat{A} \trianglelefteq N_{\hat{G}}(\hat{A}) \leqslant \hat{G}$ and $\hat{B} \leqslant N_{\hat{G}}(\hat{A})$. Since: $(e, b)^{-1}(a, \bar{e})(e, b) = (a(b^{-1}\sigma), \bar{e}) \in \hat{A}$. Hence, $\hat{G} = \hat{A}\hat{B} \leqslant N_{\hat{G}}(\hat{A})$. That is, $\hat{A} \trianglelefteq \hat{G}$. ■

3.21 *Examples (1):* The Euclidian Group in n-dimensions E(n), $n > 1$, consists of the rotations, the inversion, and the spacial translations of the n-dimensional vector space $\mathbb{R}^n$, leaving the Euclidian distance between two points invariant (see Sec. (2.3)). Let $O(n, \mathbb{R})$ be the (real) orthogonal group, $a \in O(n, \mathbb{R})$, $x \in \mathbb{R}^n$, and T(n) the group of translations. T(n) is isomorphic to $(\mathbb{R}^n, +)$. (Therefore, identify the translation $\hat{t} \in T(n) : x \mapsto x + t$ with t.) The transformation of $\mathbb{R}^n$ under $(t, a) \in E(n)$ is given by: $(t, a) : x \mapsto t(a(x)) = ax + t$. (Note, that E(n) acts from the left). Let $(t, a), (\bar{t}, \bar{a}) \in E(n)$, then $(\bar{t}, \bar{a})((t, a)x) = (\bar{t}, \bar{a})(ax + t) = \bar{a}ax + \bar{a}t + \bar{t} = (\bar{a}t + \bar{t}, \bar{a}a)x$. Hence, $(\bar{t}, \bar{a})(t, a) = (\bar{a}t + \bar{t}, \bar{a}a) = (\bar{t} + (\bar{a}\sigma)t, \bar{a}a)$. This shows that E(n) is a semidirect product

$$E(n) = T(n) \curlyvee O(n, \mathbb{R}) \tag{3.5}$$

since $\sigma : O(n, \mathbb{R}) \to \mathrm{Aut}(T(n)) : a \mapsto a\sigma : t \mapsto at$ is an homomorphism. Quite generally, all inhomogeneous matrix groups (Sec. (2.3)) are semidirect products.

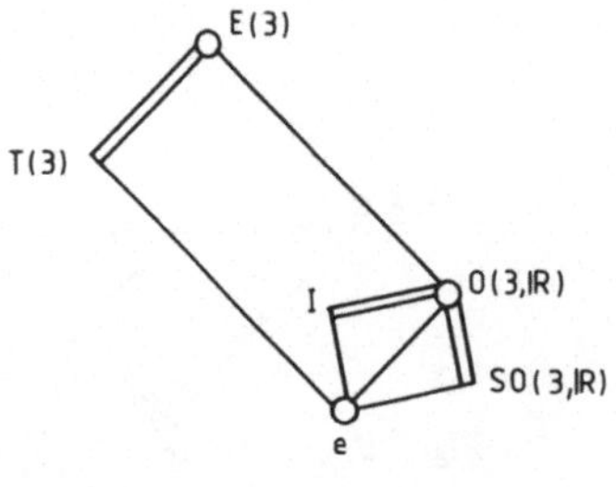

Fig. 3.10 E (3)

Let us now consider $E(3) = IO(3, \mathbb{R})$. $O(3, \mathbb{R})$ contains the inversion $i : x \mapsto -x$ and $I := \langle i \rangle$; $I \trianglelefteq O(3, \mathbb{R})$. Clearly, also the special orthogonal or rotation group $SO(3, \mathbb{R}) = \langle O(3, \mathbb{R}) - I \rangle$ is normal in $O(3, \mathbb{R})$, since $|I| = 2$ (s. 3.7). The diagram in Fig. 3.10 follows. Note, that this diagram is not complete, e.g., the group of rigid motions E(2) or the important cristallographic groups are all subgroups of E(3) not shown in this diagram [2].

(2) Fig. 3.4 indicates that the dihedral group D(4) is the (interior) semidirect product of a cyclic subgroup of order 4 by one of order 2. Let $\hat{h}$ and $\hat{k}$ be the generators of these groups, respectively. The action of $\hat{k}$ on $\hat{h}$ is given by: $\hat{k}\hat{h}\hat{k}^{-1} = \hat{h}^3 = \hat{h}^{-1}$. On the other hand, $\sigma : \hat{k} \mapsto (\hat{k}\sigma) : h \mapsto h^{-1}$ is an homomorphisn $C(2) \to \mathrm{Aut}(C(4))$ (since C(4) is abelian). Hence, $D(4) \simeq C(4) \rtimes_\sigma C(2)$.

3.22 Oberserve, that any element of a direct or a semidirect product can be uniquely factorized by a product of subgroup elements. This property can be generalized. Let $B_i (i = 1, \ldots, n)$ be subgroups of G and $B_1 B_2 \ldots B_n = \Pi B_i$. $B_i (i = 1, \ldots, n)$ factorize G if (see Fig. 3.11):

$$\begin{aligned} &\text{(i)} \quad B_i B_j = B_j B_i \text{ for all } i, j = 1, 2, \ldots, n. \\ &\text{(ii)} \quad \Pi B_i = G \text{ and} \qquad\qquad (3.4) \\ &\text{(iii)} \quad B_j \cap \prod_{i \neq j} B_i = e, \; j = 1, 2, \ldots, n. \end{aligned}$$

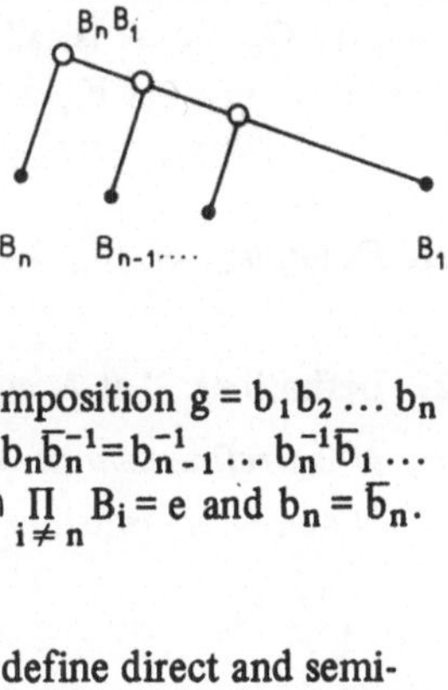

Fig. 3.11

If B_i $(i = 1, \ldots, n)$ factorize G then each $g \in G$ has a unique decomposition $g = b_1 b_2 \ldots b_n$ with $b_i \in B_i$. Suppose, $g = b_1 b_2 \ldots b_n$ and $g = \bar{b}_1 \bar{b}_2 \ldots \bar{b}_n$, then $b_n \bar{b}_n^{-1} = b_{n-1}^{-1} \ldots b_n^{-1} \bar{b}_1 \ldots \bar{b}_{n-1} \in B_{n-1} B_2 \ldots B_1 B_1 \ldots B_{n-1} = \prod_{i \neq n} B_i$. Hence, $b_n \bar{b}_n^{-1} \in B_n \cap \prod_{i \neq n} B_i = e$ and $b_n = \bar{b}_n$. The assertion follows.

3.23 *Remark:* The above definitions can easily be extended to define direct and semidirect products of n groups, $n > 2$. For discrete systems a still more complicated product, the wreath product [3], plays an important role. Its multiplication law is also "triangular". That is, the second coordinate of the product does not depend on the first coordinates of the factors (cf. 3.3). (This important property enables us to parametrize a discrete system in such a way that it decomposes into (forward) coupled subsystems.)

3.5 Subnormal chains

In this subsection we discuss briefly subgroup diagrams of the type shown in Fig. 3.12. For more details the reader is refered to [1], [4], [5].

That is, $G_0 = G$, $G_m = A \leqslant G$, $G_i \leqslant G$, and $G_{i+1} \trianglelefteq G_i$, $(i = 0, \ldots, m-1)$. This (finite) diagram is called a *subnormal chain* from G to A and A is called subnormal in G ($A \trianglelefteq\trianglelefteq G$).

Fig. 3.12

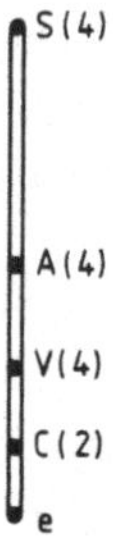

Fig. 3.13
A subnormal series of S (4)

3.24 *Exercise:* If $A \lhd\lhd G$ and $H \leqslant G$ then $A \cap H \lhd\lhd H$. If $A \lhd\lhd G$ and $H \lhd G$ then $\langle A, H\rangle \lhd\lhd G$.

3.25 Definitions: Let G be a group and $N = \{G_0, G_1, \ldots, G_m\}$ a subnormal chain from $G = G_0$ to $G_m = e$; N is called a subnormal *series* of G with the *factors* G_{i-1}/G_i $(i = 1, 2, \ldots, m)$. Let F_N denote the set of factors of N. Each group has a subnormal series, since $G = G_0 \trianglerighteq G_1 = e$; G is called *simple* if it has no normal subgroup other than G and e.

3.26 *Example:* (cf. Fig. 3.4), see Fig. 3.13

3.27 Definitions: Let N and $\hat{N}$ be two subnormal series of a group G.

(1) $\hat{N}$ is a *refinement* of N if $N \subset \hat{N}$

(2) N and $\hat{N}$ are called *isomorphic* if there is a bijection $\tau: F_N \to F_{\hat{N}}$ such that

$$\tau\left(\frac{G_{i-1}}{G_i}\right) \simeq \frac{G_{i-1}}{G_i}$$

(3) N is called a *composition* series of G if N does not have true refinements.

3.28 *Example:* The subnormal series of S(4) given above is a composition series of S(4).

3.29 Any two subnormal series of G have isomorphic refinements (O. Schreier). Any two composition series of G are isomorphic (Jordan-Hölder). All factors of a composition series are simple groups. If G is finite, G has a composition series. However, there are groups without composition series, e.g. $C\infty := \langle a\rangle \rhd \langle a^2\rangle \rhd \langle a^4\rangle \rhd \ldots$. For a proof of these statements see [4], [5], [6].

3.30 Complex simplicity: The composition series of a group is (up to isomorphism) uniquely determined. On the other hand, a group is in general not determined by its composition series. The theory of group extensions investigates, how simple groups can be fitted together (by a composition series) such that they yield new groups. Finite simple groups are the building elements from which finite groups are made up. Hence, we should have a list of all finite simple groups. For example, A(n) ($n \geqslant 3, n \neq 4$) are simple. The simplest of the simple groups are the (cyclic) groups of prime order (which have no nontrivial subgroups at all). Groups with these groups as composition factors are called *solvable.* Finite abelian groups are solvable. The symmetric groups S(n), $n \geqslant 5$, are not solvable. Composition series and solvable groups play an important role in the structure theory of discrete systems as it will be shown in Cap. 8. Many conditions for solvability are known. For example, every finite group of odd order is solvable (W. Feit, J. G. Thomson). However, the investigation of the structures of finite (and solvable) groups is complicated and is nowhere near closed.

3.6 Subgroups of semigroups

In this subsection we add a few (elementary) remarks on certain building elements of semigroups. Let S be a semigroup; $T \subseteq S$ is a subgroup of S ($T \leqslant S$) if T is a subsemigroup of S and a group. (The neutral element of T may not be the neutral element of S); S is called cyclic if there exists an $a \in S$ such that $S = \cap \{T \mid a \in T, T \text{ subsemigroup of } S\}$.

3.31 *Examples: (1)* $C_{m,r} := \{a, a^2, a^3, \ldots, a^r, a^{r+1}, \ldots, a^{r+m-1} = a^{r-1}\}$ is the cyclic semigroup of order $r + m - 1$ (r index, m period) with the cyclic subgroup C(m). If n is the multiple of m satisfying $r \leqslant n \leqslant m + r - 1$, then a^n is the identity of C(m) and C(m) is generated by a^{n+1}, if $n < m + r - 1$, and by a^r else, see Fig. 3.14; $C_{m,r \neq 1}$ is not a monoid.
(2) Let $\emptyset \neq X$ be a finite set and X^n the set of n-tuples over X. We write $x_1 x_2 \ldots x_n$ for $(x_1, x_2, \ldots, x_n)$. Define a multiplication on $X^+ := \bigcup_{n \in \mathbb{N}} X^n$ (the concatenation) as follows: $(x_1 \ldots x_n)(\bar{x}_1 \ldots \bar{x}_m) = x_1 \ldots x_n \bar{x}_1 \ldots \bar{x}_m$. Then X^+ is a semigroup, the free semigroup on X; X has no subgroups at all. In the theory of discrete systems, X^+ is called the *input semigroup* and its elements are alled input words [7]. The following statements are equivalent: (i) X^+ is commutative, (ii) $|X| = 1$, (iii) X^+ and $(\mathbb{N}, +)$ are isomorphic (Exercise).
(3) Let $\lambda \notin X$ and define $\lambda\lambda = \lambda$, $\lambda x = x\lambda = x$ for all $x \in X$. Then $X^* = \{\lambda\} \cup X^+$ is a monoid, the *input monoid;* λ is called the empty word. The only subgroup of X^* is λ.

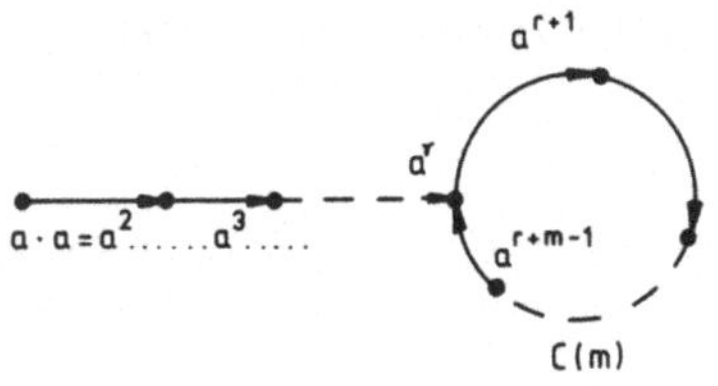

Fig. 3.14 $C_{m,r}$

3.32 Theorem: Let S be a finite semigroup and ϕ a (semigroup) homomorphism $\phi: S \to H$. If $S\phi$ is a group then there is a subgroup G of S such that $G\phi = S\phi$.

Proof: Suppose, that G is the minimal subsemigroup of S with $G\phi = S\phi$. If $e = e^2 \in G$, then $e\phi = e^2\phi = (e\phi)^2$ is the neutral element of $S\phi$. Furthermore, $(eGe)\phi = G\phi = S\phi$. Hence, $eGe = G$ and for all $g' \in G$ there is a $g \in G$ with $g' = ege$. Hence, $eg' = g'e = g'$. That is, e is the (unique) neutral element of G and G is a monoid. We started with an arbitrary idempotent e, hence, e is the unique idempotent of G. Now, let $x \in G$, then some power k of x is an idempotent (Exercise). Hence, $x^k = e$, and e is the neutral element of G. Hence, $xx^{k-1} = x^{k-1}x = e$ and x has an inverse. ∎

3.33 Definition: A subset I of a semigroup S is a *right (left) ideal* if $IS \subseteq I (SI \subseteq I)$.

If $h \in S$, then $hS(Sh)$ is a right (left) ideal of S. A subset of S which is a left and a right ideal of S is an *ideal* of S. An ideal of S is a subsemigroup of S. A group G has no (nontrivial) ideal other than $\emptyset$ and G; C(m) is an ideal of $C_{m,r}$.

3.34 Definition. Let M be a monoid with neutral element 1. An element a of M is called *invertible* if there are $b, \bar{b} \in M$ with $ab = \bar{b}a = 1$. Let U(M) denote the set of invertible elements of M, $U(M) \neq \emptyset$ since $1 \in U(M)$.

3.35 *Example:* $U((\mathbb{Z}, \cdot)) = \{+1, -1\}$.

3.36 Theorem: Let M be a monoid. Then U(M) is a subgroup of M.

Proof: Let $a \in U(M)$. That is, there are $a', a'' \in M$ such that $a'a = aa'' = 1 \in U(M)$. Hence, $a' = a'1 = a'(aa'') = (a'a)a'' = 1a'' = a''$. That is, $a' = a^{-1} \in U(M)$.
For all $a, b \in U(M)$ we have $b^{-1}a^{-1}(ab) = 1 = (ab)b^{-1}a^{-1} = 1$. Hence, $ab \in U(M)$. ∎

3.37 Theorem: Let M be a monoid and $e = e^2 \in M$ an idempotent of M. Then eMe is a monoid and a subsemigroup of M; U(eMe) is a maximal subgroup of M.

Proof:

(i) We proof that $e = eee \in eMe$ is the neutral element. If $n, m \in eMe$ and $m = e\bar{m}e$, $n = e\bar{n}e$, then $em = ee\bar{m}e = e\bar{m}e = m = me$. Furthermore, $nm = e\bar{n}e\bar{m}e \in eMe$.

(ii) $U(eMe) \leqslant eMe$. Suppose, that there is a subgroup G of M such that $U(eMe) \leqslant G \leqslant M$. Clearly, e is the neutral element in G. Hence, for all $g \in G : g \in eGe \subset U(eMe)$. Therefore, $G \leqslant U(eMe)$ and $G = U(eMe)$. ∎

Moreover, it can be shown that the maximal subgroups of a monoid M are exactly the subgroups U(eMe) with $e^2 = e \in M$. An interesting application of theorem 3.37 is given in Chap. VIII, Sec. 3.

4 Groups and topology

4.1 Topological groups

In this section we shall list some basic concepts about topology in order to be able to present the definition of a topological group. If we are given a set X such concepts as "neighboorhood", "limit", or "continuous" can be used only if X has been provided with a topology. This can be done in the following manner:

4.1 Definition: A *topology* of a set X is a family T of subsets of X with the following properties:

(1) The void set and X itself belong to T.
(2) The intersection of a finite number of elements of T belongs to T.
(3) The union of an arbitrary family of elements of T belongs to T.

A *topological space* sometimes denoted by a pair (X, T) is a set X on which a topology T is given. The elements of T are called the *open sets* of this topology. A subsystem $T' \subset T$ is called a *base* for the topology T if every element of T is the union of elements of T'. Every system of subsets of T that satisfies 1) and 2) above is a base of a certain topology. We obtain an especially simple topology on a set X if the set of subsets of X which contain only one element of X is taken as a base of the topology of X. This topology is called the *discrete* topology.

If T_1 and T_2 are two topologies on X such that $T_1 \subset T_2$, then T_1 is called *coarser* (weaker) than T_2, and T_2 is *finer* than T_1. The discrete topology is the finest topology on X.

An open set containing a point $x \in X$ is called a *neighbourhood* of that point. The complements of open sets with respect to X are called closed. For any subset $Y \subset X$ the smallest closed subset containing Y is called the *closure* of Y. A subset of X is called *dense* in X if its closure coincides with X.

4.2 Definition: An order relation on a set A is a relation "$<$" which is transitive ($x<y$ and $y<z$ implies $y<z$) and "antisymmetric" ($x<y$ and $y<x$ implies $x=y$). A set A is called a *directed* set if it is provided with an order relation with the additional property that for each pair $x, y \in A$ there exists a $z \in A$ such that $x<z$ and $y<z$.

4.3 Definition: Let $(A, <)$ be a directed set and X an arbitrary set. A mapping of A into X is called a *net* (direction) in X.

A point is called a *limit* of the net if for any neighbourhood U of the point x there exists an element $\alpha \in A$ so that $x_\beta \in U$ for all $\alpha < \beta$. Usally this will be written in the form

$$x_\alpha \to x \qquad \text{or} \qquad \lim_{\alpha \in A} x_\alpha = x \qquad \text{or} \qquad \lim x_\alpha = x\,.$$

If the topological space has a countable base, A is usually chosen to be the set $\mathbb{N}$ of natural numbers and the definition above coincides with that of elementary analysis.

A *mapping* of one topological space into another is called *continuous,* if the inverse image of every open set is open. It can be shown that a mapping is continuous if and only if the image of the limit of a net coincides with the limit of the image of a net:

$$\lim f(x_\alpha) = f(x) \qquad \text{if} \qquad \lim x_\alpha = x\,.$$

A mapping is called a *homeomorphism* if it is one to one and both the mapping and the inverse of it are continuous.

If $X = X_1 \times \ldots \times X_n = \prod_{i=1}^{n} X_i$ is the product of sets X_i each of them provided with a topology, then the set T' of n-tuples $U_1 \times \ldots \times U_n = \prod_{i=1} U_i$ is the base of a topology T on X called the *direct product topology* on X. The proof that T' generates a topology is rather trivial in the case of a finite product. If $\{(X_\alpha, T_\alpha)\}_{\alpha \in A}$ is an infinite set of topological spaces, it is possible to show that a basis of a topology of $X = \prod_{\alpha \in A} X_\alpha$ is formed by the sets $\prod_{\alpha \in A_\beta} U_\alpha \times \prod_{\alpha \in A \setminus A_\beta} X_\alpha$. A_β is any finite subset of A and the sets U_α belong to T_α.

Every subset $Y \subset X$ of a topological space X with topology T can be considered as a topological space if we define as open sets of Y the intersections of Y with open sets of X. A subset with this topology is called a *subspace* of X provided with the relative topology. A subspace with discrete relative topology is called a *discrete subset.*

Let R be an equivalence relation on X. Then the sets of equivalence classes which are projections of open sets are the open sets of a topology on $X_{(R)}$. The set $X_{(R)}$ provided with this topology is called a *factor space.* The projection mapping $X \to X_{(R)}$ is continuous by construction. The method which is applied here of inducing topology in such a way that a certain mapping is continuous is very common in topology.

An (open) *covering* of a topological space is a subset $S \subset T$ so that $\bigcup_{U \in S} U = X$. A topological space is called *compact* if and only if (iff) every covering of it by open sets admits a finite subcovering. If for example X is a subspace of an Euclidean space $\mathbb{R}^n$, this definition is equivalent to that of elementary real analysis, namely that X is compact iff it is closed and bounded (Heine-Borel theorem).

A topological space is called *separated* or *Hausdorff* iff every pair of distinct points admits disjoint neighbourhoods. In Hausdorff spaces every net admits only one limit point. This property accounts for the fact that the majority of topological spaces used are Hausdorff. But there are important classes of spaces for which the Hausdorff-separation axiom does not hold. This may happen, for example, with spaces of orbits of transformation groups on topological spaces.

4.4 Definition: A nonnegative function ρ on the product $X \times X$ of a set X is called a *distance* or *metric,* if it has the following properties:

(1) $\rho(x, y) \geqslant 0$ and $\rho(x, y) = 0$ iff $x = y$

(2) $\rho(x, y) = \rho(y, x)$

(3) $\rho(x, y) + \rho(y, z) \geqslant \rho(x, z)$.

A set X provided with a metric is called a *metric space.* Every metric space can be considered as a topological space if as a basis of open sets we take the family of open balls, that is, sets of the form $S_r(x) = \{y : \rho(x, y) < r \text{ for all } x \in X : r \in \mathbb{R}, r > 0\}$.

Here two remarks are appropriate:

(1) Different metric spaces may lead to homeomorphic topological spaces.

(2) There exist spaces which are nonmetrizable.

A topological space is called *connected* iff it can not be represented as the union of two nonvoid disjoint sets, both of which are either open or closed. A space is called *locally connected* if each of its points has a connected neighbourhood.

It would be necessary to use the concept of covering spaces and of homotopy to define multiply connectedness of general topological spaces. But for manifolds a simpler and more intuitive concept using paths will be sufficient. A *path* in a topological space is a continuous mapping of the real line into that space. If the mapping is restricted to a closed interval $[a, b]$ of the real line and if the images of a and b coincide, the *path* is called *closed.*

A space X is called *path-conneted* iff for each pair of points $x, y \in X$ there exists a path with end points x and y. A path-connected space is called simply connected iff every closed path can be continuously contracted to a single point. For example, the Euclidean spaces $\mathbb{R}^n$ or the speres, the surfaces of n-dimensional balls, are simply connected, the tube and the torus are not.

4.5 Definition: A *topological group* is a set which is both a group and a topological space in which the algebraic and the topological structures respect each other in the sense that the mapping

$$G \times G \to G : (x, y) \mapsto x y^{-1}.$$

is continuous. This requirement is equivalent to the following three conditions (which are more convenient for checking):

(1) the mapping $(x, y) \mapsto xy$ is continuous in x and y,

(2) the mapping $x \to x^{-1}$ is continuous at the point e,

(3) the mapping $(x, y) \mapsto x \cdot y$ is continuous in both variables together at the point (e, e).

Every abstract group can be regarded as a topological group if we provide it with the discrete topology. An important topological property of topological groups is the fact, that its topology is determined by the topology of an (open) neighbourhood of the unit element e. This can be seen if one takes into account that it is possible to transform this neighbourhood to every point of the group by multiplication from the left or from the right.

Most concepts of abstract groups such as subgroup, factor group, etc. as well as the important concept of a transformation group can be transferred to topological groups. It is only necessary to incorporate continuity into the definitions. Further it is possible to combine topological and algebraic concepts. Thus, a subgroup can be open, closed, connected, compact, etc. if its underlying space provided with the relative topology has this property.

A very important example which may be subsumed under the notion of a topological group is a *topological vector space (t. v. s).* A linear space V over $\mathbb{C}$ is a t.v.s if it is a topological space with the following two conditions:

(1) the mapping $V \times V \to V$ given by $(x, y) \to x + y$ is continuous (in the direct product topology of $V \times V$),

(2) the mapping $\mathbb{C} \times V \to V$ given by $(\lambda, x) \to \lambda x$ is continuous (in the direct product topology of $\mathbb{C} \times V$).

Since the first property implies in particular, that $(x, y) \to (x - y)$ is a continuous mapping, every t.v.s. is a commutative topological group.

The simplest case is that the vector space V has a *norm* (as in the case of $\mathbb{C}^n$ or of the Hilbert space), that means, the validity of the following four properties:

(i) There is a mapping $\| \; \| : V \to \mathbb{R}_+$

(ii) $x, y \in V, \; \|x + y\| \leqslant \|x\| + \|y\|$

(iii) $x \in V, \lambda \in \mathbb{C}, \; \|\lambda x\| = |\lambda| \, \|x\|$

(iv) $\|x\| = 0 \Rightarrow x = 0$.

If V is a normed space, a metrical topology can be introduced by means of the distance

$$\rho(x, y) := \|x - y\| .$$

If the property (iv) of the mapping $\| \; \|$ is dropped, then $\| \; \|$ is called a *seminorm.* A much more general method by which an appropriate topology can be given on a vector space is provided by a family of seminorms $\| \; \|_\alpha, \alpha \in A$. As a base of neighbourhoods of the neutral element $0 \in V$ one may choose the family of sets $U_{\alpha,\epsilon}, \alpha \in A, \epsilon > 0$ where

$$U_{\alpha,\epsilon} = \{x : \|x\|_\beta < \epsilon, \; \forall \beta < \alpha\} .$$

In this way, one obtains a socalled locally convex t.v.s.

Another important method for introducing a topology is achieved by means of a set W of linear functionals on V: The $\sigma(V, W)$ – topology on V is the weakest topology on V for which all the functionals in W are continuous. If the family W is so rich that $\varphi(x) = 0$ for all $\varphi \in W$, implies $x = 0$, then $\sigma(V, W)$ is a Hausdorff topology.

For a given t.v.s. V we deonte by V^* its dual space given by the set of all continuous linear forms on V. Then $\sigma(V, V^*)$ is called the weak topology on V. In general, it is coarser (weaker) than the original topology of V . $\sigma(V^*, V)$ is called the weak-$*$ topology on V^*.

If for a given t.v.s. V there exists another t.v.s. V_*, so that $(V_*)^* = V$, then we call V_* the predual of V. It is possible then to consider the $\sigma(V, V_*)$ topology on V, i.e. the weak-$*$ topology on V.

4.2 Lie groups

Many groups occuring in physical applications are topological groups whose underlying topological space can be given the structure of a manifold so that multiplication of group elements can be described by a differentiable mapping. For example, all matrix groups (cf. 2.2) belong to this class. Therefore, before we can define a Lie group we have to insert the concept of a manifold.

A manifold is a set which is locally an Euclidean space. On such a space it is possible to introduce local systems of coordinates and employ the apparatus of mathematical analysis. Let us now define a manifold more precisely in the following way:

On a topological space M a pair (U, ϕ) where U is an open subset of M and $\phi : U \to \mathbb{R}^n$ is a homeomorphism onto an open subset of a n-dimensional Euclidean space $\mathbb{R}^n$ is called a *chart*. (In the following we shall define everything only for the real case though almost everything is the same for the complex case and can be obtained simply by substituting $\mathbb{C}$ for $\mathbb{R}$.)

4.6 Definition: A C^r-atlas on a space is a family A of charts so that the following properties are satisfied:

(i) The set M is the union of all U for $(U, \phi) \in A$

(ii) If (U, ϕ) and (V, ψ) are elements of A, then on the intersection $U \cap V$ the mapping $\psi \circ \phi^{-1}$ is continuously differentiable of order r.

In the sequel we shall make no reference to the order r and therefore we shall drop it and only say that the mapping is smooth.

A *manifold* is a Hausdorff space together with an atlas. It is known from analysis that there does not exist any homeomorphism from n-dimensional Euclidiean space to m-dimensional. Therefore, if a manifold is connected the dimension of the Euclidean space in the definition of charts is the same for all charts. This number is called the *dimension* of the manifold.

Given a manifold M of dimension m, let x be an element of M and let (U, ϕ) be a chart, where U is a neighbourhood of x. A (real-valued) *function* f is called *smooth* on U if the combination $f \circ \phi^{-1}$ is a smooth (real-valued) function on the Euclidean space of the chart. We shall denote by A_x the set of all functions smooth on some open neighbourhood of x which may vary with f. By A or A_M we shall usually denote the set of all functions which are smooth everywhere on M. The sets A_x and A are commutative algebras. (An algebra is a vector space in which a distributive multiplication of elements is defined, cf. chapter II.)

We can also define *smooth mappings* from one manifold to another. A mapping f from a manifold M to a manifold N is called smooth if for every pair of charts (U, ϕ) and (V, ψ) of M and N respectively the mapping $\psi \circ f \circ \phi^{-1}$ is a smooth mapping of Euclidean spaces. Special examples of mappings are the elements of the algebra A_M which are mappings from M into the manifold $\mathbb{R}$ of real numbers. For a fixed basis in $\mathbb{R}^m$ any point of a chart (U, ϕ) is described by m coordinates. The functions on U determined by the coordinates shall be called *coordinate functions* or *coordinates* in U. Then any element of A_x for $x \in U$ can be represented as a real-valued function of m variables. Usually, if no confusion is possible, we shall denote the coordinate representation $\phi \circ f \circ \psi^{-1}$ of a mapping or the coordinate representation $g \circ \psi^{-1}$ of a function simply by the letters f and g respectively.

A smooth mapping f induces a mapping of the sets of functions which is sometimes called the pull-back (reciprocal image) of f and which we shall denote by f*. It is defined by the relation

$$(f^*(g))(x) = g(f(x))$$

for any $g \in A_{f(x)}$.

For manifolds many of the concepts of topological spaces such as subspace or cartesian product can be defined very explicitly using coordinates. For example, in some neighbourhoods, a closed submanifold can be imagined as a surface or even a plane in some appropriate coordinate system, and on the other hand, one has rather clear ideas what the direct product of Euclidean spaces is, namely again an Euclidean space. Certainly this gives only a local idea. So, for example, the direct product of two circles is represented by a torus.

One of the basic concepts in the theory of manifolds is that of a tangent vector. Its intuitive meaning is given by the velocity of a body moving on a curve on the manifold. A *smooth curve* on a manifold M is any smooth mapping of the real line $\mathbb{R}$ into M. We shall denote the parameter in $\mathbb{R}$ by the letter t and the points of M by x. In the neighbourhood of a fixed point x_0 the curve is determined by $m = \dim M$ smooth functions $x^i(t)$ the local coordinates of $x(t)$ in a chart containing $x_0 = x(0)$. Two curves $x(t)$ and $y(t)$ starting in the same point x_0 are called tangential in x_0 if

$$\left[\frac{d}{dt}(x^i(t) - y^i(t))\right]_{t=0} = 0 \qquad \text{for} \qquad i = 1, \ldots, m\,.$$

We see that two curves are tangential in a point x_0 if in some coordinate set the derivatives of their coordinate functions coincide. It is not difficult to show that tangentiality is an equivalenve relation which does not depend on the special choice of the chart.

A *tangent vector* to a manifold at the point x_0 which in the sequel we shall denote by ξ_{x_0}, η_{x_0} etc. is defined as the equivalence class of smooth curves starting at the point x_0. In each chart a tangent vector is represented by a n-tupel of numbers which coincide with the derivations of the coordinate functions of a curve at x_0. This suggests that the set of tangent vectors in a given point is a n-dimensional linear space.

The main advantage of this definition of tangent vector lies in the fact that it is concrete and coincides with the intuitive picture of velocity. But it is awkward in several aspects. For example, addition of tangent vectors and multiplication by a number require the introduction of local coordinates. We shall therefore in addition adopt a different definition.

Let φ be an element of the algebra A_{x_0} of smooth functions at x_0. Every equivalence class of curves in x_0 defines a linear mapping (functional) ξ_{x_0} from A_{x_0} to $\mathbb{R}$:

$$\begin{aligned}\xi_{x_0}(\varphi) &= \left[\frac{d}{dt}\varphi(x(t))\right]_{t=0} = \sum_i \left[\frac{dx^i}{dt}\right]_{t=0} \left[\frac{\partial\varphi}{\partial x^i}\right]_{x=x_0} \\ &= \left[\sum_i \xi^i_{x_0}\frac{\partial}{\partial x^i}\varphi\right]_{x=x_0}.\end{aligned} \tag{4.1}$$

Note that the first defining identity does not make any use of coordinates. On the other hand, the third expression shows that in a certain chart the functional is determined by the same numbers as the tangent vector to the curve defined above. Thus to every tangent vector ξ_{x_0} there corresponds a functional on A_{x_0} usually called the derivative along the vector ξ_{x_0} and which satisfies the following equation:

$$\xi_{x_0}(\varphi_1 \varphi_2) = (\xi_{x_0}(\varphi_1))\,\varphi_2(x_0) + \varphi_1(x_0)\,(\xi_{x_0}(\varphi_2))\,. \tag{4.2}$$

It can be shown that any functional on A_{x_0} satisfying (2) determines a tangent vector in a unique way. Therefore, the functional can be used to define the tangent vector. Since functionals can be added and also multiplied by numbers, the set of tangent vectors in a given point x_0 is a linear space called the tangent space which will be denoted by $T_{x_0}(M)$. Let $x^1, \dots x^n$ be coordinate functions in a neighbourhood of x_0.

The partial derivatives $\frac{\partial}{\partial x^1}, \dots, \frac{\partial}{\partial x^n}$ at the point x_0 generate a basis of the tangent space which shows again that the dimension of $T_{x_0}(M)$ coincides with that of the manifold M. Let us assume that there is given a chart $(U, \varphi(x) = (x^1, \dots, x^n))$ in some neighbourhood U of x_0. Since any tangent vector at any point $x \in U$ is defined by a functional

$$\left[\sum_i \xi^i \frac{\partial}{\partial x^i}\right]_{\varphi(x) = (x^1, \dots, x^n)}$$

it is determined by the $2n$ numbers $(x^1, \dots, x^n, \xi^1, \dots, \xi^n)$. This can be done at any chart of an atlas of M. A change of charts induces a mapping between the components of tangent vectors (which can be shown to be differentiable). This suggests to consider the collection of tangent spaces as a new manifold T(M) of dimension $2n \cdot$ T(M) has a remarkable property. Let $A = \{U_i\}$ be an atlas of M. Then $T(A) = \{U_i \times \mathbb{R}^n\}$ is an atlas of T(M) i.e. locally T(M) is a cartesian product of a neighbourhood of M with the Euclidean space $\mathbb{R}^n$. Generally, T(M) is not diffeomorph (even not homeomorph) to the cartesian product $M \times \mathbb{R}^n$. Such manifolds which are locally cartesian products are called *fibre bundles.* They have become very important in mathematical physics. (Compare also chapter VI of this volume). For completeness, we shall give one possible definition of a fibre bundle (for more details see [3] and [5].

4.7 Definition: A bundle (quasi fibre bundle) is a triple (E, B, π) consisting of two topological spaces E and B and a continuous surjective mapping $\pi : E \to B \cdot$ B is called a base of the bundle.

In other words, a bundle is nothing else than the complete description of a continuous surjective mapping π. The simplest example is the cartesian product bundle $(B_1 \times B_2, B_1, \pi_1)$, also called the trivial bundle. The simplest non-trivial example is the Möbius band [3]. The most important examples of bundles are those where for all $x \in B$ the inverse image $\pi^{-1}(x)$, called the *fibre* through x, is homeomorphic to a fixed topological space T called the *typical fibre.* π is called the *projection.*

4.8 Definition: A *fibre bundle* (E, B, π, F, G) is a bundle (E, B, π) together with a typical fibre F, a topological group G of homeomorphisms of F onto itself and an atlas $A = \{U_i\}$ of B, such that

a) for every $U_i \in A$ there is a homeomorphism $\varphi_i : \pi^{-1}(U_i) \to U_i \times F$ so that the following diagram is commutative:

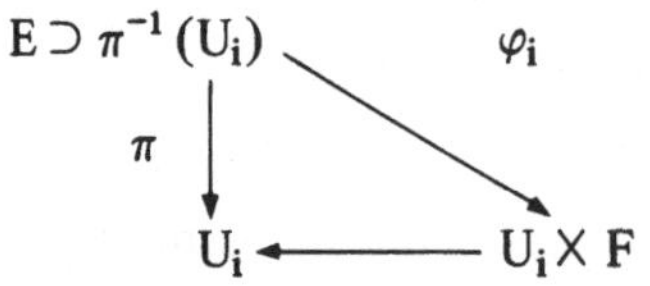

where p is the canonical projection.

b) The mapping $\varphi_i \circ \varphi_j^{-1}(x)$ for $x \in U_i \cap U_j$ is an element of G.

c) The induced mapping $g_{ik} : x \mapsto \varphi_i \circ \varphi_j^{-1}(x)$ from B into G is continuous.

If G has only one element then the bundle is trivial. If F is a vector space, the bundle is called a vector bundle. The tangent bundle is a vector bundle.

The generalisation of the graph of a mapping in the cartesian product of spaces is the concept of cross sections in bundles.

4.9 Definition: A cross section of the bundle (E, B, π) is a mapping $f : B \to E$ so that $\pi \circ f = \text{identity}$.

If at each point x of a manifold a tangent vector is chosen we call this family of tangent vectors $\{\xi_x\}_{x \in M}$ a *vector field.* Using coordinates, the vector field can be written in the form

$$\xi(x) = \Sigma \xi^i(x) \frac{\partial}{\partial x^i}. \tag{4.3}$$

For a general field, the functions $\xi^i(x)$ are neither smooth nor continuous. If they are smooth the vector field is called smooth. In the language of bundles we can say that a (smooth) vector field is a (smooth) cross section in the tangent bundle.

A vector field ξ defines a linear mapping $\xi : A_M \to A_M$ which at any point $x \in M$ associates with any function $\varphi \in A_M$ the derivative of φ along $\xi_x = \xi(x)$. It follows fron (4.2) above that

$$\xi(\varphi_1 \varphi_2) = (\xi \varphi_1)\varphi_2 + \varphi_2 (\xi \varphi_2).$$

A linear mapping with this property is called a first order linear differential operator. Again it can be shown that to every such operator on A_M there corresponds a unique vector field. By this definition it is obvious that the set of vector fields on a manifold has the structure of a vector space of infinite dimension which we shall denote by $X(M)$.

The product of two vector fields (that is the sequential application) is no longer a first order differential operator which becomes obvious if one uses coordinates. But in the same way it can be seen that the commutator

$$\zeta = [\xi, \eta] = \xi \circ \eta - \eta \circ \xi$$

is a vector field. In fact

$$\zeta = \sum_k \zeta^k \frac{\partial}{\partial x^k} = \sum_{i,k} \left(\xi^i \frac{\partial \eta^k}{\partial x^i} - \eta^i \frac{\partial \xi^k}{\partial x^i} \right) \frac{\partial}{\partial x^k} \tag{4.4}$$

has the form of (4.3). The commutator satisfies the following two identities:

$$\begin{aligned} &(a) \quad [\xi, \eta] = -[\eta, \xi] \text{ (anticommutativity)} \\ &(b) \quad [\xi[\eta, \zeta]] + [\eta, [\zeta, \xi]] + [\zeta, [\xi, \eta]] = 0 \end{aligned} \tag{4.5}$$

for any $\xi, \eta, \zeta \in \mathcal{X}(M)$ (Jacobi identity).

In addition the multiplication is distributive and therefore the space of vector fields $\mathcal{X}(M)$ has the structure of an algebra. Such an algebra, where the product satisfies the identities of (4.5) is called a *Lie algebra.*

There is also an important property of vector fields connected with the first introduction of a tangent vector. In fact, just as vectors arise from considering smooth curves, vector fields are connected with homeomorphisms of a manifold onto itself. Let $\phi_t : M \to M$ be a one parameter family of smooth mappings and let us further require that ϕ_t depends smoothly on the parameter t. This means that the mapping of $\mathbb{R} \times M \to M$ defined by $(t, x) \mapsto \phi_t(x)$ is smooth with $\phi_0(x) = x$ for all $x \in M$. Then every point moves on a smooth curve $x(t) = \phi_t(x)$ starting at the point x. This curve determines a tangent vector $\xi_x = x(0)$ for every x. In this way, we have defined a vector field called the derivative of the family $\{\phi_t\}$. Here the question arises as to whether or not every vector field is the derivative of a certain family of mappings of M onto itself. The answer to this question is affirmative [6].

We have already seen that a smooth mapping f of a manifold M into a manifold N induces a mapping f^* of A_N into A_M. We shall see that it also gives rise to a mapping of tangent spaces at every $x \in M$ which is denoted by $f_*(x)$ or $(f_x)_* : T_x \to T_{f(x)}$. This follows at once from the definition of tangent vectors by means of curves.

Let $x(t) = (x^1(t), \ldots, x^m(t))$ be a curve on M. To this there corresponds a curve $y(t) = (f^1(x(t), \ldots, f^n(x(t))$ in the manifold N. Therefore for any $\varphi \in A_{f(x)}$

$$\frac{d}{dt}\varphi(y(t)) = \Sigma \dot{y}^i(t) \frac{\partial \varphi}{\partial y^i} = \sum_{i,j} \frac{\partial y^i}{\partial x^j} \dot{x}^j(t) \frac{\partial \varphi}{\partial y^i} \tag{4.6}$$

or

$$\dot{y} = \left(\frac{\partial y}{\partial x}\right)\dot{x} \qquad \text{holds, where} \qquad \left(\frac{\partial y}{\partial x}\right) = \left(\frac{\partial y^i}{\partial x^j}\right)_{\substack{i=1,\ldots,n \\ j=1,\ldots,m}}$$

is a Jacobian matrix of this mapping.

The family $f_* = \{(f_x)_*\}_{x \in M}$ represents a mapping of smooth vector fields on M onto smooth vector fields on the image of f in N. Here the case where f is a *diffeomorphism* (that is a homeomorphism such that both f and its inverse f^{-1} are smooth) is of special importance.

With $y = f(x)$ we obtain for any $\varphi \in A_y$ and for any $\xi \in \mathcal{X}(M)$

$$((f_*\xi)\varphi)_y = (f^{-1})^*(\xi(f^*\varphi))_y = (\xi(f^*\varphi))_x \,. \tag{4.7}$$

These identities follow from the fact that for any smooth vector field η and every smooth function φ the function $\eta\,\varphi$ is smooth on M.

The mapping f_* is linear. If we apply it to the commutator and use (4.7), we have the following identities:

$$([f_*\xi, f_*\eta]\,\varphi)_y = ((f_*\xi)\circ(f_*\eta)\,\varphi)_y - ((f_*\eta)\circ(f_*\xi)\,\varphi)_y =$$
$$= ((f^{-1})^*\circ\xi\circ f^*\circ(f^{-1})^*\circ\eta\circ f^*)(\varphi))_y - ((f^{-1})^*\circ\eta\circ f^*\circ(f^{-1})^*\circ\xi\circ f^*)(\varphi))_y =$$
$$= ((f_*[\xi,\eta])(\varphi))_y$$

since $(f^{-1})^*f^*(\varphi(y))) = \varphi(f(f^{-1}(y))) = \varphi_y$.

The meaning of this identity is that f_* preserves multiplication. A linear mapping of the vector space of an Lie algebra with this property is called a *Lie algebra homomorphism*.

4.10 Definition: A topological group is called a *Lie group* if its underlying topological space is a manifold and iff the mapping $G \times G \to G : (x, y) \mapsto x\,y^{-1}$ is smooth.

The simplest example of a Lie group is provided by the additive group of real numbers, another example is the multiplicative group of complex numbers of absolute value one. (Note that this group is a Lie group over the field of real numbers.) As a simple consequence of the methods derived below, it can be seen that these are the only possible one-dimensional real Lie groups.

Mostly in physics, Lie groups occur acting as transformation groups on manifolds. Such a transformation group G is called a Lie transformation group on a manifold M if the mapping

$$\phi : G \times M \to M : (g, x) \mapsto \phi(g, x) = g(x)$$

is smooth.

Especially a Lie group acting on itself by right translations

$$r : G \times G \to G : r(g, g_0) \doteq r_g(g_0) \doteq g_0 g^{-1}$$

is a transitive Lie transformation group. Therefore the tangent space at the unit element can be carried to every element of the group by the transformations $(r_g)_*$ for any $g \in G$. In this way every tangent vector at the unit element e generates a smooth vector field and every further right translation transforms this vector field into itself. Therefore, these fields are called right invariant. Since they are generated by the tangent space at one point, they are a subspace of the space of the Lie algebra of vector fields whose dimension is equal to the dimension of the manifold of G. In addition, since every right translation is a diffeomorphism and therefore the mapping $(r_g)_*$ is an algebra homomorphism for every g, we have

$$(r_g)_*([\xi,\eta]) = [(r_g)_*(\xi), (r_g)_*(\eta)] = [\xi,\eta]\,.$$

Thus the space of right invariant vector fields is a Lie algebra which has the same dimension as the Lie group itself. It will be called the *Lie algebra of the Lie group* denoted by *g*.

There are several different ways how to obtain the Lie algebra of a Lie group, but especially one of them is very convenient to construct the differential operators. It is similar to the construction of vector fields using one-parameter sets of mappings. To do this we shall proceed in a slightly more general way.

Let G act as a Lie transformation group on a manifold M with coordinates $(x^1, \ldots, x^m)$ in a neighbourhood of a point x_0, and let $g = (\alpha^1(g), \ldots, \alpha^r(g))$ with $\alpha^\rho(e) = 0$ for all ρ be a chart of the group on a neighbourhood of e. Then an action of G on M is described by m functions of both sets of variables

$$x^i = f^i(\alpha^1(g), \ldots, \alpha^r(g), x^1, \ldots, x^m) \qquad i = 1, \ldots, m$$

If we take $\{g_\rho\}$ to be the one-parameter subset of G of those elements near the neutral element for which all but one coordinate, say α^ρ, are kept fixed to zero, the actions of the sets Γ_ρ define paths $x(\alpha_\rho)$ on M with $x(0) = x_0$. For some neighbourhood U of x_0 and $\varphi \in A_U$, $(g_\rho^{-1})^* \varphi(x_0)$ defines a smooth function

$$\alpha_\rho \mapsto \varphi(g_\rho^{-1}(x_0))$$

which can be differentiated:

$$\left[\frac{d}{d\alpha_\rho}\varphi\; g_\rho^{-1}(x_0)\right]_{\alpha_\rho=0} = -\sum_i \left[\frac{\partial\varphi}{\partial f^i}\frac{\partial f^i}{\partial\alpha^\rho}\right]_{\alpha=0} = \left(\sum_i -\left[\frac{\partial f^i(\alpha, x_0)}{\partial\alpha^\rho}\right]_{\alpha=0}\frac{\partial}{\partial x^i}\right)\varphi. \tag{4.8}$$

Finally, if we move x_0 this leads to a smooth vector field called the vector field of infinitesimal transformations. The set of vector fields we have got in this way is obviously a subspace of the space of vector fields on M. In a small neighbourhood of $(e, x) \in G \times M$ these are determined by curves g(t) on G with the initial conditions $g(0) = e$ and $\dot g(0) = (\dot\alpha(0), \ldots, \dot\alpha^r(0))$.

This implies:

$$\left[\frac{d}{dt}\varphi((g(t))^{-1}x)\right]_{t=0} = \left[\sum_{i,\rho} -\frac{d\alpha^\rho}{dt}\frac{\partial f^i}{\partial\alpha^\rho}\frac{\partial\varphi}{\partial x^i}\right]_{t=0} =$$

$$= -\sum_{i,\rho}\dot\alpha^\rho(0)\frac{\partial f^i(\alpha, x)}{\partial\alpha^\rho}\frac{\partial\varphi}{\partial x^i} = -\sum \xi^i(x)\frac{\partial}{\partial x^i}\varphi\,.$$

The commutator of two of these fields can be calculated in the usual way. In addition we shall present a different method using the commutator of elements of the Lie group. The commutator has been defined for abstract groups to be the element

$$\gamma = a^{-1}b^{-1}ab \qquad (\text{i.e. } ab = ba\gamma)$$

Now for two paths $t \mapsto a(t)$ and $t \mapsto b(t)$ starting at e with $[\frac{da}{dt}]_{t=0} = X$ and $[\frac{db}{dt}]_{t=0} = Y$, the mapping

$$s \mapsto a^{-1}(\sqrt{s})\, b^{-1}(\sqrt{s})\, a(\sqrt{s})\, b(\sqrt{s})$$

again determines a path on G starting at e (the square root is only for convenience). A straightforward but tedious calculation shows that the tangent vector corresponding to the commutator of vector fields coincides with the tangent vector to $\gamma(s)$. This implies that the space of infinitesimal transformations generated by G on M is a subalgebra of the Lie algebra of all vector fields on M. These infinitesimal transformations are also often called the generators of the transformation group. In general, this algebra is not isomorphic to the Lie algebra of the transformation group itself. It is isomorphic if the group acts freely in the neighbourhood of some point $x \in M$ i.e. if the isotropy group of x is discrete.

Both kinds of Lie algebras, the algebra of right invariant vector fields and the algebra of infinitesimal transformations can be related to each other if we consider Lie groups acting on itself. Let l_g be a left translation: $l_g(g_0) = gg_0$ and r_g a right translation defined by $r_g(g_0) = g_0 g^{-1}$. Because of the identity $r_g(l_{g'}(g_0)) = (g'g_0)g^{-1} = g'(g_0g^{-1}) = l_{g'}(r_g(g_0))$ every left translation commutes with every right translation. Let $g(t)$ be a path through the neutral element e. The family of paths $g(t)g_0$ defines a vector field of infinitesimal left translations by

$$\xi_{g_0} = \left[\frac{d}{dt}\, g(t)\, g_0\right]_{t=0} .$$

Because of $r_{g'}(g(t))g_0 = g(t)\, g_0 g^{-1} = g(t)\, g_0'$, this field is not changed by right translations. Therefore we have realized the elements of Lie algebra of right-invariant vector fields by infinitesimal left and vice versa, as it is usually done in mathematical literature.)

Let us now calculate the generators of matrix groups. Matrix groups (or linear groups) are closed Lie subgroups of the group $GL(n, \mathbb{R})$ of $n \times n$ matrices. A matrix group G acts in a natural way as a group of left translations on the general linear group in which it is embedded. Let $a = (a_{ij})$ be some fixed element of $GL(n, \mathbb{R})$ not necessarily belonging to G itself. The matrix elements may serve as coordinates of the general linear group. If $\beta_1, \ldots, \beta_r$ are the coordinates of a chart of G near the unit element e, the matrix elements of $g \in G$ are functions of these parameters β^i. If $g(\beta)$ acts as left translation on $GL(n, \mathbb{R})$, we obtain

$$a_{ij}' = \sum_j g_{ij}(\beta^\nu)\, a_{jk} .$$

Again, keeping all except one of the β^ν fixed we obtain a path in G and one in $GL(n, \mathbb{R})$ corresponding to that in G. According to (4.8), the generator of this path is given by

$$\xi_\nu = \sum_{i,j,k} \left[\frac{\partial g_{ij}(-\beta)}{\partial \beta^\nu}\right]_{\beta=0} a_{jk} \frac{\partial}{\partial a_{ij}}$$

and therefore

$$\xi_\nu = -\sum_{i,j,k} a_{ik} \left[\frac{\partial g_{ij}(\beta)}{\partial \beta^\nu}\right]_{\beta=0} \frac{\partial}{\partial a_{ij}} \doteq -\sum a_{ik}(\dot{g}_\nu)_{ij} \frac{\partial}{\partial a_{ij}} \tag{4.9}$$

With $[\xi_\rho, \xi_\sigma] = \xi_\rho \xi_\sigma - \xi_\sigma \xi_\rho$ we obtain that the commutator of two fields ξ_ρ, ξ_σ is uniquely determined by the commutator $\dot{g}_\rho \dot{g}_\sigma - \dot{g}_\sigma \dot{g}_\rho$ of the matrices corresponding to them.

Among the set of smooth curves in a Lie group there is an important subset, namely the set of (local) one parameter subgroups. *A one-parameter subgroup of a Lie group* is a curve $t \mapsto g(t)$ with the following properties:

$$g(0) = I$$
$$g(s)\, g(t) = g(s+t)$$

(the group is called local if the multiplication holds only for g(s) and g(t) close enough to I). It can be shown that the following theorem holds on any Lie group.

4.11 Theorem: For every element ξ of the Lie algebra g of a Lie group G there exists a unique smooth curve satisfying the conditions of a one-parameter subgroup above [2]. The conditions above are the same as the defining properties of the exponential function, and in fact if G is the multiplicative group $\mathbb{R}^*$ of positive real numbers different from zero, g(t) is the exponential function in the ordinary sense. In this case the mapping $g: \mathbb{R} \to \mathbb{R}^*$ can be interpreted as a homomorphism of the one-dimensional Lie algebra $\mathbb{R}$ considered as a one-dimensional commutative Lie algebra which is isomorphic to the Lie algebra of $\mathbb{R}^*$. In analogy with that we define an *exponential mapping*

$$\exp: g \to G$$

by $\exp\xi = g_\xi(1)$, where $g_\xi(t)$ is the one-parameter group that corresponds to ξ by the theorem above.
This mapping provides a local homeomorphism of an open neighbourhood of the unit element in G with an open neighbourhood of zero in the algebra g of G equipped in a natural way with the topology of an Euclidean space. For any smooth homeomorphism from a Lie group G_1 into a Lie group G_2 with the Lie algebras g_1 and g_2 respespectively, the following diagram is commutative.

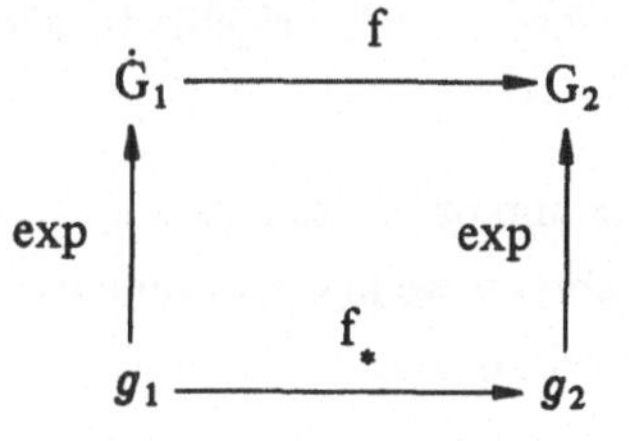

Moreover, local operations in groups can be approximated in the Lie algebra and thus in some sense they become "linearized". This is reflected by the following expansions:

4.12 Proposition:

$$\exp(t\xi)\exp(t\eta) = \exp[t(\xi+\eta) + \frac{t^2}{2}[\xi,\eta] + 0(t^3)]$$
$$\exp(-t\xi)\exp(-t\eta)\exp(t\xi)\exp(t\eta) = \exp(t^2[\xi,\eta] + 0(t^3)) \tag{4.10}$$
$$\exp(t\xi)\exp(t\eta)\exp(-t\xi) = \exp\{t\eta + t^2[\xi,\eta] + 0(t^3)\}.$$

Especially, if ξ and η commute with the commutator $[\xi, \eta]$, we obtain the important Baker-Campbell-Hausdorff formula:

$$\exp(t\xi)\exp(t\eta) = \exp(t\xi + t\eta)\exp\frac{t^2}{2}[\xi, \eta] .$$

The inverse of the exponential mapping is as usual denoted by the symbol Log. Generally, Log is defined only in a certain neighbourhood of the identity of the group. In a chart (U, Log) the coefficients of vector fields with respect to some basis in the algebra are the coordinate functions. A chart of this kind is offten called a *normal chart.* In such a coordinate system, for example multiplication along local one-parameter subgroups is performed formed by adding coordinates. A group is called *exponential* if a normal chart covers the entire group.

But even if the group is not exponential there is still a deep correspondence between the Lie groups and its Lie algebras. This is summarized in the following theorems:

4.13 Theorem: To every Lie algebra g there corresponds a unique connected and simply connected group G_0 for which g is its Lie algebra. All connected Lie groups having this algebra have the form G_0/D where D is a discrete normal subgroup contained in the center of G.

The next theorem relates subalgebras and subgroups:

4.14 Theorem: To every connected subgroup of a Lie group corresponds a subalgebra of the Lie algebra and vice versa. The invariant subgroups (normal subgroups) correspond to ideals of the algebra. (A left (right, two-sided) ideal of an algebra (ring) is a subalgebra which is invariant with respect to multiplication from the left (right, both sides). Because of anticommutativity all ideals of a Lie algebra are two-sided.

The algebraic properties of abstract groups such as simplicity, semisimplicity, solvability and nilpotence can be transferred from Lie groups to Lie algebras. A Lie algebra is called *(semi-)simple* if it contains no nontrivial (commutative) ideals.

4.15 Theorem: Every semisimple Lie algebra is the direct sum of simple ideals.

This theorem tells us that we know all semisimple Lie algebras if we know all simple ones.

In every Lie algebra we can define two sequences of subalgebras

$$g_0 = g \quad g_1 = [g, g_0] \quad g_2 = [g, g_1] \ldots g_{n+1} = [g, g_n]$$
$$g^0 = g \quad g^1 = [g^0, g^0] \quad g^2 = [g^1, g^1] \ldots g^{n+1} = [g^n, g^n]$$

A little computation which uses the Jacobi identity shows that the following inclusions hold

$$\begin{array}{ccccc} g_1 & \supset & g_2 & \supset & \ldots \\ \| & & \cup & & \\ g^1 & \supset & g^2 & \supset & \ldots \end{array}$$

If $\dim(g) < \infty$ we see that the sequences (g_n) and (g^n) must become stable and for some n_0 respectively n^0 we have $g_{n_0} = g_{n_0+1}$ respectively $g^{n^0} = g^{n^0+1}$. A Lie algebra is called nilpotent if $g_\infty = 0$ and it is called solvable if $g^\infty = 0$.
From the inclusion properties it follows at once that a nilpotent algebra is solvable. The converse is certainly not true, but it can be shown that the subalgebra $g^1 = g_1$ of a solvable algebra is nilpotent.
Finally, a Lie group is called (semi-)simple, solvable, nilpotent if its algebra has the same property. But note that these concepts differ slighlty from those of abstract groups. For example, the Lie group $SL(2, \mathbb{C})$ is simple according to our definition but is has a discrete normal subgroup, namely the group consisting of the unit element e and its negative $-e$.
Let G be a Lie group. The inner automorphisms of G are given by

$$x \mapsto g x g^{-1} = x(g)$$

If we imagine x(g) represents a curve through I on G we see at once that we can pass over to the algebra. If $\xi = [\frac{d}{dt} x(t)]_{t=0}$ determines the algebra element by this curve then

$$(\mathrm{Ad}\, g)\, \xi := \left[\frac{d}{dt} (g\, x(t)\, g^{-1})\right]_{t=0}$$

describes a linear mapping (matrix) of the algebra called the *adjoint mapping of the group*. In this way the group of inner automorphisms of a Lie group is homeomorphically mapped onto a group of linear mappings of the algebra. Now it is possible to differentiate for the second time and obtain another linear mapping of g. For $g(s) = \exp(s\eta)$ we get

$$\frac{d}{ds} (g(s))\, \xi (g(s))^{-1} = [\eta, \xi] =: (\mathrm{ad}\, \eta)\, \xi\, .$$

This mapping is called the *adjoint mapping of the algebra.*
If ξ_i is a base of the vector space of a Lie algebra we have the identity $[\xi_i, \xi_k] = \sum_j c^j_{ik}\, \xi_i$.
The constants c^j_{ik} are called *structure constants* of the algebra.
For $\eta = \Sigma\, a^i\, \xi_i$ we have

$$(\mathrm{ad}\, \eta)\, \xi_i = \sum_j a^j\, (\mathrm{ad}\, \xi_j)\, \xi_i = \sum_{j \cdot k} a^j\, c^k_{ji}\, \xi_k = \sum_k \eta^k_i\, \xi_k\, .$$

Thus we have a homeomorphism of the elements of the Lie algebra into a set of matrices which may be calculated from the structure constants. In general, this homomorphism is not an isomorphism, for – as can be seen at once – the center (that is the set of elements of the algebra which commute with the whole algebra) is mapped onto zero. But if g is semisimple the mapping is a isomorphism and that accounts for its importance.
Remark: Recently, the dual of the adjoint representation, the coadjoint representation, and especially the set of orbits of this transformation group has got some interest in the theory of representations of Lie groups (cf. [6], [7]).
For illustration, we shall present some of the concepts connected with exponential mappings for the case of matrix groups. We have already seen that the infinitesimal generators

are again matrices. As it is well-known, the exponential function of a matrix defined by its Taylor series

$$\exp tA = \sum_{n=0}^{\infty} \frac{(tA)^n}{n!} \qquad \text{converges for every } t\,.$$

From the identities $\exp(BAB^{-1}) = B(\exp A)B^{-1}$ and $(\exp A)^{-1} = -\exp A$ we get the corresponding identities for logarithms

$$\mathrm{Log}(BAB^{-1}) = B(\mathrm{Log}\,A)B^{-1}$$

and

$$\mathrm{Log}(A^{-1}) = -\mathrm{Log}\,A\,.$$

The following theorem in this context is very useful for obtaining topological properties of the manifolds of matrix groups [7].

4.16 Theorem: (Cartan decomposition of matrices) Let M be a regular $n \times n$ -matrix and put

$$B = \tfrac{1}{2}\,\mathrm{Log}(M^tM) \qquad A = \exp - B \tag{4.11}$$

then the matrices A and B exist and are real.
Moreover $M = A \exp B$

$$\text{and} \quad AA^t = 1 \qquad B^t = B \tag{4.12}$$

and the decomposition is unique. The mapping

$$M \to (A, B)$$

is a diffeomorphism of the manifold of GL(n, IR) into the manifold of the set of matrices (A, B) sufficing (4.12).
It can be shown by this theorem that, for example, the group GL(n, IR) is diffeomorphic to the manifold of $0(n) \times \mathrm{IR}^{n(n+1)/2}$.
Example: The Poincaré-group I 0 (3, 1, IR) (cf. section 2.4)
The Poincaré-group can be faithfully represented by 5×5 matrices of the form

$$P = \begin{pmatrix} & & & & c_1 \\ & & & & c_2 \\ & & L & & c_3 \\ & & & & c_4 \\ 0 & 0 & 0 & 0 & 1 \end{pmatrix}$$

where L is an element of the Lorentz group 0(3, 1, IR), a subgroup of GL(4, IR) and the c_i are real numbers. Those matrices where L is the unit metrix represent pure translations those where $c_i = 0$ homogeneous Lorentz transformations. L and P accordingly consist of four connected pieces; the connected component of the unit element is the proper orthochronous Poincaré (Lorentz) group to which we shall restrict our attention in the following. By (4.16) it can be shown that each component is two times connected (cf. section 4.1).

Since P is the semidirect product of the Lorentz group and the group of translations in $\mathbb{R}^4$ its manifold is the cartesian product of the manifolds of the corresponding pieces. Therefore every chart is the Cartesian product of charts of the pieces. An atlas of the translation group, for example, is a single chart homeomorphic to $\mathbb{R}^4$. We shall not determine an atlas of the Lorentz group (cf. [4]) for it is even not necessary to know a chart of the unit element explicitly in order to construct the Lie algebra so $(3, 1, \mathbb{R})$. We use the relation

$$^t L(t)\,(g_{ij})\,L(t) = (g_{ij})$$

with $g_{00} = 1$ and $g_{ij} = -\delta_{ik}$ for $(i, k) = (0,0)$. L(t) is a 1-parameter subgroup of L generated by an element $\frac{d}{dt} L(0) = \dot{L}$. If the defining relation of L is differentiated one obtains

$$^t\dot{L} g + g\dot{L} = 0 \qquad \text{or} \qquad ^t\dot{L} = -g\dot{L}g\,.$$

The following 5×5 matrices represent a basis of the Lie algebra of the Lorentz group

$$l_1 = \begin{pmatrix} 0 & 0 & 0 & 0 & 0 \\ 0 & 0 & 0 & 0 & 0 \\ 0 & 0 & 0 & -1 & 0 \\ 0 & 0 & +1 & 0 & 0 \\ 0 & 0 & 0 & 0 & 0 \end{pmatrix} \qquad b_1 = \begin{pmatrix} 0 & 1 & 0 & 0 & 0 \\ 1 & 0 & 0 & 0 & 0 \\ 0 & 0 & 0 & 0 & 0 \\ 0 & 0 & 0 & 0 & 0 \\ 0 & 0 & 0 & 0 & 0 \end{pmatrix}$$

$$l_2 = \begin{pmatrix} 0 & 0 & 0 & 0 & 0 \\ 0 & 0 & 0 & 1 & 0 \\ 0 & 0 & 0 & 0 & 0 \\ 0 & -1 & 0 & 0 & 0 \\ 0 & 0 & 0 & 0 & 0 \end{pmatrix} \qquad b_2 = \begin{pmatrix} 0 & 0 & 1 & 0 & 0 \\ 0 & 0 & 0 & 0 & 0 \\ 1 & 0 & 0 & 0 & 0 \\ 0 & 0 & 0 & 0 & 0 \\ 0 & 0 & 0 & 0 & 0 \end{pmatrix}$$

$$l_3 = \begin{pmatrix} 0 & 0 & 0 & 0 & 0 \\ 0 & 0 & -1 & 0 & 0 \\ 0 & 1 & 0 & 0 & 0 \\ 0 & 0 & 0 & 0 & 0 \\ 0 & 0 & 0 & 0 & 0 \end{pmatrix} \qquad b_3 = \begin{pmatrix} 0 & 0 & 0 & 1 & 0 \\ 0 & 0 & 0 & 0 & 0 \\ 0 & 0 & 0 & 0 & 0 \\ 1 & 0 & 0 & 0 & 0 \\ 0 & 0 & 0 & 0 & 0 \end{pmatrix}$$

The algebra of the translation group is represented by the matrices $t_i = (c^l_{ik}$ with $c^l_{ik} = 0$ for $k \neq 5, l \neq i)$. The action of any element of the Poincaré group on Minkowski space is given by the following matrix relations

$$\begin{pmatrix} x_0' \\ x_1' \\ x_2' \\ x_3' \\ 1 \end{pmatrix} = P \begin{pmatrix} x_0' \\ x_1' \\ x_2' \\ x_3' \\ 1 \end{pmatrix} \quad \text{or} \quad Px = Lx + c;\ x = \begin{pmatrix} x_0 \\ x_1 \\ x_2 \\ x_3 \end{pmatrix}$$

The differential operators related to the basis above are

$$l_i = -\sum_{j,k} \epsilon_{i\,jk}\, x_j \frac{\partial}{\partial x_k}; \qquad b_i = -\left(x_0 \frac{\partial}{\partial x_i} + x_i \frac{\partial}{\partial x_0} \right), \ i = 1, 2, 3$$

$$t_i = -\frac{\partial}{\partial x_i}, \ i = 0, 1, 2, 3$$

with $[a, b] = ab - ba$ for the matrix commutation relations we obtain

$$[l_i, l_j] = \epsilon^k_{ij}\, l_k, \quad [l_i, b_g] = \epsilon^k_{ij}\, b_k, \quad [b_i, b_j] = -\epsilon^k_{ij}\, b_k$$

$$[l_i, t_0] = 0 \qquad [l_i, t_j] = \epsilon^k_{ij}\, t_k$$

$$[b_i, t_0] = p_i \qquad [b_i, t_k] = \delta_{ik}\, p_0$$

$$[t_r, t_\mu] = 0 .$$

The complex Lie group $SL(2, \mathbb{C})$ can be considered as a real six-dimensional Lie group $SL(2, \mathbb{C})_{\mathbb{R}}$. Real part and imaginary part of each complex parameter are considered as independent real parameters. The Lie algebra $sl(2, \mathbb{C})_{\mathbb{R}}$ turns out to be isomorphic to the Lie algebra of the Lorentz group.

$SL(2, \mathbb{C})$ is simply connected. Therefore, according to 4.13, $SO(3, 1, \mathbb{R})$ is isomorphic to $SL(2, \mathbb{C})/_{I_2}$ where I_2 is the group consisting of the two elements

$$E = \begin{pmatrix} 1 & 0 \\ 0 & 1 \end{pmatrix} \qquad \text{and} \qquad -E = \begin{pmatrix} -1 & 0 \\ 0 & -1 \end{pmatrix}$$

The relation between $SL(2, \mathbb{C})$ and $SO(3, 1\ \mathbb{R})$ is analogous to that between $SO(3, \mathbb{R})$ and $SU(2)$ and in fact leads to the concept of spin in relativistic quantum theory (see section A. 7).

Finally we shall give some introductory remarks on the concept of the universal enveloping algebra of a Lie algebra. This concept plays an important role in theory and application of linear representations of Lie groups (cf. sect. 6 and chapter VI of this volume, and [1], [6]).

Let g be a Lie algebra. The multiplication $(\xi, \eta) \mapsto [\xi, \eta]$ is certainly not associative. We assign to g an associative algebra with unit, the universal enveloping algebra denoted by $U(g)$. This algebra is defined as the factor algebra $T(g)/I$ where $T(g)$ is the tensor algebra over the underlying vector space of g, and I is the two-sided ideal in $T(g)$ generated by the set of elements of the form $\xi \otimes \eta - \eta \otimes \xi - [\xi, \eta]$ for $\xi, \eta \in g$. It is the center of this algebra $U(g)$ which is of importance also in physics. For example, the operator J^2 of total angular momentum is an element of the center of enveloping algebra of the Lie algebra of the rotation group (cf. sect. A. 7 for similar elements corresponding to the Poincaré-group). Useful techniques for calculating basis elements of the center of $U(g)$ can be found in [1].

4.3 Invariant measures

In the theory of representations of groups the use of an integral over functions defined on the manifold (which are not necessarily elements of A_G) will be required. Certainly here cannot be the place to give an introduction into the theory of integration. Therefore we shall mainly limit ourselves to the case of Lie groups where at least locally some idea of integration is provided and we shall only indicate how to proceed to more general cases. We assume that locally the measure on the manifold which we need for integration is generated by one of the usual measures on Euclidean spaces, for example the Lebesque measure.

The operation of left multiplication transfers every sufficiently small neighbourhood U of the unit element of a Lie group G to every place of the group. So we get a measure of the whole group from a measure in a certain chart at e. But if we multiply by group elements sufficiently close to the unit element we may stay essentially in the same neighbourhood (i.e. there exists a smaller neighbourhood $V_g \subset U$ for every g which is entirely translated into U if multiplied by g). We have the intuitive idea that a translation should not alter the volume. Therefore, if f is any real (or complex) valued function, it is a reasonable reguirement for group integrals that

$$\int f(g)\, d\mu(g) = \int f(hg)\, d\mu(g) = \int f(hg)\, d\mu(hg) .$$

Since f is a rather arbitrary function, we demand of the measure defined on the group that it obeys

$$d\mu_L(g) = d\mu_L(hg)$$

Measures with this property are called *left-invariant measures.* Right-invariant measures can be similarly defined

$$d\mu_R(g) = d\mu_R(gh) .$$

The left (right) invariant measure is called a *left (right) invariant Haar measure.* Sometimes this notation is limited to the case where the measure can be defined by an invariant density

$$d\mu_L(g) = \rho_L(x)\, d^r x(g) \qquad \text{where} \qquad d^r x(g)$$

is the usual Euclidean volume element in the coordinate space of the chart. If f is the mapping defining the multiplication $z = xy$ on the Lie group in some chart, then ρ_L can be calculated according to the following identity

$$\rho_L(x) = \det \left[\frac{\partial f^i(x,y)}{\partial y^j} \right]^{-1}_{y=0}$$

where $\det\left(\frac{\partial f^i}{\partial y^j}\right)$ is the Jacobian for the mapping

$$f: G \to G : y \mapsto xy$$

for any fixed x. This formula can easily be checked for the additive group $\mathbb{R}$ where ρ_L turns out to be constant.

Left- and right-invariant measures are generally not equal and moreover, the integrals

$$\int_G \rho_L(x)\, d^r x(g) \qquad \text{and} \qquad \int_G \rho_R(x)\, d^r x(g)$$

may not converge. But if the group is a compact Lie group, the following theorem holds [4].

4.17 Theorem: The density functions ρ_L and ρ_R giving the left- and the right-invariant measures of a compact Lie group are equal.

In the following list some types of Lie groups are collected where right- and left-invariant measure are equal (cf. [4]).

(1) Abelian Lie groups (f.e. $\mathbb{R}^n$)

(2) Compact Lie groups (f.e. unitary and real orthogonal groups)

(3) Semisimple groups (f.e. the group $GL(n, \mathbb{C})$)

(4) Semidirect products of semisimple and Abelian groups (f.e. the Euclidean group $ISO(3, \mathbb{R})$ and the Poincaré-group $ISO(3, 1, \mathbb{R})$).

More general theorems on invariant measures are to be found in the literature (f.e. [6]). We shall only mention one further concept, which will be used later on.

If the group G is a general topoloical group the definition of a measure often becomes rather abstract. In this case sometimes the concept of an invariant mean can be used, the definition of which does not imply measurable sets.

Let X be a topological space and G a topological group acting as a topological transformation group on X. Let further L be a linear topological space of functions on X which is invariant under translations by elements of G. An *invariant mean* is a positive linear functional (that is a linear function on L which assumes non-negative values on non-negative functions) on L, that is invariant under the action of the group. This definition certainly applies to the case where G is the group of left-, right- or two-sided translations on its own manifold. The corresponding means are then also called left-, right- or two-sided respectively.

A group is called *amenable* if there exists a left-invariant mean on the space B(G) of continuous bounded functions on G with norm $\|f\| = \sup_{x \in G} |f(x)|$. Not all groups, even not all Lie groups are amenable. For example, any non-compact semisimple Lie group (the Lorentz group) is not amenable. On the other hand, it is known that every solvable group is amenable ([4]).

5 Representation of groups

In this section, we discuss homomorphisms of a group into linear transformation groups on linear spaces. Since most symmetry operations in quantum mechanics are implemented through linear operators, the knowledge of these homomorphisms is essential for most applications of group theory to systems described by quantum mechanics.

5.1 General properties of representations

5.1 Definition: A linear representation of a group G on a linear space E is a homomorphisr D of G into the group of all linear invertible operators on E.
In the physical applications, E may be a real or complex space of dimension m. Then, the representation is a homomorphism from G to $GL(m, \mathbb{C})$ or $GL(m, \mathbb{R})$. Alternatively, E may be the Hilbert space of a system in quantum mechanics.

5.2 Definition: If D is a linear representation of a group G and B a linear invertible operator on E, the representations D and

$$B \circ D \circ B^{-1}$$

are said to be equivalent.
For the physical applications, it should be stressed that equivalent representations of a group may be adapted to quite different physical situations.

5.3 Definition: If a representation D of the group G on E leaves a proper subspace of E invariant, it is called reducible, otherwise irreducible.
By an appropriate choice of the basis, any reducible representation may be brought to the block triangular form

$$D(g) = \begin{bmatrix} D_{11}(g) & D_{12}(g) \\ 0 & D_{22}(g) \end{bmatrix}$$

If moreover one can choose a basis of E such that $D_{12}(g) \equiv 0$, the representation D is called completely reducible.
In the following theorems 5.4–5.8 we shall assume that the group G has a left- and right-invariant measure as discussed in section 4. Hence, we shall assume that for an appropriate set of functions f on G there exists an integral

$$I(f) = \mathop{\Large\unicode{x2A0B}}_{G} f(g)\, d\mu(g)$$

such that for any $g' \in G$

$$\oint_G f(g)\, d\mu(g'g) = \oint_G f(g)\, d\mu(gg') = \oint_G f(g)\, d\mu(g)$$

For finite groups, the integral is taken as a sum over the group. The assumption holds true for compact Lie groups and for a number of other groups, compare Gilmore [1] p. 81. Under this assumption the following theorems can be proven:

5.4 Theorem: Any reducible representation D of a group G is fully reducible.

5.5 Theorem: (Schur's Lemma) Any operator M which commutes with all operators of an irreducible representation D of G is a multiple of the unit operator.

5.6 Theorem: Any representation D of G is equivalent to a unitary representation.

5.7 Theorem: Let D^a and D^b be irreducible representations of G of dimension $|a|$ and $|b|$ respectively. If a linear operator M fulfills

$$M \circ D^a(g) = D^b(g) \circ M \qquad \text{for all} \qquad g \in G$$

then

(1) for $|a| \neq |b|$ it follows that $M = 0$
(2) for $|a| = |b|$ and M invertible it follows that D^a and D^b are equivalent.

5.8 Theorem (orthogonality relations): Let D^a and D^b be irreducible representations of G and define

$$|G| = \oint_G d\mu(g)$$

The matrix elements of the two representations obey the orthogonality relations

$$\oint D^a_{ij}(g)\, D^b_{kl}(g)\, d\mu(g) = \delta(a,b)\, \delta_{il}\, \delta_{kj}\, \frac{|G|}{|a|}$$

$$\sum_{aij} \frac{|a|}{|G|} D^a_{ij}(g')\, D^a_{ji}(g^{-1}) = \delta(g', g)\,.$$

Here, a is a label which runs over all irreducible representations of G and we assume that for any irreducible representation D^a we have chosen a standard form.

The orthogonality relation may be rewritten by assuming the unitarity of the representations according to theorem 5.6. If for complex-valued functions on G we define the inner product

$$(f, h) = \oint \overline{f(g)}\, h(g)\, d\mu(g)\,,$$

then the orthogonality relations imply that the particular set of functions

$$D^a_{ij}(g) \cdot \left[\frac{|a|}{|G|}\right]^{\frac{1}{2}}$$

are an orthonormal and complete set on G.
We shall not discuss the proofs of theorems 5.4–5.8 as they are given in any standard book on representations of groups, and we prefer to consider specific groups in more detail.
An important tool in the analysis of a given representation is the character.

5.9 Definition: The character χ of a representation D is the function on G given by

$$\chi(g) = \text{trace } D(g) .$$

5.10 Proposition: The character is a function on conjugacy classes of G.

Proof: From its definition, the character obeys

$$\begin{aligned}\chi(g'g(g')^{-1}) &= \text{trace } [D(g') \circ D(g) \circ D^{-1}(g')] \\ &= \text{trace } D(g) \\ &= \chi([g])\end{aligned}$$

where [g] is the class of elements conjugate to g. ∎

5.11 Proposition: The characters χ^a, χ^b of pairs of irreducible representations obey the orthogonality relations

$$\oint\!\!\!\!\!\!\sum_G \chi^a(g)\chi^b(g^{-1})\,d\mu(g) = \delta(a, b)\ |G| .$$

Proof: The orthogonality follows by the introduction of traces into the general orthogonality theorems 5.8. ∎

5.12 Proposition: Suppose that the reducible representation D^α contains the irreducible representation D^a of G with multiplicity $m(\alpha, a)$. Then $m(\alpha, a)$ is given by

$$m(\alpha, a) = |G|^{-1} \oint\!\!\!\!\!\!\sum_G \chi^\alpha(g)\,\chi^a(g^{-1})\,d\mu(g) .$$

Proof: Imagine that D^α has been brought into explicitly reduced form. Then, from the orthogonality relations proposition 5.11, the inner product of the characters counts the number $|G|$ times the multiplicity of the irreducible representation D^a. But since the character $\chi^\alpha(g)$ is the same for equivalent representations, the result may be obtained from the original representation D^α without explicit reduction. ∎

Somer other aspects of the general theory representations are treated in later sections. The Kronecker product of representations and the adjoint and complex conjugate representation are discussed in section 5.3, subduction and induction of representations in section 5.5.

5.2 Irreducible representations of the symmetric group

In the present section, we describe the irreducible representations of the symmetric group S(n). We stress some aspects which will be used in many-body applications.

5.13 Definition: An (ordered) partition

$$f = [f_1 f_2 \dots f_j]$$

of n is a set of j integers fulfilling

$$\sum_{i=1}^{j} f_i = n\,, \qquad f_1 \geqq f_2 \geqq \dots \geqq f_j > 0$$

To any partition we associate a Young diagram by arranging f_i boxes into the row i for $i = 1\,2 \dots j$.

5.14 Proposition: The irreducible representations of the symmetric group S(n) are in one-to-one correspondence to the partitions of n. The representation with partition $f = [n]$ is the identity representation of S(n). The representation with the partition $f = [1\,1 \dots 1]$ is the one-dimensional antisymmetric representation.

Example: The group S(4) has five irreducible representations charaterized by the partitions

$$f = [4], [31], [22], [211], [1111]$$

Consider now the subgroup

$$S(w_1) \times S(w_2) < S(n), \qquad w_1 + w_2 = n$$

Let d^{f_1}, d^{f_2}, d^f be the irreducible representations of the groups $S(w_1), S(w_2)$ and S(n). The irreducible representations of $S(w_1) \times S(w_2)$ are then given by the Kronecker products $d^{f_1} \times d^{f_2}$. We are interested in the reduction of the irreducible representation d^f of S(n) under restriction to the subgroup $S(w_1) \times S(w_2)$. The corresponding multiplicities as used in proposition 5.12 are denoted as $m(f, f_1 \times f_2)$.

5.15 Proposition: The irreducible representation d^f of S(n) when restricted to its subgroup $S(w_1) \times S(w_2)$ contains the subgroup representation $d^{f_1} \times d^{[w_2]}$ with multiplicity $m(f, f_1 \times [w_2]) = 1$ iff $f = [f_1 f_2 \dots f_j]$ and $f_1 = [f_{11} f_{21} \dots f_{j1}]$ obey the inequalities

$$f_1 \geqq f_{11} \geqq f_2 \geqq f_{21} \geqq f_3 \geqq \dots \geqq f_{j-1} \geqq f_{j-11} \geqq f_j \geqq f_{j1}$$

Note that we selected the special representation $f_2 = [w_2]$. The content of proposition 5.15 is best seen from the Young diagrams f and f_1. The diagram f_1 is obtained from the diagram f by removal of w_2 boxes from the shaded region in Fig. 5.1. Consider now a weight of n,

$$w = (w_1 w_2 \dots w_j)\,, \qquad w_1 + w_2 + \dots w_j = n\,,$$

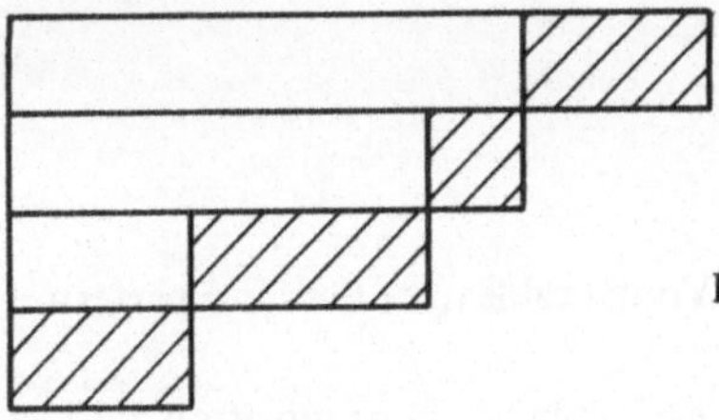

Fig. 5.1

the corresponding subgroup

$$S(w_1) \times S(w_2) \times \ldots \times S(w_j) < S(n)$$

and the reduction of the irreducible representation d^f of $S(n)$ to the identity representation of this subgroup. The multiplicity in this reduction may be obtained by repeated application of proposition 5.15 according to the scheme

$$\begin{aligned} S(n) &> S(n-w_j) \times S(w_j) \\ &> S(n-w_{j-1}-w_j) \times S(w_{j-1}) \times S(w_j) \\ &\vdots \\ &> S(w_1) \times S(w_2) \times \ldots \times S(w_j)\,. \end{aligned}$$

The algorithm implied by this stepwise procedure is expressible by the concept of a generalized Young tableau or a Gelfand pattern.

5.16′ Definition: A generalized Young tableau is a Young diagram f whose n boses are labelled by the j numbers $1\ 2 \ldots j$ with repetition frequencies $w_1\, w_2 \ldots w_j$. The labelling is such that the numbers increase with increasing row or column number.

5.16″ Definition: A Gelfand pattern q associated with the partition f is a triangular arrangement of j partitions of the numbers $n-w_j, n-w_{j-1}-w_j, \ldots, w_1$ of the form

$$\begin{array}{ccccccccc} f_{1j-1} & & f_{2j-1} & & \cdots & & f_{j-2j-1} & & f_{j-1j-1} \\ & f_{1j-2} & & f_{2j-2} & \cdots & & & f_{j-2j-2} & \\ & & \ddots & & & & \iddots & & \\ & & & f_{12} & & f_{22} & & & \\ & & & & f_{11} & & & & \end{array}$$

where the inequalities given in proposition 5.15 are valid for the top partition with respect to the partition f and for any other partition with respect to the next one on top of it.

5.17 Proposition: If the irreducible representation d^f of $S(n)$ is restricted to the subgroup

$$S(w_1) \times S(w_2) \times \ldots \times S(w_j) < S(n)\,,$$

each occurence of the identity representation

$$d^{[w_1]} \times d^{[w_2]} \times \ldots \times d^{[w_j]}$$

of this subgroup is labelled by one allowed generalized Young tableau or Gelfand pattern.

Proof: We use the process by reduction in steps and in each step apply proposition 5.15. The w_j boxes removed in the first step from f we label by the number j, the w_{j-1} ones removed in the second step by $j-1$, etc. The overall reduction through these steps is described by a generalized Young tableau, and any generalized Young tableau yields one possible reduction. The Gelfand pattern q associated with a generalized Young tableau is found by reading off the intermediate partitions. ■

Example: A Young tableau and corresponding Gelfand pattern for the weight $w = (2\,3\,2\,3)$. We put the partition f on top of the Gelfand pattern:

Young tableau

1	1	2	3
2	2	4	
3	4		
4			

Gelfand pattern

$$\begin{array}{ll} f: & 4 \quad 3 \quad 2 \quad 1 \\ q: & \left\{ \begin{array}{c} 4 \quad 2 \quad 1 \\ 3 \quad 2 \\ 2 \end{array} \right. \end{array}$$

The interest in these particular subgroups of S(n) and the particular representation will become apparent in section 5.3 and in the physical application. Consider now the matrix elements of the representation d^f of S(n) where two weights

$$\tilde{w} = (\tilde{w}_1 \tilde{w}_2 \ldots \tilde{w}_j), \qquad w = (w_1 w_2 \ldots w_j)$$

have been chosen along with corresponding Gelfand patterns $\tilde{q}, q$ to label the rows and columns. Any permutation $p \in S(n)$ has a double coset decomposition with respect to the subgroups associated with these two weights and may be written as

$$p = \tilde{h}\, z_k h\,,$$

$$\tilde{h} \in S(\tilde{w}_1) \times S(\tilde{w}_2) \times \ldots \times S(\tilde{w}_j)$$

$$h \in S(w_1) \times S(w_2) \times \ldots \times S(w_j)\,.$$

But then, the matrix elements of p must depend only on z_k,

$$d^f_{\tilde{q}q}(p) = d^f_{\tilde{q}q}(z_k)$$

since the actions of $\tilde{h}$ and h on these functions on the group only involve the identity representation of the subgroups.

For the weights $\tilde{w} = w = (1\,1 \ldots 1)$ one obtains the so-called Young tableau and Young's orthogonal representation. The numerical values of the representation matrices are determined by the axial distances of boxes.

5.18 Definition: Denote by (α_i, β_i) and (α_l, β_l) the row and column positions of two boxes in a Young diagram f. The axial distance of the two boxes is given by

$$\tau_{il} = \beta_i - \beta_l + \alpha_l - \alpha_i = -\tau_{li}$$

We now give the numerical value of the representation d^f for the permutation

$$p = \begin{bmatrix} 1\,2\ldots n-2\;n-1\;n \\ 1\,2\ldots n-2\;n\;n-1 \end{bmatrix}$$

Since this permutation commutes with the subgroup S(n−2), its representation depends at most on the position of the numbers n−1 and n in the standard Young tableau. Let m and $\bar{m}$ be two standard Young tableaus obtained from one another by interchange of the numbers n−1 and n. Then, for any such pair, the representation d^f has the form

	m	$\bar{m}$
m	τ_{nn-1}^{-1}	$[1-\tau_{nn-1}^{-2}]^{\frac{1}{2}}$
$\bar{m}$	$[1-\tau_{nn-1}^{-2}]^{\frac{1}{2}}$	$-\tau_{nn-1}^{-1}$

where τ_{nn-1} is the axial distance computed from the standard tableau m. If $\bar{m}$ as described above is not an allowed Young tableau, the form given above reduces to the single entry τ_{nn-1}^{-1} which has the value ± 1.

Example: To the representations f = [n] and f = [11 ... 1] there corresponds a single Young tableau respectively. One easily verifies that

$$d^{[n]}(p) = 1 \qquad d^{[11\ldots1]}(p) = -1 .$$

Since arbitrary permutations may be generated from the interchanges of j−1 with j for j = 2 3 ... n, the full representation of S(n) may be generated from the values given above. The standard representation theory of S(n) is treated in much detail by Hamermesh [2]. The computation of the general coefficients $d^f_{\tilde{q}q}(z_k)$ on the basis of axial distances is treated in [3, 4]. The concept of Gelfand patterns was introduced by Gelfand and Zetlin [5] in relation to the unitary group. Its use for the theory of irreducible representations of the symmetric group is explained in [4].

5.3 Finite irreducible representations of the general linear group

The finite irreducible representations of the general linear group GL(j, ℂ) are constructed by reduction of Kronecker product representations.

5.19 Definition: Let D^a, D^b be two representations of a group G. The Kronecker product representation $D^a \times D^b$ is the representation obtained by defining

$$(D^a \times D^b)(g) = D^a(g) \times D^b(g) .$$

Note that the right-hand side is just the Kronecker product of the matrices of the representations D^a and D^b.

If $\varphi = (\varphi_\beta)$ is a basis of the complex linear space E underlying the action of GL (j, ℂ), we have for $g \in GL(j, \mathbb{C})$

$$D(g)\,\varphi_\beta = \sum_\alpha \varphi_\alpha D_{\alpha\beta}(g)$$

where the representation D consists of the elements of GL(j, ℂ). Then, by taking n copies of this basis labelled by an upper index, we obtain a new basis with elements

$$\varphi^{(i)}_{\beta(i)} = \varphi^{1}_{\beta_1}\varphi^{2}_{\beta_2}\dots\varphi^{n}_{\beta_n} \qquad \beta_i = 1\,2\dots j$$

of the tensor product space $E \times E \times \dots \times E = E^n$. The representation of GL(j, ℂ) obtained by the action on each factor is the nth Kronecker power representation $D^{(n)}$. Now it is easy to verify that the action of the symmetric group S(n) on the upper tensor indices commutes with the action of the group GL(j, ℂ). This observation allows one to break the space E^n into subspaces characterized by irreducible representations f of S(n). Since the index β can take at most j values, the partitions f are restricted to at most j non-vanishing components. This decomposition can be shown to yield irreducible representations of GL(j, ℂ).

5.20 Theorem: The irreducible representations of the group GL(j, ℂ) contained in the Kronecker power $D^{(n)}$ are characterized by the partitions f of n with at most j components.

A canonical form of the representations of GL (j, ℂ) arises by demanding that the representation be explicitly reduced with respect to the subgroups

$$\begin{aligned} GL(j,\mathbb{C}) &> GL(j-1,\mathbb{C}) \times GL(1,\mathbb{C}) \\ &> GL(j-2,\mathbb{C}) \times GL(1,\mathbb{C}) \times GL(1,\mathbb{C}) \\ &\dots \end{aligned}$$

It is found that the partitions which characterize the sequence representations of the subgroups GL(j−1, ℂ), GL(j−2, ℂ), ... are related to each other by the rules underlying a Gelfand pattern q according to definition 5.16″. In other words, the irreducible representations of GL(j, ℂ) may be given as matrices whose rows and columns are labelled by Gelfand patterns $\tilde{q}$ and q respectively. An explicit form of these representations can be given by the following theorem:

5.21 Theorem: If $g = (\epsilon_{ij})$ is an arbitrary element of GL(j, ℂ), the matrix elements of the irreducible representation D^f of GL(j, ℂ) are given by

$$D^f_{\tilde{q}q}(\epsilon) = \left[\prod_i \tilde{w}_i!\ \Pi w_l!\right]^{\frac{1}{2}}$$

$$\sum_k d^f_{\tilde{q}q}(z_k)\left[\prod_{i,l=1}^{j} k_{il}!\right]^{-1} \prod_{i,l=1}^{j} (\epsilon_{il})^{k_{il}}$$

where the sum runs over all double cosets of the subgroups of S(n) associated with the weights $\tilde{w}$ and w.
The proof of this theorem is given in [4].
This description does not exhaust the finite irreducible representations of the group $GL(j, \mathbb{C})$. Additional representations are obtained by considering automorphisms of $GL(j, \mathbb{C})$.

5.22 Proposition. Consider the group $GL(j, \mathbb{C})$ with elements g. The maps $GL(j, \mathbb{C}) \to GL(j, \mathbb{C})$ defined by

the complex conjugation operation	:	$g \to \bar{g}$
the contragredient operation	:	$g \to {}^t g^{-1}$
the complex conjugate contragredient conjugation	:	$g \to {}^t \bar{g}^{-1}$

are automorphisms of $GL(j, \mathbb{C})$.

5.23 Definition: Given a representation D of a group G on a complex linear space, the new representations obtained by the three automorphisms specified in proposition 5.22 are called the complex conjugate, the contragredient or adjoint, and the complex conjugate contragredient representation.
For the general linear group $GL(j, \mathbb{C})$, the process of taking the nth Kronecker power and reducing with respect to the symmetric group commutes with the operations given in proposition 5.22. This leads to the equations

$$\bar{D}(g) = D(\bar{g})$$
$${}^tD^{-1}(g) = D({}^tg^{-1})$$
$${}^t\bar{D}^{-1}(g) = D({}^t\bar{g}^{-1}) \,.$$

The most general finite irreducible representation of the group $GL(j, \mathbb{C})$ may now be characterized as a Kronecker product with factors from all four possible types of representations. If subgroups are considered, there may occur additional equivalences. For the subgroup

$$SL(j, \mathbb{C}) < GL(j, \mathbb{C})$$

it is found that

$${}^t(D^f)^{-1}(h) \sim D^{f_c}(h), \qquad h \in SL(j, \mathbb{C})$$

where the partitions f and f_c are related by the diagram given in Fig. 5.2. In this case the contragredient representation is eliminated. Moreover, we may restrict the partitions to those with $j-1$ nonvanishing components since all partitions of the type $[f_1 + f_j \; f_2 + f_j \dots f_{j-1} + f_j f_j]$ lead to the same partition f_c.
For the unitary group

$$U(j) < GL(j, \mathbb{C})$$

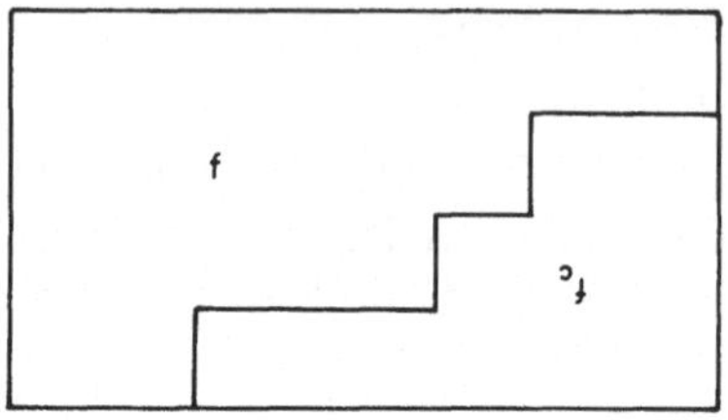

Fig. 5.2

we have

$$^{t}(\bar{D}^{f})^{-1}(h) = D(h)$$

which shows that the representation itself is unitary and eliminates the complex conjugate contragredient representation. Finally, for the group SU(j) both equivalences are valid and we end up with a single type of unitary representations.

Example 1: Irreducible representations of the group SU(2)

The irreducible representations are characterized by partitions $f = [n0]$. The Gelfand patterns are identical to the weights $\tilde{w} = (\tilde{w})$ and $w = (w)$. Application of proposition 5.21 yields

$$D^{[n0]}_{\tilde{w}w}(u) = [\tilde{w}!\,w!]^{\frac{1}{2}} \sum_{k} \left[(k_{11}!\,k_{12}!\,k_{21}!\,k_{22}!)^{-1}\; u_{11}^{k_{11}}\, u_{12}^{k_{12}}\, u_{21}^{k_{21}}\, u_{22}^{k_{22}}\right]$$

since $d^{[n0]}_{\tilde{w}w}(z_k) \equiv 1$ as we are dealing with the identity representation of S(n). The sum over double cosets may be reduced to a single summation index since the numbers k_{il} must fulfill the equalities

$$k_{11} + k_{21} = w \qquad k_{12} + k_{22} = n - w$$
$$k_{11} + k_{12} = \tilde{w} \qquad k_{21} + k_{22} = n - \tilde{w}\,.$$

The correspondence to the usual quantum numbers associated with angular momentum is obtained by introduction of

$$j = \tfrac{1}{2}\,n \qquad \tilde{m} = \tfrac{1}{2}\,(2\,\tilde{w} - n) \qquad m = \tfrac{1}{2}\,(2\,w - n)$$

Then, the representation matrices become the usual functions $D^{j}_{\tilde{m}m}$ discussed for example by Edmonds [8].

Example 2: Irreducible representations of SL(2, ℂ)

The finite irreducible representations are Kronecker products of the representations D^{f} and $\bar{D}^{f}$. The four-dimensional representation $D^{[1]} \times \bar{D}^{[1]}$ yields the two-to-one correspondence to the proper orthochronous Lorentz group acting in Minkowski space. The spinors with four components transform according to the reducible representation $D^{[1]} + \bar{D}^{[1]}$ which induces an irreducible representation upon introduction of parity as a linear operator.

For the finite representations of the general linear group we refer in particular to Bacry [6]. Theorem 5.21 is discussed in [7]. This reference contains many algebraic properties derived from this theorem.

5.4 Group algebras and representations

A complex linear (vector) space V involves elements (vectors) $v_\alpha \in V$, an operation + for vector addition, and an operation $\cdot$ for multiplication with complex numbers $z \in \mathbb{C}$, we denote it more explicitly by $(V, \mathbb{C}, +, \cdot)$.

5.24 Definition: A *linear algebra* $A = (V, \mathbb{C}, +, \cdot, \Delta)$ over the complex field is a complex linear space with a second internal composition law Δ. The composition

$$v_\alpha \,\Delta\, v_\beta$$

is required to be compatible with vector addition according to

$$(v_\alpha + v_\beta)\,\Delta\, v_\gamma = (v_\alpha \,\Delta\, v_\gamma) + (v_\beta \,\Delta\, v_\gamma)$$
$$v_\alpha \,\Delta\, (v_\beta + v_\gamma) = (v_\alpha \,\Delta\, v_\beta) + (v_\alpha \,\Delta\, v_\gamma)\,.$$

Particular cases of linear algebras arise if, with respect to Δ, the associative law holds, the Lie properties hold or if there exists a unit element.
If we introduce the formal sums

$$a = \sum_{g \in G} f(g)\, g$$

where f is a function $f : G \to \mathbb{C}$, these elements can be taken as vectors of a linear complex space of dimension $|G|$. Moreover, by using the group multiplication law we may define for

$$a' = \sum_{g \in G} f'(g)\, g, \qquad a'' = \sum_{g \in G} f''(g)\, g$$

the internal composition

$$a'a'' = \sum_{g \in G} f(g)\, g$$

where f(g) is obtained by convolution

$$f(g) = \sum_{g' \in G} f'(g')\, f''(g'^{-1} g)\,.$$

5.25 Definition: The construction outlined above renders a linear associative algebra A_G called the group algebra with an identity element

$$e = \sum_{g \in G} f_e(g) g \qquad \text{where } f_e(g) = \delta(g, e).$$

5.26 Definition: The Young operators for irreducible representations D^a of G are the algebra elements

$$c_{ij}^{a} = \frac{|a|}{|G|} \sum_{g \in G} D_{ji}^{a}(g^{-1})g \,. \tag{Y0}$$

5.27 Proposition: The Young operators have the following properties:

$$c_{ij}^{a} c_{kl}^{b} = \delta(a,b)\, \delta_{jk} c_{il}^{a} \tag{Y1}$$

$$g\, c_{ij}^{a} = \sum_{m} c_{mj}^{a} D_{mi}^{a}(g) \tag{Y2}$$

$$c_{ij}^{a} g = \sum_{m} c_{im}^{a} D_{jm}^{a}(g) \tag{Y3}$$

$$g = \sum_{aij} c_{ij}^{a} D_{ij}^{a}(g) \,. \tag{Y4}$$

Eq. (Y1) determines the multiplication law. The other equations may be interpreted as follows: If the group elements g are taken as basis of the algebra A_G, eq. (Y0) yields a linear decomposition into linear subspaces which, according to eq. (Y2) and (Y3), are invariant under the action of G from right and left. Eq. (Y4) yields the decomposition of the original basis into the invariant parts.

Any representation D of G provides a representation of the group algebra A_G by

$$D: g \to D(g), \; a = \sum_{g \in G} f(g) g \to D(a) = \sum_{g \in G} f(g)\, D(g) \,.$$

Suppose now that the representation D(g) has matrix elements $D_{\alpha\beta}(g)$ in some basis and consider its explicit irreducible decomposition by an operator M,

$$D_{\alpha\beta}(g) = \sum_{aij} M_{\alpha,ai} D_{ij}^{a}(g)\, M_{aj,\beta}^{-1} \,.$$

Then the elements of M may be obtained from

$$D_{\alpha\beta}(c_{ij}^{a}) = M_{\alpha,ai} M_{aj,\beta}^{-1}$$

if no irreducible representation appears more than once or

$$D_{\alpha\beta}(c_{ij}^{a}) = \sum_{q} M_{\alpha,qai} M_{qaj,\beta}^{-1}$$

if the index q labels multiplicities.

5.5 Induced and subduced representations

In this section, we consider some relations between the representations of a group and a subgroup.

5.28 Definition: Let D^a be a representation of a group G. Its restriction to a subgroup $H < G$ we call the subduced representation $D^{a\downarrow}$ of H.
Suppose now that a representation D^α of $H < G$ is given and that a representation D of G has the property that for $h \in H$

$$D(gh) = D(g) \circ D^\alpha(h)\,.$$

Choosing a set of left coset generators c_i of H in G and decomposing $g \in G$ in the form $g = c_i h$ detemines D(g) essentially as a function of the coset generators c_i.

5.29 Definition: The representation of G induced by D^α is denoted by $D^{\alpha\uparrow}$ and defined by

$$D(g') \circ D(c_i) = D(g'c_i) = \sum_l D(c_l)\, D^{\alpha\uparrow}_{li}(g')\,.$$

This representation is given explicitly by

$$D^{\alpha\uparrow}_{li}(g') = \delta(g'c_i, c_l h')\, D^\alpha(c_l^{-1} g' c_i)\,.$$

Induced and subduced representations are linked by the following propositions:

5.30 Proposition (reciprocity theorem): Let D^a and D^α be irreducible representations of the groups G and $H < G$ respectively. Then the multiplicities in induction and subduction are related by

$$m(\alpha\uparrow, a) = m(a\downarrow, \alpha)\,.$$

5.31 Proposition. For D^a and D^α as in proposition 5.30, the induced representation has the explicit reduction

$$D^{\alpha\uparrow}_{l\lambda, i\nu}(g') = \sum_{qq'ars} \left[\frac{|a|}{|G|}\,\frac{|H|}{|\alpha|}\right]^{\frac{1}{2}} D^a_{q\alpha\lambda, r}(c_l^{-1})\, D^a_{rs}(g')$$

$$\left[\frac{|a|}{|G|}\,\frac{|H|}{|\alpha|}\right]^{\frac{1}{2}} D^a_{s q'\alpha\nu}(c_i)$$

where the indices q,q′ label the multiplicities in the subduction of D^a.

Proof: The proposition follows by application of Young operators to the induced representation.

As an application consider the representation of the group S(n) induced by the identity representation of the group $S(w_1) \times S(w_2) \times \ldots \times S(w_j)$ associated with the weight $w = (w_1\, w_2\, \ldots\, w_j)$. The subduction was shown in proposition 5.17 to be expressible through all possible Gelfand patterns compatible with a fixed partition f and the weight w. By proposition 5.30 it is clear that the same Gelfand patterns serve to label the reduction of the induced representation into its irreducible parts.

There are cases in which the irreducible representations of a group are obtained by induction from an appropriate subgroup. This applies in particular to the semidirect product groups. An application to the Poincaré-group is given in section 6.

We conclude this section by giving some general references. For the representations of groups we mention the books by Hamermesh [2], Jansen and Boon [8], Bacry [6] and Barut and Razka [9].

6 Induced representations of the Poincaré group

In this section we will construct the irreducible representations of the Pincaré group by means of the inducing procedure, and discuss the connection between irreducible representations and wave equations. As we want only to give an outline of the main ideas, we relax mathematical rigour. The mathematical details can be found in the literature [1] ... [5].

6.1 Poincaré transformations

The Poincare transformations P are defined as a group of transformations (a, Λ) on the 4-dimensional Euclidean space $\mathbb{R}^4$, acting in the following way:

$$(a, \Lambda): x_\mu \to x'_\mu = a_\mu + \Lambda^\nu_\mu x_\nu, \mu, \nu = 0 \ldots 3$$

with a_μ as translations and Λ as homogeneous Lorentz transformations which leave the indefinite form $x_\mu x^\mu := x_0^2 - (x)^2 := x_\nu g^{\nu\mu} x_\mu$ invariant. $g^{\mu\nu}$ is the metric tensor, $g^{00} = -g^{11} = -g^{22} = -g^{33} = 1, g^{\mu\nu} = 0, \mu \neq \nu$. The Λ fulfill the relation

$$\Lambda^T g \Lambda = g \tag{6.1}$$

We deduce the following composition law for Poincaré transformations P:

$$(a', \Lambda')(a, \Lambda) = (a' + \Lambda' a, \Lambda' \Lambda) \tag{6.2}$$

P is a group because $(a' + \Lambda' a, \Lambda'\Lambda)$ is again an element of P. P is a 10-parametric, non compact Lie group. Furthermore, P is a semidirect product of the translation group T and the homogeneous Lorentzgroup L written as

$$P = T \curlyvee L$$

To see this, we recall the definition of the semidirect product. Let T and L be two groups. Consider the pairs (a, Λ), $a \in T, \Lambda \in L$. There exists a homomorphism from L in the group of automorphisms of T:

$$\sigma : L \to \mathrm{Aut}\, T\,.$$

If we define a composition law for the pairs (a, Λ) by

$$(a', \Lambda')(a, \Lambda) = (a' \sigma_{\Lambda'}(a), \Lambda'\Lambda) \tag{6.3}$$

with $\sigma_{\Lambda'}$ the image of Λ' under σ, then the set $\{(a, \Lambda)\}$ becomes a group called the semi-direct product of T and L. By comparing (6.3) and (6.2) we see that for the Poincaré group $\sigma_{\Lambda'}$ is Λ' itself. For later applications we notice that the translation subgroup is an abelian invariant subgroup of P.

As is easily seen, P acts transitively on $\mathbb{R}^4$, i.e. $\mathbb{R}^4$ is a homogeneous space of P. Then we know that $\mathbb{R}^4$ must be realizable as a coset space of P modulo an isotropy group. We chose as the isotropy group the Lorentzgroup L, then the coset space P/L contains essentially only the translations which, regarded as a vectorspace, are isomorphic to the Euclidean space $\mathbb{R}^4$. Therefore we can write:

$$P/L \cong \mathbb{R}^4$$

which is the representation of the $\mathbb{R}^4$ as the coset space of P with respect to L.

6.2 Representations

The Poincaré group is a noncompact group, therefore no irreducible unitary finite-dimensional representations exist. We look for the infinitesimal generators of an unitary irreducible representation. For a Poincaré transformation close to the identity we get

$$x'_\mu = (g^\nu_\mu + \omega^\nu_\mu)\, x_\nu + a_\mu \; .$$

From (6.1) follows $\omega^{\mu\nu} = -\,\omega^{\nu\mu}$. Consider a representation $U(a, \Lambda)$. Expanding $U(a, \Lambda)$ near the identity, one gets

$$U(a, \Lambda) = 1 + i\,(P_\mu a^\mu + \tfrac{1}{2}\,\omega_{\mu\nu}\, I^{\mu\nu})$$

where P_μ, $I^{\mu\nu}$ are hermitean operators. P_μ are the generators of the translations, $I^{\mu\nu}$ of the four-rotations with $I^{\mu\nu} = -\,I^{\nu\mu}$. From

$$U(0, \Lambda)\; U(a, \Lambda)\; U(0, \Lambda^{-1}) = U(\Lambda a, \Lambda)$$

we get by taking 'a' infinitesimal

$$U(\Lambda)\, P^\mu\, U(\Lambda^{-1}) = P^\nu \Lambda^\mu_\nu \, .$$

From

$$U(0, \Lambda)\; U(0, \Lambda')\; U(0, \Lambda^{-1}) = U(0, \Lambda\Lambda'\Lambda^{-1})$$

we get by taking Λ' close to the identiy:

$$U(\Lambda)\, I^{\mu\nu}\, U(\Lambda^{-1}) = I^{\varphi\sigma}\, \Lambda^\mu_\varphi\, \Lambda^\nu_\sigma$$

i.e. P^μ transforms like a vector and $I^{\mu\nu}$ like a tensor under Lorentz transformations. Taking now Λ infinitesimal, we get the Lie-algebra of the Poincaré group

$$[I_{\mu\nu}, I_{\varphi\sigma}]_- = i\,(g_{\mu\varphi}\, I_{\nu\sigma} - g_{\nu\varphi}\, I_{\mu\sigma} + g_{\nu\sigma}\, I_{\mu\varphi} - g_{\mu\sigma}\, I_{\nu\varphi})$$

$$[P_\mu, P_\nu]_- = 0$$

$$[I_{\mu\nu}, P_\lambda]_- = i(g_{\mu\lambda}\, P_\nu - g_{\nu\lambda}\, P_\mu) \, .$$

It is easily verified that the operators $P_\mu P^\mu$ and $\Gamma_\mu \Gamma^\mu$ with $\Gamma_\mu = \frac{1}{2} \epsilon_{\mu\nu\varphi\sigma} I^{\nu\varphi} P^\sigma$ commute with all generators of the Poincaré group, i.e. $P_\mu P^\mu$ and $\Gamma_\mu \Gamma^\mu$ are the Casimir operators of the Poincaré group. By Schur's Lemma, they are multiples of the identity for an irreducible representation and can therefore serve as good quantum numbers. $P_\mu P^\mu$ and $\Gamma_\mu \Gamma^\mu$ in this case are interpreted as mass and spin squared:

$$P_\mu P^\mu = m^2$$

$$\Gamma_\mu \Gamma^\mu = s(s+1)$$

P is then interpreted as operator of the total four momentum. Because the Casimir operators are c-numbers in an irreducible representation, they are constants of motion. Because of this we may say, the U.I.R.'s describe elementary physical systems characterized by mass and spin.

We should mention a topological peculiarity: The Poincaré group is not simply connected, i.e. there exist closed curves which cannot be contracted to a point. This is, because P has the rotation group $O(3, \mathbb{R})$ as a subgroup which itself is not simply connected because the points of $O(3, \mathbb{R})$ can be identified with the points of a sphere where the north- and the southpole are regarded as equal. A curve connecting these two points is closed, but cannot be contracted to a point. From a representation theoretic point of view it is necessary to regard the so called universal covering group of P, i.e. the group which is simply connected and locally isomorphic to P (that means, they have the same infinitesimal generators). The universal covering group P is the $ISL(2, \mathbb{C})$, i.e. the semidirect product of T with $SL(2, \mathbb{C})$:

$$ISL(2, \mathbb{C}) = T \rtimes SL(2, \mathbb{C}) .$$

There exists a two-to-one homomorphism from $ISL(2, \mathbb{C})$ to P, this gives raise to spin-representations. To simplify our notation, we ignore this, but the reader should have in mind that if we say P, we mean $ISL(2, \mathbb{C})$.

Let us make a final remark concerning relativistic quantum field theory. There exists the possibility to get finite results from a nonrenormalizable theory by the introduction of an indefinite metric in the state space. Because of the required invariance of the observable quantities, one then is forced to introduce nonunitary representations of the Poincaré group. Then the difficulty arises that the probability interpretation of the theory is destroyed and has to be restored by a procedure called 'unitarization'. An example for such a situation is Heisenberg's nonlinear spinor-theory of elementary particles.

6.3 Induced representations

We want to construct the UIR's of the Poincaré group by means of the inducing construction. This method is very old and was first used by Frobenius. In his famous paper [7], Wigner applied it to the Poincare group. The theory of induced representations was worked out in full mathematical rigour by Mackey [1]. In the following we summarize the main features of induced representations and give as an application the UIR's of the Poincaré group. Let G be a group with "good" topological properties, i.e. locally compact, sepa-

rable. Let K be a closed subgroup of G, $K \to L(K)$ an unitary representation of K in a Hilbert space H(L).
Consider maps from G to H(L):

$$f: g \to f(g) \in H(L), g \in G$$

with the following covariance property

$$f(g \cdot k) = f(g) L(k), g \in G, k \in K. \qquad (6.4)$$

For the scalar product in H(L) we have:

$$(f(g \cdot k), f(g \cdot k)) = (f(g) L(k), f(g) L(k)) = (f(g), f(g)) \qquad (6.5)$$

because of the unitary of L. (6.5) shows that the scalar product of H(L) is constant on the cosets gK, $g \in G$. Thus (f(g), f(g)) defines a function on the cosetspace N = G/K. Now let $d\mu(x)$ be an invariant measure on N = G/K, i.e. $d\mu(x) = d\mu(gx)$, $\forall g \in G$. Then we also require for the set $Z := \{f, \text{ which fulfill (6.4)}\}$ that

$$\int_N d\mu(x)\,(f(x), f(x)) < \infty.$$

Defining now a scalar product in Z by

$$\langle f_1, f_2 \rangle := \int_N (f_1(x), f_2(x))\, d\mu(x), f_1, f_2 \in Z,$$

Z becomes a Hilbert space after completion. Now we define:

$$(U_x^L f)(y) := f(xy), x, y \in G. \qquad (6.6)$$

(6.6) defines a representation U^L of G which depends on the choice of L. We call U^L the representation induced by the unitary representation L of K. U is itself unitary:

$$\int_N ((U_y^L f)(x), (U_y^L g)(x))\, d\mu(x) = \int_N (f(yx), g(yx))\, d\mu(x) = \int_N (f(x), g(x))\, d\mu(x).$$

We have assumed that the topological properties of G/K are nice enough to guarantee the existence of the invariant measure $d\mu(x)$. In this case, $d\mu(x)$ is essentially unique (apart from a factor). This applies for the Poincaré group, therefore we shall not treat here the more complicated case where only quasi invariant measures exist on G/K [1].

6.4 UIR's of the Poincaré group

Now we want to construct the UIR's of the Poincaré group P by means of the inducing procedure.

Let $a \to e^{i\mathring{p}a}$ be the irreducible representations of the translation subgroup T of P, labelled by four-vectors $\mathring{p}$. Now we look for the isotropy group of $\mathring{p}$, i.e. for all transformation $\{r\}$ which leave $\mathring{p}$ fixed:

$$R_{\mathring{p}} := \{r ; r\mathring{p} = \mathring{p}\}$$

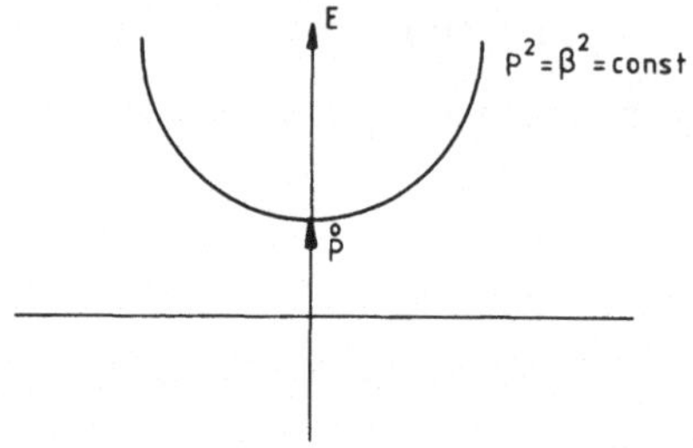

Fig. 6.1

$R_{\mathring{p}}$ is called the little group of $\mathring{p}$. Now we form the semidirect product $T \rtimes R_{\mathring{p}}$ and consider the cosetspace $P/T \rtimes R_{\mathring{p}}$. What are the representatives of $P/T \rtimes R_{\mathring{p}}$? This is most easily visualized for the case $\mathring{p}^2 > 0$.

As one can show that the whole construction does not depend on the choice of a special vector $\mathring{p}$, we choose the vector $(\mathring{p}, 0, 0, 0)$. Then evidently $R_{\mathring{p}}$ is the 3-dimensional rotation group $O(3, \mathbb{R})$, and the representatives of $P/T \times R_{\mathring{p}}$ are all rotation-free boosts, i.e. all homogeneous Lorentz transformations L_p which carry the point $\mathring{p}$ to any point p on the hyperboloid $p^2 = \text{const}$ along a hyperbel $L_p \mathring{p} = p$. Clearly the set L_p can be identified with the points of the hyperboloid.

The subgroup $T \rtimes R_{\mathring{p}}$ is represented by

$$T \rtimes R_{\mathring{p}} = \{a, r_{\mathring{p}}\} \to e^{i\mathring{p}a} D(R_{\mathring{p}})$$

where $D(R_{\mathring{p}})$ is some irreducible unitary representation of $R_{\mathring{p}}$.

Now we choose an element L_p of $P/T \rtimes R_{\mathring{p}}$; it suffices to take any representative because the effect of any other (a, Λ) is completely specified by means of the covariance law. Therefore, the definition of the representation (6.6) reads in our case

$$(U(a, \Lambda) f)(0, L_p) = f\{(a, \Lambda)(0, L_p)\} = f(a, \Lambda L_p) = f\{(0, L_{\Lambda p})(L_{\Lambda p}^{-1} a, L_{\Lambda p}^{-1} \Lambda L_p)\}.$$

Λp denotes the point into which the four-vectors p is carried by the transformation Λ. The transformation $L_{\Lambda p}^{-1} \Lambda L_p$ leaves $\mathring{p}$ fixed:

$$L_{\Lambda p}^{-1} \Lambda L_p \mathring{p} = L_{\Lambda p}^{-1} \Lambda p = \mathring{p}.$$

Therefore $\{L_{\Lambda p}^{-1} a, L_{\Lambda p}^{-1} \Lambda L_p\}$ belongs to the little group of $\mathring{p}$:

$$\{L_{\Lambda p}^{-1} a, L_{\Lambda p}^{-1} \Lambda L_p\} \in T \rtimes R_{\mathring{p}}.$$

Then the covariance law (6.4) gives

$$\begin{aligned}(U(a, \Lambda) f)(0, L_p) &= f(0, L_{\Lambda p})\, e^{i(\mathring{p}, L_{\Lambda p}^{-1} a)} D(L_{\Lambda p}^{-1} \Lambda L_p) \\ &= f(0, L_{\Lambda p})\, e^{i(\Lambda p, a)} D(L_{\Lambda p}^{-1} \Lambda L_p)\end{aligned} \tag{6.7}$$

because of the invariance of (,):

$$(p_0, L_{\Lambda p}^{-1} a) = (L_{\Lambda p} p_0, L_{\Lambda p} L_{\Lambda p}^{-1} a) = (\Lambda p, a) .$$

As explained above for the special case $p = (\mathring{p}, 0, 0, 0)$, the homogeneous spaces $P/T \rtimes R_{\mathring{p}}$ are isomorphic to the hyperboloids $p^2 = \mathring{p}^2 = \text{const}$. This holds for any choice of $\mathring{p}$. In a given hyperboloid $p^2 = \text{const}$ the construction does not depend on a special choice of $\mathring{p}$, the corresponding little groups are all isomorphic. Therefore, the invariant measures on $P/T \rtimes R_{\mathring{p}}$ can be identified with the invariant measures on the hyperboloids $p^2 = \text{const}$, depending on the value of p^2. We get the following types of representations:

Orbit	Standard choice of $\mathring{p}$	Little group	Invariant measure
$p^2 = m^2, p_0 \gtrless 0$	$(p_0, 0, 0, 0)$	Rotation group	$\theta(\pm p_0)\,\delta(p_2^2 - m^2)\, d^4p$
$p^2 = -\sigma^2 < 0$	$(0, 0, 0, \sigma)$	$0(2, 1, \mathbb{R})$	$\delta(p^2 + \sigma^2)\, d^4p$
$p^2 = 0$	$(p_0, 0, 0, \pm p_0)$	$SU(2)$	$\theta(\pm p_0)\,\delta(p^2)\, d^4p$
$p = 0$	$(0, 0, 0, 0)$	$0(3, 1, \mathbb{R})$	$\delta(p)\, d^4p$.

The question remains open whether one finds all irreducible representations of the Poincaré group by taking all irreducible representations of the little groups. This is answered by a general theorem of Mackey [1] which applies to the Poincaré-group:

Let $G = M \rtimes K$, M, K closed subgroups of G, M an abelian normal subgroup of G. Then one gets *all* unitary irreducible representations of G by taking all unitary irreducible representations of a special subgroup K' of G.

6.5 The Dirac equation

Because of the identification of the homogeneous spaces $P/T \times R_{\mathring{p}}$ with the hyperboloids $p^2 = \text{const}$ it is possible to regard the functions $f(0, \Lambda_p)$ in the transformation law (6.7) as wave functions. Usually the transformation properties of wave functions are derived by means of invariance requirements imposed on the corresponding wave equations. From the group theoretical point of view, the natural way is to derive the wave equations from the corresponding transformation laws. This can be done quite generally. We will illustrate the general method by the special example of a wave function for particles of real non-zero mass, $p^2 = m^2 > 0$ and spin $\frac{1}{2}$.

In this case, the little group is the rotation group and according to the possible two spin orientations we have to take for the representation D(R) the two-dimensional representation of SU(2) – the universal covering group of $0(3, \mathbb{R})$ – that is $D^{1/2}(SU(2))$.

It can be shown that the representations $D(SL(2, \mathbb{C}))$ of the universal covering group of the homogeneous Lorentz group can be obtained by analytic continuation of the representations $D(SU(2, \mathbb{C}))$, so that the symbol $D(\Lambda)$ for $\Lambda \in SL(2, \mathbb{C})$ makes sense. The realization of $SL(2, \mathbb{C})$ is given by two matrices, therefore we have

$$D(\Lambda) = \Lambda, \Lambda \in SL(2, \mathbb{C}) .$$

The rotation-free Wigner boosts $\Lambda_p \in SL(2, \mathbb{C})$ are represented by

$$\Lambda_p = (2m(m + p_0))^{-1/2} \left\{ (p_0 + m)\, 1 + \sum_{\kappa=1}^{3} p_\kappa \sigma_\kappa \right\}$$

where 1 denotes the unit matrix, and the σ_κ are the Pauli matrices. We have

$$\Lambda_p = \Lambda_p^+$$

and

$$\Lambda_p^2 = \frac{1}{m}\tilde{p} = \frac{1}{m}(p_0 1 + \sum_{k=1}^{3} p_\kappa \sigma_\kappa) .$$

With

$$\frac{1}{m}\underset{\sim}{p} = \frac{1}{m}(p_0 1 - \sum_{\kappa=1}^{3} p_\kappa \sigma_\kappa)$$

we have

$$\frac{1}{m}\underset{\sim}{p}\,\frac{1}{m}\tilde{p} = 1 . \tag{6.8}$$

We define new amplitudes by

$$\varphi^+(p) = f(0, \Lambda_p)\, D(\Lambda_p^{-1})$$

and (6.9)

$$\hat{\varphi}^+(p) = f(0, \Lambda_p)\, D(\Lambda_p^+) .$$

φ^+ and $\hat{\varphi}^+$ obey the transformation laws

$$U(a, \Lambda)\, \varphi^+(p) = e^{i(\Lambda p, a)}\, \varphi^+(\Lambda p)\, D(\Lambda)$$
$$U(a, \Lambda)\, \hat{\varphi}^+(p) = e^{i(\Lambda p, a)} \hat{\varphi}^+(\Lambda p)\, D(\Lambda^{+-1})$$

and are not independent. From (6.9) we derive

$$\begin{aligned} \hat{\varphi}^+(p) &= \varphi^+(p)\, D^{-1}(\Lambda_p^{-1})\, D(\Lambda_p^+) \\ &= \varphi^+(p)\, D(\Lambda_p)\, D(\Lambda_p^+) \\ &= \varphi^+(p)\, D(\Lambda_p \Lambda_p) \\ &= \varphi^+(p)\, \frac{1}{m}\tilde{p} \end{aligned} \tag{6.10}$$

by means of $D^+(\Lambda) = D(\Lambda^+)$ and $D^{-1}(\Lambda) = D(\Lambda^{-1})$.

From (6.10) we get for the 2 x 2 dimensional spinor $\binom{\hat{\varphi}}{\varphi} =: \psi$ the equation

$$\psi = \begin{pmatrix} 0 & \frac{1}{m}\tilde{p} \\ \frac{1}{m}\underset{\sim}{p} & 0 \end{pmatrix} \psi$$

where we have used (6.8). Multiplication with m gives

$$m\,\psi = p_\mu \gamma^\mu \psi \tag{6.11}$$

which is the usual Dirac equation in a special representation with

$$\gamma_0 = \begin{pmatrix} 0 & 1 \\ 1 & 0 \end{pmatrix}, \; \gamma_\kappa = \begin{pmatrix} 0 & \sigma_\kappa \\ -\sigma_\kappa & 0 \end{pmatrix}.$$

So we have demonstrated that the free Dirac equation is a direct consequence of group theory. The two parts of (6.11) (corresponding to each two-spinor φ, $\hat{\varphi}$ respectively) are interrelated with each other by the linear realization of the parity transformation. For this and for more general relativistic wave equations cf. chapter VI in this volume.

References

Section 1:

[1] P. Roman, Some modern mathematics for physicists and other outsiders, Pergamon, New York 1975, vol. 1.

[2] H. Bacry, Lectures on Group Theory and Particle Theory, Gordon and Breach, London 1977.

Section 2:

[3] M. Hamermesh, Group Theory and its Application to Physical Problems, Reading and London, 1962.

[4] P. Kramer and T. H. Seligman, Nucl. Phys. **A136**, 545 (1969) and **A186**, 49 (1972).

[5] R. Gilmore, Lie Groups, Lie Algebras, and Some of Their Applications, Wiley, New York 1974.

[6] M. Brunet and P. Kramer, Rep. Math. Phys., to be published.

[7] P. Kramer, G. John and D. Schenzle, Group Theory and the Interaction of Composite Nucleon Systems, Vieweg, Braunschweig 1979.

Section 3:

[1] H. W. Wielandt, Topics in the Theory of Composite Groups, Lecture Notes University of Wisconsin and University of Tübingen, 1967/68.

[2] W. Miller, Symmetry Groups and Their Applications, Academic Press, New York 1972.

[3] S. Eilenberg, Automata, Languages, and Machines, Academic Press, New York, vol. A and B, 1974.

[4] B. Huppert, Endliche Gruppen I, Springer Verlag, Berlin-Heidelberg-New York 1967.

[5] A. Machi, Introduzione alla Theoria dei Gruppi, Feltrinelli, Milano 1974.

[6] J. J. Rotman, The Theory of Groups: An Introduction, Allyn and Bacon, London 1965.

[7] H. Jürgensen, Some Applications of the Theory of Semigroups to Automata, in Group Theoretical Methods in Physics, Eds. P. Kramer and A. Rieckers, Springer Lecture Notes in Physics 79, 307–322, Springer Verlag, Berlin-Heidelberg-New York 1978.

Section 4:

[1] A. O. Barut, R. Raczka, Theory of Group Representations and Applications, PWN – Polish Scientific Publishers, Warszawa 1977.

[2] P. M. Cohn, Lie groups, Cambridge University Press 1961.

[3] Y. Choquet-Bruhat, C. DeWitt-Morette, M. Dillard-Bleick, Analysis, Manifolds and Physics, North Holland 1977.

[4] R. Gilmore, Lie Groups, Lie Algebras and Some of Their Applications, John Wiley 1974.

[5] D. Husemoller, Fibre Bundles, Springer 1975.

[6] A. A. Kirillov, Elements of the Theory of Representations, Springer 1976.

[7] J. M. Souriau, Structure de Systèmes Dynamiques, Dunod 1970.

Section 5:

[1] R. Gilmore, Lie Groups, Lie Algebras, and Some of Their Applications, Wiley, New York 1974.

[2] M. Hamermesh, Group Theory and its Application to Physical Problems, Reading and London, 1962

[3] P. Kramer, Z. Phys. **205**, 181 (1967); **216**, 68 (1968).

[4] P. Kramer, G. John and D. Schenzle, Group Theory and the Interaction of Composite Nucleon Systems, Vieweg Braunschweig 1980.

[5] I. M. Gelfand and M. L. Zetlin, Dokl. Akad. Nauk USSR **71**, 1017 (1950).

[6] H. Bacry, Lectures on Group Theory and Particle Theory, Gordon and Breach, New York 1977.

[7] P. Kramer, in Group Theoretical Methods in Physics, ed. R. T. Sharp and B. Kolman, Academic Press, New York 1977, p 173.

[8] A. R. Edmonds, Angular Momentum in Quantum Mechanics, Princeton University Press, Princeton 1957.

[9] A. O. Barut and R. Raczka, Theory of Group Representations and Applications, PWN – Polish Scientific Publishers, Warszawa 1977.

Section 6:

[1] G. Mackey, Induced Representations of Groups and Quantum Mechanics, Benjamin, New York 1968.

[2] R. Hermann, Lie Groups for Physicists, Benjamin, New York 1966.

[3] W. Klink, Induced Representation Theory and the Poincaré Group, in Boulder Lectures in Theoretical Physics, Vol. XI-D, Gordon and Breach, New York 1969.

[4] S. Strohm, Introduction to the Theory of Groups and Group Representations, Lectures Notes CPT-120, Austin 1971.

[5] W. Rühl, The Lorentz Group and Harmonic Analysis, Benjamin, New York 1970.

[6] U. H. Niederer, L. O'Raifeartaigh, Fortschr. Phys. **22**, 111 (1974).

[7] E. Wigner, Ann. Math. **40**, 39 (1939).

Chapter II
Fundamentals of Algebraic Quantum Theory

A. Rieckers

Introduction

Many body physics is concerned with macroscopic systems consisting of a very large number of microscopic particles ($N \approx 10^{20}$). In this case the direct dynamical treatment of the complete system is impossible. A quite useful and frequently applied approximation consists in the neglection of terms which are relatively small as the volume of the systems tends to infinity, the particle density remaining constant. For the concise theoretical description of the typical many body phenomena such as phase transitions and spontaneous symmetry breakdown the use of the thermodynamical limit is even necessary, since no existing theory for the finite systems shows these effects. As a general rule one can say that a quantum system behaves like a thermodynamical system only in the infinite-volume limit.

In the thermodynamical limit the many body system has – as well as a quantized field – infinitely many degrees of freedom, a fact which changes the microdynamical aspects radically compared with the usual description of quantum systems of finitely many particles. In the latter formalism, which occasionally we shall refer to as "traditional quantum theory", a well known theorem of von Neumann tells us that all irreducible representations of canonical commutation relations are unitarily equivalent. For infinitely many degrees of freedom there are continuously many inequivalent representations of the basic algebraic structure, that means that the physical statements depend on the special choice of the representation space. Since one does not know from the outset which representation space is appropriate for a given dynamical system one is urged to develop the fundamental principles of the microscopic description without representing the observables as operators in a Hilbert space, that is, one treats the problem first of all in an algebraic stage. For this one must define an abstract algebra of observables as the starting point for all later developments and investigations. It turns out that the natural algebra of observables for an infinite many body system – the so-called quasilocal algebra – is essentially different from an algebra B (H), the set of all bounded operators in a Hilbert space H, which is characteristic for traditional quantum theory. It is exactly this fact that causes great mathematical complications but which on the other hand allows for the characteristic many body effects. All basic notions as observables, states, symmetries and dynamics have then to be reconsidered from a new and a more general point of view. These lectures are meant as a short introduction into the mathematical and physical fundamentals of the algebraic quantum theory with special emphasis on symmetry transformations for macroscopic systems. They are also meant as a preparatory step for Prof. Emch's lecture in this volume, although the latter is self-contained for those readers who are familiar with the most basic concepts of algebraic quantum theory. We have laid emphasis on general definitions and structural theorems, but have also given representative mathematical argumentations in which the introduced notions are involved. It seemed, however, not useful to prove all mentioned assertions. We hope that also some arguments could be made plausible, which enforce the abstract algebraic approach in the dynamical treatment of many body phenomena.

1 Algebra of observables

Since the very beginning algebraic aspects played a very important role in the formulation of quantum mechanics. In order to apply rigorous mathematical theorems some restrictions and precautions in comparison to common physical text book presentations are required, above all that one deals only with bounded observables. Without this limitation a rigorous formulation of algebraic quantum mechanics would not be possible, at least not at the present stage of the mathematical developments. The restriction is not so severe as it appears at first sight since in representations the unbounded operators can be approximated by bounded ones with arbitrary accuracy. This being accepted the algebraic structure of traditional quantum mechanics as presented in [1] can be characterized as being isomorphic to an operator algebra $B(H)$, i.e. isomorphic to the set of *all* bounded operators in a separable Hilbert space H. Here we follow the usual convention that also non-observables as, e.g., commutators of observables, are included into the algebraic structure. Strictly speaking constitute the observables alone only a real, commutative but not associative algebra, a so-called Jordan algebra. (Jordan algebras were first studied in [3] and further developed in [4]. A very interesting algebraic approach based on the Jordan algebraic structure has been worked out in [8].) Nevertheless, let us call for brevity $B(H)$ the algebra of observables of traditional quantum mechanics. If we remember that the states usually are constructed by means of density operators (the pure states being given by one dimensional projections resp. by unit rays in H) then we have already described the main features of traditional quantum mechanics we are interested in.

A first step to a generalization of traditional quantum mechanics has been the introduction of superselection rules, cf., e.g., [2]. If there are superselection rules, then H decomposes into a direct sum of (coherent) subspaces

$$H = \sum_{i=1}^{n} \oplus H_i$$

and the algebra of observables is given by $\sum_{i=1}^{n} \oplus B(H_i)$, which is strictly smaller than $B(H)$ (and no more irreducible, cf. Def. 1.9) and has many new properties. The investigation of many body systems has even brought abstract algebras into play (cf. below) which are not connected with a special representation in a physically distinguished Hilbert space, and which are in general very different from $B(H)$. In order to work with these abstract algebras let us introduce some mathematical notions.

1.1 Definition:

a) A is a distributive and associative algebra over $\mathbb{C}$ if A is a linear vector space over $\mathbb{C}$ and has an associative inner product which satisfies for $\lambda_i \in \mathbb{C}$ and $a_i, b \in A$

$$(\lambda_1 a_1 + \lambda_2 a_2) b = \lambda_1 a_1 b + \lambda_2 a_2 b$$
$$b(\lambda_1 a_1 + \lambda_2 a_2) = \lambda_1 b a_1 + \lambda_2 b a_2 .$$

b) A is a Banach *-algebra if it is an algebra with a norm, i.e. there is a mapping

$$\| \ \| : A \to \mathbb{R}_+$$

with $(a, b \in A, \lambda \in \mathbb{C})$

(i) $\|a\| = 0 \Leftrightarrow a = 0$
(ii) $\|a + b\| \leqslant \|a\| + \|b\|$
(iii) $\|\lambda a\| = |\lambda| \, \|a\|$
(iv) $\|ab\| \leqslant \|a\| \, \|b\|$.

Furthermore, A has to be complete w.r.t. the norm topology. Finally, there exists a mapping

$$* : A \to A,$$

called involution, with

(i) $(a^*)^* = a$
(ii) $(a + b)^* = a^* + b^*$
(iii) $(ab)^* = b^* a^*$
(iv) $(\lambda a)^* = \lambda^* a^*$.

1.2 Definition: A is a C*-algebra if

a) A is a Banach*-algebra
b) $\|a^* a\| = \|a\|^2, \quad \forall a \in A$.

1.3 Definition: A is a W*-algebra, if

a) A is a C*-algebra
b) There is a Banach space (i.e. a complete normed space) A_*, the so-called predual of A with

$$(A_*)^* = A.$$

(The notion of the dual space V^* of a topological vector space $V^*$is defined in Chapter I.4 of the fundamental course.)

We observe that a W*-algebra has the richest structure. It is known that $B(H)$ is a W*-algebra with predual $T(H)$, the set of all trace class operators in H. The abstract algebra of observables for infinite many body systems is, however, introduced merely as a C*-algebra. We shall always assume that our C*-algebras possess a unit element.

1.4 Definition: Let A_1 and A_2 be two C*-algebras. A mapping

$$\alpha: A_1 \rightarrow A_2$$

is called a C*-morphism if it is complex linear, adjoint preserving ($\alpha a^* = (\alpha a)^*$), and multiplicative ($\alpha(ab) = (\alpha a)(\alpha b)$). α is then automatically norm-continuous. α is called a C*-isomorphism if it is a one-to-one morphism. It is then automatically norm-preserving.

Remark: The norm-continuity of a C*-morphism α is a direct consequence of the property b) of Def. 1.2.

For a full understanding of the algebraic description of an infinite quantum system the following construction is basic.

Let be I a directed set, i.e. I is partially ordered and for every pair $i, j \in I$ there exists a $k \in I$ with $i \leqslant k$ and $j \leqslant k$. Let be $\{A_i; i \in I\}$ a family of C*-algebras with identity. If for all $i, j \in I$, with $i \leqslant j$, there exists a C*-isomorphism α_{ji} of A_i into A_j such that $\alpha_{ji}(1_i) = 1_j$ and that

$$\alpha_{ki} = \alpha_{kj} \circ \alpha_{ji}, \quad i \leqslant j \leqslant k$$

then one can construct a new C*-algebra A, in which all algebras A_i, $i \in I$, are contained. For this define A_0 as the set of all functions $\{a_i\}$ on I with $a_i \in A_i$ such that there exists an index $k \in I$ depending on $\{a_i\}$ with

$$a_i = \alpha_{ik} a_k, \quad \forall i \geqslant k.$$

A_0 is easily seen to be an algebra w.r.t. the following operations:

$$\lambda \{a_i\} := \{\lambda a_i\}, \quad \lambda \in \mathbb{C}$$
$$\{a_i\} + \{b_i\} := \{a_i + b_i\}$$
$$\{a_i\} \{b_i\} := \{a_i b_i\}.$$

Furthermore, we define

$$\{a_i\}^* := \{a_i^*\}$$

and $\| \{a_i\} \| := \lim_i \| a_i \|$.

Then $\| \ \|$ is a semi-norm on A_0. In order to introduce a normed algebra we consider the two-sided ideal J of A_0:

$$J := \{\{a_i\}; \| \{a_i\} \| = 0\}.$$

Then the quotient algebra A_0/J consisting of all $\{a_i\} + J =: a$, with $\| a \| := \| \{a_i\} \|$ is a normed algebra, i.e. $\| a \| = 0$ implies $a = 0$ $(= \{0\} + J = J)$.

1.5 Definition:

a) Let be $\{A_i; i \in I\}$ a family of C*-algebras as described above. We call then $\{A_i; i \in I\}$ an increasing family of C*-algebras.

b) For an increasing family $\{A_i; i \in I\}$ of C*-algebras let be A_0 and J constructed as above. We call then

$$A := \overline{A_0/J}^{\text{norm}}$$

the inductive limit of $\{A_i; i \in I\}$.

By construction is the inductive limit A of $\{A_i\}$ a Banach*-algebra. It is not hard to show, that also property b) of Def. 1.2 is fulfilled by A. Thus A is a C*-algebra. Every $a_j \in A_j$ can be identified with a function $\{a_i\}$ on I given by $a_i = \alpha_{ij} a_j$ for $j \leqslant i$ and $a_i = 0$ for $j \nleqslant i$. This leads to an isomorphic embedding

$$\alpha_j : A_i \to A.$$

One can show, that $A = \overline{\bigcup_j \alpha_j(A_j)}^{\text{norm}}$. So A is the smallest C*-algebra which contains all A_i as subalgebras.

After these mathematical preliminaries we come to the central point of the whole investigation: How can we formulate a quantum theory for an infinitely extended system? Of course is the infinite system to be approximated by finite systems. We thus assume that for every bounded region $\Lambda \subset \mathbb{R}^3$ we have a traditional quantum mechanical description of that part of the system which is confined to the region Λ, i. e. for every Λ there is a Hilbert space H_Λ and the corresponding set of observables $B(H_\Lambda) =: A_\Lambda$. With increasing Λ there exists no such a concept of a limiting Hilbert space, to which the H_Λ would converge in some sense, and which would contain the wave functions of an infinitely extended system. Consider, e.g., the characteristic function χ_Λ of the region Λ which is in $L^2(\Lambda)$ for all bounded Λ. For $\Lambda \to \mathbb{R}^3$, $\chi_\Lambda \to 1$, and this function is not square-integrable over $\mathbb{R}^3$.

On the other hand we learn from the construction of the C*-inductive limit, that we have a natural C*-algebra A containing all A_Λ, if we take $\{\Lambda; \Lambda \subset \mathbb{R}^3, \Lambda \text{ bounded}\} =: L$ as directed index set I with the inclusion as partial order, and if the appropriate isomorphisms $\alpha_{\Lambda'\Lambda}$ exist for all $\Lambda \subset \Lambda'$. The latter condition must obviously be valid, if every A_Λ is to describe a part of one and the same infinite system. There must hold even stronger conditions which result from the spacial homogeneity of the infinite system.

1.6 Definition:

a) Let be $\{A_\Lambda; \Lambda \in L\}$ an increasing family of C*-algebras. Then

$$A_0 := \bigcup_{\Lambda \in L} A_\Lambda$$

is called the local algebra.

b) The increasing family $\{A_\Lambda; \Lambda \in L\}$ is called spacially homogeneous if for every $x \in \mathbb{R}^3$ and every $\Lambda \in L$ there exists a C*-isomorphism

$$\tau_x : A_\Lambda \xrightarrow{\text{onto}} A_{\Lambda + x}$$

such that

(i) $\tau_{x+y} = \tau_x \circ \tau_y, \quad \tau_{-x} = \tau_x^{-1}$

(ii) For $\Lambda \subset \Lambda'$

$$\tau_x \circ \alpha_{\Lambda'\Lambda} = \alpha_{\Lambda'+x,\Lambda+x} \circ \tau_x$$

(iii) For $\Lambda \cap \Lambda' = \emptyset$ and $a \in A_\Lambda$, $a' \in A_{\Lambda'}$

$$[\alpha_{\Lambda \cup \Lambda',\Lambda}(a), \alpha_{\Lambda \cup \Lambda',\Lambda'}(a')]_- = 0$$

(cf., e.g., [10]).

c) The inductive limit A of a spacially homogeneous increasing family $\{A_\Lambda; \Lambda \in L\}$ of C*-algebras A_Λ is called the quasilocal algebra.

The meaning of b) in the above definition is to guarantee the algebraic isomorphy of the algebras of observables localized in Λ and $\Lambda + x = \{y; y = z + x, z \in \Lambda\}$ in such manner that this is compatible with the embedding isomorphisms $\alpha_{\Lambda'\Lambda}$. One easily sees that b) (i) (ii) lead to a representation of the translation group first in A_0 by means of automorphisms (that are isomorphisms $A_0 \xrightarrow{\text{onto}} A_0$), and then also in A. Property b) (iii) gives the compatibility of observables in disjoint regions.

As described at the introduction of the inductive limit of an increasing family of C*-algebras the quasilocal algebra A is the smallest C*-algebra which comprises the local algebra A_0. A is by construction an abstract algebra, also if all A_Λ are concrete operator algebras. Since in the limit $\Lambda \to \mathbb{R}^3$ we could not build up a Hilbert space which had some connection with the localized Hilbert spaces H_Λ, there is no distinguished realization of A by operators acting in a Hilbert space. We see that the performance of the thermodynamical limit leads necessarily to an algebraic formulation of the quantum theory.

There remains, of course, the question if the quasilocal algebra is suited for physical purposes. First let us indicate that many operations known for operators can be taken over to an abstract C*-algebra. We give for illustration some definitions: $a \in A$ is self-adjoint, if $a^* = a$. a is positive if there is a $b \in A$ such that $a = b^*b$. It is sometimes useful to know that w.r.i.g. we can choose b self-adjoint in the positivity definition. A positive a is obviously self-adjoint. $P \in A$ is a projection if it is self-adjoint and idempotent (i.e. $P^2 = P$). $u \in A$ is unitary if $u^*u = uu^* = 1$. $v \in A$ is a partial isometry if v^*v is a projection. The spectrum $\sigma(a)$ of a is the set of all $\lambda \in \mathbb{C}$, such that $a - \lambda 1$ is not invertible.

In spite of the mathematical structure of a C*-algebra being almost as rich as that of an operator algebra it is certainly easier in many cases to work with a representation of the algebra (cf. Def. 1.7). We shall see later (Prop. 2.17) that every C*-algebra has an isomorphic universal representation. This representation can then be extended to a (concrete) W*-algebra. One could think that this "envelopping W*-algebra" of the quasilocal A would be the appropriate observable algebra of an infinite quantum system. It has turned out, however, that the envelopping W*-algebra is too large. Using this bigger algebra one looses a concept which is very important for the treatment of many body systems, namely the distinction between local (resp. quasilocal) excitations and global changes of the system. These notions can be appropriately formulated by means of the quasilocal C*-algebra but not with the envelopping W*-algebra.

Thus we have to get acquainted with the properties of abstract C*-algebras, their representations, states, and isomorphisms. Let us start with the general study of their representations.

1.7 Definition: A representation of a C*-algebra A is a C*-morphism

$$\pi : A \to B(H),$$

where H is some Hilbert space.

The image $\pi(A) \subset B(H)$ is then a C*-subalgebra of $B(H)$, i.e. a norm-closed operator algebra. The kernel $\pi^{-1}(0) \subset A$ is a two sided norm-closed ideal in A. A subalgebra B of an algebra A is a left (sided) ideal, if $AB \subset B$, a right (sided) ideal if $BA \subset B$, and a two sided ideal if it is a left and right ideal. If $a \in \pi^{-1}(0)$, then $\pi(b_1 a b_2) = \pi(b_1)\pi(a)\pi(b_2) = 0$, so that $b_1 a b_2 \in \pi^{-1}(0)$, too. π is a C*-isomorphism if, and only if $\pi^{-1}(0)$ is trivial. Every C*-algebra has an isomorphic representation. Two isomorphic representatives $\pi(A)$ and $\pi'(A)$ of A are isomorphic to each other. If A is simple, i.e. has no nontrivial ideals, then all representations are isomorphic.

Since positivity in A is defined purely algebraically, it is preserved under C*-morphisms:

$$a \in A, \quad a \geqslant 0 \Leftrightarrow a = b^*b \Rightarrow \pi(a) = \pi^*(b)\,\pi(b) \geqslant 0. \tag{1.1}$$

In the subset of all self-adjoint elements of A an order relation can be introduced:

$$a \leqslant b \Leftrightarrow 0 \leqslant b - a. \tag{1.2}$$

Every C*-morphism is in virtue of (1.1) order perserving.

1.8 Definition: Let A_1 and A_2 be two concrete C*-algebras (acting in H_1 and H_2 respectively). A_1 and A_2 are called unitarily equivalent if there exists a unitary operator

$$U : H_1 \to H_2$$

(i.e. U is bijective and isometric), so that the mapping π_u, defined by

$$\pi_u(a) = U a U^{-1}$$

maps A_1 onto A_2. π_u is then automatically a C*-isomorphism.

Thus we have, that unitarily equivalent representatives of A are isomorphic to each other. The converse, however, is not true: Two isomorphic representatives of A are in general not unitarily equivalent. The physical content of an algebra of observables is invariant only against unitary equivalence and not against C*-isomorphy. Thus the introduction of an abstract algebra of observables A does not fix the physical properties of the observables, if there are isomorphic representations of A which are not unitarily equivalent, a case which is typical for infinite systems.

Some notions are still needed for later occasions.

1.9 Definition: Let R be a subset of $B(H)$, then R' is defined as the set of all $a \in B(H)$ which commute with every $r \in R$. R' is called the commutant of R (in $B(H)$).

(i) A representation $\pi: A \to B(H)$ is cyclic if there is a $\Phi \in H$, so that $\pi(A)\Phi$ is dense in H. Φ is called a cyclic vector of the representation π.

(ii) π is said to be irreducible if

$$\pi(A)' = \mathbb{C}1$$

(iii) π is said to be factorial if

$$\pi(A)' \cap \pi(A)'' = \mathbb{C}1$$

(iv) $\Phi \in H$ is a separating vector for $\pi(A)$, if $\pi(a)\Phi = 0$ implies $\pi(a) = 0$.

In spite of the fact that our algebra of observables A is assumed to be a C*-algebra, we are frequently concerned with concrete W*-algebras (von Neumann-algebras) if we study representations of A. The weak operator topology on $B(H)$ is defined as the weakest topology, so that all functions

$$B(H) \ni a \to |(\Psi, a\Phi)| \in \mathbb{C},\ \Psi, \Phi \in H, \tag{1.3}$$

are continuous. There is a famous characterization of von Neumann-algebras.

1.10 Theorem: Let M be a subalgebra of $B(H)$ containing the unit operator and suppose $M^* = \{a^*; a \in M\} = M$. Then M is a von Neumann algebra

(i) if and only if it is weakly closed (this by definition)

(ii) if and only if $M'' = M$ (von Neumann's commutation theorem) ([5] a).

If $\pi(A)$ is a represented C*-algebra in the Hilbert space H, then from $\pi(A') = \pi(A)'''$ and from $\pi(A)'' = \pi(A)''''$ one concludes by means of Theorem 1.10 that $\pi(A)'$ as well as $\pi(A'')$ are von Neumann-algebras.

The W*-algebras are abstractions of the von Neumann algebras. If A is an abstract W*-algebra, then the existence of the predual A_* allows for the introduction of the $\sigma(A, A_*)$-topology in A which has some similarity to the weak operator topology. The mathematical structure of a W*-algebra is even closer to an operator algebra than that of a C*-algebra. Self-adjoint elements in a W*-algebra have a spectral resolution ([7]), and the set of all projections of a W*-algebra constitutes a complete orthomodular lattice (not necessarily atomic) ([9]). Thus one would expect from general axiomatic considerations that a W*-algebra is the appropriate set up for an algebraic treatment of quantum systems. An interesting quantum mechanical system theory based on W*-algebras is presented in [35]. It remains, however, the fact that typical many body effects are better described by means of the quasilocal C*-algebra.

2 States

Let us consider the state concept in a statistical theory. In general the physical preparation of a state does not fix the observable values itself but only their expectation values. The system's dwelling in a well defined state is thus indicated by the association of every observable the system has with a real number, the expectation value. Since it is not useful to ask, how the system accomplishes this fixation of expectation values, one has to accept the expectation value functional itself as the appropriate formalization of the state concept. In traditional quantum mechanics the state vector (or, more precisely, a unit ray of vectors) plays a predominant role. That is, there one has not only the statistical state concept as mentioned above, but a richer structure which has the purpose to describe interference phenomena. The mixed states are mostly considered to be less fundamental since they result either from an outside disturbance of an otherwise closed system or from insufficient information on the system.

One has, however, not in all physically meaningful situations a non-trivial superposition principle, which leads to interference phenomena. Already in the purely microscopic domain exist (charge like) superselection rules which prevent the interference of two vector states from different sectors. In this case the vector addition between the two vectors is equivalent to an incoherent superposition of the two states. That means, that in this case the additional structure of the vector states in comparison to the purely statistical state concept is partly lost. In many body physics the macroscopic character of the states leads to a breakdown of the unrestricted superposition principle: two macroscopically different states cannot interfere coherently with each other. So one cannot in general expect that the pure states are given by vector states, and what remains as the general structure is the purely statistical state concept.

In the case of the quasilocal algebra A we loose by the abstraction process itself contact with a (distinguished) Hilbert space. So if we consider the limit of a sequence of states on the localized algebras A_Λ we know not more about the limiting state on A – if it exists – than that it must fulfill the requirements of a general statistical state. The discussions of infinite quantum systems show as a rule, that such a limiting state is qualitatively different from vector states and from mixed states of traditional quantum mechanics (i.e. it is not normal, see below). Thus we have to employ the general statistical state concept.

2.1 Definition: Let A be the (C*-)algebra of observables for a quantum system. A state of the system is then (in the theoretical description) a functional

$$\varphi: A \to \mathbb{C}$$

which is linear, positive and normalized, i.e. for the "expectation values" $\langle \varphi; a \rangle$ of the elements $a \in A$ in the state φ holds

(i) $\langle \varphi; \lambda_1 a_1 + \lambda_2 a_2 \rangle = \lambda_1 \langle \varphi; a_1 \rangle + \lambda_2 \langle \varphi; a_2 \rangle$

$$a_i \in A, \lambda_i \in \mathbb{C}$$

(ii) $\langle \varphi; a^* a \rangle \geqslant 0, \quad \forall\, a \in A$

(iii) $\langle \varphi; 1 \rangle = 1$.

The set of all states is denoted by S.

If $\varphi_1, \varphi_2 \in S$ and $\lambda \in [0, 1]$ then also $\lambda \varphi_1 + (1 - \lambda) \varphi_2 \in S$; S is a convex set. Because of the rich structure of A much more can be said about states.

2.2 Proposition: Let be $\varphi \in S$. Then

a) $\langle \varphi; a\rangle^* = \langle \varphi; a^*\rangle, \forall a \in A$

b) $|\langle \varphi; a_1^* a_2\rangle|^2 \leqslant \langle \varphi; a_1^* a_1\rangle \langle \varphi_i a_2^* a_2\rangle, a_i \in A$

c) $\|\varphi\| = \langle \varphi; 1\rangle = 1.$

Proof:

a) The mapping $A \times A \to \mathbb{C}$ defined by $\langle \varphi; a^* a'\rangle$ is conjugate linear in a and linear in a′. Because of positivity

$$\begin{aligned} &\langle \varphi; (a_1 + \lambda a_2)^* (a_1 + \lambda a_2)\rangle \\ &= \langle \varphi; a_1^* a_1\rangle + \lambda \langle \varphi; a_1^* a_2\rangle + \lambda^* \langle \varphi; a_2^* a_1\rangle + \lambda^* \lambda \langle \varphi; a_2^* a_2\rangle \geqslant 0 \end{aligned} \tag{2.1}$$

The sum of the two terms in the middle of the r.h.s. must be real for arbitrary $\lambda \in \mathbb{C}$, thus

$$\langle \varphi; a_1^* a_2\rangle = \langle \varphi; a_2^* a_1\rangle^*,$$

which proves the assertion.

b) If $\langle \varphi; a_2^* a_2\rangle = 0$ and $\langle \varphi; a_1^* a_2\rangle \neq 0$ a large negative value of λ would lead to a contradiction in (2.1). Thus in this case $\langle \varphi; a_1^* a_2\rangle = 0$, too. For $\langle \varphi; a_2^* a_2\rangle \neq 0$ choose $\lambda = -\langle \varphi; a_1 a_2^*\rangle / \langle \varphi; a_2^* a_2\rangle$ and multiply by $\langle \varphi; a_2^* a_2\rangle$.

c) For $a \in A$ define $b = a - \|a\|\, 1$ and deduce from $b^* b \geqslant 0$

$$a^* a \leqslant \|a\|^2\, 1. \tag{2.2}$$

Apply b) with $a_2 = 1$

$$\begin{aligned} &|\langle \varphi; a\rangle| \leqslant \langle \varphi; 1\rangle^{1/2} \langle \varphi; a^* a\rangle^{1/2} \leqslant \langle \varphi; 1\rangle\, \|a\|^2. \\ &\|\varphi\| := \sup_{a \in A} |\langle \varphi; a/\|a\|\rangle| = \langle \varphi; 1\rangle. \end{aligned} \tag{2.3}$$

From 2.2 a) follows that self-adjoint a have real expectations, as it should be.

2.3 Definition: $\varphi \in S$ is a pure state if a decomposition of the form

$$\varphi = \lambda_1 \varphi_1 + \lambda_2 \varphi_2, \quad \lambda_i \in (0, 1), \quad \varphi \neq \varphi_1 \tag{2.4}$$

is impossible.

In traditional quantum mechanics states are constructed by means of density operators ρ, which are positive operators with trace 1:

$$\langle \varphi; a\rangle := \mathrm{tr}\,[\rho\, a], \; \rho, a \in B(H). \tag{2.5}$$

The pure states are exactly those, for which ρ is a (one-dimensional– projection. Not all states on $B(H)$ – in the sense of 2.1 – are of the form (2.5). There are even pure states which cannot be expressed as in (2.5). These are e.g. those states which ascribe to a certain observable a dispersion-free mean value in its continuous spectrum. The inclusion of such cases is not an undesirable feature of our state definition; in many body physics most of the physically relevant states cannot be written by means of a density operator. But it is indeed true, and important to note, that for a given class of systems in general not all of S is needed for the theoretical description. That subset $S_0 \subset S$ which plays a relevant role may, however, change from theory to theory. The only thing one can generally expect for these various state sets S_0 is completeness in a certain sense. We develop the appropriate definition.

2.4 Proposition: The three following assertions are equivalent:

(i) $a \geqslant 0$

(ii) $\pi(a) \geqslant 0$ for all representations π of A

(iii) $\langle \varphi; a \rangle \geqslant 0$ for all $\varphi \in S$.

Proof: (i) ⇒ (iii): Follows from the definition of a state. (iii) ⇒ (ii): Let π represent A in H, and let $\Phi \in H$, $\|\Phi\| = 1$. Then

$$\langle \varphi; b \rangle := (\Phi, \pi(b)\Phi), \; b \in A, \tag{2.6}$$

is a state on A and gives positive values to our fixed a by assumption. This being true for all unit vectors Φ, we have $\pi(b) \geqslant 0$.
(ii) ⇒ (i): Let π be an isomorphic representation of A onto $\pi(A) \subset B(H)$, which is known to exist for every C*-algebra (cf. 2.17). If $\pi(a)$ is positive then $a = \pi^{-1} \circ \pi(a)$ is positive by (1.1). ■

Prop. 2.4 shows that positivity of an observable a can be tested by means of states; it shows also that only a subclass for this is really necessary, namely e.g. the set of states of the form (2.6) where π is some isomorphic representation of A.

2.5 Definition: A convex subset S_0 of S is full (w.r.t. A), if $a \in A$ and $\langle \varphi; a \rangle \geqslant 0$, for all $\varphi \in S_0$, implies $a \geqslant 0$. One can show that a convex set of states is full, iff it is $\sigma(S, A)$-dense in $S(A)$ [17].
Of course is the set of states which are constructed via a density operator a full set on $B(H)$.

2.6 Definition: The pair (A, S_0), A a C*-algebra, S_0 a full set of states, is called a description of a (class of) quantum system (s).

For traditional quantum mechanics (of finite systems) the use of the pair $(B(H), S_n)$, S_n the set of all density operator states, is typical.
It is interesting that the set S_n can be characterized in a way capable of generalizations.

2.7 Definition: Let f be a linear positive functional on the C*-algebra A. f is called normal if

$$\langle f; \sup_i a_i \rangle = \sup_i \langle f; a_i \rangle$$

for all uniformly bounded, directed, increasing families of self-adjoint elements a_i, $i \in I$.

2.8 Proposition: Let A be a W*-algebra. Then a state φ on A is normal if and only if it is $\sigma(A, A_*)$-continuous. The normal states on $B(H)$ are exactly all density-operator states. For every W*-algebra is the set of normal states full (but not for every C*-algebra) [6].

For infinite systems normality is too severe a restriction.

2.9 Definition: Let A be the quasilocal algebra of $\{A_\Lambda\}$. A state φ on A is said to be locally normal, if $\varphi | A_\Lambda$ is a normal state for all Λ.

2.10 Proposition: Let A be the quasilocal algebra of an increasing family of W*-algebras A_Λ. Then the set of all locally normal states on A is full.

Proof: The essential idea of the rather technical proof is that every normal state on $\{A_\Lambda\}$ can be considered as a partial state of a quasilocal state on A [26]. ∎

With every state $\varphi \in S$ is associated a cyclic representation π_φ of A. Many properties of φ are expressible by properties of π_φ and vice versa.

2.11 Proposition (GNS-construction): For every $\varphi \in S$ exists a cyclic representation

$$\pi_\varphi : A \to B(H_\varphi)$$

with cyclic vector Φ, so that

$$\langle \varphi; \mathrm{a} \rangle = (\Phi, \pi_\varphi(\mathrm{a})\,\Phi), \qquad \forall\, \mathrm{a} \in A. \tag{2.7}$$

Every other cyclic representation π_1 with cyclic vector Φ_1 satisfying the analogue of (2.7) is unitarily equivalent to π_φ.

Proof:

(i) Define

$$I_\varphi := \{\mathrm{a}; \mathrm{a} \in A, \langle \varphi; \mathrm{a}^*\mathrm{a} \rangle = 0\}. \tag{2.8}$$

If $\mathrm{a} \in I_\varphi$ and $\mathrm{b} \in A$, then

$$0 \leqslant \langle \varphi; \underbrace{\mathrm{a}^*\mathrm{b}^*}_{\mathrm{c}^*}\mathrm{b}\,\mathrm{a} \rangle \overset{2.2\mathrm{b})}{\leqslant} \langle \varphi; \mathrm{c}^*\mathrm{c} \rangle^{1/2} \langle \varphi; \mathrm{a}^*\mathrm{a} \rangle^{1/2} = 0$$

thus $\mathrm{b\,a} \in I_\varphi$. I_φ is, therefore, a left ideal (but not a right ideal). If $\mathrm{a_i} \xrightarrow{\text{norm}} \mathrm{a}$, $\mathrm{a_i} \in I_\varphi$, then $\langle \varphi; \mathrm{a_i^*\,a_i} \rangle \to \langle \varphi; \mathrm{a^*\,a} \rangle = 0$, which shows that I_φ is norm closed.

(ii) We write

$$\Phi_\mathrm{a} := \mathrm{a} + I_\varphi. \tag{2.9}$$

The set $A/I_\varphi = \{\Phi_\mathrm{a}; \mathrm{a} \in A\}$ equipped with the scalar product

$$(\Phi_\mathrm{a}, \Phi_\mathrm{b}) := \langle \varphi; \mathrm{a}^*\mathrm{b} \rangle \tag{2.10}$$

is a pre-Hilbert space. We define

$$H_\varphi := \overline{A/I_\varphi} \tag{2.11}$$

the closure being taken in the norm associated with the scalar product (2.10). The mapping

$$\pi_\varphi : A \to B(A/I_\varphi)$$

given by $a \to \pi_\varphi(a)$,

$$\pi_\varphi(a)\,\Phi_b := \Phi_{ab} \tag{2.12}$$

connects every a with a bounded operator on A/I_φ, which can be extended to a bounded operator $\pi_\varphi(a)$ on the whole of H_φ. π_φ is easily seen to be a C*-morphism. Set $\Phi_1 =: \Phi$ and calculate

$$\langle \varphi; a\rangle = \langle \varphi; 1^* a\rangle = (\Phi_1, \Phi_a) = (\Phi, \pi_\varphi(a)\,\Phi)$$

in order to prove (2.7).
(iii) The action of $\pi_\varphi(A)$ on Φ gives A/I_φ which is dense in H_φ by construction: π_φ is cyclic with cyclic vector Φ.
(iv) Let H_1 be the Hilbert space corresponding to π_1! Define

$$u : A/I_\varphi \to H_1$$

by

$$u\,\Phi_a = \pi_1(a)\,\Phi_1 .$$

u is densely defined and invertible on a dense set in H_1. Furthermore

$$\|u\,\Phi_a\|^2 = (\Phi_1, \pi_1(a^* a)\,\Phi_1) = (\Phi_1 \pi_\varphi(a^* a)\,\Phi) = \|\Phi_a\|^2, \quad \forall\, a \in A.$$

Thus u can be extended to a unitary mapping $H \xrightarrow{\text{onto}} H_1$. ∎

The GNS-representation – a shorthand for the names Gelfand-Naimark-Segal – is one of the most useful tools in algebraic quantum theory. It is in some sense a generalization of the Fock-space construction, where the Hilbert space is also "generated" by applying an algebra – the algebra of creation and annihilation operators – to a cyclic vector, the no-particle state.

2.12 Lemma: Let ψ and φ be two states, for which there is a number $\lambda \geq 1$ with

$$\langle \psi; a^* a\rangle \leq \lambda \langle \varphi; a^* a\rangle, \quad \forall\, a \in A. \tag{2.13}$$

We say, ψ is dominated by φ. Let $\pi_\varphi : A \to B(H_\varphi)$ be the GNS-representation associated with φ. Then there is a unique operator $Z \in B(H_\varphi)$ satisfying

(i) $\langle \psi; a\rangle = (Z\Phi, \pi_\varphi(a)\,Z\Phi), \quad \forall\, a \in A,$

where Φ is the cyclic vector of the GNS-construction.

(ii) $Z \geq 0$

(iii) $Z \in \pi_\varphi(A)'$

(iv) $\|Z\Phi\| = 1.$

Proof: From (2.13), (2.10) and 2.2b)

$$|\langle\psi; a^* b\rangle| \leqslant \lambda \, \|\Phi_a\| \, \|\Phi_b\|. \tag{2.14}$$

If $b - b' \in I_\varphi$, i.e. $b' \in \Phi_b$, then (2.14) gives $\langle\psi; a^*(b - b')\rangle = 0$. Thus

$$B(\Phi_a, \Phi_b) := \langle\psi; a^* b\rangle$$

is a well defined, bounded sesquilinear form on $H_\varphi \times H_\varphi$, for which we can apply the Riesz representation theorem [13] to conclude that

$$\langle\psi; a^* b\rangle = (Y\Phi_a, \Phi_b) \tag{2.15}$$

for a uniquely defined positive $Y \in B(H_\varphi)$. We calculate

$$\begin{aligned}(Y\pi_\varphi(a)\Phi_b, \Phi_c) &= (Y\Phi_{ab}, \Phi_c) = \langle\psi; b^* a^* c\rangle \\ &= (Y\Phi_b, \pi_\varphi^*(a)\Phi_c) = (\pi_\varphi(a) Y\Phi_b, \Phi_c).\end{aligned} \tag{2.16}$$

Since Φ_b and Φ_c may vary independently in the dense set $A/I_\varphi \subset H_\varphi$ it follows that

$$Y\pi_\varphi(a) = \pi_\varphi(a) Y, \quad \forall a \in A,$$

i.e. $Y \in \pi_\varphi(A)'$. Define

$$Z := Y^{1/2} > 0, \tag{2.17}$$

which is unique, since there is exactly one positive square root of Y, and furthermore $Z \in \pi_\varphi(A)'$, too. Taking this into account, formula (i) is deduced from (2.15) by setting $b = c = 1$, and (ii) as well as (iii) have already been proved. For (iv) consider

$$\|Z\Phi\|^2 = (Z\Phi, Z\Phi) = (Y\Phi, \Phi) = \langle\psi; 1\rangle = 1,$$

since ψ is normalized. ∎

2.13 Proposition: φ is a pure state if and only if π_φ is irreducible.

Proof:

a) Let π_φ be irreducible. Then Z of Lemma 2.12 must be equal to 1 and every ψ dominated by φ is equal to φ itself. Assume

$$\varphi = \lambda\varphi_1 + (1 - \lambda)\varphi_2, \quad \lambda \in (0, 1)$$

then both states φ_i are dominated by φ and thus coincide with it.

b) Let π_φ be reducibel and p a non-trivial projection in $\pi_\varphi(A)'$. Then

$$\langle\varphi; a\rangle = \lambda_1 \underbrace{\frac{(\pi_\varphi(a)\, p\,\Phi, \Phi)}{\lambda_1}}_{\langle\varphi_1; a\rangle} + \lambda_2 \underbrace{\frac{(\pi_\varphi(a)(1-p)\Phi, \Phi)}{\lambda_1}}_{\langle\varphi_2; a\rangle}$$

with

$$\lambda_1 = (p\Phi, \Phi), \quad \lambda_2 = ((1 - p)\Phi, \Phi)$$

is a non-trivial decomposition of φ. Hence φ is not pure. ∎

2.14 Proposition: Let φ be a state and $\Phi \in H_\varphi$ the corresponding cyclic vector of the GNS-representation π_φ. If Φ is cyclic for $\pi_\varphi(A)'$, too, then Φ is separating for $\pi_\varphi(A)$. If Φ is separating for $\pi_\varphi(A)''$, then Φ is cyclic for $\pi_\varphi(A)'$. If Φ is separating for $\pi_\varphi(A)''$ and $H_\varphi \neq \mathbb{C}$ then $\pi_\varphi(A)$ is reducible and φ not pure.

Proof:

(i) Let Φ be cyclic for $\pi_\varphi(A)'$. Then $\pi_\varphi(\mathrm{a})\,\Phi = 0$ implies

$$\pi_\varphi(\mathrm{a})\,\pi_\varphi(A)'\Phi = 0,$$

and thus $\pi_\varphi(\mathrm{a}) = 0$, because $\pi_\varphi(A)'\Phi$ is dense in H_φ.

(ii) Let Φ be separating for $\pi_\varphi(A)''$, and denote by p the projection on the subspace $\pi_\varphi(A)'\,\Phi \subset H_\varphi$. Then for every $\mathrm{b} \in \pi_\varphi(A)'$ holds $\mathrm{pbp} = \mathrm{bp}$. Thus

$$\mathrm{pb} = (\mathrm{b}^*\mathrm{p})^* = (\mathrm{pb}^*\mathrm{p})^* = \mathrm{pbp} = \mathrm{bp}$$

and $\mathrm{p} \in \pi_\varphi(A)''$. From $(1-\mathrm{p})\,\Phi = 0$ then follows $\mathrm{p} = 1$. Therefore, $\pi_\varphi(A)'\,\Phi = H_\varphi$ and Φ is cyclic for $\pi_\varphi(A)'$.

(iii) If Φ is separating for $\pi_\varphi(A)''$, then $\pi_\varphi(A)'\,\Phi$ spans an at least two dimensional vector space, so $\pi_\varphi(A)' \neq \mathbb{C}\,1$. Because of 2.13 φ is then not pure. ∎

Remark: The Tomita-Takesaki theory [12] shows that in fact $\pi_\varphi(A)'$ is isomorphic to $\pi_\varphi(A)''$ if Φ is separating for $\pi_\varphi(A)''$.

2.15 Proposition: Let A be the quasilocal algebra of a countable increasing family $\{A_{\Lambda \mathrm{n}}\}$ of von Neumann algebras in separable Hilbert spaces H_n and φ a locally normal state on A. Then the GNS-representation space H_φ is separable.

This result, the proof of which is given in [8], p. 282, makes clear that also for infinite quantum systems separable Hilbert spaces may be suitable for representing the theoretical description (A, S_0).
For structural investigations the existence of the so-called universal representation $\pi_\mathrm{u}(A)$ of an arbitrary C*-algebra is of fundamental importance.

2.16 Definition: Let for every $\varphi \in S(A)$ be π_φ the corresponding GNS-representation of A in the Hilbert space H_φ and define $H_\mathrm{u} := \sum_{\varphi \in S} \oplus H_\varphi$. Then

$$\pi_\mathrm{u} : A \to B(H)$$

given by

$$\pi_\mathrm{u}(\mathrm{a}) := \sum_{\varphi \in S} \pi_\varphi(\mathrm{a})$$

is called the universal representation of A.

2.17 Proposition: The universal representation π_u of A is a C*-isomorphism of A onto $\pi_\mathrm{u}(A) \subset B(H_\mathrm{u})$ [6].

From Prop. 2.17 follows that every C*-algebra A is isomorphic to an operator algebra. We shall have need alsc for the following fact.

2.18 Proposition: Let $\overline{\pi_u(A)}$ the weak closure of $\pi_u(A)$ in $B(H_u)$. Then $\overline{\pi_u(A)}$ is a W*-algebra with predual A^* [6].

The importance of such general structural insights is illustrated by the following norm expression which again shows that it is sufficient to deal with a full set of states S_0 only.

2.19 Proposition: Let be $A_0^* \subset A^*$ the linear span of a full set of states S_0. Then

$$\|a\| = \sup_{\substack{\varphi \in A_0^* \\ \|\varphi\| = 1}} |\langle \varphi; a \rangle| \tag{2.18}$$

for all $a \in A$.

Proof: In virtue of Prop. 2.17 we have $\|a\| = \|\pi_u(a)\|$, and from 2.18 we conclude that $\pi_u(a)$ is an element of A^{**}. Thus

$$\begin{aligned}\|\tau_u(a)\| &= \sup_{\substack{\varphi \in A^* \\ \|\varphi\| = 1}} |\langle \pi_u(a); \varphi \rangle| \\ &= \sup_{\substack{\varphi \in A^* \\ \|\varphi\| = 1}} |\langle \varphi; a \rangle| \\ &= \sup_{\substack{\varphi \in A_0^* \\ \|\varphi\| = 1}} |\langle \varphi; a \rangle|\end{aligned}$$

where in the last step we used the fact that A_0 is $\sigma(A^*, A)$-dense in A^*. ∎

The norm of an algebra element $a \in A$ can be associated with its spectrum: For $a \in A$ we define the spectrum by

$$\sigma(a) = \{\lambda; \lambda \in \mathbb{C}, (a - \lambda 1)^{-1} \text{ does not exist}\}.$$

Then one has

$$\|a\| = \sup_{\lambda \in \sigma(a)} |\lambda| \tag{2.19}$$

(cf. [6], p. 4). Thus the norm-topology in A may be considered as not too an artificial concept from the physical point of view. This is somewhat different for the norm topology in the state space, since for $\varphi, \psi \in S$ we have

$$\|\varphi - \psi\| = \sup_{a \in A} |\langle \varphi - \psi; a \rangle| / \|a\| \tag{2.20}$$

and the simple formula from Pro. 2.2c) is no more valid. The supremum construction in (2.20) requires first the determination of the norms of all observables. A little bit simpler is the $\sigma(S_0, A)$-topology defined as the weakest (coarsest) topology, for which the functions

$$S_0 \ni \varphi \to \langle \varphi; a \rangle \in \mathbb{C}$$

are continuous for all $a \in A$. We shall call this the "physical topology" in S_0.

2.20 Proposition: The state space $S(A)$ is compact in the physical topology [6].

The special choice of the full set S_0 singles out certain representations of A as appropriate.

2.21 Definition: A representation $\pi: A \to B(H)$ is said to be a representation of (A, S_0), if for every unit vector $\Phi \in H$ the state φ defined by

$$\langle \varphi; a \rangle := (\Phi, \pi(a)\Phi) \tag{2.21}$$

is in S_0.

Let us add that model studies of nonrelativistic many body systems [27], [28], [29] have enforced an even more general set up for the description of a quantum system than outlined above. According to [30] the full convex set S_0 of Def. 2.6 should be substituted by a convex, norm-closed set, which is invariant under local excitation, i.e., it should hold

$$\varphi \in S_0 \Rightarrow \varphi_a \in \varphi_0, \quad \forall a \in A, \ \langle \varphi; a^* a \rangle > 0,$$

where

$$\langle \varphi_a; b \rangle := \langle \varphi; a^* b a \rangle / \langle \varphi; a^* a \rangle, \quad b \in A.$$

Such a set is called folium in [31] and [33]. Because this generalization can be reduced to our set up in most cases of interest [32] we stick to Def. 2.6 in this article.

3 Symmetry transformations

One should discern between different notions of symmetry transformations. The most general ones are those which preserve the physically relevant structures of a theory; we call them structural symmetries. They may be induced by purely formal alterations in the theoretical language without any specific physical interpretation. The time translations should always be special cases of structural symmetries, because as time proceeds the theoretical language should not alter its relevant structure. Another important subclass of structural symmetries consists of all transformations which are induced by changes of the spatial coordinate system. This class reflects our conception of physical space and is independent of the special system under consideration. The dynamical symmetries, on the other hand, depend on the forces in a special system and are defined as those structural symmetries which commute with the time translations.

In traditional quantum mechanics the structural symmetries can be defined as bijective mappings of the set V of vector states, which leave all transition probabilities invariant [14], [15]. This definition is quite popular but nevertheless not very natural, since it is by no means evident that all of the physical relevant structure can be reduced to transition probabilities. In the consequent vector-state language which is commonly used in talking about these Wigner symmetries, some aspects appear even paradoxical. According to this definition a spatially localized state of a system may equally well be described by a localized or a spreadout wave function, provided that the wave functions of all other states are also properly transformed. It is clear from this that, in order to give the attribute "localized" an invariant meaning, one has to transform the observables simultaneously with the states. So one can compensate for the state variations. Beside that one must give a prescription how to transform mixed states. Only after all this being concisely formulated one knows in which way the basic elements of the theoretical language change under a structural symmetry. But then arises the question if the condition of the conserved transition probabilities is still needed.

It is clear, especially in the algebraic formulation, that the physically relevant structure of a quantum theory is the set of all expectation values $\{\langle\varphi; a\rangle; \varphi \in S_0, a \in A, a^* = a\}$. (The transition probabilities are only a subset of expectation values, where φ are (pure) vector states and a one dimensional projections.) In order to leave these quantities invariant, a structural symmetry has to act on both the state space and the observable space $A^s \subset A$ ($s \equiv$ selfadjoint). One can restrict the considerations to one of these two aspects and work either in the state picture or in the observable picture if one has the guarantee that the one entails the other. If one starts again in the state picture a structural symmetry should certainly act bijectively in S_0 and should furthermore, in accordance with the above reasoning, ensure the existence of the adjoint mapping in A^s. Including also continuity aspects and remembering that it is easier to deal with $A = A^s + iA^s$, we arrive thus at the following definition.

3.1 Definition: A structural symmetry transformation of the description (A, S_0) of a quantum system is a bijection $\nu: S_0 \to S_0$, which is continuous in the physical topology of S_0 and for which there exists a linear mapping $\alpha: A \to A$, such that $\alpha A^s = A^s$.

If ν is a structural symmetry according to Def. 3.1 then the simultanous action of ν in S_0 and of ν^{*-1} in A^s, the existence of which can be easily concluded from the properties of ν, leave all measurable quantities invariant. This is the most direct characterization of a structural symmetry, but it is not common to put the definition in this way and it is not clear at all, how this definition corresponds to a Wigner symmetry. The first step is now to formulate the symmetry in one picture alone, and this we will do again first in the state picture.

3.2 Proposition: The structural symmetries of (A, S_0) are exactly the $\sigma(S_0, A)$-continuous, affine bijections of S_0

Proof: Since $\sigma(S_0, A)$ is by definition the "physical" topology of S_0, the only thing we have to show is that affinity – in connection with the other properties of ν-– entails the existence of ν^* and vice versa.

Assume first that $\nu^* =: \alpha$ exists. Then for $\lambda \in [0, 1]$ we have

$$\langle \nu(\lambda\varphi_1 + (1-\lambda)\varphi_2); a\rangle = \langle \lambda\varphi_1 + (1-\lambda)\varphi_2; \alpha a\rangle$$
$$= \lambda\langle\varphi_1; \alpha a\rangle + (1-\lambda)\langle\varphi_2; \alpha a\rangle = \langle \lambda\nu\varphi_1 + (1-\lambda)\nu\varphi_2; a\rangle.$$

This being true for all $a \in A$ we obtain affinity of ν, i.e.

$$\nu(\lambda\varphi_1 + (1-\lambda)\varphi_2) = \lambda\nu\varphi_1 + (1-\lambda)\nu\varphi_2 \tag{3.1}$$

as a necessary condition.

Assume now that ν fulfills all suppositions of the Proposition.

(i) We extend ν to a linear mapping $\tilde{\nu}$ on the linear span $A_0^* \subset A^*$ of S_0, which is clearl clearly $\sigma(A_0^*, A)$-continuous. For all $a \in A$ the mapping

$$A_0^* \ni f \to \langle \tilde{\nu} f, a\rangle \in \mathbb{C}$$

is linear and $\sigma(A_0^*, A)$-continuous. Thus one knows ([13], p. 114), that there is a unique $a' \in A$ with $\langle \tilde{\nu} f; a\rangle = \langle f; a'\rangle$. Since the mapping $\alpha: a \to a'$ is linear, we have proved the existence of α as a linear operator in A.

(ii) It holds

$$\langle\varphi; \alpha a^*\rangle = \langle\nu\varphi; a^*\rangle = \langle\nu\varphi; a\rangle^* = \langle\varphi; (\alpha a)^*\rangle, \quad \forall \varphi \in S_0.$$

Because of fulness of S_0 we thus have

$$\alpha a^* = (\alpha a)^*$$

and, therefore

$$\alpha A^s \subset A^s$$

As mentioned earlier, $\alpha^{-1} = \nu^{-1\,*}$ exists, hence

$$\alpha A^s = A^s. \qquad \blacksquare$$

Remark: The proof of Prop. 3.2 shows that beside affinity also $\sigma(S_0, A)$-continuity is essential for ν having an adjoint $\nu^*: A \to A$.

The characterization of symmetry transformations as in the Proposition was first given by Kadison [16], [17] (in the case of time translations), we call them Kadison symmetries. In fact is it quite natural to assume from the beginning – as Kadison did – that a decomposition of a state into pure states should be preserved under a symmetry transformation (this is a verbal circumscription of affinity). It is only that one does not immediately realize the structural invariance of the theory under a Kadison symmetry, why we preferred our Definition 3.1.

We can now join Kadison's interesting investigations [17] on the transition to the observable-language (cf. also [8], Chapter 2.2).

3.3 Definition: A Jordan automorphism of a C*-algebra A is a linear, adjoint-preserving bijection α of A with

$$\alpha(ab + ba) = (\alpha a)(\alpha b) + (\alpha b)(\alpha a) \tag{3.2}$$

for all a, b $\in A$.
Since

$$ab + ba = (a + b)^2 - a^2 - b^2$$

(3.2) is equivalent with

$$\alpha a^2 = (\alpha a)^2, \quad \forall a \in A. \tag{3.3}$$

3.4 Theorem: The correspondence $\langle \nu \varphi; a\rangle = \langle \varphi; \alpha a\rangle$, $\forall \varphi \in S_0$, $\forall a \in A$, associates with every Kadison symmetry ν a $\sigma(A, S_0)$-continuous Jordan automorphism α of A, and conversely.

Proof:

(i) We have already proved in Prop. 3.2 that with every ν is associated a linear adjoint-preserving bijection $\alpha: A \to A$.

(ii) $a \geq 0 \Leftrightarrow \langle \varphi; a\rangle \geq 0, \quad \forall \varphi \in S_0,$

(since S_0 is full, cf. Def. 2.5)

$\Leftrightarrow \langle \nu^{-1} \varphi; \alpha a\rangle \geq 0, \quad \forall \varphi \in S_0,$

$\Leftrightarrow \alpha a \geq 0.$

(iii)
$$\|\alpha a\| \overset{(2.18)}{=} \sup_{f \in A_0^*} |\langle f; \alpha a\rangle| / \|f\| = \sup_{f \in A_0^*} |\langle \tilde{\nu} f; a\rangle| / \|f\| = \|a\|$$

(iv) For mappings α satisfying (i)–(iii) one has Kadison's generalized Schwartz inequality [18]

$$(\alpha a)^2 \leq \alpha a^2$$

for all self-adjoint $a \in A^s$. Since α^{-1} fulfills also (i)–(iii) we obtain

$$(\alpha^{-1} b)^2 \leq \alpha^{-1} b^2$$

for $b \in A^s$. Insert $b = \alpha a$ which is in A^s if a is so by (i), to get

$$a^2 \leq \alpha^{-1} (\alpha a)^2.$$

Thus

$$\alpha a^2 \leq (\alpha a)^2 \leq \alpha a^2$$

and the equality sign has to hold for all $a \in A^s$.

(v) Since $a = a_1 + i a_2$, with $a_i \in A^s$ for every $a \in A$, we have

$$a^2 = a_1^2 - a_2^2 + i(a_1 a_2 + a_2 a_1).$$

From (iv) follows $\alpha a_i^2 = (\alpha a_i)^2$ as well as

$$\alpha(a_1 a_2 + a_2 a_1) = (\alpha a_1)(\alpha a_2) + (\alpha a_2)(\alpha a_1),$$

since

$$\alpha(a_1 + a_2)^2 = (\alpha a_1 + \alpha a_2)^2 .$$

Linearity of α gives then

$$\alpha a^2 = (\alpha a)^2, \ \forall a \in A. \tag{3.4}$$

(vi) Let $a_i \to a$ in the $\sigma(A, S_0)$-topology, which is equivalent with $\langle\varphi; a_i\rangle \to \langle\varphi; a\rangle$, $\forall \varphi \in S_0$, and is further equivalent with $\langle\nu\varphi; a_i\rangle \to \langle\nu\varphi; a\rangle$, $\forall \varphi \in S_0$. Thus $\alpha a_i \to \alpha a$ in the $\sigma(A, S_0)$-topology, which proves the continuity of α w.r.t. this topology.

(vii) Let on the other hand α be a $\sigma(A, S_0)$-continuous Jordan automorphism of A. There exists then evidently the adjoint linear mapping $\alpha^* : A^* \to A^*$. For $f \in A_0^*$ we have that $\alpha^* f = f \circ \alpha$ is a $\sigma(A, A_0^*)$-continuous linear functional on A, and we may apply again Th. IV.20 of [13] to conclude that $\alpha^* f \in A_0^*$. Thus $\alpha^* A_0^* = A_0^*$, α^* being invertible. We further observe that every Jordan automorphism α is a positive map, since from $a \geqslant 0 \Leftrightarrow a = b^2$, $b \in A^s$ it follows $\alpha a = (\alpha b)^2 \geqslant 0$. Thus, if $f \in A^*$ is positive, we have $\langle\alpha^* f, a\rangle = \langle f, \alpha a\rangle \geqslant 0$ for all $a \geqslant 0$, and $\alpha^* f$ is again a positive linear form on A. Since $\alpha(1) = 1$, α^* preserves also the normalization of a linear form. From this follows that $\nu := \alpha^* | S_0$ is an affine, $\sigma(S_0, A)$-continuous bijection of S_0. ■

We have now a simple characterization of a structural symmetry in the observable picture as a $\sigma(A, S_0)$-continuous Jordan automorphism of A. This is intuitive appealing since the proper algebraic structure for observables is just the Jordan algebra. Nevertheless it is important to know how much of the C*-algebraic structure is also conserved under symmetry transformations. This can be investigated most conveniently if the C*-algebra A is realized as an operator algebra (isomorphically).

3.5 Theorem (Generalized Wigner's Theorem): Let A be a concrete C*-algebra. Then α is a structural symmetry in the observable picture if and only if it is a linear adjoint-preserving bijection of A, so that there exists a projection $p \in A'' \cap A'$ with

$$\begin{aligned} &\alpha(ab)\, p = (\alpha a)(\alpha b)\, p \\ &\alpha(ab)(1-p) = (\alpha b)(\alpha a)(1-p) \\ &\forall a, b \in A. \end{aligned} \tag{3.5}$$

The proof will be given in the Appendix. Observe that p is not necessarily in A!

3.6 Corollary: Let $A = B(H)$. Then every structural symmetry α in the observable picture is either a C*-automorphism or a C*-anti-automorphism.

Proof: Since in this case $A'' \cap A' = \mathbb{C}\,1$, p of (3.5) is either 1 or 0, that means that the linear, adjoint-preserving bijection α is either multiplicative (automorphism) or anti-multiplicative (anti-automorphism) on the whole of A. ∎

In this simple case $A = B(H)$ one knows ([5], Chapter III) that every (anti-)automorphism is inner and can be implemented by an (anti-)unitary operator $U(V) \in B(H)$, i.e.

$$\alpha a = \begin{cases} UaU^{-1}, \text{ U unitary (automorphism)} \\ Va^*V^{-1}, \text{ V anti-unitary (anti-automorphism)} \end{cases} \tag{3.6}$$

Observe that U and V are unique up to a phase factor only. Thus we have

3.7 Corrolary (Wigner's Theorem): In traditional quantum mechanics every structural symmetry is associated uniquely with either a unitary or an anti-unitary operator ray $\{\lambda U; |\lambda| = 1\}$ and vice versa.

Since Wigner proved the same statement for his kind of symmetries, the Corollary shows the equivalence of Wigner symmetries and Kadison symmetries in classical quantum mechanics. This correspondence, however, can be shown more directly [19]. The three assumptions on ν to be a Kadison symmetry imply indeed directly that ν restricted to the vector states V keeps the transition probabilities invariant, and the latter condition guarantees that a bijection ν_0: $V \to V$ can be uniquely extended to a Kadison symmetry ν: $S_0 \to S_0$ where $S_0 = S_n$, the set of normal states on $B(H)$. That means that in Wigner's definition all of the relevant structure for a symmetry is comprised and condensed to only two conditions, but it is not so easy to realize the full content of this formulation.

In applications one is concerned mostly not with a single symmetry transformation, but with a whole group of such transformations. Let us introduce on the set Γ^* of all (Kadison) symmetries of (A, S_0) the topology of which is given by the family of seminorms

$$\{|\langle \nu\varphi; a\rangle|; \nu \in S_0, a \in A\}. \tag{3.7}$$

Γ^* may be viewed at as a convex subset of the vector space constituted of all linear transformations in $A_0^* \supset S_0$. The topology on Γ^* induced by (3.7) is locally convex and Hausdorff. We call it the weak topology in Γ^*. It is weaker than the metric topology induced by the operator norm of ν.

3.8 Proposition: Γ^* equipped with the usual composition of mappings and with the weak topology is a topological group.

Proof:

(i) If ν and μ are in Γ^* then $\nu \circ \mu$ is clearly bijective, affine. Since the composition $\circ$ of mappings preserves continuity, $\nu \circ \mu$ is also $\sigma(S_0, A)$-continuous, hence $\nu \circ \mu \in \Gamma^*$. If $\nu \in \Gamma^*$ then ν^{-1} is also an affine bijective mapping of Γ^*. Continuity of ν^{-1} can be shown as follows: S_0 is $\sigma(S, A)$-dense in S ([17], Theorem (2.2)), and S is $\sigma(S, A)$-compact ([13], Theorem IV.2 (Banach-Alaoglu-Theorem). ν can be extended to a $\sigma(S, A)$-continuous mapping $S \to S$. Then by [13], Theorem IV.4, ν^{-1} is $\sigma(S, A)$-continuous on S, and the restriction to S_0 is then $\sigma(S_0, A)$-continuous, since S_0 is a topological subspace of S. Since also $1 \in \Gamma^*$, we have shown that Γ^* is a group.

(ii) The continuity of the group operations in Γ^* is stated in [8], [1] c) and [20]. ■

For the further reasoning it is preferable to use the observable language. The set of all Jordan automorphisms in A is denoted by Γ. The bijection

$$\delta : \Gamma^* \leftrightarrow \Gamma, \quad \delta(\nu) = \alpha, \tag{3.8}$$

where

$$\langle \nu\varphi; a \rangle = \langle \varphi; \delta(\nu)\, a \rangle$$

(cf. Theorem 3.4) is a continuous group anti-isomorphism, where Γ is to be equipped with the coarsest topology, for which all mappings $\alpha \to \langle \varphi; \alpha a \rangle$ ($\varphi \in S_0$, $a \in A$) are continuous. Under the assumption that Proposition 3.8 is valid, Γ is a topological group, too.

3.9 Definition: A topological group G is a symmetry group of (A, S_0) if there is a continuous group-morphism

$$\begin{aligned} &\gamma : G \to \Gamma \\ &\gamma(g) = \alpha_g. \end{aligned} \tag{3.9}$$

One says that G acts in A via Jordan automorphisms.

Given γ one can construct a dual continuous group morphism

$$\begin{aligned} &\gamma^* : G \to \Gamma^* \\ &\gamma^*(g) := \delta^{-1} \circ \gamma(g^{-1}). \end{aligned} \tag{3.10}$$

Thus a symmetry group G acts also in S_0 via affine mappings.
Now, there is a natural subgroup Γ_1 of Γ consisting of all C*-automorphisms of A. For many important cases one can show that necessarily $\gamma(G) \subset \Gamma_1$.

3.10 Proposition: Let G be a symmetry group of (A, S_0) and $\pi : A \to B(H)$ a representation of (A, S_0). Then $\pi \circ \alpha_g a$ is strongly continuous in g for all $a \in A$.

Proof: With $\Phi \in H$, a unit vector in H, we define the state φ by

$$\langle \varphi; a \rangle := (\Phi, \pi(a)\, \Phi)$$

where $\varphi \in S_0$ because of the definition of a representation of (A, S_0) (Def. 2.17). Since $g \to \alpha_g$ is weakly continuous, we have

$$g \to g' \Rightarrow (\Phi, \pi(\alpha_g a)\, \Phi) \to (\Phi, \pi(\alpha_{g'} a)\, \Phi)$$

and by polarization

$$(\Phi, \pi(\alpha_g a)\, \Psi) \to (\Phi, \pi(\alpha_{g'} a)\, \Psi), \ \forall\, \Phi, \Psi \in H.$$

We calculate for selfadjoint a

$$\|\pi(\alpha_g a)\Phi - \pi(\alpha_{g'} a)\Phi\|^2 = (\pi(\alpha_g a)\Phi, \pi(\alpha_g a)\Phi) - (\pi(\alpha_g a)\Phi, \pi(\alpha_{g'} a)\Phi)$$
$$- (\pi(\alpha_{g'} a)\Phi, \pi(\alpha_g a)\Phi) + (\pi(\alpha_{g'} a)\Phi, \pi(\alpha_{g'} a)\Phi).$$

Since, if a is selfadjoint, it is also $\pi(\alpha_g a)$, and because of Jordam homomorphy: $(\alpha_g a)(\alpha_g a) = \alpha_g a^2$, we have

$$(\pi(\alpha_g a)\Phi, \pi(\alpha_g a)\Phi) = (\Phi, \pi(\alpha_g a^2)\Phi).$$

We can now apply weak continuity for showing

$$g \to g' \Rightarrow \|\pi(\alpha_g a)\Phi - \pi(\alpha_{g'} a)\Phi\|^2 \to 0.$$

This is the strong (operator) continuity of $\pi(\alpha_g a)$. ∎

3.11 Theorem: If G is a connected symmetry group of (A, S_0), and if (A, S_0) has a separating family of factor representations, then G acts in A via C*-automorphisms.

Proof:

(i) Let $\pi: A \to B(H)$ be one of the factor representations of (A, S_0) and p_g the maximal element of those $p \in \pi(A)' \cap \pi(A)''$ for which (3.5) is valid. Since $\pi(A)' \cap \pi(A)''$ is a von Neumann algebra, p_g is again in $\pi(A)' \cap \pi(A)''$. One can then show ([17], Lemma (4.8)) that $g \to p_g \Phi$ is a continuous mapping $G \to H$ for all $\Phi \in H$.

(ii) Since π is a factor representation $\pi(A)' \cap \pi(A)'' = \mathbb{C}\,1$, and p_g is either 1 or 0. But for $g = e$ p_g must be 1. Because of connectedness of G every other p_g can be reached from p_e on a continuous path in G and must be 1, too.

(iii) According to (ii) we have for all factor representations π that

$$\pi(\alpha_g(ab) - (\alpha_g a)(\alpha_g b)) = 0. \tag{3.11}$$

The fact that the set of factor representations is separating means exactly that one may conclude from (3.11) the validity of

$$\alpha_g(ab) = (\alpha_g a)(\alpha_g b),$$

so that α_g is a C*-isomorphism for all $g \in G$. ∎

The assumptions of Theorem 3.11 are e.g. fulfilled if already the set of pure states in S_0 is full, i.e. if the observables can be identified by their values in pure states alone. For, the GNS-representation π_φ of a pure state is irreducible and thus factorial. And from $\pi_\varphi(a) = 0$ follows $\langle\varphi; a\rangle = (\Phi, \pi_\varphi(a)\Phi) = 0$ for a full set of states φ, so $a = 0$.

4 Represented systems and symmetries

In the preceding section we dealt with symmetries of a theoretical description (A, S_0) without considering representations. Now we investigate what can be said about symmetries in connection with representations.

4.1 Definition: Let G be a symmetry group of (A, S_0) and $\pi : A \to B(H)$ a representation of (A, S_0). We say that π is compatible with α_g if

$$\alpha_g\, \pi^{-1}(0) \subset \pi^{-1}(0). \tag{4.1}$$

If π is compatible with α_g, then

$$\alpha_g^\pi\, \pi(a) := \pi(\alpha_g a) \tag{4.2}$$

is a well defined Jordan automorphism of $\pi(A)$. For, if $a - a' \in \pi^{-1}(0)$ then $\pi(a) = \pi(a')$ and

$$\alpha_g^\pi(\pi(a) - \pi(a')) = \pi(\alpha_g(a - a')) = 0$$

since $\alpha_g(a - a') \in \pi^{-1}(0)$. Thus α_g^π is well defined and easily shown to be a Jordan automorphism.

4.2 Proposition:

a) Let α be an (anti-) automorphism of A and the state φ be invariant under α^*. Then the GNS-representation π_φ is compatible with α and it holds

$$\alpha^\pi\, \pi_\varphi(a) = \pi_\varphi(\alpha a) = U_\varphi \pi_\varphi(a)\, U_\varphi^{-1}, \forall\, a \in A \tag{4.3a}$$

resp.

$$\alpha^\pi\, \pi_\varphi(a) = \pi_\varphi(\alpha a) = V_\varphi \pi_\varphi^*(a)\, V_\varphi^{-1}, \forall\, a \in A \tag{4.3b}$$

where V_φ (U_φ) is the some (anti-)unitary operator in H_φ.

b) Let the state $\varphi \in S_0$ be invariant under the connected symmetry group G of (A, S_0), then there exists a strongly continuous unitary representation $\{U_\varphi(g); g \in G\}$ of G in H_φ such that

$$\alpha_g^\pi\, \pi_\varphi(a) = U_\varphi(g)\, \pi_\varphi(a)\, U_\varphi^{-1}(g). \tag{4.4}$$

Proof:

(i) α has an inverse on A. Into $\langle\varphi; \alpha a\rangle = \langle\varphi; a\rangle$, $\forall\, a \in A$, insert $a = \alpha^{-1} b$ in order to show that φ is also invariant against α^{-1}. $c \in \pi_\varphi^{-1}(0)$ iff

$$(\Phi_a, \pi_\varphi(c)\, \Phi_b) = \langle\varphi; a^* c b\rangle = 0, \ \forall\, a, b \in A.$$

But then also

$$0 = \langle\alpha^* \varphi; a^* b c\rangle = (\Phi_{\alpha a}, \pi_\varphi(\alpha c)\, \Phi_{\alpha b})$$

so that $\alpha \pi^{-1}(0) \subset \pi^{-1}(0)$ is valid. Since the same reasoning holds for α^{-1} we have

$$\alpha \pi^{-1}(0) = \pi^{-1}(0)$$

and α^π according to (4.2) is well defined.

(ii) Let α be a C*-automorphism; define on A/I_φ

$$U_\varphi \Phi_a := \Phi_{\alpha a}. \tag{4.5}$$

Then

$$(U_\varphi \Phi_a, U_\varphi \Phi_b) = \langle \varphi; \alpha(a^* b)\rangle = (\Phi_a, \Phi_b),$$

so that U_φ can be extended to an unitary operator on H_φ. Furthermore

$$(\alpha^\pi \pi_\varphi(a))\, \Phi_b = \pi_\varphi(\alpha a)\, \Phi_b = \Phi_{\alpha(a\alpha^{-1} b)} = (U_\varphi \pi_\varphi(a)\, U_\varphi^{-1})\, \Phi_b, \quad \forall a, b \in A.$$

(iii) Let α be a C*-anti-automorphism; define on A/I_φ

$$V_\varphi \Phi_a := \Phi_{\alpha a^*}. \tag{4.6}$$

Then

$$(V_\varphi \Phi_a, V_\varphi \Phi_b) = \langle \varphi; (\alpha a)(\alpha b^*)\rangle = \langle \varphi; \alpha(b^* a)\rangle = (\Phi_b, \Phi_a).$$

Thus V_φ can be extended to an anti-unitary operator on H_φ. Furthermore

$$(\alpha^\pi \pi_\varphi(a))\, \Phi_b = \pi_\varphi(\alpha a)\, \Phi_b = \Phi_{(\alpha a) b} = \Phi_{\alpha((\alpha^{-1} b) a)} = V_\varphi \Phi_{a^* \alpha^{-1} b^*}$$
$$= (V_\varphi \pi^*(a)\, V_\varphi^{-1})\, \Phi_b, \ \forall a, b \in A.$$

With (i), (ii) and (iii) we have shown the assertion a).

(iv) According to Theorem 3.11 all α_g, $g \in G$ are C*-automorphisms, so that (4.2) is valid for all α_g^π. The strong continuity follows from Proposition 3.10. This proves b). ∎

Prop. 4.2 shows the importance of invariant states for group representations. It is interesting that there exists a general characterization of those groups which allow for invariant states if they act via C*-automorphisms in a C*-algebra.

4.3 Definition:

a) Let $C(G)$ be the set of all bounded, continuous, complex-valued functions on the topological group G. A mapping

$$\eta: C(G) \to \mathbb{C},$$

which is linear, positive and normalized (i.e. a state on the commutative algebra $C(G)$) is called a mean on G.

b) Let G be a locally compact topological group. G is said to be amenable if it has an invariant mean η, i.e. a mean which satisfies

$$\eta(g_0 f) = \eta(f), \ \forall g_0 \in G \tag{4.7}$$

where $g_0 f(g) := f(g_0 g)$ (cf. [21]).

Remark: Relation (4.7) is equivalent with

$$\eta(f g_0) = \eta(f), \ \forall g_0 \in G; \ f g_0(g) := f(g g_0) \tag{4.8}$$

and with

$$\eta(g f h) = \eta(f), \ \forall g, h \in G. \tag{4.9}$$

Every abelian or solvable locally compact group is amenable, and every compact group is trivially amenable by the existence of the finite, invariant Haar measure. Likewise is the Euclidean group amenable. However, no non-compact semi-simple Lie group is amenable. So are SL(2, R) and the Lorentz group not amenable.

For non-compact groups amenability has nothing to do with the existence of the Haar measure: The Haar measure corresponds to a functional on $C_0(G)$, the set of continuous functions which vanish at infinity. If η is a mean, then $\eta(f) = 0$, for $f \in C_0(G)$, if G is non-compact.

In the subsequent investigations we specialize S_0 to S since otherwise the formulation of the assumptions would be too complicated.

4.4 Proposition: Let G be an amenable symmetry group of (A, S), then there exists a G-invariant state.

Proof: Let φ be an arbitrary state. Then the function $f(g) := \langle \varphi; \alpha_g a \rangle$ for a fixed $a \in A$ is in $C(G)$, since it is continuous by the definition of a symmetry group and bounded by $\|a\|$. Thus ηf exists, and we may define

$$\langle \eta\varphi; a \rangle := \eta(\langle \varphi; \alpha_g a \rangle). \tag{4.10}$$

It is easy to show that (4.10) defines a state $\eta\varphi \in S$. It is invariant since

$$\langle \eta\varphi; \alpha_h a \rangle = \eta(\langle \varphi; \alpha_{gh} a \rangle) = \langle \eta\varphi; a \rangle.$$ ■

Let us denote by S_G the set of all G-invariant state, where G is a symmetry group of (A, S). S_G is clearly convex. It is weakly closed since α_g^* is weakly continuous by definition of a structural symmetry. Thus S_G is a closed subset of a compact set S and is, therefore, itself compact. The extreme points of S_G are called G-ergodic.

4.5 Proposition: Let φ and ψ be in S_G and $\psi \leqslant \lambda\varphi$ for a $\lambda \geqslant 1$. Then

$$\langle \psi; a \rangle = (Z\Phi, \pi_\varphi(a) Z\Phi) \tag{4.11}$$

where π_φ is the GNS-representation with cyclic vector Φ associated with φ and where Z is a positive operator in $\pi_\varphi(A)' \cap U_\varphi(G)' =: N_\varphi$.

Proof: We conclude from Lemma 2.12 that (4.11) is valid with $Z \in \pi_\varphi(A)'$! Consider

$$(\Phi, \pi_\varphi(b^*) Z^2 U_\varphi(g) \pi_\varphi(a) U_\varphi^{-1}(g) \Phi) = (Z\Phi, \pi_\varphi(b^*) \pi_\varphi(\alpha_g a) Z\Phi)$$
$$= \langle \psi; b^* \alpha_g a \rangle = \langle \psi; \alpha_g^{-1}(b^* \alpha_g a) \rangle = \langle \psi; (\alpha_g^{-1} b^*) a \rangle$$
$$= (\Phi, U_\varphi^{-1}(g) b^* U_\varphi(g) Z^2 \pi_\varphi(a) \Phi).$$

Comparing the first and the last term we obtain

$$(\Phi_b, Z^2 U_\varphi(g) \Phi_a) = (\Phi_b, U_\varphi(g) Z^2 \Phi_a), \forall a, b \in A.$$

Thus

$$Z^2 U_\varphi(g) = U_\varphi(g) Z^2,$$

so that $Z^2 \in N_\varphi$, which is a von Neumann algebra. So $Z \in N_\varphi$, too. ■

4.6 Proposition: A state φ is G-ergodic, if and only if N_φ (defined as in Prop. 4.5) is trivial.

Proof:

(i) Let $N_\varphi = \mathbb{C}\,1$ and $\varphi = \lambda\varphi_1 + (1-\lambda)\,\varphi_2$, $\varphi_i \in S_G$. Then the operators Z_i which correspond to φ_i according to Prop. 4.5, are both equal to 1. Thus $\varphi_i = \varphi$, i = 1, 2, and φ is an extremal point of S_G.

(ii) Assume now $N_\varphi \neq \mathbb{C}\,1$. Then there is a projection $p \in N_\varphi$ which is different from 1 and 0. Use formula in (2.13b) to construct a nontrivial decomposition of φ! ■

This proposition is an analogue of the classical statement that for an ergodic system the constants of motion are $\mathbb{C}\,1$. Every invariant state can be decomposed into ergodic states. Such a decomposition is unique if S_G is a simplex. In fact a simplex can be defined by the unique decomposition of its elements into extreme points.

4.7 Definition: A description (A, S) with symmetry group G is said to be G-abelian if $E_\varphi \pi_\varphi(A) E_\varphi$ is commutative for all $\varphi \in S_G$, where E_φ is the projection onto the subspace of $U_\varphi(G)$-invariant vectors in the GNS-representation space H_φ.

4.8 Proposition: if (A, S) is G-abelian, then S_G is a simplex.

The proof is given in [22].

The properties of an arbitrary G-invariant state φ can be reduced to that of the G-ergodic states ψ by means of the decomposition

$$\varphi = \int_{S_G} \psi \, d\mu(\psi) \tag{4.12}$$

where μ is a measure on S_G which is carried by the G-ergodic states ($\mu(E_G) = 1$, E_G = extreme points of S_G). If Prop. 4.8 is applicable, then the decomposition (4.12) is unique.

In the case that A is a quasilocal algebra, the translation group $T \simeq \mathbb{R}^3$ is a symmetry group of (A, S), since (ii) of Def. 1.5 leads to a group $\{\tau_x; x \in T\}$ of C*-automorphisms in A. From (iii) of Def. 1.5 one concludes

$$\lim_{x \to \infty} \|[\tau_x a, b]\| = 0 \tag{4.13}$$

for all $a, b \in \bigcup_\Lambda A_\Lambda$. But this relation can be generalized to all $a, b \in A$, since every element of A can uniformly be approximated by local "observables". Therefore, we can state

4.9 Proposition: A quasilocal algebra A is T-abelian.

Proof:

(i) One can show that for amenable groups, as in the case of the group T, G-abelianess is equivalent to

$$\eta(\langle\varphi; (\alpha_g a)\, b - b\,(\alpha_g a)\rangle) = 0 \tag{4.14}$$

for some mean η of G, and all $\varphi \in S_G$ [8].

(ii) Because of norm-boundedness of φ and positivity of η

$$\eta(\langle\varphi; (\tau_x a)\, b - b\,(\tau_x a)\rangle) \leqslant \|(\tau_x a)\, b - b\,(\tau_x a)\|.$$

Thus (4.13) ensues T-abelianess. ■

4.10 Proposition: Let φ be a translation invariant state on a quasilocal algebra A. Then φ can uniquely be decomposed into T-ergodic states ψ_λ, $\lambda \in I$, according to

$$\varphi = \int_I \psi_\lambda\, d\mu(\lambda). \tag{4.15}$$

The GNS-representations π_λ, associated with ψ_λ are mutually inequivalent.

Formula (4.15) with unique measure follows from Props. 4.9 and 4.8, and the proof of the last assertion can be found in [23].

For an infinite system it requires an infinite amount of energy to bring the system from one translation invariant state to another; this is the meaning of the mutual inequivalence of the π_λ.

4.11 Proposition: Let φ be a T-ergodic state on the quasilocal algebra A and $U_\varphi(T)$ the representation of T in the associated GNS-space H_φ. Since $U_\varphi(x)$ is unitary for all $x \in T$ one has the spectral representation

$$U_\varphi(x) = \sum_{p_n \in S_d} e^{ip_n x}\, E(p_n) + \int_{S_c} e^{ipx}\, dE(p), \tag{4.16}$$

where S_d is the discrete and S_c the continuous spectrum of the generator P_φ (with 3 components) of U_φ. Then S_d is simple and an additive group [23].

If $S_c = \emptyset$, then φ is a lattice state. The symmetry of the lattice is connected with the momentum spectrum and with certain cluster properties. All this can be investigated in a systematic fashion only in the algebraic approach, because one has to work with several representation spaces at the same time (cf. [7], Ch. IV).

5 Dynamical symmetries and equilibrium states

As we already mentioned before, time translations should constitute a group of structural symmetries of a description (A, S_0). The group to be represented is the (simply connected) additive group $\mathbb{R}$. With regard to Theorem 3.11 we thus define:

5.1 Definition: The dynamics of a quantum system is given by a weakly continuous representation of $\mathbb{R}$ by C*-automorphisms $\{\tau_t; t \in \mathbb{R}\}$ acting $\sigma(A, S_0)$-continuously in A. We write (A, S_0, τ_t) for the description of a dynamical quantum system.

In traditional quantum mechanics it is well known how to construct $\{\tau_t\}$. One starts with an ansatz for the potentials effecting the interactions between the particles of which the considered system is composed. The guiding line is usually the correspondence principle (a step to be thought over again in future theoretical developments). Having now a formal expression for the Hamiltonian H one must specify a dense domain $D(\mathrm{H}) \subset H$, on which H is self-adjoint and on which the correct physical boundary conditions are valid. Since every self-adjoint H is the generator of a strongly continuous group $\{\mathrm{U}(t) = \exp[it\mathrm{H}]; t \in \mathbb{R}\}$ of unitary operators ([13], p. 265), τ_t can be defined as

$$\tau_t a = \mathrm{U}(t)\, a\, \mathrm{U}(-t) \tag{5.1}$$

and possesses all required properties. The conjugate transformations τ_t^* leave the set S_n of normal states invariant.

The new aspect for infinite systems is not a dynamical one, but only the simultaneous consideration of a compatible family of finite systems. In dealing now with dynamical systems the compatibility condition 1.5 must be supplemented by assumptions on the dynamical transformations τ_t^{Λ} of every local system in the region Λ. For this one uses an absorbing sequence $\{\Lambda_n\}$ of geometrically similar, bounded regions, so that the boundary conditions for $\mathrm{H}_{\Lambda n}$ can analogously be formulated for every Λ_n. For quite a considerable class of model systems the sequence of time automorphisms $\{\tau_t^n = \tau_t^{\Lambda n}; n \in \mathbb{N}\}$ can be formulated as to converge for $n \to \infty$ to a well defined C*-automorphism τ_t in the quasilocal algebra A, for $\forall t \in \mathbb{R}$. We shall restrict our considerations in the following onto this class.

On the other hand let us emphasize that model studies in non-relativistic many body physics show quite definitely that the formalism has to be enlarged [7], [27], [28], [29], [34]. In physically relevant cases one has realized, that the time automorphism could be defined in certain (GNS-) representations only and not for the quasilocal algebra itself. In these cases one can nevertheless work in the abstract algebraic set up if one uses the state picture: time translations are then affine bijections in a folium of states (cf. the remark at the end of Chapter 2). So the state picture is no more equivalent to the observable picture, and the dynamics depends on a selected set of states. The concept of a fundamental microscopic dynamics becomes problematical for infinite systems.

Also if one assumes the existence of the time automorphism group one has to be careful with the physical interpretation. If (A, S_0, τ_t) is a dynamical description of an infinite quantum system according to Def. 5.1, then every time-invariant state φ leads in virtue of Prop. 4.2 to a so-called implementation of the dynamics by unitary operators $U_\varphi(t)$:

$$\pi_\varphi(\tau_t a) = U_\varphi(t)\, \pi_\varphi(a)\, U_\varphi(-t). \tag{5.2}$$

The generator of $U_\varphi(t)$ is a selfadjoint operator acting in the representation space H_φ. Now, one must be aware of the fact that the implemented dynamics describes only partial aspects of the full abstract dynamics, namely the excitation and relaxation of local changes of φ. If φ is macroscopically different from the ground state of the infinite system then the generator of $U_\varphi(t)$ is unbounded from below. The system has no absolute energy but only energies with reference to a certain state.

In spite of the mentioned difficulties the algebraic approach has led to some general structural insights into the dynamical symmetries and their spontaneous breakdown of many body systems, to which we shall now turn our attention. For simplicity we set in the following $S_0 = S$ and assume thus the dynamics to be given by abstract C*-automorphisms in A without extra continuity assumptions. Since $\mathbb{R}$ is an amenable group we know from Prop. 4.4 that there are for every (A, S, τ_t) time-invariant states. There exists a very important class of invariant states: the thermodynamical equilibrium states. The first question is, how to characterize an equilibrium state of an infinite system being in general not able to work with a density operator? Let us begin with finite systems, that is with systems in a finite volume Λ but with varying particle number N_Λ. For this situation we have the well understood Fock representation with Hamiltonian H_Λ and the so called reduced Hamiltonian $\overline{H}_\Lambda := H_\Lambda - \mu N_\Lambda$ where μ is the chemical potential. The equilibrium states for the natural temperature $\beta = 1/kT$ are the Gibbs states (all quantities referring to a selected Λ)

$$\langle\varphi; a\rangle := \mathrm{tr}\,[\exp(-\beta\overline{H})\, a]\, Z^{-1}, \tag{5.3}$$

which are the most "chaotic" states for a given value of the internal energy and the average particle number. The states (5.3) have the following property:

$$\begin{aligned}\langle\varphi; a\tau_{t+i\beta} b\rangle &= \mathrm{tr}\,[e^{-\beta\overline{H}}\, a\, e^{i(t+i\beta)\overline{H}}\, b\, e^{-i(t+i\beta)\overline{H}}]\, Z^{-1} = \mathrm{tr}\,[a\, e^{it\overline{H}}\, e^{-\beta\overline{H}}\, b\, e^{-it\overline{H}}]\, Z^{-1} \\ &= \mathrm{tr}\,[e^{-\beta\overline{H}}\, (\tau_t b)\, a]\, Z^{-1} = \langle\varphi; (\tau_t b)\, a\rangle.\end{aligned} \tag{5.4}$$

It turns out that this purely formal property (5.4) is characteristic for equilibrium states in general and can also be used for infinite systems.

5.2 Definition: A state φ of (A, S, τ_t) is called a β-KMS-state if for all a, b $\in A$ there is a bounded continuous function $F_{ab}(z)$ on the strip $S_\beta = \{z; z \in \mathbb{C}, 0 \leqslant \mathrm{Im}\, z \leqslant \beta\}$ analytic in the interior $\mathring{S}_\beta$, so that

$$\langle\varphi; a\tau_t b\rangle = F_{ab}(t)$$

and (5.5)

$$\langle\varphi; (\tau_t b)\, a\rangle = F_{ab}(t + i\beta).$$

The set of all β-KMS-states is denoted by S_β.

Remark: The shorthand KMS stands for Kubo, Martin and Schwinger, who investigated the analogue of the property (5.4) for thermal Green-functions. This condition was first introduced into the algebraic formulation of many body physics by [24]. Observe that in (5.5) no use is made of complex time translations!

That the KMS-states can indeed be interpreted as thermodynamical equilibrium states is supported by the following propositions.

5.3 Proposition: Every KMS-state is time invariant.

Proof: Set $a = 1$ in (5.5) and obtain

$$F_{1b}(t) = F_{1b}(t + i\beta), \quad \forall t \in \mathbb{R}. \tag{5.6}$$

Let now b be self-adjoint, so that $F_{1b}(t)$ is real. Then we may apply the reflection principle to the lines $z = ni\beta$, $n \in \mathbb{Z}$ to construct an entire analytic function which is bounded by periodicity and thus a constant. Hence, $F_{1b}(t)$ is constant for all self-adjoint b and, therefore, constant for all $b \in A$. ■

5.4 Proposition: In traditional quantum mechanics the Gibbs state is the only one which satisfies the KMS-condition.

Proof: Here every state of interest is normal and we can apply the KMS-condition in the direct form (5.4). For $t = 0$ we get

$$\mathrm{tr}\,[\rho\, a\, e^{-\beta \bar{H}}\, b\, e^{\beta \bar{H}}] = \mathrm{tr}\,[\rho\, b\, a], \; \forall a, b \in B(H).$$

By a cyclic permutation and the arbitrariness of a we conclude that

$$e^{-\beta \bar{H}}\, b\, e^{\beta \bar{H}}\, \rho = \rho\, b, \; \forall b \in B(H),$$

or, that $e^{\beta \bar{H}} \rho$ commutes with all $b \in B(H)$ and thus is a constant. Hence

$$\rho = e^{-\beta \bar{H}}\, Z^{-1} \tag{5.7}$$

where $Z \in \mathbb{R}$ is determined by normalization. ■

By Prop. 5.4 the possibility of non-normal KMS-states on $B(H)$ – different from the Gibbs state – is not excluded, but such states are foreign to usual quantum statistics; there is no natural procedure which leads from the formalism of density operators to non-normal states as long as the thermodynamical limit is not involved.

The next assertion seems rather technical but is in fact the basis for many properties of physical relevance of KMS-states.

5.5 Proposition: Let φ be β-KMS and ψ be β'-KMS, and assume $\psi \leqslant \lambda\varphi$, for a $\lambda \geqslant 1$. Then there is a unique positive operator $Z \in \pi_\varphi(A)' \cap \pi_\varphi(A)'' =: Z_\varphi$ with $\|Z\Phi\| = 1$ so that

$$\langle \psi; a \rangle = (Z\Phi, \pi_\varphi(a)\, Z\Phi), \; \forall a \in A. \tag{5.8}$$

Proof: Since φ dominates ψ and both states are τ_t-invariant, Prop. 4.5 ensues the relation (5.8) with $Z \in \pi_\varphi(A)' \cap U_\varphi(\mathbb{R})'$. The fact that $Z \in Z_\varphi$ cannot be proved by elementary reasoning (cf. e.g. [8]). ■

Remark: Prop. 5.5 can immediatly be generalized to the case the ψ is not normalized. Then all assertions remain valid with the only exception of $\|Z\Phi\| = 1$.

5.6 Proposition: Let φ be β-KMS and

$$\varphi = \lambda\varphi_1 + (1-\lambda)\varphi_2, \lambda \in (0,1) \tag{5.9}$$

where the φ_i, i = 1, 2, are β_i-KMS. Then $\beta_i = \beta$.

Proof:

(i) We have need for the following lemma which is again somewhat technical: If φ is β-KMS, then the state $\tilde{\varphi}$ on $\pi_\varphi(A)''$ defined by

$$\langle\tilde{\varphi}; c\rangle := (\Phi, c\Phi), \forall c \in \pi_\varphi(A)'', \tag{5.10}$$

is β-KMS w.r.t the dynamics $\tau_t^\varphi c := U_\varphi(t)\, c\, U_\Psi(-t)$, i.e. is β-KMS for the description $(\pi_\varphi(A)'', S_n^\varphi, \tau_t^\varphi)$ where S_n^φ stands for the normal states on $\pi_\varphi(A)''$ [8].

(ii) From (5.9) follows that φ dominates φ_1 and φ_2, so that Prop. 5.5 can be applied: There are positive operators $Z_i \in Z_\varphi$ with

$$\langle\varphi_i, a\rangle = (Z_i\Phi, \pi_\varphi(a) Z_i\Phi), \ \forall a \in A. \tag{5.11}$$

Since $Z_i \pi_\varphi(a) Z_i \in \pi_\varphi(A)''$ we can now easily show by means of (i) that the φ_i are both β-KMS. ■

Prop. 5.6 can be expressed in the form, that an equilibrium state φ to the natural temperature β can only be decomposed into such equilibrium states which have the same temperature.

5.7 Theorem: S_β is convex, $\sigma(S, A)$-compact and a simplex.

Proof:

(i) If $\varphi_1, \varphi_2 \in S_\beta$ with analytic functions F_{ab}^1 and F_{ab}^2 then the analytic function which makes $\varphi = \lambda\varphi_1 + (1-\lambda)\varphi_2, \lambda \in (0,1)$ a KMS-state according to (5.5) is $F_{ab} = \lambda F_{ab}^1 + (1-\lambda) F_{ab}^2$.

(ii) Since S is $\sigma(S, A)$-compact and contains S_β we have to show that S_β is closed. If the net $\{\varphi_\gamma\} \subset S_\beta$ converges to a $\varphi \in S$, then the corresponding net $\{F_{ab}^\gamma\}$ of analytic functions on S_β converges to an analytic function F_{ab} on S_β, since the derivatives of the F_{ab}^γ are uniformly bounded. F_{ab} has the properties which make φ a β-KMS-state.

(iii) According to Prop. 5.6 every positive functional ψ with $0 \leqslant \psi \leqslant \varphi, \varphi \in S_\beta$ and with ψ satisfies the KMS-condition (necessarily with β), corresponds to a unique operator $Z \in Z_\varphi = \pi_\varphi(A)' \cap \pi_\varphi(A)''$ with $0 \leqslant Z \leqslant 1$. This correspondence is easily seen to be order preserving. Since Z_φ is commutative it is a lattice. So all ψ, with $0 \leqslant \psi \leqslant \varphi$

constitute a lattice. Thus S_β is the basis of a cone with apex 0 which is a lattice. But this is equivalent with S_β being a simplex. ■

From Theorem 5.7 follows that every $\varphi \in S_\beta$ can be uniquely decomposed into extremal β-KMS-states $\psi \in E_\beta$ (E_β shall signify the extremal points of S_β), i.e.

$$\varphi = \int_{S_\beta} \psi \, d\mu(\psi) \tag{5.12}$$

where $\mu(E_\beta) = 1$. The unicity of the measure μ in (5.12) is the basis for the interpretation of extremal KMS-states as pure phases. Could φ likewise be decomposed by means of another measure μ' – concentrated on E_β – then the kind and the amount of a certain phase in φ would not be uniquely definable, the concept of phases would not be applicable at all. On the other hand there are some models for which the decomposition (5.12) is in accordance with more conventional phase definitions by means of thermodynamic functions.

5.8 Proposition: $\varphi \in S_\beta$ is a pure phase, if and only if $\pi_\varphi(A)$ is a factor.

Proof: Every decomposition of φ corresponds uniquely to a decomposition of $1_\varphi \in Z_\varphi$ into a sum of two projections in Z_φ (see Prop. 5.5) which can be effected in a non-trivial manner if and only if $Z_\varphi \neq \mathbb{C}\,1$. ■

Let us furthermore mention the so called commutant theorem for KMS-states which leads beside its mathematically interesting structure also to useful physical results.

5.9 Theorem: Let $\varphi \in S_\beta$ be faithful, i.e. $\pi_\varphi^{-1}(0) = 0$. Then there is an antiunitary operator C in H_φ with

(i) $C^2 = 1$

(ii) $C\Phi = \Phi$

(iii) $C\,U_\varphi(t)\,C = U_\varphi(t), \ \forall t \in \mathbb{R}$

(iv) $C\,\pi_\varphi(A)'\,C = \pi_\varphi(A)''$.

Furthermore Φ is separating for $\pi_\varphi(A)''$.

For the proof cf. [8]!

Let us remark that in Theorem 5.9 the faithfulness of φ plays an important role; in a certain sense it is more important than the KMS-condition. Compare also Prop. 2.14.

Now we consider dynamical symmetries of (A, S, τ_t), that are symmetry-transformations α_g, $g \in G$, which commute with τ_t.

5.10 Proposition: Let G be a connected dynamical symmetry group of (A, S, τ_t). Then S_β as well as E_β are invariant under the action of α_g^*, for all $g \in G$.

Proof: Let $\varphi \in S_\beta$, a, b $\in A$, and F_{ab} the analytic function associated with these three quantities according to (5.5). Then we infer from the fact that every α_g is a C*-automorphism which commutes with τ_t that

$$\langle \alpha_g^* \varphi; a \tau_t b \rangle = \langle \varphi; a' \tau_t b' \rangle = F_{a'b'}(t),$$

where $a' = \alpha_g a$ and $b' = \alpha_g b$. Because φ is β-KMS, we have

$$\langle \varphi; (\tau_t b') a' \rangle = \langle \alpha_g^* \varphi; (\tau_t b) a \rangle = F_{a'b'}(t + i\beta).$$

Thus $\alpha_g^* \varphi \in S_\beta$. The same can be done with α_g^{*-1}. Therefore α_g^* is an affine bijection of S_β onto itself. Such a transformation leaves E_β invariant. ■

Consider now the following situation: $\varphi \in S_\beta$ is not extremal KMS but is decomposed into pure phases according to (5.12), and assume φ to be G-ergodic, for a connected dynamical symmetry group (Fig. 2.1). Then the pure phases occuring in φ cannot be G-invariant, they can be only invariant against the action of a subgroup $H \leqslant G$. This situation is referred to as a spontaneous breakdown of the G-symmetry. It is typical for phase transitions of the second kind. It can only occur in the thermodynamical limit since for finite systems S_β consists of one state only (cf. Prop. 5.4). The pure phases ψ of a state φ with broken G-symmetry are transformed into each other by the action of G. They remind us of a multiplett in usual quantum mechanics. This similarity is confirmed by the following result.

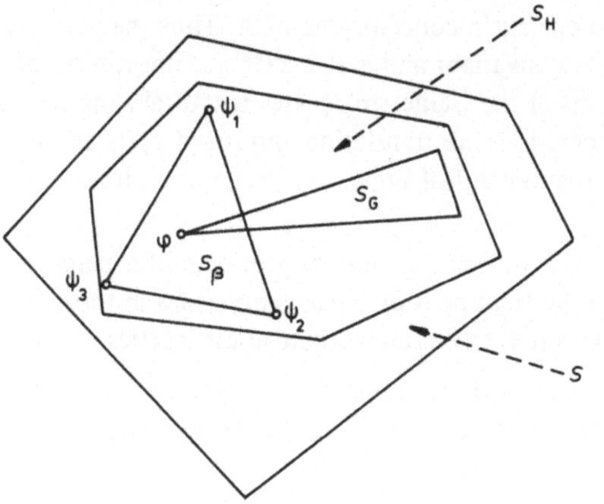

Fig. 2.1

5.11 Proposition: Let G be a connected dynamical symmetry group, and let φ be G-ergodic and

$$\varphi = \int_I \psi_\lambda \, d\mu(\lambda) \tag{5.13}$$

its decomposition into pure phases. Let $U_\varphi(g)$ be the representation operators of G in H_φ. Then the ψ_λ can be extended to vector states Ψ_λ on $\pi_\varphi(A)''$, and the subspace $K_\varphi \subset H_\varphi$ spanned by the $\{\Psi_\lambda; \lambda \in I\}$ is $U_\varphi(G)$-invariant.

Proof: The assertions follow immediately if one has convinced oneself that every Ψ_λ, $\lambda \in I$, is in virtue of (5.13) dominated by φ. Then according to Prop. 5.5 we have a family of positive operators $Z_\lambda \in \mathcal{Z}_\varphi$, $\lambda \in I$, so that

$$\langle \psi_\lambda; a\rangle = (Z_\lambda \Phi, \pi_\varphi(a) Z_\lambda \Phi), \quad \forall a \in A. \tag{5.14}$$

But (5.14) has the form of vector states with vectors $\Psi_\lambda = Z_\lambda \Phi$, $\|\Psi_\lambda\| = 1$, and can be extended to $\pi_\varphi(A)''$. Since φ and E_β are invariant under α_g^*, $g \in G$, the Ψ_λ are transformed into each other by $U_\varphi(g)$, $g \in G$. ∎

So the pure phases of φ constitute a G-multiplett of vector states Ψ_λ in the GNS representation of φ.

Let us finally discuss the symmetries of an infinite nonrelativistic many body system. We have already mentioned before (4.13) that the translations T act as structural symmetries in the quasilocal algebra A. The rotations can be introduced as automorphisms of A if one chooses as absorbing sequence $\{\Lambda_n\}$ a sequence of spheres centered at the origin of $\mathbb{R}^3$. The special Galilei-transformations with velocity $\vec{v}$ are represented in a N-particle space as

$$U(\vec{v}, t) = \exp[i\vec{v} \cdot (M\vec{R} - \vec{P}t)], \tag{5.15}$$

where M is the total mass, $\vec{R}$ the center of mass coordinate and $\vec{P}$ the total momentum of the system. The term $\exp[-i\vec{v}t \cdot \vec{P}]$ is a space translation and thus a structural symmetry also in A. The operation $\exp[i\vec{v} \cdot M\vec{R}]$ – a momentum translation – maps every wave function concentrated in the region Λ into one again concentrated in Λ. Thus the corresponding C*-automorphism leaves every A_Λ invariant and is also a C*-automorphism of A. We have the result that all transformations of the Galilei group are structural symmetries of the quasilocal algebra A. Since most interactions are translation and rotation invariant, we may assume that the Galilei group is also a dynamical symmetry group of an infinitely extended system.

Another very important but conceptually more difficult symmetry group are the gauge transformations. We cannot enter here into the treatment of gauge symmetries in the algebraic framework, which is connected with non-trivial mathematical difficulties.

The spontaneous symmetry breakdown of the mentioned groups in equilibrium states of many particle systems is the origin of spectacular macroscopic quantum effects, some of which are listed in the following table:

Effect	*Spontaneously broken symmetry group*
cristallization	translation and rotation group
structural phase transition	a cristallographic group
ferromagnetism	rotation group
superfluidity	special Galilei group
Bose-Einstein condensation	gauge group
superconductivity	gauge group

For some algebraic models with spontaneous symmetry breakdown compare Emch's lecture in this volume.

Appendix: Jordan homomorphisms

This Appendix is devoted to the derivation of the statement that every Jordan homomorphism of a C*-algebra can be written as a sum of a C*-homomorphism and a C*-antihomomorphism. The complete proof of this physically important theorem requires nontrivial results of ring theory, classification of von Neumann algebras and modern topology, and cannot be found in the textbook literature. We have worked out a deduction from the ring theoretic origins, but we saw no way how to provide all ab initio proofs of the necessary C*-algebraic tools. For some of these we have to refer to [5], [6]. Nevertheless, we think that the presentation of the whole line of thought may give valuable structural insights.

A.1 Definition: A ring is a set R with two inner compositions "+" and "." for which the following rules are valid: $(q, r, s \in R)$

(i) $(q + r) + s = q + (r + s)$

$$r + s = s + r$$

Given r and s there exists a q, such that

$$r + q = s$$

(ii) $(q \cdot r) \cdot s = q \cdot (r \cdot s)$

(iii) $q \cdot (r + s) = q \cdot r + q \cdot s$

$$(r + s) \cdot q = r \cdot q + s \cdot q$$

A.2 Definition: Let R and $\hat{R}$ be two rings and α a mapping

$$\alpha : R \to \hat{R}$$

(i) α is a ring homomorphism if

$$\alpha(r + s) = \alpha r + \alpha s$$
$$\alpha(r \cdot s) = \alpha r \cdot \alpha s$$

(ii) α is a ring anti-homomorphism if

$$\alpha(r + s) = \alpha r + \alpha s$$
$$\alpha(r \cdot s) = \alpha s \cdot \alpha r.$$

(iii) Let $r \circ s := \frac{1}{2}(r \cdot s + s \cdot r)$

and the analogue in $\hat{R}$. α is a ring Jordan homomorphism if

$$\alpha(r + s) = \alpha r + \alpha s$$
$$\alpha(r \circ s) = \alpha r \circ \alpha s$$

Observe that the last condition is equivalent with α preserving squares.

A.3 Definition: Let R be a ring with identity e (i.e. $e \cdot r = r \cdot e = r$ for all $r \in R$). A set $\{e_{ij}; e_{ij} \in R, 1 \leqslant i, j \leqslant n, 2 \leqslant n < \infty\}$ is called a set of matrix units if

$$e_{ij} \cdot e_{kl} = \delta_{jk} e_{il}, \quad \sum_{i=1}^{n} e_{ii} = e.$$

Let furthermore D denote the subring of R whose elements commute with all e_{ij}. Then R is called an $(n \times n)$-matrix ring over D.

If R is a matrix ring over D, every $r \in R$ can be written as

$$r = \sum_{ij} d_{ij} \cdot e_{ij}$$

where the $d_{ij} \in D$ are uniquely determined. From A.3 follows then

$$r \cdot r' = \sum_{ij} \left(\sum_{k} d_{ik} \cdot d'_{kj} \right) \cdot e_{ij}.$$

Thus the ring product is expressed by a matrix multiplication.

A.4 Example: Let $M^{(n)}$ be the set of all $(n \times n)$-matrices with complex entries. $M^{(n)}$ is an $(n \times n)$-matrix ring over D for the following definitions:

$$\begin{aligned} e &:= (\delta_{mm'}), && \text{where } m, m' \text{ are dummies} \\ e_{ij} &:= (\delta_{mi}\delta_{jm'}), && 1 \leqslant i, j \leqslant n \\ D &:= \{\lambda e; \lambda \in \mathbb{C}\} \end{aligned}$$

A.5 Example: Let be $M^{(n)}$ as in A.4 with $n = 2l$. Then $M^{(n)}$ is a (2×2)-matrix ring over D for the following definitions: Write for $r \in M^{(n)}$

$$r = \begin{pmatrix} \rho_{11} & \rho_{12} \\ \rho_{21} & \rho_{22} \end{pmatrix} \qquad \text{with } \rho_{ij} \in M^{(l)}$$

and let ϵ be the unit-matrix of $M^{(l)}$. Define

$$e_{11} = \begin{pmatrix} \epsilon & 0 \\ 0 & 0 \end{pmatrix}, \; e_{22} = \begin{pmatrix} 0 & 0 \\ 0 & \epsilon \end{pmatrix}, \; e_{12} = \begin{pmatrix} 0 & \epsilon \\ 0 & 0 \end{pmatrix}, \; e_{21} = \begin{pmatrix} 0 & 0 \\ \epsilon & 0 \end{pmatrix}$$

and

$$D = \left\{ \begin{pmatrix} \rho_{11} & 0 \\ 0 & \rho_{22} \end{pmatrix}; \rho_{ii} \in M^{(l)} \right\}$$

Observe that the identity $e \in M^{(n)}$ can be "halved":

$$e = e_{11} + e_{22}$$

where the projections e_{11} and e_{22} are of equal dimension.

In general a ring R with identity is an (n × n)-matrix ring iff it is a direct sum of n isomorphic left ideals I_i, $1 \leq i \leq n$. That this is a necessary condition can be seen by setting

$$I_i := R\, e_{ii}.$$

A.6 Theorem: Any Jordan homomorphism α of an (n × n)-matrix ring R over D into a second ring $\hat{R}$ is the sum of a ring homomorphism and a ring anti-homomorphism [25].

Proof:

(i) One shows that (for $d \in D$) αd commutes with every αe_{ij} and that

$$\alpha\Big(\sum_{ij} d_{ij} \cdot e_{ij}\Big) = \sum_{ij} \alpha d_{ij} \cdot \alpha e_{ij}$$

(ii) Define for $i \neq j$

$$g_{ij} := \alpha e_{ii} \cdot \alpha e_{ij} \cdot \alpha e_{jj}, \;\; h_{ij} := \alpha e_{ii} \cdot \alpha e_{ji} \cdot \alpha e_{jj}$$

and

$$g_{ii} := g_{ij} \cdot g_{ji}, \;\; h_{ii} := h_{ij} \cdot h_{ji}$$

the last two definitions depending not of the choice of $j\,(\neq i)$. Then $\{g_{ij}\}$ and $\{h_{ij}\}$ constitute two (n × n)-systems of matrix units in $\hat{R}$ and one has

$$\alpha e_{ij} = g_{ij} + h_{ij}$$

and

$$g_{ij} \cdot h_{kl} = h_{kl}\, g_{ij} = 0, \;\; \forall\, i, j, k, l.$$

(iii) We introduce

$$g := \sum_i g_{ii}, \;\; h := \sum_i h_{ii}.$$

Let $\hat{R}_\alpha$ be the smallest subring of $\hat{R}$ containing αR and $\alpha e = \hat{e}$ the identity of $\hat{R}_\alpha$. Then

$$\hat{e} = g + h.$$

Furthermore

$$g^2 = g, \;\; h^2 = h, \;\; gh = hg = 0$$

so that g and h are disjoint idempotents. According to (ii) g and h belong to the center of $\hat{R}_\alpha$. Thus $\hat{R}_\alpha$ splits into two subrings

$$\hat{R}_\alpha = \hat{R}_\alpha g \oplus \hat{R}_\alpha h =: \hat{R}_1 \oplus \hat{R}_2.$$

(iv) We define

$$\alpha_1 : R \to \hat{R}_1, \alpha_1 r := (\alpha r) \cdot g$$

$$\alpha_2 : R \to \hat{R}_2, \alpha_2 r := (\alpha r) \cdot h.$$

Then obviously

$$\alpha = \alpha_1 + \alpha_2 .$$

(v) Let in the following calculations alsways $i \neq j$, and consider two arbitrary elements a, b of D. According to the multiplication law of A.3

$$a \cdot b \cdot e_{ii} + b \cdot a \cdot e_{jj} = (a \cdot e_{ij} + b \cdot e_{ji})^2 .$$

Since α preserves squares and in virtue of (i) we may evaluate

$$\alpha(a \cdot b) \cdot \alpha e_{ii} + \alpha(ba) \cdot \alpha e_{jj} = \alpha[(a \cdot e_{ij} + b \cdot e_{ji})^2] = [\alpha a \cdot \alpha e_{ij} + \alpha b \cdot \alpha e_{ij}]^2$$
$$= \alpha a \cdot \alpha b \cdot \alpha e_{ij} \cdot \alpha e_{ji} + \alpha b \cdot \alpha a \cdot \alpha e_{ji} \cdot \alpha e_{ij} \qquad (*)$$

since

$$\alpha e_{ij} \cdot \alpha e_{ij} = \alpha e_{ij}^2 = 0 \quad (i \neq j!).$$

Multiplying (*) on the right by g_{ij} and observing

$$\alpha e_{ij} \cdot g_{ij} = (g_{ij} + h_{ij}) \cdot g_{ij} = 0$$

as well as

$$\alpha e_{ij} \cdot \alpha e_{ji} \cdot g_{ij} = \alpha e_{ij} \cdot g_{jj} = g_{ij}$$

we arrive at

$$\alpha(a \cdot b) \cdot g_{ij} = \alpha a \cdot \alpha b \cdot g_{ij} .$$

On the other hand, multiplication of (*) on the right by h_{ij} gives with

$$\alpha e_{ij} \cdot h_{ij} = 0$$

and

$$\alpha e_{ji} \cdot \alpha e_{ij} \cdot h_{ij} = h_{ij}$$

the result

$$\alpha(a \cdot b) \cdot h_{ij} = \alpha b \cdot \alpha a \cdot h_{ij} .$$

(vi) For arbitrary elements $r, s \in R$

$$r = \sum_{ij} a_{ij} \cdot e_{ij}, \quad s = \sum_{ij} b_{ij} \cdot e_{ij}, \quad a_{ij}, b_{ij} \in D$$

we have

$$\alpha_1 r = \alpha r \cdot g = \sum_{ij} \alpha a_{ij} \cdot g_{ij}$$

$$\alpha_1 s = \alpha s \cdot g = \sum_{ij} \alpha b_{ij} \cdot g_{ij}$$

and

$$\alpha_1 r \cdot \alpha_1 s = \sum_{ij} \sum_{k} \alpha a_{ik} \cdot \alpha b_{kj} \cdot g_{ij}.$$

Similarly

$$\alpha_2 r = \alpha r \cdot h = \sum_{ij} \alpha a_{ij} \cdot \alpha e_{ij} \cdot \sum_{l} h_{ll} = \sum_{ij} \alpha a_{ij} \cdot h_{ji}$$

$$\alpha_2 s = \alpha s \cdot h = \sum_{ij} \alpha b_{ij} \cdot h_{ji},$$

so that

$$\alpha_2 s \cdot \alpha_2 r = \Big(\sum_{kj} \alpha b_{kj} \cdot h_{jk}\Big) \cdot \Big(\sum_{il} \alpha a_{il} \cdot h_{li}\Big) = \sum_{ij} \sum_{k} \alpha b_{kj} \cdot \alpha a_{ik} \cdot h_{ji}.$$

Observing the formulas of (v), we deduce finally

$$\alpha_1 (r \cdot s) = \sum_{ij} \alpha \Big(\sum_{k} a_{ik} \cdot b_{kj}\Big) \cdot g_{ij} = \sum_{ij} \sum_{k} \alpha a_{ik} \cdot \alpha b_{kj} \cdot g_{ij} = \alpha_1 r \cdot \alpha_1 s$$

and

$$\alpha_2 (r \cdot s) = \sum_{ij} \alpha \Big(\sum_{k} a_{ik} \cdot b_{kj}\Big) \cdot h_{ji} = \sum_{ij} \sum_{k} \alpha b_{kj} \cdot \alpha a_{ik} \cdot h_{ji} = \alpha_2 s \cdot \alpha_2 r. \qquad \blacksquare$$

A.7 Definition: A linear mapping $\alpha : R \xrightarrow{\text{onto}} \hat{R}$, where R and $\hat{R}$ are two rings, is said to be a direct sum of the linear mappings α_1 and α_2, i.e.

$$\alpha = \alpha_1 \oplus \alpha_2,$$

if there are direct linear space decompositions $R = R_1 \oplus R_2$ and $\hat{R} = \hat{R}_1 \oplus \hat{R}_2$ so that

$$\alpha_1 R_1 = \alpha R_1 = \hat{R}_1, \alpha_1 R_2 = 0$$

$$\alpha_2 R_2 = \alpha R_2 = \hat{R}_2, \alpha_2 R_1 = 0.$$

A.8 Corollary: Any Jordan isomorphism α of an $(n \times n)$-matrix ring R onto a ring $\hat{R}$ is the direct sum of a ring isomorphism and a ring anti-isomorphism.

Proof: Now $\hat{R}_\alpha$ in the proof of A.6 is equal to $\hat{R} = \hat{R} g \oplus \hat{R} h =: \hat{R}_1 \oplus \hat{R}_2$. Define

$$R_i := \alpha^{-1} (\hat{R}_i), \quad i = 1, 2$$

and observe that

$$R = R_1 \oplus R_2$$

as well as $\alpha_i R_i = \alpha R_i = \hat{R}_i$

and $\alpha_i R_j = 0, \; i \neq j.$ ■

In order to transfer the results on Jordan homomorphisms of $(n \times n)$-matrix rings to C*-algebras, we first show that W*-algebras may be split into direct sums of matrix rings. Some definitions are required.

A.9 Definition: Let M be a W*-algebra.

(i) $p \in M$ is called a projection if $p^* = p$ and $p^2 = p$.

(ii) Let $p_1, p_2 \in M$ be projections. We write

$$p_1 \leqslant p_2$$

and say "p_2 contains p_1" if

$$p_1 p_2 = p_2 p_1 = p_1 .$$

(iii) A projection $p \in M$ is called abelian if $p > 0$ and pMp is abelian.

(iv) M is called discrete if every non-zero projection $p \in M' \cap M$ contains an abelian projection. M is said to be continuous if it has no discrete W*-algebra as a direct summand.

A.10 Definition: Let M be a W*-algebra.

(i) $v \in M$ is called a partial isometry if v^*v is a projection.

(ii) Let q and p be projections in M. We say "q is equivalent to p" and write $q \sim p$, if there exists a partial isometry $v \in M$, such that

$$p = v^*v, \quad q = vv^*.$$

The meaning of Def. A.10 can best be understood in terms of operators acting on a Hilbert space H. If we consider a partial isometry v then the projection $v^*v =: p$ projects on the subspace of those vectors which are not annihilated by v; pH is called the initial space of v, it is also the range of v^*. Now $vv^* =: q$ is also a projection since it is hermitean and since $vv^*vv^* = vpv^* = vv^*$. And q is now the projection on the range of v, qH being called the final space of v. Since v maps pH isometrically onto qH these two subspaces have the same dimension. One says also that p and q are of the same dimension.

On the other hand, let p and q be two projections of the same dimension and $\{\Phi_m\}$ and $\{\Psi_n\}$ two complete orthonormal systems of pH and qH respectively. Define

$$v := \sum_n \Psi_n \otimes \Phi_n^*$$

where $(\Psi_n \otimes \Phi_n^*)\,\Phi := \Psi_n (\Phi_n, \Phi)$. Then v is easily seen to be a partial isometry with initial space pH and final space qH. Thus we observe that the abstract equivalence relation of two projections is a generalization of the property that two concrete projections are of the same dimension.

A.11 Theorem: Every W*-algebra M can be written as the direct sum of matrix rings.

Proof:

(i) Consider the set $P_a \subset M' \cap M$ of central projections which contain an abelian projection and let be z the supremum of P_a, w.r.t. the ordering A.9 (ii). Then every central projection z' with $z' \leqslant 1 - z$ does not contain an abelian projection (because otherwise we had also $z' \leqslant z$ and $z' \leqslant z(1-z) = 0$). From this follows that the decomposition

$$M = Mz \oplus M(1-z) =: M_d \oplus M_c$$

splits M into a discrete and into a continuous W*-algebra.

(ii) According to [5], pp. 239, 240, M_d can be written as

$$M_d = \sum_k {}_\oplus C_k \otimes B(H_k) =: \sum_k {}_\oplus M_k$$

where k runs through the ordinal numbers, C_k is an abelian W*-algebra and H_k is a Hilbert space of dimension k. If k is finite, i.e. $k < \aleph_0$, M_k is a $(k \times k)$-matrix ring with $D_k := C_k \otimes 1_k$. If $k \geqslant \aleph_0$ we decompose $H_k = H_k' \oplus H_k''$ into two Hilbert spaces of dimension k. The projections p_k' and p_k'' on H_k' and H_k'' are then equivalent, i.e. there is a partial isometry v_k with

$$v_k^* v_k = p_k', \quad v_k v_k^* = p_k''.$$

Observe also

$$p_k' \cdot p_k'' = 0, \quad p_k' + p_k'' = 1_k.$$

(iii) Since M_c has no abelian projections we may apply Prop. 2.2.13 of [6] to conclude that there are projections $p_c', p_c'' \in M_c$ with

$$p_c' p_c'' = 0, \quad p_c' + p_c'' = 1_c$$

and

$$p_c' \sim p_c''.$$

(iv) Define

$$\bar{M} := \sum_{k \geqslant \aleph_0} M_k \oplus M_c$$

and the projections $p', p'' \in \bar{M}$ by

$$p' := \sum_{k \geqslant \aleph_0} p_k' + p_c'$$

$$p'' := \sum_{k \geqslant \aleph_0} p_k'' + p_c''.$$

Then $p' \sim p''$ (cf. [6], p. 79)
and $p'p'' = 0,\ p' + p'' = \bar{1}$.

Let $v \in \bar{M}$ be the partial isometry with

$$v^*v = p',\ vv^* = p''.$$

We define

$$e_{11} := p',\quad e_{22} := p''$$
$$e_{12} := v^*,\quad e_{21} := v$$

Then the set $\{e_{ij}\}$, $1 \leqslant i, j \leqslant 2$ is a set of matrix units according to Def. A.3. For we have

$$v = vp'$$

and

$$v = p''vp',\ v^* = p'v^*p''$$

and thus

$$e_{12}e_{11} = v^*p' = v^*p''p' = 0 \qquad e_{12}e_{21} = v^*v = p' = e_{11}$$
$$e_{12}e_{22} = v^*p'' = v^* = e_{12} \qquad e_{11}e_{11} = e_{11},\ e_{22}e_{22} = e_{22}$$
$$e_{12}e_{12} = v^*v^* = v^*p''p'v^* = 0$$

Thus we have proved the composition law

$$e_{ij}\,e_{kl} = \delta_{jk}\,e_{il}$$

for some cases and the remaining cases follow by taking the star operation.

We see that $\bar{M}$ is a (2×2)-matrix ring over the set D consisting of all elements $d \in \bar{M}$ which commute with the set of matrix units $\{e_{ij}\}$.

(v) From (i) and (ii) follows that

$$M = \sum_{k < \aleph_0} \oplus M_k \oplus \bar{M}$$

where all direct summands have now been shown to be matrix rings. ∎

A.12 Theorem: Every Jordan-isomorphism α of a W*-algebra M onto a C*-algebra A is the direct sum of a C*-isomorphism and a C*-anti-isomorphism [16].

Proof:

(i) From A.11 we know that M is the direct sum

$$M = \sum_{n \in \mathbb{N}} \oplus\, M_n$$

of $(n \times n)$-matrix rings M_n.

(ii) For every n let α_n be the restriction of α to M_n. From A.8 we know that there exist idempotents g_n and h_n in the center of $\alpha(M_n) \subset A$, such that

$$\alpha_n = \alpha'_n \oplus \alpha''_n$$

with α'_n a ring-isomorphism from M_n onto $\alpha(M_n)\, g_n$ and α''_n a ring-anti-isomorphism from M_n onto $\alpha(M_n)\, h_n$.

(iii) Taking the direct sum over n we conclude that there exist the idempotents

$g := \sum_n g_n$ and $h := \sum_n h_n$ in the centre of $\sum_n{}_{\oplus}\ \alpha(M_n) = \alpha(M) = A$, such that

$$\alpha = \alpha' \oplus \alpha''$$

with α' a ring-isomorphism from M onto $A\,g$ and α'' a ring-anti-isomorphism from M onto $A\,h$. Now g and h commute especially also with their adjoint elements which shows that they are normal idempotents and which again ensues that they are projections. But then we get for all $a \in M$

$$(\alpha' a)^* = (\alpha a g)^* = (\alpha a)^* g.$$

Since Jordan homomorphisms of C*-algebras are by definition adjoint-preserving (cf. Def. 3.3) we derive

$$(\alpha' a)^* = (\alpha a^*)\, g = \alpha' a^*$$

and similarely for α''. This shows that α', α'' are also C*-isomorphisms resp. C*-anti-isomorphisms. ■

In order to extend Theorem A.12 to C*-homomorphisms of C*-algebras, we have to associate a W*-algebra with a given C*-algebra. This can be done in a canonical way.

A.13 Definition: Let A be a C*-algebra.

(i) A may be represented isomorphically as an operator algebra by means of the so-called universal representation

$$\pi_u : A \to B\left(\sum_{\varphi \in S} \oplus H_\varphi\right)$$

$$A \ni a,\ \pi_u(a) := \sum_{\varphi \in S} \oplus\ \pi_\varphi(a)$$

where π_φ are the GNS-representations associated with the states φ.

(ii) The envelopping W*-algebra of A is defined as $\pi_u(A)''$. It can be shown to be isometrically isomorphic with the bidual A^{**} of A. Henceforth we shall identify the envelopping W*-algebra and A^{**}.

For the statements smuggled into Def. A.13 cf. [6], pp. 41–43.

A.14 Lemma: Let $\alpha: A \to B$ be a linear, norm-continuous mapping from a C*-algebra into some operator algebra B. Then α can uniquely be extended to a $\sigma(A^{**}, A^*)$-continuous mapping

$$\tilde{\alpha}: A^{**} \to B''.$$

If α is a Jordan homomorphism, then so is its extension $\tilde{\alpha}$.

Proof: The proof is given in [17], p. 182. ■

A.15 Theorem: Let A be a C*-algebra and B a C*-subalgebra of $B(H)$, H some Hilbert space. A linear, adjoint preserving mapping

$$\alpha: A \xrightarrow{\text{onto}} B$$

is a Jordan homomorphism, iff there is a projection $p \in B' \cap B''$ such that

$$\alpha(ab)\,p = (\alpha a)(\alpha b)\,p$$
$$\alpha(ab)(1-p) = (\alpha b)(\alpha a)(1-p)$$

for all $a, b \in A$ ([17], p. 183).

Proof:

(i) If such a p exists, then α is evidently a Jordan homomorphism.

(ii) If α is a Jordan homomorphism, we extend it to the $\sigma(A^{**}, A^*)$-continuous Jordan homomorphism

$$\tilde{\alpha}: A^{**} \to B''.$$

The kernel of $\tilde{\alpha}$ is then a two-sided $\sigma(A^{**}, A^*)$-closed ideal $I \subset A^{**}$ which can be written as

$$I = A^{**}(1-r)$$

r being a central projection in A^{**} (cf. [6], p. 24). Thus the restriction of $\tilde{\alpha}$ onto $A^{**}r$, denoted by the same symbol $\tilde{\alpha}$, is a Jordan isomorphism of $A^{**}r$ onto B'' and Theorem A.12 may be applied.

(iii) According to the proof of A.12, there exist the central projections g and h in B'' so that the assertion is valid with $p := g$ for all $a, b \in A^{**}r$. For $a \in A^{**}(1-r)$ the relations are trivially satisfied. ■

Since the projection p is in general not in B, one has to assume an operator representation for the algebra B. Nevertheless is the decomposition of α into C*-isomorphic resp. C*-anti-isomorphic parts quite general. Theorem 3.5 is a special case of Theorem A.15.

References

[1] a) von Neumann, J.: Grundlagen der Quantenmechanik, Springer, Berlin (1932)
b) Mackey, G. W.: Mathematical Foundations of Quantum Mechanics, Benjamin, New York (1963)
c) Jauch, J. M.: Foundations of Quantum Mechanics, Addison-Wesley, Reading Mass. (1968)

[2] Wick, G. C., A. S. Wightman and E. P. Wigner: Intrinsic Parity of Elementary Particles, Phys. Rev. **88**, 101–105

[3] a) Jordan, P.: Über eine Klasse nicht assoziativer hyperkomplexer Algebren, Nachr. Ges. Wiss. Göttingen **569** (1932)
b) Jordan, P., J. von Neumann and E. Wigner: On an Algebraic Generalization of the Quantum Mechanical Formalism, Ann. Math. **35**, 29 (1934)

[4] Braun, H. und M. Koecher: Jordan Algebra, Springer, Berlin (1966)

[5] a) Dixmier, J.: Les algèbres d'opérateurs dans l'espace hilbertien, Gauthier-Villars, Paris (1957)
b) Dixmier, J.: Les C*-algèbres et leurs réprésentations, Gauthier-Villars, Paris (1964)

[6] Sakai, I.: C*-Algebras and W*-Algebras, Springer, Berlin (1971)

[7] Emch, G. G.: An Algebraic Approach for Spontaneous Symmetry Breaking in Quantum Statistical Mechanics, this volume (1977)

[8] Emch, G. G.: Algebraic Methods in Statistical Mechanics and Quantum Field Theory, Wiley, New York (1972)

[9] Holland, S. S.: The Current Interest in Orthomodular Lattices, in: C. A. Hooker: The Logico-Algebraic Approach to Quantum Mechanics, D. Reidel Publ. Comp. (1973)

[10] Ruelle, D.: Statistical Mechanics, Benjamin, New York (1969)

[11] Segal, I. E.: Postulates for General Quantum Mechanics, Ann. Math. **48**, 930 (1947)

[12] Takesaki, M.: Tomita's Theory of Modular Hilbert Algebras and its Application, Springer, Berlin (1970)

[13] Reed, M. and B. Simon: Functional Analysis, Academic Press, New York (1972)

[14] Wigner, E. P.: Gruppentheorie und ihre Anwendung, Vieweg, Braunschweig (1931), Reprint Vieweg, Braunschweig (1977)

[15] Bargmann, V.: Note on Wigner's Theorem on Symmetry Operations, J. Math. Phys. **5**, 862 (1964)

[16] Kadison, R. V.: Isometries of Operator Algebras, Ann. Math. **54**, 325 (1951)

[17] Kadison, R. V.: Transformations of States in Operator Theory and Dynamics, Topology 3, Suppl. **2**, 177 (1965)

[18] Kadison, R. V.: A Generalized Schwartz Inequality and Algebraic Invariants for Operator Algebras, Ann. Math. **56**, 494 (1952)

[19] Rieckers, A.: Equivalence of Kadison and Wigner Symmetries in Traditional Quantum Mechanics, Contribution to the VIth International Colloquium on Group Theoretical Methods in Physics, Tübingen (1977)

[20] Emch, G. G.: Méchanique quantique quaternionienne et rélativité restreinte I, Helv. Phys. Acta **36**, 739 (1963)

[21] Hewitt, E. and K. A. Ross: Abstract Harmonic Analysis, Springer, Berlin (1963)

[22] Lanford, O. E. and D. Ruelle: Integral Representations of Invariant States on B*-Algebras, J. Math. Phys. **8**, 1460 (1967)

[23] Kastler, D. and D. W. Robinson: Invariant States in Statistical Mechanics, Comm. Math. Phys. **3**, 151 (1966)

[24] Haag, R., N. M. Hugenholtz and M. Winnink: On the Equilibrium States in Quantum Statistical Mechanics, Comm. Math. Phys. **5**, 215 (1967)

[25] Jacobson, N. and C. E. Rickart: Jordan Homomorphisms of Rings, Trans. Amer. Math. Soc. **69**, 47 (1950)

[26] Takesaki, M. and M. Winnink: Local Normality in Quantum Statistical Mechanics, Comm. Math. Phys. **30**, 129 (1973)

[27] Emch, G. G. and H. J. Knops: J. Math. Phys. **11**, 3008 (1970)

[28] Dubin, D. A. and G. Sewell: J. Math. Phys. **11**, 2990 (1970)

[29] Jelinek, F.: Comm. Math. Phys. **9**, 169 (1968)

[30] Roos, H.: KMS Condition without Time Automorphism of the Observable Algebra

[31] Haag, R., R. V. Kadison and D. Kastler: Nets of C*-Algebras and Classification of States, Comm. Math. Phys. **10**, 81 (1970)

[32] Rieckers, A.: On Symmetries of Infinite Macrosystems in a Standard Representation, Contribution to the Integrative Conference on Group Theory and Mathematical Physics, Austin (1978)

[33] Sewell, G. L.: States and Dynamics of Infinitely Extended Physical Systems, Comm. Math. Phys. **33**, 43 (1973)

[34] Battle, G. A.: Dynamics and Phase Transitions for a Continuous System of Quantum Particles in a Box, J. Math. Phys. **19**, 39 (1978)

[35] Primas, H. and U. Müller-Herold: Quantum Mechanical System Theory, Adv. Chem. Phys.

Chapter III
Pauli Principle and Indirect Exchange Phenomena in Molecules and Solids

L. Jansen, R. W. J. Roël, R. Block

1 Permutation symmetry and chemical bonding in molecules and solids

1.1 Permutation symmetry and exchange interactions

All analyses of interactions in molecular and solid-state systems, however diverse they may be in method of approach and degree of sophistication, must conform to basic principles imposed by what is called the quantum-mechanical *"symmetry"* of the system. The concept of symmetry of a system characterized by a Hamilton operator H is embodied by the group of Hilbert-space operators which have the following two important properties: a) they preserve the absolute value of the scalar product, $|(f, g)|$, between any two vectors f, g in the Hilbert space. Let A be such an operator. Then, either $(Af, Ag) = (f, g)$ or $(Af, Ag) = (g, f) = (f, g)^*$. From this it follows that A must be *unitary,* or an *anti-unitary* operator, respectively. Further, b) any A must *commute* with H: $AHf = HAf$, for all $\underline{f}$. We call them *"symmetry operators"* for the system. It can indeed be easily verified, and it is left as an exercise to the Reader, from the properties a) and b), that these operators form a group in the abstract sense of the word, the symmetry group of the system under consideration. We denote this group by $\mathcal{G}_H$.

From the properties of unitary and anti-unitary operators it readily follows that $\mathcal{G}_H$ is of the form

$$\mathcal{G}_H = \{G_H, \Theta G_H\}, \tag{1.1}$$

where G_H is the (normal) subgroup of all *unitary* operators and Θ is an *arbitrary* anti-unitary operator. Specifically, since $\mathcal{G}_H$ is a group, and since the product of a unitary and anti-unitary operator is anti-unitary, (the product of two (anti-)unitary operators is unitary) it follows that any anti-unitary operator $\hat{A}_i$ can be written as ΘA_j, for some $A_j \in G_H$. Suppose $\hat{A}_j = \hat{A}_i A_k$, then $\hat{A}_j = \Theta A_j A_k = \Theta A_l$, etc. This shows that G_H possesses the structure indicated. For a system with real H, the operator K of complex conjugation is an obvious choice for θ. It is convenient to study first the effect of the unitary operators on the solutions of the Schrödinger equation and to treat the anti-unitary operators separately.

Restricting ourselves then to the subgroup G_H, the next step lies in observing that any eigenfunction of H transforms under G_H as a basis vector for a (matrix) *representation* of G_H. This is seen as follows:

Let $A \in G_H$. Suppose $\{\phi_1, \phi_2, \ldots, \phi_l\}$ denotes an orthonormal basis of eigenvectors of H belonging to the eigenvalue λ (l-fold degenerate). Then $H(A\phi_i) = A(H\phi_i) = \lambda(A\phi_i)$. It thus follows that $A\phi_i$ is a linear combination of the basis functions, i.e., $A\phi_i = \sum_{j=1}^{l} D_{ji}(A)\,\phi_j$.

As is the case with all linear transformations, we have

$$D(AB) = D(A)\,D(B), \quad \text{for all} \quad A, B \in G_H.$$

From this it follows, in particular, that (E is the identity of G_H)

$$D(A)\,D(E) = D(AE) = D(A) = D(EA) = D(E)\,D(A) \quad \text{for all} \quad A \in G_H$$

so that D(E) is the unit matrix. Further we have

$$D(E) = D(AA^{-1}) = D(A)\,D(A^{-1}),$$

i.e. $D(A^{-1}) = [D(A)]^{-1}$.

Since $A^{-1} \in G_H$, $D(A^{-1}) = [D(A)]^{-1}$ exists, so that all matrices D(A) are non-singular. In total, this means that $\Gamma = \{D(A)\}$ is a homomorphic (matrix) image of G_H and thus, by definition, a (matrix) representation of G_H.

The matrix of H in the basis $\{\phi_i\}$ is, of course, $\lambda \mathbb{1}$, where $\mathbb{1}$ is the unit matrix of dimension l. Now, if $\Gamma = \{D(A)\}$ were an *irreducible* representation of G_H, then this form for the matrix of H would be imposed by Schur's (Second) Lemma. Note, however, that the *converse* is not true: If H is represented by $\lambda \mathbb{1}$ ($\lambda \mathbb{1}$ commutes trivially with Γ) then Γ is not necessarily irreducible.

Physical experience has taught us that in all known cases the representation $\Gamma = \{D(A)\}$ is indeed *irreducible* under G_H. In well-known examples (hydrogen atom, 3-dimensional harmonic oscillator) where this rule seemingly did not apply, *additional* symmetry operators could be found such that Γ did become irreducible under the enlarged G_H. Although from a group-theoretical point-of-view irreducibility of Γ cannot be imposed, it is now believed to hold in general. If not, we speak of *accidental* degeneracy. We call this the *Irreducibility Postulate.* In the second Part, Dr. Roël will deal with the concept of symmetry-and-degeneracy in more detail.

The consequences of the above postulate are of considerable weight: the possible *degeneracies* of eigenlevels are known when G_H is known (they can only be equal to the dimensions of irreducible representations of G_H), and the *transformation properties* under G_H of an eigenfunction are determined, once the irreducible representation associated with the eigenlevel has been identified. Additional degeneracies may, of course, be due to anti-unitary symmetries; the rules governing this eventuality are well-known [1].

To be more specific, let us consider a system of atoms or molecules with a total of N electrons and a number of clamped nuclei with arbitrary charges. The Hamilton operator H is invariant under permutations of an arbitrary number of electron labels, i.e. the permutation group (symmetric group) S_N is a subgroup of G_H. We denote all other variables on which H depends, collectively by r (in configuration- and spin-space of the N electrons). We collect all the transformations of r under which H is invariant, noting that single-electron variables which occur coupled in H (Coulomb interactions e.g.) must be transformed *simultaneously* to ensure invariance of H. This subgroup of G_H is denoted by F_H. Clearly, each element of S_N commutes with each element of F_H and their intersection contains the identity only. Also, each element of G_H is uniquely determined as the product of an element of F_H and of S_N. It follows that G_H is the *direct* product of its two subgroups, i.e.

$$G_H = F_H \times S_N, \tag{1.2}$$

in which F_H contains all unitary symmetry operators of H other than permutations.

The definition of symmetry operator in the Hilbert space has still to be made more precise, since such an operator acts on vectors, which are *functions* of the variables occurring in H. Suppose A is an invariance transformation of $H(r)$, i.e. $H(Ar) = H(r)$, for all r. We define a symmetry operator A in the Hilbert space as follows

$$Af(r) = f'(r) \equiv f(A^{-1}r) \text{ for all } f(r).$$

This definition establishes an isomorphism between the group $\{A\}$ of symmetry operators and the group $\{A\}$ of invariance transformations of H. On the other hand, the definition $Af(r) \equiv f(Ar)$ does *not* lead to homomorphism between $\{A\}$ and $\{A\}$, as is easily verified.

We continue with $G_H = F_H \times S_N$ $(= S_N \times F_H)$. It is a property of direct-product groups that all of its irreducible representations can be obtained by multiplying the irreducible representations of its "direct factors" (in this case, S_N and F_H) as Kronecker products. Specifically, the Irreducibility Postulate then states that any eigenfunction for the system must transform under permutations of electron labels as a basis vector for an irreducible representation of S_N and under the "rest-symmetries" as a basis vector for an irreducible representation of F_H.

As far as S_N is concerned, however, we now impose a much stricter limitation, *not* of group-theoretical origin. Electrons are "indistinguishable" ("identical"), implying that Hilbert-space vectors differing only by a permutation of electron labels must represent the *same* physical state. No measurement of any physical observable can yield different results for wavefunctions in which electron labels are permuted in different ways. We now *define* the concept of indistinguishability as *implying* that $P\psi = (\text{const})\,\psi$, for any $P \in S_N$, the constant being just a phasefactor, for any physical state ψ. Note that this definition holds for any basisvector ϕ_i of a degenerate state, i.e. $P\phi_i = (\text{const})\,\phi_i$, not just that $P\phi_i$ is a linear combination of basis vectors. This implies immediately that every physical state of the system must transform as a basis vector for a *one-dimensional* representation of S_N. Since one-dimensional representations are by definition irreducible, the Irreducibility Postulate is automatically satisfied. There are just *two* one-dimensional representations of S_N: the totally symmetric (also called "trivial") representation, in which every permutation is represented by the number + 1, and the totally antisymmetric (also called "alternating") representation, in which even permutations are represented by + 1, odd permutations by − 1.

The *Pauli principle* states that any electron wavefunction must belong to the *antisymmetric* representation of S_N, i.e. $P\psi = (-1)^P\,\psi$, with $(-1)^P = +1$ when P is an even permutation,

$(-1)^P = -1$ when P is an odd permutation. As a consequence, we can limit the search for solutions of the Schrödinger equation for a system of N electrons to those vectors of Hilbert space belonging to the totally antisymmetric representation of S_N. Further restrictions are then imposed by the condition that such functions must *also* belong to one of the irreducible representations of the "rest-symmetry" group F_H. Later on we will, for a variety of systems in which "*weak* chemical bonding" occurs, concentrate our attention on permutation symmetry *alone*. Formally, this means that in such systems the rest-symmetry group F_H may in first approximation be replaced by the identity. For a more explicit discussion of symmetry, in particular permutation symmetry and electron spin, we refer to Part 2.

On our way to properties of molecular and solid-state systems, we next discuss the somewhat artificial, principally non-physical, concepts of *exchange effects* and *exchange interactions*. The "exchange" phenomenon has no connection with any physically observable aspect of a system containing identical particles. It "corrects", so-to-speak, in the sense discussed above, our erroneous assumption that identical particles can be distinguished through different labels. Thus, it has no classical analogue.

The "exchange energy" of a system of N electrons would, ideally, be defined as the difference between its actual energy and the energy it would have if there were no restrictions whatever on the wavefunction due to permutation symmetry. Without these restrictions, the symmetry group of the system is simply the restgroup F_H. *Note* that we are here throwing overboard not only the indistinguishability of electrons, but also the Irreducibility Postulate. Namely, considering electrons as somehow distinguishable would *still* leave S_N as symmetry group $\subset G_H$. Retaining the Irreducibility Postulate would then *still* entail restrictions on the wavefunctions under permutations. The non-physical nature of this concept is clear from the above definition.

However, such a definition of exchange energy is unmanageable because it does not lead to simple algebraic expressions. Instead, we follow a more heuristic approach, which will be illustated for the simple case of two hydrogen atoms, hopefully without obscuring its general context. Suppose A and B are two protons, $\underline{a}$ is an electron orbital centered on A, $\underline{b}$ an orbital centered on B; 1 and 2 are the two electrons. We calculate the expectation value of the Hamilton operator H *first* with the correct antisymmetric singlet (non-degenerate) and triplet (three-fold degenerate) wavefunctions constructed from $\underline{a}$, $\underline{b}$ and spin functions, and *then* for the simple-product-type function a(1) b(2) or a(2) b(1), and substract the results. The differences are:

$$\begin{aligned} &\textit{triplet}: \quad -(J + 2\,\Delta K)/(1 - \Delta^2), \\ &\textit{singlet}: \quad +(J + 2\,\Delta K)/(1 + \Delta^2), \end{aligned} \tag{1.3}$$

where

$$J \equiv \iint a^*(1)\, b^*(2)\, (e^2/r_{12})\, a(2)\, b(1)\, d\tau_1 d\tau_2\,;$$

$$K \equiv \int a^*(1)\, (-e^2/r_{1B})\, b(1)\, d\tau_1\,,$$

and where

$$\Delta \equiv \int a^*(1)\, b(1)\, d\tau_1, \quad \text{the overlap integral.}$$

The integral J is called the exchange Coulomb repulsion between the two electrons and K the exchange part of the electron-proton interaction. With a and b orbitals on two widely separated hydrogen atoms, the differences (1.3) are "in some sense" first-oder perturbation energies for the system of two interacting hydrogen atoms. The "heuristic approach" involves replacing the rest-symmetry group F_H by $F_{H_A} \times F_{H_B}$, where F_{H_A} and F_{H_B} are the (rest-)symmetry groups for the separated atoms. This means that we have not compared the system with-and-without-permutation symmetry for the definition of exchange, as we ideally planned to do, but we have coupled with "no permutation symmetry" the condition "and no interactions".

Although this is not necessary for the general argument, we consider the special case of such small overlap that $\Delta^2 \ll 1$. Then we can combine the expressions (1.3) in the following form

$$-\tilde{J}\,[1/2 + 2\langle S_1 \cdot S_2 \rangle], \tag{1.4}$$

where $\tilde{J} \equiv J + 2\Delta K$, S_1 and S_2 are the spin operators for the two electrons, and where $\langle --- \rangle$ denotes the expectation value in *spin space.* The equations (1.3) and (1.4) are indeed the same, since for the triplet state $\langle S_1 \cdot S_2 \rangle = +1/4$ and for the singlet state $\langle S_1 \cdot S_2 \rangle = -3/4$. The *difference* singlet-triplet, equal to $2\tilde{J}$, is called the *exchange energy of the system.* It should be kept in mind that, in spite of the above notation in terms of spin variables, the exchange energy is entirely *orbital* in nature, i.e. it is of *electrostatic* origin, as is seen from the definitions of J and K.

Expressions for exchange energies, in the sense defined above, can be given for general N-electron systems, without basing ourselves on explicit spin notations similar to (1.4). Since the Pauli principle applies to the total wavefunction for the system, i.e. including spin variables, the symmetry properties under permutations of the *orbital* part of the wavefunction depend upon the spin state of the system (cf. the hydrogen molecule in the singlet or triplet state). Thus, exchange energies are "spin-dependent". This dependence, however, is caused by symmetry, not by spin dynamics: exchange energies do not involve the interaction of the magnetic moments associated with spin. *The differences in energy for different spin states are typically what we call exchange energies.*

After this brief excursion into "symmetry" and "exchange energy", let us see first what the above has to do with phenomena of chemical bonding in general.

1.2 Exchange interactions and chemical bonding [2]

The interpretation of chemical bonding in molecules on the basis of the concept of *"valence"* of the atoms involved, has, of course, a long history: already Gassendi ("Syntagma Philosophicum" 1658) suggested that atoms could attach themselves to one another by means of "hooks" on their surface. The number of such hooks for each particular kind of atom constitutes what later became known as the (maximum) valence of the atom in question.

A more thorough analysis of the concept of valence remained dormant until about one century ago, although Berzelius (1819) had already suggested that chemical combination between two atoms arises from electrostatic interactions between differently charged parts of the atoms. However, even in the days of Kekulé (second half of the nineteenth century) the picture of an atom was much like that of a billiard ball with a certain number of fixed knobs on the surface, by means of which other atoms could be attached. It appears, therefore, that until towards the end of the last century, the interpretation of the concept of valence had remained practically stagnant for two-and-one-half centuries. At the turn of the 19th century, rapid changes in the field of valence started to occur from different, and opposite, sides. Following J. J. Thomson's first attempt (1897) towards an electron theory of valence, Kossel (1915) and Lewis (1916) proposed practically simultaneously two seemingly very different models for chemical bonding. These ideas and, in fact, subsequent developments until, say, the mid-thirties, are so well documented in textbooks that it would be quite superfluous to dwell upon them in the present context. Lewis' idea of the electron-pair bond (covalent, or homopolar, bond) seemed to be validated by Heitler and London's [3] analysis (1927) of the hydrogen molecule; this marked the beginning of a promising attempt to transform the field of chemistry from a largely intuitive into an exact science. The method (valence-bond method, VB) of describing a molecule as an assembly of *atoms-plus-bonds* incorporates elements of intuitive chemical thinking and forms the basis *par excellence* for further development of the notion of valence as a characteristic of each chemical element.

In parallel with the development of the VB-method, Hund [4] (1928), Mulliken [5] (1928) and others developed a "physical" description of molecules in which the system of nuclei and electrons was treated as a whole and the electrons were arranged in "molecular orbitals" (MO). In this MO-method the identity of a bond between two atoms was not maintained, and the concept of valence as an elemental characteristic disappeared to a large extent from the description.

Next to the necessity of explaining why bonding does indeed occur between atoms, or why it does not occur, a satisfactory theory of chemical bonding must account for the observed steric configuration of molecules. In 1931 Pauling [6] developed the concept of "hybridisation" of atomic orbitals on the carbon atom ("valence state"); this marked the beginning of the stereochemistry of carbon compounds. In the years following, the hybridisation concept was generalized to include atoms other than carbon. For heavier atoms, other hybridisation schemes are possible, generally several ones for one and the same central atom, so that *varying* valence and varying geometric configurations of molecules could in principle be explained. The high stereospecificity of hybridised electron-pair bonding is not reproduced by the MO-approach; this is to be expected since here one starts from the molecule as a whole.

In spite of the fact that the VB electron-pair method, with its stereospecificity and its chemically intuitive picture of a molecule as atoms-plus-bonds, did score a large number of at least qualitative successes, serious criticisms have arisen against its interpretation of valence and chemical bonding. *First* of all, we note that for an atom to bind $\underline{n}$ ligands in a molecule, $\underline{n}$ *orbitals are needed.* If $\underline{n}$ is not a small number, then promotion of electrons from deep-lying states to excited configurations is necessary to create the valence state as a precursor to molecule formation. It is often more than doubtful whether this formation energy can be overcome by bonding. *Secondly,* the assumption of electron-pair bonding in general molecules is an extrapolation from the case of molecular hydrogen. From a MO point-of-view this extrapolation is dangerous: if we start building a general molecule from an assembly of nuclei and electrons, then it would be very surprising indeed if the state of lowest energy of the system did correspond to the VB picture, in view of the multitude of other possibilities. *Thirdly*, the VB electron-pair language is, however appealing intuitively, largely descriptive; it does not readily lend itself to numerical verification in any but the simplest cases.

The MO-method does not suffer from the first two defects. In numerical calculations it is in principle considerably less cumbersome, more *a priori*, than the VB procedure, although it as well can mostly be applied only in an approximated version, of which there now exists a confusingly large number of variants. It suffers from relatively low stereospecificity and it effaces, to a large extent, the notion of valence as an elemental characteristic, in spite of the very large body of chemical evidence to the contrary. One might be tempted to state that the VB method is too "atomic", the MO method too "molecular" in its approach, and that chemical bonding should be treated from a basis which is more local than the MO approach, but less local than the VB electron-pair method.

We will in this paper concentrate our attention on a class of phenomena characterized by *"weak (chemical) bonding"*, i.e. ~ 1 eV or less. A convenience of this restriction is that the bonding may be evaluated on the basis of perturbation theory, with the separated atoms as zeroth order of approximation. Phenomena of weak bonding include: bonding in rare-gas halides such as in XeF_4 (binding energy ~ 1.3 eV per bond), hydrogen bonding (mostly a few tenths of 1 eV) and magnetic interactions in insulating solids with paramagnetic cations, such as NiO and MnS (interaction energy 10^{-2} to 10^{-3} eV). *Chemical* bonding implies the assumption that exchange interactions must be taken into account in the analysis.

For our purpose it is most convenient to start from the example of magnetic interactions in solids. In the following Section we will briefly review the status of our present understanding regarding such phenomena.

1.3 The Heisenberg (effective spin-) Hamiltonian approach to exchange interactions

Much of our understanding of magnetic coupling between unpaired spins in solids containing paramagnetic cations crucially depends on the validity of an approach characterized by a socalled "Heisenberg (effective spin-) Hamiltonian". This concept is defined in close connection with the discussion on the hydrogen molecule given in Section 1. There it was found that (for $\Delta^2 \ll 1$) the expectation value of the Hamilton operator can be calculated, without taking account of permutation symmetry, by adding a spin operator

$$-\tilde{J}\,[1/2 + 2\,\mathbf{S}_1 \cdot \mathbf{S}_2]$$

to the corresponding expression obtained on the basis of an orbital *product* function for the electrons. It is again to be noted that this operator has been derived from a "first-order" perturbation expression, so that we should call it a *first-order operator.* Concerning the *sign* of $\tilde{J} = J + 2\Delta K$, it can be easily shown that the integral J is always positive. This follows readily if we rearrange the integrand in J as $(a(1)\, b^*(1))^*\, (e^2/r_{12})\, (a(2)\, b^*(2))$; the "exchange charge densities" left and right are just each other's complex conjugates. Consequently, if the orbitals $\underline{a}$ and $\underline{b}$ for the electrons happen to be orthogonal, then $\tilde{J} = J > 0$, and the triplet state ($\langle S_1, S_2 \rangle = 1/4$) lies lower than the singlet ($\langle S_1, S_2 \rangle = -3/4$). For $\underline{a}$ and $\underline{b}$ two 1 s-orbitals, on the other hand, it is found, with $K < 0$, that $|2\Delta K| > J$ up to very large distances, i.e. the singlet lies lower. In more general two-electron cases, $\tilde{J}$ can be either positive or negative.

The generalisation of these results to more-than-two electrons was given by Dirac [7] (1929). He showed that, for the special case of localized, *orthogonal*, orbital functions, in first order of approximation, the effect of the Pauli principle may be replaced by adding to the expectation value of the Hamilton operator, in the space of orbital product functions, an operator in spin space

$$-\sum_{i \leq j} J_{ij}\,(1/2 + 2\, S_i \cdot S_j); \tag{1.5}$$

$\underline{i}$ and $\underline{j}$ index the different electrons and J_{ij} is defined as for the hydrogen molecule. Suppose we consider the *interaction* between two atoms A and B (with n_A and n_B unpaired spins, respectively). In the literature, one often writes the spin-dependent part of (1.5) for this case as

$$-2\sum_{i=1}^{n_A}\sum_{j=1}^{n_B} J_{ij}\, S_i \cdot S_j \equiv -2 J_{AB} \sum_{i=1}^{n_A} S_i \cdot \sum_{j=1}^{n_B} S_j = -2 J_{AB}\, S_A \cdot S_B, \tag{1.6}$$

with the total-spin operators $S_A = \sum_{i=1}^{n_A} S_i$, $S_B = \sum_{j=1}^{n_B} S_j$, and with J_{AB} an "average exchange integral". The expression (1.6) is generally called the *Heisenberg (effective spin-) Hamiltonian,* although it was Dirac who first derived (1.5) and Van Vleck [8] (1932) who was the first to consider applications to magnetic materials.

There are several difficulties regarding validity of the Heisenberg Hamiltonian (1.6). In the *first* place, the total spins of A and B must remain good quantum numbers for the coupled system. This means that we must limit ourselves to cases of weak interactions (validity of Russell-Saunders coupling). *Secondly,* each (scalar) coefficient J_{ij} on the left of (1.6) depends on the spatial orientations of the orbitals for electrons $\underline{i}$ and $\underline{j}$. The expression on the right-hand side presupposes that the sum of all these orientation-dependent terms can be written as the product of a *scalar* ($S_A \cdot S_B$) and an *isotropic* quantity (J_{AB}). In the general case this is not a valid assumption.

The expression (1.6) *is* strictly valid (under Russell-Saunders coupling) if the incomplete shells on A and B are just half-filled (e.g. Mn^{2+} for 3d−; Eu^{2+}, Gd^{3+} for 4f-electrons). In that case J_{AB} is the isotropic average

$$J_{AB} = (1/n_A n_B) \sum_{i,j} J_{ij}. \tag{1.7}$$

The trivial special case when J_{ij} is independent of $\underline{i}, \underline{j}$ is, of course, included in (1.7).

A further difficulty with the Heisenberg Hamiltonian is that it strictly holds only for *orthogonal* orbitals. Since then every J_{ij} is positive, this implies that in a magnetic solid the unpaired spins of all cations will be parallel to each other, with the consequence of *ferromagnetic* spin ordering as the state of lowest energy (at 0 K). As is well known, however, ferromagnetism is not a common phenomenon in nature: the majority of solids with paramagnetic cations such as NiO, MnS, etc. are *anti*ferromagnetic.

Many efforts have been undertaken, in the course of the years, toward placing the Heisenberg Hamiltonian on a more realistic basis. Primarily, these efforts were aimed at removing the condition of *orthogonality* between singly-occupied orbitals of the interacting systems. This is a most serious handicap indeed, since in actual calculations it is often not possible to use orthogonal orbitals for the electrons. This applies, in particular, to a perturbation analysis of magnetic interactions, starting from isolated ions as zeroth approximation. The wavefunctions of different anions or cations are then not orthogonal. They can, of course, be orthogonalized, but these modified wavefunctions are no longer eigenfunctions of the unperturbed Hamiltonian. It appears that, in the case of non-orthogonality, and in the absence of orbital degeneracy[1]), the exchange interactions can be written as the expectation value in spin space of a series of positive integral powers of the operator $S_A \cdot S_B$, the highest power occurring being equal to the number of unpaired electrons of the magnetic atom (cation). In case A and B are different cations, then the highest power is equal to the number of unpaired electrons for the cation with the smaller number of such electrons. Thus, in the case of $A = B = Mn^{2+}$, the power series ends with $\langle (S_A \cdot S_B)^5 \rangle$, the same as for $A = Mn^{2+}$, $B = Eu^{2+}$. (Mn^{2+} has 5 3d$^-$, Eu^{2+} 7 4f-electrons.) If we consider only the term proportional to $\langle S_A \cdot S_B \rangle$, then the *form* of the interaction is the same as that of the Heisenberg Hamiltonian, but the proportionality constant J_{AB} is *not* defined as in (1.7). In addition, terms with higher powers of $S_A \cdot S_B$ can become more important if the spin quantum numbers are high (5/2 for Mn^{2+}, 7/2 for Eu^{2+}).

The explicit evaluation of the coefficients of the different terms $\langle (S_A \cdot S_B)^n \rangle$ can be accomplished in a number of ways. An elegant method is based on the double-coset decomposition of permutation groups [9]. For an explicit discussion we refer to Chapter 3.2.

It should also be realized that, even in the event that only bilinear spin terms must be considered in good approximation, the spin-Hamiltonian for a system of *three* cations A, B and C must be properly written as $-2[J_{AB(C)} S_A \cdot S_B + J_{AC(B)} S_A \cdot S_C + J_{BC(A)} S_B \cdot S_C]$. This is because exchange interactions are not additive in pairs, so that, as indicated in the notation, the exchange integrals depend on the positions *of all three cations simultaneously.*

The situation is still further aggravated by the fact that the Heisenberg Hamiltonian refers to *direct* exchange between paramagnetic atoms or ions. This exchange decreases rapidly with increasing separation, and it is impossible to explain the observed high Néel-temperatures on the basis of such direct exchange interactions, in view of the relatively large dis-

1) e.g. cations with half-filled shells; also three d-electrons in a cubic field so strong that the (threefold degenerate) T_{2g} level is by far the lowest one, thus giving effectively a half-filled shell.

tance between nearest-neighbour paramagnetic atoms. One now knows that the coupling between two cations usually involves a (one or more) diamagnetic anion(s). This phenomenon is called *indirect* or *super*exchange. Also in this case, the double-coset decomposition method can be applied to obtain expressions for the coupling coefficients [10].

All in all, it has appeared over the years that the accurate determination of exchange fields in magnetic materials is one of the most difficult (albeit one of the most important!) problems in the theory of the solid state. It is now believed that the Heisenberg Hamiltonian (1.6), but *not* with J_{AB} defined following (1.4), constitutes a valid first approximation for the interpretation of properties of magnetic solids. However, first-principle theoretical values of exchange coupling constants, and in many cases even their *sign*, are so uncertain that an accurate theory of magnetic interactions is still remote at present. This is the more disappointing, as the form of the Heisenberg Hamiltonian is very suitable for use in the determination of macroscopic properties of magnetic materials on the basis of methods of statistical thermodynamics.

In the literature on superexchange phenomena, semi-empirical rules have filled the gap between first-principle theory and experiment; they were developed by Goodenough [11] and Kanamori [12] (so-called "Goodenough-Kanamori rules"); for a clear review, and a concise formulation of these rules, see Anderson [13]. Still, it must be said that, with the enormous supply of precise experimental data now available, also semi-empirical theory lags very far behind the experimental facts.

References

[1] *Jansen, L., Boon, M.:* "Theory of Finite Groups. Applications to Physics", N.-H. Publ. Cy., Amsterdam, 1967.

[2] Taken in part from
Jansen, L.: Research Futures (4-th Quarter, 1973) 11.

[3] *Heitler, W., London, F.:* Z. Phys., **44** (1927) 455.

[4] *Hund, F.:* Z. Phys. **51** (1928) 739; **73** (1931) 1.

[5] *Mulliken, R. S.:* Phys. Rev. **32** (1928) 186, 761; **41** (1932) 49.

[6] *Pauling, L.:* J. Am. Chem. Soc. **53** (1931) 1367.

[7] *Dirac, P. A. M.:* Proc. Roy. Soc. **A123** (1929) 714.

[8] *Van Vleck, J. H.:* "The Theory of Electric and Magnetic Susceptibilities", University Press, Oxford, 1932.

[9] *Block, R.:* Physica **70** (1973) 397.

[10] *Block, R.:* Physica **73** (1974) 312.

[11] *Goodenough, J. B.:* Phys. Rev. **100** (1955) 564; J. Phys. Chem. Solids **6** (1958) 1287.

[12] *Kanamori, J.:* J. Phys. Chem. Solids **10** (1959) 87.

[13] *Anderson, P. W.:* in "Solid State Physics", Ed. F. Seitz, D. Turnbull, Vol. 14, Ac. Press, New York, 1963.

2 Group-theoretical-aspects pertaining to the quantum-mechanical N-particle system

2.1 Introduction

2.1.1 Hamilton operator and Hilbert space

Consider a system which contains N identical particles. Let H denote the Hamilton operator pertaining to this system and let V (N) denote the N-particle Hilbert space on which the Hamilton operator acts. Then the Schrödinger equation must be solved. This is a problem of such extreme complexity that simplifying assumptions must be introduced. We define:

a) *A model Hamilton operator* H,

1. which is spinfree, i.e., the operator does not contain terms which explicitly refer to the spin coordinates of the particles. Most of chemistry and part of physics can be discussed on this basis

2. which looks like $H = \sum_{R=0} H_R$

   where H_R, the complete R-particle part, R = 0, 1, 2, . . . , can be expressed as

3. $H_R = \binom{N}{R} \frac{1}{N!} \sum_{g \in G} g h_R \bar{g}$, $g \in G = S_N$ the symmetric group permuting the N identical particles,

   where h_R denotes a prototype R-particle operator which contains the coordinates of R selected particles, say 1, 2, . . . , R.

Thus:

$$H = \sum_{R=0} \binom{N}{R} \frac{1}{N!} \sum_{g \in G} g h_R \bar{g} \qquad h_R \text{ is spinfree, } R = 0, 1, 2, \ldots .$$

b) *A model Hilbert space* V (N),

1. which is a product space

   $$V(N) = V^\rho(N) \otimes V^\sigma(N)$$

   where ρ and σ refer to spatial- and spin coordinates, respectively. The N-particle orbital space is chosen as

2. $V^\rho(N) = V_n^\rho \otimes^N$, the N-fold tensor product

   of V_n^ρ, an n-dimensional 1-particle orbital space.

   The N-particle spin space is given by

3. $V^\sigma(N) = V_t^\sigma \otimes^N$, the N-fold tensor product of 1-particle spin space with $t = 2s + 1$ for spin-s-particles.

Thus:

$$V(N) = (V_n^\rho \otimes V_t^\sigma)\otimes^N = (V_n^\rho \otimes^N)\otimes(V_t^\sigma \otimes^N).$$

Our aim is to discuss some group-theoretical aspects of the model Schrödinger equation defined by (a, b).

2.1.2 Symmetry and Wigner's theorem. The irreducibility postulate, the identity postulate and the Pauli postulate

Again we consider a quantum-mechanical system with hamiltonian H. Let V denote its Hilbert space which is spanned by a basis $\{\phi_j | j = 1, 2, \ldots\}$ and let G denote a group of symmetry operators on V, i.e., operators $g: V \to V$ which preserve the modulus of scalar products and which commute with H. Such a group is called a symmetry group of H.

Relative to the basis $\{\phi\}$ of V we represent the operator H by a matrix H with elements $H_{jk} = (\phi_j | H | \phi_k)$ and the group of symmetry operators by a matrix group $\{\Gamma(g) | g \in G\}$, with $\Gamma_{jk}(g) = (\phi_j | g | \phi_k)$. This matrix group is a representation of G.
Suppose its decomposition reads

$$\Gamma = \bigoplus_\Lambda f(\Lambda)\,\Lambda$$

where $f(\Lambda)$ denotes the number of times that the irreducible representation (IR) Λ of G occurs in Γ. Then there exists a non-singular matrix S such that

$$\forall g \in G \qquad S\,\Gamma(g)\,\bar{S} = \bigoplus_\Lambda \{\mathbb{1}_{f(\Lambda)} \otimes \Lambda(g)\}$$

in which the right-hand side is a matrix where $\Lambda(g)$ occurs $f(\Lambda)$ times along the diagonal. ($\oplus$ is direct sum and $\otimes$ is direct product of matrices.)
Under this decomposition one has

$$S H \bar{S} = \bigoplus_\Lambda \{\tilde{H}^\Lambda \otimes \mathbb{1}_{|\Lambda|}\}$$

where $\tilde{H}^\Lambda$ denotes an $f(\Lambda)$-square matrix and $\mathbb{1}_{|\Lambda|}$ a unit matrix of dimension $|\Lambda|$, the dimension of the IR.
Let $\Gamma(g)$ and H be partitioned into blocks with a row label $\Lambda'k'$, $k' = 1, 2, \ldots f(\Lambda')$ and a column label $\Lambda''k''$, $k'' = 1, 2, \ldots, f(\Lambda'')$.
Then $\{S\,\Gamma(g)\,\bar{S}\}_{\Lambda'k', \Lambda''k''} = \delta(\Lambda'\Lambda'')\,\delta(k'k'')\,\Lambda'(g)$

and from

$$H\,\Gamma(g) = \Gamma(g)\,H \qquad \forall g \in G$$

it follows that

$$\forall \Lambda', \Lambda'', k', k'' \text{ and } g \in G$$

the following expression

$$(S H \bar{S})_{\Lambda' k', \Lambda'' k''} \Lambda''(g) = \Lambda'(g) (S H \bar{S})_{\Lambda' k', \Lambda'' k''}$$

holds.

This means that the block $(S H \bar{S})_{\Lambda' k', \Lambda'' k''}$ intertwines the IR Λ' and Λ''. Thus:

Schur's first lemma $\Rightarrow (S H \bar{S})_{\Lambda' k', \Lambda' k''} = \tilde{H}^{\Lambda'}_{k' k''} \, \mathbb{1}_{|\Lambda'|}$

with $\tilde{H}^{\Lambda'}_{k' k''}$ a proportionality constant, and

Schur's second lemma $\Rightarrow (S H \bar{S})_{\Lambda' k', \Lambda'' k''} = \mathbb{0}$.

Then let T be a matrix of the form

$$T = \bigoplus_{\Lambda} (U^{\Lambda} \otimes \mathbb{1}_{|\Lambda|}) \qquad \text{(which commutes with all } S\,\Gamma(g)\,\bar{S})$$

such that

$$U^{\Lambda} \tilde{H}^{\Lambda} \bar{U}^{\Lambda} = \epsilon^{\Lambda}, \qquad \text{a diagonal } f(\Lambda)\text{-square matrix.}$$

Then:

The entries of ϵ^{Λ}, ϵ^{Λ}_{k} $k = 1, 2 \ldots f(\Lambda)$, Λ ranging, are the eigenvalues of H on V. Thus, as a consequence of Schur's lemmas we can state *Wigner's Theorem.*

The eigenspaces of the Hamilton operator H (subspaces of V associated with different eigenvalues of H) are the carrier space for *representations* of G. More explicitly: With any eigenvalue ϵ^{Λ}_{k} that occurs *once* there is associated

a) a set of eigenvectors which constitute a basis for the IR Λ,

b) a degree of degeneracy which is equal to the dimension of the IR Λ.

However: On the basis of group theory one cannot possibly exclude the occurrence of equal eigenvalues $\epsilon^{\Lambda}_{k} = \epsilon^{\Lambda}_{m} = \ldots$.

Quoting MacIntosh from "Symmetry and Degeneracy" (Applications Group Theory, Loebl II, 75–144):

"Every symmetric system will show characteristic degeneracies whose multiplicity is prescribed by the dimensions of the irreducible representations of its symmetry group. Yet, there is no restriction arising from group-theoretical reasoning which prevents there from being a higher multiplicity of degeneracy that that required by Schur's lemmas, but any degeneracy so arising is commonly called "accidental" degeneracy due to *a presumption as to its unlikelihood*".

"There has always been a feeling that accidental degeneracy might not be so much of an accident after all, in the sense that *there might actually have been a larger group* which would incorporate several different degenerate representations of the overt symmetry group in a single one of its own irreducible representations".

"Since the mathematical understanding of these conditions (the conditions which govern the occurrence of accidental degeneracy) has *never been completely decisive,* there has grown up a considerable folklore about the nature of accidental degeneracy, ... Mainly we know of certain conditions under which accidental degeneracy will arise but *it is largely a matter of faith* and our limited range of experience, that there will be no accidental degeneracy when these conditions are absent. ... One of the more attractive tenets of the folklore has been that accidental degeneracy arises from hidden symmetry, as has been amply demonstrated by the major examples (hydrogen atom, harmonic oscillator, symmetric top)". (Italics and bracketed remarks by the present author.)

In our understanding one generally accepts the (very convenient but not group-theoretically rationalisable) presumption that, provided the group G of Hilbert space operators that commute with H is "large" enough, one may replace Wigner's Theorem by the sharpened statement:

The eigenspaces of H carry *irreducible* representations of G.

And according to MacIntosh:

"In a formal sense this (irreducibility) is clearly true, since one can simply *postulate* such a group".

In our opinion one thus has to introduce the *postulate of irreducibility* in either one of the following two ways:

a) There exists a group G^H of Hilbert space operators that commute with H such that the eigenspaces of H carry irreducible representations of G^H.

Whether or not the elements of such a group will have physical significance (we recall that elements of G^H are actually state transformations to be interpreted as physical operations performed on the system) is a question which cannot be easily answered.

b) The eigenstates of H span irreducible representations of its symmetry group G_S^H defined as any group of Hilbert space operators which a) commute with H, b) predict the spectrum of H and c) have a physical interpretation. It may be very difficult to find such groups.

Let us now determine such a symmetry group for a Hamilton operator H pertaining to a system which contains N identical particles.

First: Consider h, a proto-type 1-particle operator and let G_S^h denote its symmetry group. Then obviously, the direct product $G_S^{h_1} \otimes G_S^{h_2} \otimes \ldots \otimes G_S^{h_N} \equiv G_S^{h\otimes}$ commutes with H_I, the complete 1-particle part of H, but not necessarily with H_{II}, etc.

Secondly: With H totally symmetric in the coordinates of the N identical particles the group $G = S_N$ will be a subgroup of the symmetry group of H. It is easily seen that $g G_S^{h\otimes} \bar{g} = G_S^{h\otimes}$ $\forall g \in G = S_N$, and also that $G \cap G_S^{h\otimes} = e$, the identity.

Thus $G_S^{h\otimes}$ G is a semi direct product group and the symmetry group of H, G_S^H will be one of its subgroups.

Let $F_S^h \subseteq G_S^h$ denote the largest subgroup such that the inner direct product $F_S^H =$ $= F_S^{h_1} \boxtimes F_S^{h_2} \boxtimes \ldots$ commutes with H.

The inner direct product is chosen in order to ensure that the operations effects the *same* symmetry transformation on each particle separately. This is necessary since although we do not specify the explicit form of the interaction (H_{II}, etc.) between two, three etc. particles, the complete interaction operators (H_{II}, etc.) have been defined totally symmetric.

Since F_S^H and G commute even elementwise, the symmetry group so defined is a direct product group

$$G_S^H = F_S^H \times G, \quad G = S_N.$$

Obviously, the symmetric group permuting identical particles is always present whatever system we consider. The group $G = S_N$ always occurs on account of the *principle of identity* (which is just a definition of identity of particles): The interchange of identical particles cannot be observed or, what amounts to the same, identical particles are indistinguishable. Thus no experiment exists under which states of N identical particles $g\psi$, $g \in S_N$, can be distinguished, i.e. all $g\psi$ represent the same state.
The irreducible representations of S_N which enter the game are severely restricted by the *identity postulate* however:
Physically realizable N-identical particle states transform according to 1-dimensional irreducible representations of S_N. According to this postulate there exist only two possibilities since the group S_N possesses only two 1-dimensional irreducible representations, the totally symmetric one, denoted by [N], and the totally antisymmetric one, $[1^N]$.

The actual realization of these two possibilities is given by the *Pauli-postulate*: The states of a system which contains N-identical integer-spin particles (bosons) transform according to the symmetric representation [N].
The states of a system which contains N-identical half-integer-spin particles (fermions) transform according to the antisymmetric representation $[1^N]$.
Thus:

$$t = 2s + 1 = \text{odd} \Leftrightarrow [N]$$
$$t = 2s + 1 = \text{even} \Leftrightarrow [1^N].$$

We note that if the symmetry group G_S^H is restricted to the symmetric group, the postulate of irreducibility is not essential. The identity postulate is more fundamental and its consequences are more restrictive since 1-dimensionality implies irreducibility.

2.1.3 The spin free Pauli-principle

The model Hamilton operator defined above is spin-free. Physically this means that we neglect all dynamical effects caused by spin. Group-theoretically this assumption has far-reaching consequences for the description of the N-particle system. Actually, in combination with the Pauli principle it is possible to select subspaces of $V_n^\rho \otimes^N$ which are physically "allowed" and we can then solve the Schrödinger equation within these "allowed" subspaces.
Consider $F_h \subseteq G_h$, the 1-particle symmetry group, again. Let $F_h = F_h^\rho \times F_h^\sigma$ denote the factorization into a group affecting only spatial coordinates, and a group on spin coordinates. Since here we are not interested in F_h^ρ we put $F_h^\rho = E^\rho$, the identity. F_h^σ is the group of non-singular transformations GL_t, $t = 2s + 1$, on 1-particle spin-s-space. The factor $G = S_N$ will be written $G = G^\rho \boxtimes G^\sigma$ an inner product the elements of which permute spatial- and spin coordinates simultaneously.

Then the symmetry group

$$G_H \cong GL_t^\sigma \times (G^\rho \boxtimes G^\sigma)$$

is a group of operators acting on N-particle Hilbert space

$$V(N) = (V_n^\rho \otimes V_t^\sigma) \otimes^N = (V_n^\rho \otimes^N) \otimes (V_t^\sigma \otimes^N).$$

The *Schur-Weyl theorem* (App. A) states that: Symmetry adaptation of spin space $V_t^\sigma \otimes^N$ with respect to either of the two groups GL_t^σ or G^σ induces symmetry adaption with respect to the other.

It then follows that: Symmetry-adapted subspaces of $V_t^\sigma \otimes^N$ are labelled by

1. Partitions of N, $(\Lambda) = \dots j^{\lambda_j} \dots$, such that
2. $\sum_j \lambda_j \leqslant t = 2s + 1$, i.e.,

symmetry-adapted subspaces are characterized by Young diagrams with at most 2s + 1 rows.

The Pauli (fermion) principle states that only the antisymmetric subspace of $V(N) = (V_n^\rho \otimes^N) \otimes (V_t^\sigma \otimes^N)$ is physically realizable. This in combination with the following facts:

1. The antisymmetric IR $[1^N]$ only occurs in the product of an IR and its associate, in which product it occurs once,
2. The IR of G are characterized by Young diagrams,
3. Associate IR are characterized by transposed Young diagrams, yields *the spin-free exclusion principle:*

Physically allowed spin-free states pertaining to a system of N identical spin-s-fermions transform according to IR of $G = S_N$ which are characterized by Young diagrams with at most 2s + 1 columns.

Appendix A

The Schur-Weyl Theorem (Schur 1906, Weyl 1930)

With V_n a linear n-dimensional vector space, $V_n \otimes^N$ the corresponding N-fold tensor product space, defined as the product of N identical spaces $V_n^{(k)}$, k = 1, 2, ..., N, $V_n \otimes^N$ is stable with respect to:

1. The symmetric group S_N if the action of $g \in S_N$ on $V_n \otimes^N$ is defined on the labels (k) in $V_n^{(k)}$, k = 1, 2, ..., N,
2. The general linear group GL_n if the action of $\pi \in GL_n$ on $V_n \otimes^N$ is defined as $\pi: V_n^{(k)} \to V_n^{(k)}$ for all k = 1, 2, ... N simultaneously.

   Moreover:
3. The groups S_N and GL_n, the elements of which act on $V_n \otimes^N$ as defined above, commute elementwise.

Thus, a complete decomposition of $V_n \otimes^N$ into irreducible subspaces with respect to either of these two groups (subsequently called symmetry adaptation) induces a complete decomposition with respect to the other.

More explicitly (IR = irreducible representation, SA = symmetry adaptation or symmetry-adapted):

SA of $V_n \otimes^N$ to S_N yields

$$V_n \otimes^N = \bigoplus_{(\Lambda)} V^{(\Lambda)}$$

where (Λ) ranges over the partitions of N, which characterize the IR $[\Lambda]$ of S_N. SA of $V_n \otimes^N$ to GL_n yields

$$V_n \otimes^N = \bigoplus_{(\Lambda)} V^{(\Lambda)}$$

where (Λ) ranges over n-tuples of non-increasing integers, which characterize IR of GL_n, $\langle\Lambda\rangle$.

Then:

$V^{(\Lambda)}$ carries the IR $[\Lambda]$ of S_N with multiplicity $d_{\langle\Lambda\rangle}$, the dimension of the IR $\langle\Lambda\rangle$ of GL_n.

$V^{(\Lambda)}$ carries the IR $\langle\Lambda\rangle$ of GL_n with multiplicity $d_{[\Lambda]}$, the dimension of the IR $[\Lambda]$ of S_N.

Thus:

$$\bigoplus_{M}^{d_{\langle\Lambda\rangle}} V_M^{[\Lambda]} = V^{(\Lambda)} = \bigoplus_{W}^{d_{[\Lambda]}} V_W^{\langle\Lambda\rangle}.$$

With

$$\{\phi_{MW}^{(\Lambda)} \mid M = 1, 2, \ldots d_{\langle\Lambda\rangle};\ W = 1, 2, \ldots d_{[\Lambda]}\}$$

a SA basis for the subspace $V^{(\Lambda)} \subset V_n \otimes^N$ it then follows that

$V_W^{\langle\Lambda\rangle}$, the W-th carrier space for the IR $\langle\Lambda\rangle$ of GL_n that occurs in the decomposition of $V_n \otimes^N$, $W = 1, 2, \ldots d_{[\Lambda]}$, is spanned by the basis

$$\{\phi_{MW}^{(\Lambda)} \mid W \text{ fixed}, M = 1, 2, \ldots d_{\langle\Lambda\rangle}\} \equiv \{|\phi(W); \langle\Lambda\rangle M\rangle \mid M \text{ ranging}\}$$

$V_M^{[\Lambda]}$, the M-th carrier space for the IR $[\Lambda]$ of S_N that occurs in the decomposition of $V_n \otimes^N$, $M = 1, 2, \ldots d_{\langle\Lambda\rangle}$, is spanned by the basis

$$\{\phi_{MW}^{(\Lambda)} \mid M \text{ fixed}, W = 1, 2, \ldots d_{[\Lambda]}\} \equiv \{|\phi(M); [\Lambda]\, W\rangle \mid W \text{ ranging}\}.$$

With (Λ) any partition of N we can associate (App. C) a Young diagram $D(\Lambda)$, an h-graph number $h(\Lambda)$ and an n-graph number $n(\Lambda)$.

Then from

$$d_{[\Lambda]}\, h(\Lambda) = |G|, \quad d_{\langle\Lambda\rangle}\, h(\Lambda) = n(\Lambda)$$

the pertaining dimensions can be easily calculated.

A short explanation pertaining to the Schur-Weyl Theorem may be appreciated. Let G denote a group of operators acting on an n-dimensional linear space V and suppose this space decomposes into irreducible subspaces (with respect to G) in the following way:

$$V = \bigoplus_{\Lambda} V^{(\Lambda)} = \bigoplus_{\Lambda} \bigoplus_{j=1}^{f(\Lambda)} V_j^{(\Lambda)}$$

where $V_j^{(\Lambda)}$ denotes the j-th carrier space for the IR $[\Lambda]$ of G.
Let

$$\{\phi_{jk}^{(\Lambda)} \mid k = 1, 2, \ldots n(\Lambda)\}$$

denote the basis for $V_j^{(\Lambda)}$ and let the bases for $j = 1, 2, \ldots f(\Lambda)$ be chosen such as to afford identical matrix representations.
Then:

$$g\phi_{jk}^{(\Lambda)} = \sum_{k'}^{n(\Lambda)} \phi_{jk'}^{(\Lambda)} \Lambda_{k'k}(g) \qquad \text{(a)}$$

i.e., $g \in G$ is represented in V by

$$D(g) = \bigoplus_{\Lambda} \{\mathbb{1}_{f(\Lambda)} \otimes \Lambda(g)\}.$$

Let F denote a group of operators on V such that

$$\forall f \in F \quad fg = gf \quad \forall g \in G.$$

Then (as in Wigner's Theorem, (2.1.2)) we have

$$D(f) = \bigoplus_{\Lambda} \{\tilde{D}^{\Lambda}(f) \otimes \mathbb{1}_{|\Lambda|}\}$$

Thus:

$$f\phi_{jk}^{(\Lambda)} = \sum_{j'=1}^{f(\Lambda)} \phi_{j'k}^{(\Lambda)} \tilde{D}_{j'j}^{\Lambda}(f) \qquad \text{(b)}.$$

Comparing (a) and (b) we see that in the following arrangement

$$\begin{pmatrix} \phi_{11}^{(\Lambda)} & \phi_{12}^{(\Lambda)} & \ldots & \ldots & \phi_{1n}^{(\Lambda)} \\ \phi_{21}^{(\Lambda)} & \cdot & \ldots & \ldots & \cdot \\ \vdots & \vdots & & & \vdots \\ \phi_{f1}^{(\Lambda)} & \phi_{f2}^{(\Lambda)} & \ldots & \ldots & \phi_{fn}^{(\Lambda)} \end{pmatrix}$$

rows are stable under the action of $g \in G$ (i.e., (a))
columns are stable under the action of $f \in F$ (i.e., (b)).

Each row constitutes a basis for an IR of G (which we denote by $[\Lambda]$) and one can prove, that under the assumption that the group F contains *all* operators that commute element-wise with G, that each column constitutes a basis for an IR of F (which we denote by $\langle\Lambda\rangle$).

It is also clear that dimension and multiplicity change roles, i.e.,

$$f([\Lambda]) = n(\langle\Lambda\rangle) \quad \text{and} \quad n([\Lambda]) = f(\langle\Lambda\rangle).$$

Electron spin and permutation symmetry

The simplest non-trivial example of the above statements is specified by n = 2 which corresponds to electron spin.

V_2: 1-electron spin space is spanned by $\{\sigma_+ = \alpha, \sigma_- = \beta\}$

$V_2 \otimes^N$: N-electron spin space is spanned by the 2^N dimensional orthornormal basis $\{\sigma_\pm(1)\,\sigma_\pm(2)\ldots\sigma_\pm(N)\}$.

SA of $V_2 \otimes^N$ to S_N yields

$$V_2 \otimes^N = \bigoplus_p V^{(p,N-p)}$$

and

$$\bigoplus_M^{d\langle p,N-p\rangle} V_M^{[p,N-p]} = V^{(p,N-p)} = \bigoplus_W^{d[p,N-p]} V_W^{\langle p,N-p\rangle}$$

and

$$d_{\langle p,N-p\rangle} = N - 2p + 1, \quad d_{[p,N-p]} = \frac{N-2p+1}{N+1}\binom{N+1}{p}.$$

There exists a unique relation between p in (p, N − p) and S in S(S + 1), the eigenvalue of the N-electron spin operator S^2. The relation reads $p = \frac{1}{2}N - S$.

This is very plausible since for given S there exist 2S + 1 eigenvalues M for S_Z to be compared with the dimension N − 2p + 1 of ⟨p, N − p⟩.

Hence, with $\Xi(N; W, S, M)$ the W-th linearly independent N-electron spin (S, M) eigenfunctions we have:

1. $S \Leftrightarrow (\frac{1}{2}N - S, \frac{1}{2}N - S) = (\Lambda_S) \Rightarrow [\Lambda_S]$ and $\langle\Lambda_S\rangle$.
2. $\Xi(N; W, S, M)$, S and M fixed, span $[\Lambda_S]$ of S_N.
3. $\Xi(N; W, S, M)$, S and W fixed, span $\langle\Lambda_S\rangle$ of GL_2

2.2 Symmetry adaptation

2.2.1 Matric bases

Consider some linear space V with basis $\{\phi \mid \phi \text{ ranging}\}$.

Symmetry adaptation of V to a group G consists of:

1. The construction of a basis for the group algebra of G, i.e., a set of |G| operators which are linearly independent

$$Q_{JK}^{\Lambda} = \sum_{g \in G} Q_{JK}^{\Lambda}(g)\, g \tag{1}$$

Λ ranging over the IR of G J, K = 1, 2, ... $|\Lambda|$, the dimension of Λ, $Q^{\Lambda}_{JK}(g)$ algebraic coefficients.

2. The construction of a symmetry-adapted basis for V by projecting the basis $\{\phi\}$ with the operators Q. This is a non-trivial problem since linear dependencies wil always occur if the complete range of ϕ is used.

With the following definition:

$$\{Q^{\Lambda}_{JK}|\phi\rangle \equiv |\phi(K); \Lambda J\rangle | K \text{ fixed}, J = 1, 2, \ldots |\Lambda|\} \tag{2}$$

is a basis (the K-th) for the IR Λ, it can be easily shown that general matric bases elements must be of the following type

$$Q^{\Lambda}_{JK} = \frac{|\Lambda|}{|G|} \sum_{g \in G} (A\Lambda^*(g)\bar{B})_{JK}\, g \tag{3}$$

Λ ranging, J, K = 1, 2, ... $|\Lambda|$, where $\Lambda(g)$ denotes a unitary matrix representing g in Λ and A and B are arbitrary non-singular.

Alternative expressions, sometimes easier to handle, can be derived by introducing

$$\begin{aligned} A\Lambda^*(g)\bar{A} &= \Lambda^L(g) \quad \text{the left matrix representation for the IR } \Lambda \\ B\Lambda^*(g)\bar{B} &= \Lambda^R(g) \quad \text{the right matrix representation for the IR } \Lambda \end{aligned} \tag{4}$$

and

$$A\bar{B} = \frac{|G|}{|\Lambda|} Q^{\Lambda}(e)$$ Substitution into (3) yields

$$Q^{\Lambda}_{JK} = \sum_{g \in G} \{\Lambda^L(g)\, Q^{\Lambda}(e)\}_{JK}\, g = \sum_{g \in G} \{Q^{\Lambda}(e)\, \Lambda^R(g)\}_{JK}\, g \tag{5}$$

Some special choices of (3):

1. $A = B = \Lambda(e)$.

   The operators Q^{Λ}_{JK} constitute a unitary matric basis (UMB). We then use the notation

$$Q(J\Lambda K) = \frac{|\Lambda|}{|G|} \sum_{g \in G} \langle g|J\Lambda K\rangle^* g = \frac{|\Lambda|}{|G|} \sum_{g \in G} \langle \bar{g}|K\Lambda J\rangle\, g.$$

2. A or B or both A and B $\neq \Lambda(e)$ but unitary such that A decomposes $\Lambda(g)$ with respect to $G^{\alpha} \subseteq G$, B decomposes $\Lambda(g)$ with respect to $G^{\beta} \subseteq G$. The pertaining UMB

$$Q(J^{\alpha}\Lambda K^{\beta}) = \frac{|\Lambda|}{|G|} \sum_{g \in G} \langle g|J^{\alpha}\Lambda K^{\beta}\rangle^* g = \frac{|\Lambda|}{|G|} \sum_{g \in G} \langle g|K^{\beta}\Lambda J^{\alpha}\rangle\, \bar{g},$$

   are called sequence-adapted UMB (App. B).

3. $A = B \neq \Lambda(e)$, $A^+ \neq \bar{A}$,

   yields non-orthogonal matric bases for which notation (3) will be maintained.

Note: The set of matrices

$$\{A\,\Lambda(g)\,\overline{B}\,|\,g \in G\}$$

is not a representation in general, $A \neq B$. However,

$$A\,\Lambda(g')\,\overline{B}\,(B\overline{A})\,A\,\Lambda(g'')\,\overline{B} = A\,\Lambda(g'g'')\,\overline{B} \tag{6}$$

and one speaks of the set as a skew representation with metric $B\overline{A} = (A\,\Lambda(e)\,\overline{B})^{-1}$.

2.2.2 Some fundamental expressions pertaining to matric bases

Some expressions will be given for future reference, mostly without derivation.

$$f\,Q^{\Lambda}_{J'K} = \sum_{J''} \Lambda^{L}_{J'J''}(\overline{f})\,Q^{\Lambda}_{J''K} \tag{7}$$

Thus, the set

$$\{Q^{\Lambda}_{JK}\,|\,K \text{ fixed, } J = 1, 2, \dots |\Lambda|\}$$

spans a minimal left ideal.

$$Q^{\Lambda}_{JK'}\,f = \sum_{K''} Q^{\Lambda}_{JK''}\,\Lambda^{R}_{K''K'}(\overline{f}). \tag{8}$$

Thus: The set

$$\{Q^{\Lambda}_{JK}\,|\,J \text{ fixed, } K = 1, 2, \dots |\Lambda|\}$$

spans a minimal right ideal.

The basic multiplication rule:

$$Q^{\Lambda'}_{J'K'}\,Q^{\Lambda''}_{J''K''} = (A\overline{B})^{\Lambda'}_{J''K'}\,\delta(\Lambda'\Lambda'')\,Q^{\Lambda'}_{J'K''} = \frac{|G|}{|\Lambda'|}\,Q^{\Lambda'}_{J''K'}(e)\,Q^{\Lambda'}_{J'K''}\,\delta(\Lambda'\Lambda''). \tag{9}$$

Expansion of the elements $g \in G$:

$$g = \sum_{\Lambda JK} Q^{\Lambda}_{JK}\,C^{\Lambda}_{JK}(g) \tag{10}$$

We will prove that

$$C^{\Lambda}(g) = \overline{A}^{t}\,\Lambda(g)\,B^{t} = \frac{|\Lambda|}{|G|}\,\{\Lambda^{R}(\overline{g})\,\overline{Q}^{\Lambda}(e)\}^{t} = \frac{|\Lambda|}{|G|}\,\{\overline{Q}^{\Lambda}(e)\,\Lambda^{L}(\overline{g})\}^{t}. \tag{11}$$

Multiply both sides of (3) from the left by $(A\,\Lambda^{*}(f)\,\overline{B})^{-1}_{KJ}$ and perform the sum over J and K.

$$\sum_{JK} Q^{\Lambda}_{JK}\,(A\,\Lambda^{*}(f)\,\overline{B})^{-1}_{KJ} = \frac{|\Lambda|}{|G|}\sum_{g \in G} \langle g|\Lambda\rangle^{*}\,gf$$

where $\langle g|\Lambda\rangle$ denotes the character of g in the IR Λ. Summing over Λ then yields

$$\sum_{\Lambda JK} Q^{\Lambda}_{JK}\,(A\Lambda^*(f)\,\bar{B})^{-1}_{KJ} = \frac{1}{|G|}\sum_{g\in G}\Big(\sum_{\Lambda}|\Lambda|\langle g|\Lambda\rangle\Big)^* gf.$$

Since the part between brackets on the right-hand side equals $|G|\,\delta(e, g)$ we find

$$f = \sum_{\Lambda JK} Q^{\Lambda}_{JK}\,(A\Lambda^*(f)\,\bar{B})^{-1}_{KJ}.$$

Comparison with (10) yields the desired result.

In particular, the resolution of the identity:

$$e = \sum_{\Lambda JK} Q^{\Lambda}_{JK}\,(B\bar{A})^{\Lambda}_{KJ} = \frac{|\Lambda|}{|G|}\sum_{\Lambda JK} Q^{\Lambda}_{JK}\,\bar{Q}^{\Lambda}_{KJ}\,(e). \tag{12}$$

Since subsequently we will discuss matrix elements over symmetry-adapted bases, we are interested in the hermitian adjoint operators as well.

$$(Q^{\Lambda}_{JK})^+ = \frac{|\Lambda|}{|G|}\sum_{g\in G}(A^*\Lambda(g)\,\bar{B}^*)_{JK}\,\bar{g}. \tag{13}$$

Transformation (substitute (10, 11) into (13)):

$$(Q^{\Lambda}_{JK})^+ = \sum_{M}\sum_{N}(AB^+)^{-1}_{KM}\,Q^{\Lambda}_{MN}\,(AB^+)^+_{NJ} \tag{14}$$

Multiplication rules:

$$\begin{aligned}(Q^{\Lambda'}_{J'K'})^+\,(Q^{\Lambda''}_{J''K''}) &= (AA^+)_{J''J'}\,\delta(\Lambda'\Lambda'')\sum_{K}(AB^+)^{-1}_{K'K}\,(Q^{\Lambda'}_{KK''}) = \\ &= (AA^+)_{J''J'}\,\delta(\Lambda'\Lambda'')\sum_{K}(A^*B^t)^{-1}_{K''K}\,(Q^{\Lambda'}_{KK'})^+\end{aligned} \tag{15}$$

$$\begin{aligned}(Q^{\Lambda'}_{J'K'})\,(Q^{\Lambda''}_{J''K''})^+ &= (BB^+)^{-1}_{K''K'}\,\delta(\Lambda'\Lambda'')\sum_{J}(Q^{\Lambda'}_{J'J})\,(AB^+)^+_{JJ''} = \\ &= (BB^+)^{-1}_{K''K'}\,\delta(\Lambda'\Lambda'')\sum_{J}(Q^{\Lambda'}_{J''J})^+\,(A^*B^t)^+_{JJ'}.\end{aligned} \tag{16}$$

In particular we define:

$$\begin{aligned}\sigma^{\Lambda}_{K'K''} \equiv \frac{(Q^{\Lambda}_{JK'})^+\,(Q^{\Lambda}_{JK''})}{(AA^+)_{JJ}} &= \sum_{K}(AB^+)^{-1}_{K'K}\,(Q^{\Lambda}_{KK''}) = \\ &= \frac{|\Lambda|}{|G|}\sum_{g\in G}(\bar{B}^+\Lambda^*(g)\,\bar{B})_{K'K''}\,g,\end{aligned} \tag{17}$$

from which

$$\sigma^{\Lambda}_{K'K''}(e) = \frac{|\Lambda|}{|G|}(\bar{B}^{+}\bar{B})^{\Lambda}_{K'K''} \quad \text{i.e.,} \quad (BB^{+})^{\Lambda} = \frac{|\Lambda|}{|G|}\bar{\sigma}^{\Lambda}(e) \tag{18}$$

and

$$(\sigma^{\Lambda}_{K'K''})^{+} = \sigma^{\Lambda}_{K''K'}$$

and

$$f\,\sigma^{\Lambda}_{K'K''} = \sum_{K} \Lambda^{R}_{KK'}(f)^{*}\,\sigma^{\Lambda}_{KK''}. \tag{19}$$

Also:

$$\tau^{\Lambda}_{J'J''} \equiv \frac{(Q^{\Lambda}_{J'K})(Q^{\Lambda}_{J''K})^{+}}{(\bar{B}^{+}\bar{B})_{KK}} = \sum_{J}(Q^{\Lambda}_{J'J})(AB^{+})^{+}_{JJ''} =$$

$$= \frac{|\Lambda|}{|G|}\sum_{g\in G}(A\Lambda^{*}(g)A^{+})_{J'J''}\,g, \tag{20}$$

from which

$$\tau^{\Lambda}_{J'J''}(e) = \frac{|\Lambda|}{|G|}(AA^{+})^{\Lambda}_{J'J''} \quad \text{i.e.} \quad (AA^{+})^{\Lambda} = \frac{|G|}{|\Lambda|}\tau^{\Lambda}(e) \tag{21}$$

and

$$(\tau^{\Lambda}_{J'J''})^{+} = \tau^{\Lambda}_{J''J'}$$

and

$$\tau^{\Lambda}_{J'J''}\,f = \sum_{J}\tau^{\Lambda}_{J'J}\,\Lambda^{L}_{J''J}(f)^{*}, \tag{22}$$

By means of (5),

$$(A\bar{B})^{\Lambda} = \frac{|G|}{|\Lambda|}Q^{\Lambda}(e),$$

so that from (18, 21),

$$(AB^{+})^{\Lambda} = Q^{\Lambda}(e)\,\bar{\sigma}^{\Lambda}(e), \quad (BA^{+})^{\Lambda} = \bar{Q}^{\Lambda}(e)\,\tau^{\Lambda}(e). \tag{18', 21'}$$

Therewith

$$(Q^{\Lambda'}_{J'K'})^{+}(Q^{\Lambda''}_{J''K''}) = \frac{|G|}{|\Lambda|}\tau^{\Lambda'}_{J''J'}(e)\,\delta(\Lambda'\Lambda'')\,\sigma^{\Lambda'}_{K'K''} \tag{15'}$$

and

$$(Q^{\Lambda'}_{J'K'})(Q^{\Lambda''}_{J''K''})^{+} = \frac{|G|}{|\Lambda|}\sigma^{\Lambda'}_{K''K'}(e)\,\delta(\Lambda'\Lambda'')\,\tau^{\Lambda'}_{J'J''}. \tag{16'}$$

The matrices $Q^\Lambda(e)$, $\sigma^\Lambda(e)$ and $\tau^\Lambda(e)$ will be called the structure matrices of the matric bases (3). They play an important role in all transformations: E.g.

$$(5) \to \Lambda^L(g) = Q^\Lambda(g)\,\bar{Q}^\Lambda(e),\quad \Lambda^R(g) = \bar{Q}^\Lambda(e)\,Q^\Lambda(g)$$

$$(11) \to C^\Lambda(g) = \frac{|\Lambda|}{|G|}\{\bar{Q}^\Lambda(e)\,Q^\Lambda(\bar{g})\,\bar{Q}^\Lambda(e)\}^t \qquad (11')$$

$$(14) \to \{Q^\Lambda_{JK}\}^+ = \sum_{MN} (\sigma^\Lambda(e)\,\bar{Q}^\Lambda(e))_{KM}\,\{Q^\Lambda_{MN}\}\,(\bar{Q}^\Lambda(e)\,\tau^\Lambda(e))_{NJ} \qquad (14')$$

Many interesting relations among the matrices $\Lambda^L(g)$, $\Lambda^R(g)$, $Q^\Lambda(g)$, $\sigma^\Lambda(g)$, $\tau^\Lambda(g)$ and $C^\Lambda(g)$ can be derived by means of the properties of the structure matrices $Q^\Lambda(e)$, $\sigma^\Lambda(e)$ and $\tau^\Lambda(e)$. In particular, the transformations between matric bases Q, Q^+, σ and τ are completely governed by the structure matrices. We will not treat this interesting part of the theory here.

We only note

$$\left.\begin{aligned} &\frac{|\Lambda|}{|G|}\sum_{g\in G} Q^\Lambda(g)^+\,Q^\Lambda(g) = \sigma^\Lambda(e)\,\mathrm{Sp}\,[\tau^\Lambda(e)] \\ &\frac{|\Lambda|}{|G|}\sum_{g\in G} Q^\Lambda(g)\,Q^\Lambda(g)^+ = \tau^\Lambda(e)\,\mathrm{Sp}\,[\sigma^\Lambda(e)] \end{aligned}\right\} \qquad (22')$$

from which it follows that the structure matrices are directly connected with the invariant means of skew irreducibele matrix representations.

In the way discussed above a variety of matric bases can be constructed. Which basis we choose to perform the symmetry adaptation of space V to group G is principally irrelevant. However subsequently matrix elements of the Hamilton operator over symmetry-adapted bases must be evaluated and there we have to decide which bases can be handled most efficiently.

2.2.3 Some special choices of matric bases, $G = S_N$.

The two unitary matric bases well-known and used in N-particle theory are: (23)

1. Young-Yamanouchi orthogonal (= real unitary).

   Eq. (3) with $\Lambda(g)$ unitary, $A = B$ unitary, and $A\,\Lambda(g)\,\bar{A}$ is sequence adapted to the chain of subgroups

   $$S_N \supset S_{N-1} \supset S_{N-2} \ldots \supset S_2 \supset S_1.$$

2. Jahn-Serber orthogonal.

   Eq. (3) with $\Lambda(g)$ unitary, $A = B$ unitary, and $A\Lambda(g)\,\bar{A}$ is sequence adapted to the chain of subgroups

   $$S_N \supset S_{N-2} \supset S_{N-4} \ldots \supset S_2(S_1).$$

For these two cases the basic expressions given above simplify according to

$$\frac{|G|}{|\Lambda|} Q^{\Lambda}(e) = A\bar{B} = AB^{+} = A^{*}B^{t} = AA^{+} = BB^{+} = \Lambda(e).$$

Non-orthogonal bases used in N-particle theory are: (24)

1. Young operators type NP: Y^{Λ}_{JK}
2. Young operators type PN: $(Y^{\Lambda}_{JK})^{+}$
3. Sandwich operators type PNP = σ; $(Y^{\Lambda}_{JK'})^{+}(Y^{\Lambda}_{JK''})$
4. Sandwich operators type NPN = τ; $(Y^{\Lambda}_{J'K})(Y^{\Lambda}_{J''K})^{+}$,

collectively called Tableau operators. (App. C). They can be specified from (5) by Young's natural representations for Λ^{L} and Λ^{R} and a structure matrix for $Q^{\Lambda}(e)$.

2.2.4 Symmetry adaptation of $V_n \otimes^N$ to S_N.

Let 1-particle orbital (or spin) space V_n have a basis $\{\phi_i / i = 1, 2, \ldots n\}$. Then the basis of $V_n \otimes^N$ consists of all n^N products

$$\{\phi_{i_1}\phi_{i_2} \ldots \phi_{i_N} \mid i_k \in \{1, 2, \ldots n\}\ \forall k = 1, 2, \ldots N\}.$$

Any basis vector can be represented by an index set $(i, i_2 \ldots i_N)$.
Any index set can be generated from an ordered index set $I = (i_1 \leqslant i_2 \leqslant \ldots \leqslant i_N)$ by applying $g \in G = S_N$.
Let the basisfunction pertaining to the ordered index set I be the denoted by ϕ_I. Then:
The set

$$\{g\phi_I \mid g \in G\}$$

spans a stable subspace $V(I) \subset V_n \otimes^N$ and obviously, linear dependencies that occur after projection can only occur within each such subspace.
For the choices mentioned above (2.2.3): If there exist J, K such that $Q^{\Lambda}_{JK}\phi_I \neq 0$ the set

$$\{Q^{\Lambda}_{JK}\phi_I \mid J = 1, 2, \ldots |\Lambda|\}$$

is a basis (the K-th) for the IR Λ.
All bases (ranging K), for which the index tableau I^{Λ}_{K} pertains to the ordered set I and the Young tableau T^{Λ}_{K}, are linearly independent.
Hence: A symmetry-adapted basis for $V_n \otimes^N$ is given by: (25)

$\{Q^{\Lambda}_{JK}\phi_I \mid \Lambda$ ranging over the IR of G,

$J = 1, 2, \ldots |\Lambda|$,

I ranging over ordered index sets,

K ranging over Young tableaux T^{Λ}_{K} such that the index tableau I^{Λ}_{K} is standard$\}$.

Again consider any ordered N-particle orbital product function

$$\phi = \phi_{i_1}\phi_{i_2} \ldots \phi_{i_N}, \quad i_1 \leqslant i_2 \leqslant \ldots \leqslant i_N.$$

Group theoretically we are not interested in the actual labelling $i_1, i_2 \ldots i_N$ but only in how many times each label occurs. Therefore we rewrite ϕ as

$$\phi = \ldots \cdot \underbrace{\phi_1 \ldots \phi_1}_{j} \; \underbrace{\phi_2 \ldots \phi_2}_{j} \; \cdots \; \underbrace{\ldots \phi_{\lambda_j} \ldots \phi_{\lambda_j}}_{j} \cdot \ldots \quad \text{times}$$

$$= \ldots \cdot \phi_1^j \phi_2^j \ldots \phi_{\lambda_j}^j \cdot \ldots = \ldots \phi_\lambda^j \ldots$$

$$\phi = \ldots \phi_\lambda^j \ldots$$

with $j = 1, 2, \ldots N$, $\lambda = 1, 2, \ldots \lambda_j$ defines an N-particle orbital product which contains λ_j different orbitals each occurring j times, $j = 1, 2, \ldots$.
The number j is called the occupation number of the orbitals ϕ_λ, $\lambda = 1, 2, \ldots \lambda_j$. In this way each ϕ is characterized by a partition of N namely $(\lambda) = \ldots j^{\lambda_j} \ldots$, $\sum_j j \lambda_j = N$. We write $\phi = \phi(\lambda)$.
The invariance group of $\phi(\lambda)$ is defined by

$$H^\lambda = \{g \in S_N \mid g\phi(\lambda) = \phi(\lambda)\}.$$

Thus

$$H^\lambda = \underset{j}{\mathsf{X}} \; \overset{\lambda_j}{\underset{\lambda}{\mathsf{X}}} \; S_{j_\lambda} \tag{26}$$

the direct product of symmetric groups S_{j_λ} $j = 1, 2, \ldots, \lambda = 1, 2, \ldots \lambda_j$ where S_{j_λ} acts on the particles that occupy the λ-th orbital with occupation number j.
By definition $\phi(\lambda)$ transforms according to the trivial 1-dimensional IR of its invariance group. Henceforth these functions will therefore be denoted by $|\phi; \lambda_t\rangle$. The number of linearly independent projections of given final symmetry ΛJ originatin from $|\phi; \lambda_t\rangle$ is given by $\langle \Lambda | \lambda_t \uparrow \rangle = \langle \lambda_t | \Lambda \downarrow \rangle$ i.e. the number of times the IR Λ occurs in the induced representation $\lambda_t \uparrow$.
A particularly convenient matric basis is then of the sequence-adapted type

$$Q(J \Lambda k \lambda_t) = \frac{|\Lambda|}{|G|} \sum_{g \in G} \langle g | k \lambda_t \Lambda J \rangle \, \bar{g}$$

$$\Lambda J \text{ fixed}; \; k = 1, 2, \ldots \langle \Lambda | \lambda_t \uparrow \rangle.$$

We define

$$Q(J \Lambda k \lambda_t) \, |\phi; \lambda_t\rangle = |\phi(k \lambda_t); \Lambda J\rangle \tag{27}$$

in accordance with (2).

Note: $\langle \Lambda | \lambda_t \uparrow \rangle = 0 \quad \forall \Lambda < \lambda$ (28)

where $\Lambda = \ldots j^{\Lambda_j} \ldots < \lambda = \ldots j^{\lambda_j} \ldots$ if the first non-vanishing difference $\Lambda_j - \lambda_j$, starting with the highest j-value, is negative.

With $\lambda > \Lambda$ there is at least one occupation number j the corresponding index of which, say λ, is distributed over at least two rows in Λ such that there is at least one column in which it occurs twice. Thus no standard index tableau exists.

From the characterization of SA bases for $V_n \otimes^N$ by means of standard index tableaux it follows that:

$V_n \otimes^N$ carries IR Λ of S_N with at most n-rows.

Suppose this is not true and consider the first column of a standard index tableau pertaining to the Young diagram with n + k rows. The numbers in this column should increase whereas we only have 1,2, . . . n at our disposal. Thus ↯ .

In particular:

$V_{2s+1} \otimes^N$ carries IR Λ of S_N with at most 2s + 1 rows.

In combination with the Pauli (fermion) principle we already inferred that the physically allowed subspace of $V_n \otimes^N$ then carries IR $\tilde{\Lambda}$ of S_N with at most 2s + 1 columns.

Together with (28) we arrive at the *Occupation number exclusion principle*: Any spatial orbital can accommodate at most 2s + 1 spin-s-fermions. (29)

2.2.5 Antisymmetrization

The construction of antisymmetric N-particle functions is a coupling problem. Namely one has to couple SA bases for orbital space $V_n^\rho \otimes^N$ to SA bases for spin space $V_t^\sigma \otimes^N$ in such a way that the resulting combinations are antisymmetric. If, for the moment, we omit all non group-theoretical labels, and we denote by $|(\Lambda^\rho)\, 1^N\rangle$ any antisymmetric N-particle function with orbital transformation properties given by $\Lambda^\rho \in S_N^\rho$, then:

$$|(\Lambda^\rho)\, 1^N\rangle = \sum_J \sum_{\Delta K} |\Lambda K\rangle^\rho\, |\Delta K\rangle^\sigma \langle \Lambda J \otimes \Delta K|\, 1^N\rangle$$

where the symbols $\langle \Lambda J \otimes \Delta K|\, 1^N\rangle$ denote the appropriate coupling coefficients.

Since the antisymmetric IR $[1^N]$ only occurs in the product of associate IR Λ and $\tilde{\Lambda}$, in which products it occurs once only, we have

$$|(\Lambda^\rho)\, 1^N\rangle = \sum_{JK} |\Lambda J\rangle^\rho\, |\tilde{\Lambda}\tilde{K}\rangle^\sigma \langle \Lambda J \otimes \tilde{\Lambda}\tilde{K}|\, 1^N\rangle \qquad (30)$$

The value of the coupling coefficients depends on the actual definition of the irreducible matrix representations, i.e., on the explicit form of the operators which have been used to SA the two spaces V^ρ and V^σ.

Since an antisymmetric function must change sign if any two particles are permuted, it follows easily that the coupling matrix

$$C^\Lambda \text{ with elements } \quad C^\Lambda_{J\tilde{K}} = \langle \Lambda J \otimes \tilde{\Lambda}\tilde{K}|\, 1^N\rangle$$

should satisfy

$$(-)^g C^\Lambda = D^\Lambda(g)\, C^\Lambda D^{\tilde\Lambda}(g)^t \quad \forall g \in G$$

where $D^\Lambda(g)$ and $D^{\tilde\Lambda}(g)$ are the matrix representations afforded by the bases $\{|\Lambda J\rangle\}$ and $\{|\tilde\Lambda\tilde K\rangle\}$ of the IR Λ and $\tilde\Lambda$, respectively.

When we use matric bases (3) to perform the SA of the spaces V^ρ and V^σ, then it follows from (2, 4, 7) that $D^\Lambda(g) = (A^\Lambda \Lambda^*(g)\, \bar A^\Lambda)^t$ and $D^{\tilde\Lambda}(g)^t = A^{\tilde\Lambda} \tilde\Lambda^t(g)\, \bar A^{\tilde\Lambda}$ where $\Lambda(g)$ and $\tilde\Lambda(g)$ are associate unitary matrix representations.

We *always* choose $\Lambda(g)$ in the Young-Yamanouchi orthogonal matrix representation,

$$\Lambda(g) = \Lambda^t(\bar g) = \Lambda^*(g) \quad \text{and} \quad (-)^g \tilde\Lambda(g) = \mu^\Lambda \Lambda(g)\, \bar\mu^\Lambda \tag{31}$$

where μ^Λ is a matrix with elements $\mu^\Lambda_{JK} = \delta(JK)\,(-)^{\mu^\Lambda_J}$.

The numbers μ^Λ_J depend on the labels J which, as we recall from (2.2.3), stand for a series of IR from the groups in the chain $S_{N-1} \supset S_{N-2} \supset \ldots \supset S_1$.

By means of (31):

$$(A^{\Lambda^t} C^\Lambda A^{\tilde\Lambda} \mu^\Lambda)\, \Lambda(g)\, (\bar\mu^\Lambda \bar A^{\tilde\Lambda} \bar C^\Lambda \bar A^{\Lambda^t}) = \Lambda(g).$$

Application of Schur's theorem yields:

$$\bar\mu^\Lambda \bar A^{\tilde\Lambda} \bar C^\Lambda \bar A^{\Lambda^t} \sim \Lambda(e) \quad \text{i.e.,} \quad \bar C^\Lambda = A^{\tilde\Lambda} \mu^\Lambda A^{\Lambda^t}.$$

We choose the proportionality constant equal to 1 and define the coupling matrix by

$$\bar C^\Lambda = A^{\tilde\Lambda} \mu^\Lambda A^{\Lambda^t}. \tag{32}$$

In the following section we will frequently consider antisymmetric functions which are obtained from spin-adapted functions

$$\phi Q^\Lambda_{JK} \theta = [(E)^\rho \otimes (Q^\Lambda_{JK})^\sigma](\phi\theta), \quad \phi \in V^\rho,\ \theta \in V^\sigma, \tag{33}$$

by the action of the antisymmetrizer

$$Q^{\rho\sigma}(1^N) = \frac{1}{|G|} \sum_{g \in G} g^\rho (-)^g g^\sigma. \tag{34}$$

Therefore we are interested in the operator $Q^{\rho\sigma}(1^N)$ and in the operator products

$$Q^{\rho\sigma}(1^N)\,[(E)^\rho \otimes (Q^\Lambda_{JK})^\sigma].$$

First, with Q^Λ_{JK} defined by (3),

$$Q^{\tilde\Lambda}_{\tilde J\tilde K} = \frac{|\Lambda|}{|G|} \sum_{g \in G} (A^{\tilde\Lambda} \tilde\Lambda^*(g)\, \bar B^{\tilde\Lambda})_{\tilde J\tilde K}\, g$$

and

$$P^\Lambda_{JK} = \sum_L Q^\Lambda_{JL} \bar Q^\Lambda_{LK}(e); \quad P^{\tilde\Lambda}_{\tilde J\tilde K} = \sum_L Q^{\tilde\Lambda}_{\tilde J\tilde K} \bar Q^{\tilde\Lambda}_{\tilde L\tilde K}(e),$$

we derive:

$$\left.\begin{aligned} |G|^2\, Q^{\rho\sigma}(1^N) &= \sum_{\Lambda} |\Lambda| \sum_{JK} (C^{\Lambda} P^{\widetilde{\Lambda}} \overline{C}^{\Lambda})^{\rho}_{JK} (P^{\Lambda}_{JK})^{\sigma} = \\ &= \sum_{\Lambda} |\Lambda| \sum_{JK} (C^{\Lambda} P^{\widetilde{\Lambda}})^{\rho}_{J\widetilde{K}} (P^{\Lambda} \overline{C}^{\Lambda^t})^{\sigma}_{J\widetilde{K}} = \\ &= \sum_{\Lambda} |\Lambda| \sum_{JK} (P^{\widetilde{\Lambda}}_{\widetilde{J}\widetilde{K}})^{\rho} (C^{\Lambda^t} P^{\Lambda} \overline{C}^{\Lambda^t})^{\sigma}_{\widetilde{J}\widetilde{K}}. \end{aligned}\right\} \qquad (35)$$

Substitute (11) for g^{σ} into (34). This yields

$$|G|\, Q^{\rho\sigma}(1^N) = \sum_{g} \sum_{\Lambda JK} ((-)^g (\overline{A}^{\Lambda^t} \Lambda(g)\, B^{\Lambda^t})_{JK}\, g)^{\rho} (Q^{\Lambda}_{JK})^{\sigma}.$$

Replace $(-)^g \Lambda(g)$ by $\overline{\mu}^{\Lambda} \widetilde{\Lambda}(g) \mu^{\Lambda}$. Insert $\overline{A}^{\widetilde{\Lambda}} A^{\widetilde{\Lambda}}$ and $\overline{B}^{\widetilde{\Lambda}} B^{\widetilde{\Lambda}}$ in the appropriate positions and use the definition of $Q^{\widetilde{\Lambda}}_{\widetilde{J}\widetilde{K}}$. This yields

$$Q^{\rho\sigma}(1^N) = \sum_{\Lambda JK} \frac{1}{|\Lambda|} \sum_{ST} (\overline{A}^{\Lambda^t} \overline{\mu}^{\Lambda} \overline{A}^{\widetilde{\Lambda}})_{J\widetilde{S}} (Q^{\widetilde{\Lambda}}_{\widetilde{S}\widetilde{T}})^{\rho} (B^{\widetilde{\Lambda}} \mu^{\Lambda} B^{\Lambda^t})_{\widetilde{T}\widetilde{K}} (Q^{\Lambda}_{JK})^{\sigma}.$$

With $\overline{A}^{\Lambda^t} \overline{\mu}^{\Lambda} \overline{A}^{\widetilde{\Lambda}} = C^{\Lambda}$ and $B^{\widetilde{\Lambda}} \mu^{\Lambda} B^{\Lambda^t} = B^{\widetilde{\Lambda}} \overline{A}^{\widetilde{\Lambda}} A^{\widetilde{\Lambda}} \mu^{\Lambda} A^{\Lambda^t} \overline{A}^{\Lambda^t} B^{\Lambda^t} =$

$$= (B\overline{A})^{\widetilde{\Lambda}} \overline{C}^{\Lambda} (B\overline{A})^{\Lambda^t} = \left(\frac{|\Lambda|}{|G|}\right)^2 \overline{Q}^{\widetilde{\Lambda}}(e)\, \overline{C}^{\Lambda}\, \overline{Q}^{\Lambda}(e)^t$$

where we used (4) to arrive at the last line, we get the desired result after introduction of the operators P^{Λ}_{JK} defined above.

Secondly,

$$|G|\, Q^{\rho\sigma}(1^N)\, [(E)^{\rho} \otimes (Q^{\Lambda}_{ST})^{\sigma}] = \sum_{J} (C^{\Lambda} P^{\widetilde{\Lambda}} \overline{C}^{\Lambda})^{\rho}_{JS} (Q^{\Lambda}_{JT})^{\sigma} \qquad (36)$$

is then easily derived.

Two special cases will be treated by way of an example as well as for future reference.

A. Young-Yamanouchi orthogonal matric bases

Here $\Lambda(g)$ is the Young-Yamanouchi matrix which represents $g \in G = S_N$ in the IR Λ.

$$A^{\Lambda} = \Lambda(e) \;\; \forall \Lambda, \quad \text{i.e.,} \quad C^{\Lambda} = \overline{\mu}^{\Lambda}$$

$$\overline{Q}^{\Lambda}(e) = \frac{|G|}{|\Lambda|} \Lambda(e), \quad \text{i.e.,} \quad P^{\widetilde{\Lambda}} = \frac{|G|}{|\Lambda|} Q^{\widetilde{\Lambda}}.$$

Hence:

$$Q^{\rho\sigma}(1^N)\, [E^{\rho} \otimes Q^{\sigma}(S\Lambda T)] = \frac{(-)^{\mu^{\Lambda}_S}}{|\Lambda|} \sum_{J} Q^{\rho}(\widetilde{J}\widetilde{\Lambda}\widetilde{S})\, (-)^{\mu^{\Lambda}_J}\, Q^{\sigma}(J\Lambda T). \qquad (37)$$

B. Young unit type NP *matric bases (App. C).*

From (1, 5)

$$Y^{\Lambda}(g) = \Lambda^L(g)\, Y^{\Lambda}(e).$$

Thus,

$$\Lambda^L(g) = Y^{\Lambda}(g)\, \overline{Y}^{\Lambda}(e)$$

and similarly

$$\tilde{\Lambda}^L(g) = Y^{\tilde{\Lambda}}(g)\, \overline{Y}^{\tilde{\Lambda}}(e).$$

Further,

$$(-)^g\, Y^{\Lambda}_{JK}(\overline{g}) = (-)^{\phi_J}\, Y^{\tilde{\Lambda}}_{\overline{K}\overline{J}}(g)\, (-)^{\phi_K},\ \phi_J, \phi_K \text{ phases.}$$

This is

$$(-)^g\, Y^{\Lambda}(\overline{g}) = (\phi\, Y^{\tilde{\Lambda}}(g)\, \overline{\phi})^t$$

and in particular

$$Y^{\Lambda}(e) = (\phi\, Y^{\tilde{\Lambda}}(e)\, \overline{\phi})^t$$

from which

$$(-)^g\, \tilde{\Lambda}^L(g) = (\phi\, Y^{\Lambda}(e)^t)\, \Lambda^L(\overline{g})^t\, (\phi\, Y^{\Lambda}(e)^t)^{-1} \tag{38}$$

is then easily derived.

By definition (4)

$$\Lambda^L(g) = A^{\Lambda}\, \Lambda^*(g)\, \overline{A}^{\Lambda},\quad \tilde{\Lambda}^L(g) = A^{\tilde{\Lambda}}\, \tilde{\Lambda}^*(g)\, \overline{A}^{\tilde{\Lambda}}$$

and by choice (31)

$$\Lambda^*(g) = \Lambda^t(\overline{g}) = \Lambda(g) \quad \text{and} \quad (-)^g\, \tilde{\Lambda}(g) = \mu^{\Lambda}\, \Lambda(g)\, \overline{\mu}^{\Lambda}.$$

One then easily shows that

$$\tilde{\Lambda}^L(g) = \overline{C}^{\Lambda}(\phi\, Y^{\Lambda}(e)^t)^{-1}\, \tilde{\Lambda}^L(g)\, (\phi\, Y^{\Lambda}(e)^t)\, C^{\Lambda}$$

with $\overline{C}^{\Lambda} = A^{\tilde{\Lambda}}\mu^{\Lambda}A^{\Lambda^t}$ the coupling matrix (32). Then, according to Schur's theorem

$$\phi\, Y^{\Lambda}(e)^t\, C^{\Lambda} \sim \Lambda(e).$$

We choose the proportionality constant equal to 1 so that

$$\overline{C}^{\Lambda} = \phi\, Y^{\Lambda}(e)^t = Y^{\tilde{\Lambda}}(e)\, \phi \tag{39}$$

from which

$$C^{\Lambda}\, P^{\tilde{\Lambda}}\, \overline{C}^{\Lambda} = \overline{Y}^{\Lambda}(e)^t\, \overline{\phi}\, Y^{\tilde{\Lambda}}\, \phi.$$

Substitution into (36) yields:

$$Q^{\rho\sigma}(1^N)\,[(E)^\rho \otimes (Y^\Lambda_{ST})^\sigma] = \frac{(-)^{\phi^\Lambda_S}}{|G|} \sum_K (Y^{\tilde{\Lambda}}_{\tilde{K}\tilde{S}})^\rho\,(-)^{\phi^\Lambda_K}\Big(\sum_J \bar{Y}^\Lambda_{KJ}(e)\,Y^\Lambda_{JT}\Big)^\sigma. \qquad (40)$$

Matrix elements of the Hamilton operator H over antisymmetrized spin-adapted N-particle functions will contain the operator product

$$\{Q^{\rho\sigma}(1^N)\,[(E)^\rho \otimes (Q^\Lambda_{ST})^\sigma]\}^+\,H\,\{Q^{\rho\sigma}(1^N)\,[(E)^\rho \otimes (Q^\Delta_{MN})^\sigma]\}. \qquad (41)$$

According to the discussion in 2.1.1 the Hamilton operator is totally symmetric. Thus the antisymmetrizer commutes with H. Since the antisymmetrizer is hermitian as well as idempotent (41) can be written

$$\{[(E)^\rho \otimes (Q^\Lambda_{ST})^\sigma]\}^+\,H\,\{Q^{\rho\sigma}(1^N)\,[(E)^\rho \otimes (Q^\Delta_{MN})]\}. \qquad (42)$$

From here on one has to distinguish between an operator H which is spinfree or not. Our Hamilton operator is spinfree by choice with the advantage that the spin-space operators Q^σ commute with H. Thus the previous expression reads

$$|G| \sum_J \{(E)\,H(C^\Lambda P^{\tilde{\Lambda}} \bar{C}^\Lambda)_{JM}\}^\rho\,\{(Q^\Lambda_{ST})^+\,(Q^\Delta_{MN})\}^\sigma.$$

Application of (15, 18) yields

$$|G|\,\delta(\Lambda\Delta)\,H\,\{(\tau^\Lambda(e)^t C^\Lambda P^{\tilde{\Lambda}} \bar{C}^\Lambda)_{SM}\}^\rho\,\{\sigma^\Lambda_{TN}\}^\sigma. \qquad (43)$$

Special cases:

A. Young Yamanouchi

In this case $C^\Lambda = \mu^\Lambda$, $P^{\tilde{\Lambda}} = Q^{\tilde{\Lambda}}$ and $\tau^\Lambda(e) = \Lambda(e)$
so that (43) reads

$$\delta(\Lambda\Delta)\,H(-)^{\mu^\Lambda_S + \mu^\Lambda_M}\,Q^\rho(\tilde{S}\tilde{\Lambda}\tilde{M})\,Q^\sigma(T\Lambda N). \qquad (44)$$

B. Young unit type NP

In this case

$$C^\Lambda P^{\tilde{\Lambda}} \bar{C}^\Lambda = \bar{Y}^\Lambda(e)^t\,\phi\,Y^{\tilde{\Lambda}}\phi$$

so that

$$\tau^\Lambda(e)^t C^\Lambda P^{\tilde{\Lambda}} \bar{C}^\Lambda = \{\bar{Y}^\Lambda(e)\,\tau^\Lambda(e)\}^t\,\phi\,Y^{\tilde{\Lambda}}\phi.$$

From (18′, 21′)

$$\bar{Y}^\Lambda(e)\,\tau^\Lambda(e) = BA^t)^\Lambda, \quad Y^\Lambda(e)\,\bar{\sigma}^\Lambda(e) = (AB^t)^\Lambda$$

so that

$$\{\bar{Y}^{\Lambda}(e)\,\tau^{\Lambda}(e)\}^{t} = Y^{\Lambda}(e)\,\bar{\sigma}^{\Lambda}(e).$$

One can prove that

$$\bar{Y}^{\tilde{\Lambda}}(e)\,\tau^{\tilde{\Lambda}}(e) = \phi\,\bar{\tau}^{\Lambda}(e)\,Y^{\Lambda}(e)\,\phi \quad \text{and}$$

$$Y^{\tilde{\Lambda}}(e)\,\bar{\sigma}^{\tilde{\Lambda}}(e) = \phi\,\sigma^{\Lambda}(e)\,\bar{Y}^{\Lambda}(e)\,\phi.$$

Therewith:

$$\tau^{\Lambda}(e)^{t}\,C^{\Lambda}\,P^{\tilde{\Lambda}}\,\bar{C}^{\Lambda} = \{\bar{Y}^{\Lambda}(e)\,\tau^{\Lambda}(e)\}^{t}\,\phi\,Y^{\tilde{\Lambda}}\phi = Y^{\Lambda}(e)\,\bar{\sigma}^{\Lambda}(e)\,\phi\,Y^{\tilde{\Lambda}}\,\phi =$$

$$= \phi\,\sigma^{\tilde{\Lambda}}(e)\,\bar{Y}^{\tilde{\Lambda}}(e)\,Y^{\tilde{\Lambda}}\phi = \phi(A^{\tilde{\Lambda}}B^{\tilde{\Lambda}^{+}})^{-1}\,Y^{\tilde{\Lambda}}\phi = \phi\,\sigma^{\tilde{\Lambda}}\,\phi$$

where the last line follows by means of (15, 17). Thus (43) reads

$$\delta(\Lambda\Delta)\,H(-)^{\phi^{\Lambda}_{S}+\phi^{\Lambda}_{M}}\,(\sigma^{\tilde{\Lambda}}_{\tilde{S}\tilde{M}})^{\rho}\,(\sigma^{\Lambda}_{TN})^{\sigma}. \tag{45}$$

From the above expressions (41–45) it follows that the combination of a spin-free operator and the Pauli principle leads to a spin-free formulation.

Explictly: starting from operator products with definite spin symmetry $(\Lambda\,S)^{\sigma}$ on the left and $(\Delta\,M)^{\sigma}$ on the right, the resulting operator exhibits $(\tilde{\Lambda}\,\tilde{S})^{\rho}$ and $(\tilde{\Lambda}\tilde{M})^{\rho}$ orbital symmetry on the left and right, respectively. Thus, spin transformation properties have been transformed to orbital transformation properties. This possibility arises from the fact that the operator H has been chosen spin-free.

Appendix B

Sequence adaptation. Double coset decomposition of unitary matric bases

Consider a group G and any two subgroups $H^{\alpha}, H^{\beta} \subset G$.

Let $\Lambda, \Lambda', \Lambda'', \Delta, \Delta', \Delta'', \ldots$ denote the IR of G.

$\lambda^{\alpha}, \mu^{\alpha}, \ldots, \lambda^{\beta}, \mu^{\beta}, \ldots$ denote the IR of H^{α}, H^{β}.

Let $\Lambda(g)$ denote unitary matrices representing $g \in G$ in Λ.

The matrix elements will be denoted by

$$\langle g|J\,\Lambda K\rangle = \langle\bar{g}|K\,\Lambda\,J\rangle^{*}.$$

Consider subduction $G\downarrow H$, H either H^{α} or H^{β}.

Then

$$\Lambda\downarrow = \bigoplus_{\lambda}\,\lambda\,\langle\lambda|\Lambda\downarrow\rangle$$

i.e., if the IR Λ of G is restricted to $H \subset G$ then the representation

$$\Lambda\downarrow = \{\Lambda(h)\,|\,h \in H\}$$

is a representation of H which is reducible in general.

It decomposes into IR λ of H each of which occurs with multiplicity $\langle\lambda|\Lambda\downarrow\rangle$ given by

$$\langle\lambda|\Lambda\downarrow\rangle = \frac{1}{|H|}\sum_{h\in H}\langle h|\lambda\rangle\langle h|\Lambda\rangle^*$$

where $\langle h|\lambda\rangle$ and $\langle h|\Lambda\rangle$ denote the character of $h\in H$ in λ of H and in Λ of G respectively.

The decomposition given above implies the existence of a unitary matrix U (which depends on Λ and H), such that:

$$U\Lambda(h)U^+ = \bigoplus_{\lambda}\lambda(h)\langle\lambda|\Lambda\downarrow\rangle.$$

I.e.: The matrix $U\Lambda(h)U^+$ is block diagonalized. There occur $\langle\lambda|\Lambda\downarrow\rangle$ identical blocks $\lambda(h)$ along the diagonal.

Rows and columns of $U\Lambda(h)U^+$ are most naturally labelled by a triplet of indices $(t\lambda j)$ where

λ refers to the IR of H that occur in $\Lambda\downarrow$,

t counts the number of times λ occurs, $t = 1, 2, \ldots\langle\lambda|\lambda\downarrow\rangle$,

$j = 1, 2, \ldots|\lambda|$ indicates a row/column within the diagonal block specified by λ and t.

Definition: A sequence-adapted irreducible representation (SAIR) is an irreducible matrix representation which elements

$$\langle g|t\lambda j\,\Lambda\, n\mu m\rangle \quad \lambda,\mu \text{ ranging} \quad \begin{array}{ll} t=1,2,\ldots\langle\lambda|\Lambda\downarrow\rangle, & n=1,2,\ldots\langle\mu|\Lambda\downarrow\rangle \\ j=1,2,\ldots|\lambda|, & m=1,2,\ldots|\mu|, \end{array}$$

such that

$$\langle h|t\lambda j\,\Lambda\, n\mu m\rangle = \delta(tn)\,\delta(\lambda\mu)\,\langle h|j\mu m\rangle \quad \forall h\in H.$$

Thus: A matrix representation $\Lambda(g)$ is $H\subset G$-sequence-adapted if $\Lambda(h)\ \forall h\in H$ is completely decomposed.

More general: With U^α the matrix which accomplishes sequence adaptation of $\Lambda(g)$ with respect to $H^\alpha\subset G$, and similarly U^β for $H^\beta\subset G$, we consider

$$\langle g|(t\lambda j)^\alpha\,\Lambda\,(n\mu m)^\beta\rangle$$

elements of the matrix $U^\alpha\Lambda(g)U^{\beta^+}$.

These matrices do not longer constitute an ordinary representation. They constitute a skew representation with metric $(U^\alpha U^{\beta^+})^{-1}$, i.e.,

$$U^\alpha\Lambda(g')U^{\beta^+}(U^\alpha U^{\beta^+})^{-1}U^\alpha\Lambda(g'')U^{\beta^+} = U^\alpha\Lambda(g'g'')U^{\beta^+}.$$

The matrix $U^\beta U^{\alpha^+}$ is properly called a recoupling matrix since it governs the transformation between SAIR defined with respect to H^α and the SAIR defined with respect to H^β, namely

$$(U^\beta U^{\alpha^+})\, U^\alpha \Lambda(g)\, U^{\alpha^+} (U^\alpha U^{\beta^+}) = U^\beta \Lambda(g)\, U^{\beta^+}.$$

The skew representations so defined possess all properties of the ordinary unitary representations.

In particular: Let

$$\Lambda^\alpha(g) = U^\alpha \Lambda(g)\, U^{\alpha^+}, \quad \Lambda^{\alpha\beta}(g) = U^\alpha \Lambda(g)\, U^{\beta^+}$$

$$\Lambda^\beta(g) = U^\beta \Lambda(g)\, U^{\beta^+}, \quad \Lambda^{\beta\alpha}(g) = U^\beta \Lambda(g)\, U^{\alpha^+}.$$

Then:

1. $\{\Lambda^{\alpha\beta}(g)\}^{-1} = \{\Lambda^{\alpha\beta}(g)\}^+ = \Lambda^{\beta\alpha}(\overline{g})$

   $\Lambda^{\alpha\beta}(\overline{g}) = \{\Lambda^{\beta\alpha}(g)\}^+ = \{\Lambda^{\beta\alpha}(g)\}^{-1}$

2. $\Lambda^{\alpha\beta}(g'gg'') = \Lambda^\alpha(g')\, \Lambda^{\alpha\beta}(g)\, \Lambda^\beta(g'')$

   $\Lambda^{\beta\alpha}(g''gg') = \Lambda^\beta(g'')\, \Lambda^{\beta\alpha}(g)\, \Lambda^\beta(g')$

   which relations are obviously of interest for $g' \in H^\alpha$, $g'' \in H^\beta$ in which case the matrices $\Lambda^\alpha(g')$ and $\Lambda^\beta(g'')$ are block diagonalized.

3. With $A^\alpha, B^\beta, C^\gamma$ and D^δ denoting sequence-adapted labels with respect to $H^\alpha, H^\beta, H^\gamma, H^\delta \subset G$ and $g', g'' \in G$ any two fixed elements, the following column-orthogonality theorem holds:

$$|\Lambda| \sum_{g \in G} \langle g | A^\alpha \Lambda B^\beta \rangle \langle g'gg'' | C^\gamma \Delta D^\delta \rangle^* = |G|\, \delta(\Lambda\Delta)\, \langle \overline{g}' | A^\alpha \Lambda C^\gamma \rangle \langle g'' | B^\beta \Lambda D^\delta \rangle^*.$$

The corresponding row-orthogonality theorem:

$$\sum_\Lambda |\Lambda| \sum_A \sum_B \langle g' | A^\alpha \Lambda B^\beta \rangle \langle g'' | A^\alpha \Lambda B^\beta \rangle^* = |G|\, \delta(g', g'').$$

The matric bases corresponding to the above defined SAIR are given by

$$Q(k'\lambda'j'\Lambda k''\lambda''j'') = \frac{|\Lambda|}{|G|} \sum_{g \in G} \langle g | k'\lambda'j'\Lambda k''\lambda''j'' \rangle^*\, g =$$

$$= \frac{|\Lambda|}{|G|} \sum_{g \in G} \langle \overline{g} | k''\lambda''j''\Lambda k'\lambda'j' \rangle\, g.$$

with $(k'\lambda'j')$ referring to $H' \subset G$, $(k''\lambda''j'')$ to $H'' \subset G$.

For the symmetric group $G = S_N$ one can always choose real orthogonal matrices. We now briefly discuss the double coset (DC) decomposition of this type of operator (See App. D).

Consider the DC decomposition of G with respect to H′ and H″ on the left and right, respectively. Any element $g \in G$ is written $h'Zh''$, $Z \in G$ called the generator. A complete summation over all h′ and h″ yields the DC $H'ZH''$ and each element $g \in H'ZH''$ will occur with a certain frequency given by $d_Z = |H' \cap ZH''\bar{Z}|$.

The quantity $l_Z = |H'||H''|\bar{d}_Z$ will be called the length of the DC. Then:

$$Q(k'\lambda'j'\Lambda k''\lambda''j'') = \frac{|\Lambda|}{|G|} \sum_Z \bar{d}_Z \sum_{h'} \sum_{h''} \langle h'Zh''|k'\lambda'j'\Lambda k''\lambda''j''\rangle\, h'Zh''.$$

Factorize:

$$\langle h'Zh''|k'\lambda'j'\Lambda k''\lambda''j''\rangle =$$

$$\sum_{t'\zeta's'} \sum_{t''\zeta''s''} \langle h'|k'\lambda'j'\Lambda t'\zeta's'\rangle \langle Z|t'\zeta's'\Lambda t''\zeta''s''\rangle \langle h''|t''\zeta''s''\Lambda k''\lambda''j''\rangle =$$

$$= \sum_{t'\zeta's'} \sum_{t''\zeta''s''} \delta(k't')\,\delta(\lambda'\zeta')\,\langle h'|j'\zeta's'\rangle \langle Z|t'\zeta's'\Lambda t''\zeta''s''\rangle \cdot$$

$$\cdot \delta(t''k'')\,\delta(\zeta''\lambda'')\,\langle h''|s''\lambda''j''\rangle =$$

$$= \sum_{s'} \sum_{s''} \langle h'|j'\lambda's'\rangle \langle Z|k'\lambda's'\Lambda k''\lambda''s''\rangle \langle h''|s''\lambda''j''\rangle.$$

Substitution, collecting terms, and application of

$$\frac{|\lambda|}{|H|} \sum_{h \in H} \langle h|j\lambda s\rangle\, h = q(j\lambda s)$$

a UMB element pertaining to the subgroup H, yields

$$Q(k'\lambda'j'\Lambda k''\lambda''j'') =$$

$$= \frac{|\Lambda|}{|\lambda'||G||\lambda''|} \sum_Z l_Z \sum_{s'}^{|\lambda'|} \sum_{s''}^{|\lambda''|} \langle Z|k'\lambda's'\Lambda k''\lambda''s''\rangle \{q(j'\lambda's')\, Z\, q(s''\lambda''j'')\}$$

the DC decomposition of a general sequence-adapted UMB element.

Each term refers to a particular DC. Each term is factorized into UMB elements pertaining to the groups $H', H'' \subset G$.

Of some importance are the following three cases:

a) The multiplicity-free case:

$$\langle \lambda'|\Lambda\downarrow\rangle = \langle \lambda''|\Lambda\downarrow\rangle \leqslant 1.$$

The indices k′ and k″ are superfluous.

b) The 1-dimensional case:

$$|\lambda'| = |\lambda''| = 1.$$

The summation over s′ and s″ drop out and the indices j′, j″, s′ and s″ are superfluous.

c) The multiplicity-free 1-dimensional case.

Of considerable interst in quantum mechanical applications are subgroups of $G = S_N$, defined by partitions of N:

$$S(\lambda) = \mathop{\times}_{j} \mathop{\times}_{\lambda}^{\lambda_j} S_{j_\lambda}$$

where S_{j_λ} is the λ-th, $\lambda = 1, 2, \dots \lambda_j$ symmetric group of degree j in this product, $(\lambda) = \dots j^{\lambda_j} \dots$ with $\sum_j^N j\lambda_j = N$, a partition of N.

With

$$S(\lambda) = \mathop{\times}_{j} \mathop{\times}_{\lambda}^{\lambda_j} S_{j_\lambda} \quad \text{and} \quad S(\mu) = \mathop{\times}_{k} \mathop{\times}_{\mu}^{\mu_k} S_{k_\mu}$$

the intersection group pertaining to the DC $S(\lambda)\ ZS(\mu)$ is given by

$$S(\lambda) \cap ZS(\mu)\bar{Z} = \mathop{\times}_{j} \mathop{\times}_{\lambda}^{\lambda_j} \mathop{\times}_{k} \mathop{\times}_{\mu}^{\mu_k} (S_{j_\lambda} \cap Z S_{k_\mu} \bar{Z})$$

where $S_{j_\lambda} \cap Z S_{k_\mu} \bar{Z}$ is a symmetric group defined on the numbers common to S_{j_λ} and S_{μ_k}.

Let $V(j,\lambda)$, $V(k,\mu)$ and $ZV(k,\mu)$ denote the sets of numbers on which the groups S_{j_λ}, S_{k_μ} and $ZS_{k_\mu}\bar{Z}$ act.

Then $V(j,\lambda) \cap ZV(k,\mu)$ is the set of numbers on which the intersection group $S_{j_\lambda} \cap ZS_{k_\mu}\bar{Z}$ is defined.

We introduce

$$D_{j_\lambda, k_\mu} = |V(j,\lambda) \cap ZV(k,\mu)|$$

the degree of the symmetric group $S_{j_\lambda} \cap ZS_{k_\mu}\bar{Z}$.

The matrix: $D(Z)$ with elements $D_{j_\lambda,k_\mu}(Z)$ is called a DC symbol (DCS) (See App. D).

It can be proven that: the correspondence

$$S(\lambda)\ ZS(\mu) \Leftrightarrow D(Z)$$

is unique. The elements of $D(Z)$ satisfy, by definition:

a) $D_{j_\lambda,k_\mu}(Z) \geqslant 0$, integer,

b) $\sum_{j\lambda} D_{j_\lambda,k_\mu}(Z) = k, \quad \sum_{k\mu} D_{j_\lambda,k_\mu} = j, \quad \sum_{j\lambda}\sum_{k\mu} D_{j_\lambda,k_\mu}(Z) = N.$

The concept of a DCS and the unique correspondence stated above enables one to characterize DC decompositions through algorithms based on "matrices with special properties" rather than on the DC themselves.

As an example, the frequency of the DC $S(\lambda)\,Z\,S(\mu)$ is given by

$$d_Z = \prod_j^{\lambda_j} \prod_\lambda \prod_k^{\mu_k} \prod_\mu D_{j_\lambda, k_\mu}(Z)!$$

i.e. the factorial product of the entries of the DCS.

Appendix C

Tableau operators

Consider any positive integer N and its partitions (λ).

A *partition* (λ) is defined as a set of positive integers 1, 2, ... N each of which occurs $\lambda_1, \lambda_2 \ldots \lambda_N$ times such that

$$\sum_j^N j\lambda_j = N, \quad (\lambda) = \ldots \lambda^{\lambda_j} \ldots.$$

A partition (λ) characterizes:

a) An IR Λ of S_N, $\Lambda = [\ldots j^{\lambda_j} \ldots]$

b) A class Λ of S_N, $\Lambda = (\ldots j^{\lambda_j} \ldots)$, the elements of which have the following cycle structure,

$$\ldots \cdot \underbrace{(\ldots)(\ldots)\ldots(\ldots)}_{\lambda_j \text{ cycles of length } j} \cdot \ldots$$

Let $\sum_j^N \lambda_j = N_\lambda$ denote the number of cycles.

c) A *Young Diagram*, YD, denoted D^Λ.

A YD is an arrangement of N_λ rows ordered according to non-increasing length from top to bottom with each row starting at the left in the first column.

Corresponding to any YD, D^Λ, *Young Tableaux*, YT, are defined by filling D with the integers 1, 2, ... N, each integer occurring once only. There are N! tableaux.

Those YT which satisfy:

numbers ascend in rows from left to right, and
numbers ascend in columns from top to bottom,

are called *standard YT*. Their number equals $d_{[\Lambda]}$, the dimension of the IR Λ of S_N pertaining to the YD, D^Λ.

The number is inversely proportional to the h-graph number $h(\Lambda)$.

The h-graph pertaining to D^Λ has entries $h_{jk} = l_j + l_k - (j + k) + 1$ on the intersection of the j-th row (length l_j) and the k-th column (length l_k).
Then:

$$h(\Lambda) = \prod_{jk} h_{jk} \quad \text{and} \quad d_{[\Lambda]}\, h(\Lambda) = N!$$

Corresponding to any YD, D^Λ, *Index Tableaux*, IT, are defined by filling D with integers 1, 2, ... n, *not* restricted by the condition that any number occurs once only. There are n^N such tableaux.

Those IT which satisfy:

numbers do not descend in rows from left to right,
numbers ascend in columns from top to bottom,

are called *standard IT*. Their number equals the dimension $d_{\langle\Lambda\rangle}$ of the IR of GL_n pertaining to the YD, D^Λ. This number is proportional to the n-graph number $n(\Lambda)$. The n-graph pertaining to D^Λ has an entry $n_{jk} = n - j + k$ on the intersection of the j-th row and k-th column. Then:

$$n(\Lambda) = \prod_{jk} n_{jk} \quad \text{and} \quad d_{\langle\Lambda\rangle}\, h(\Lambda) = n(\Lambda).$$

Consider any YT, T_s^Λ.

The *Row group* is the direct product of symmetric groups defined on the numbers in the row of T_s. Thus

$$R_s^\Lambda = \mathop{\times}_{j} \mathop{\times}_{\lambda}^{\lambda_j} S_{j\lambda} \qquad S_{j\lambda} \Leftrightarrow \lambda\text{-th row of length } j.$$

Similarly the *Column group* C_s^Λ is defined.

Let $\mathcal{T}^\Lambda$ denote the "first" standard YT. Then any tableau T_s^Λ can be characterized by a permutation $f_s \in G$ which, acting on $\mathcal{T}^\Lambda$ yields T_s^Λ.

With R^Λ and C^Λ the row- and column group pertaining to $\mathcal{T}^\Lambda$ we have

$$R_s^\Lambda = f_s R^\Lambda \bar{f}_s \quad C_s^\Lambda = f_s C^\Lambda \bar{f}_s.$$

Let:

P^Λ denote the *row symmetrizer* for $\mathcal{T}^\Lambda$
N^Λ denote the *column symmetrizer* for $\mathcal{T}^\Lambda$

$$P^\Lambda = \sum_{h \in R^\Lambda} h; \quad N^\Lambda = \sum_{v \in C^\Lambda} (-)^v v$$

where $h \in R^\Lambda$ are called horizontal permutations and $v \in C^\Lambda$ vertical permutations for obvious reasons.

Then the operators

$$P^\Lambda N^\Lambda, N^\Lambda P^\Lambda, P^\Lambda N^\Lambda P^\Lambda \quad \text{and} \quad N^\Lambda P^\Lambda N^\Lambda$$

are called tableau operators pertaining to T^Λ.

More general: For any pair of YT, T_s^Λ and T_t^Λ tableau operators are defined by

$$Y_{st}^\Lambda = f_s Y^\Lambda \bar{f}_t = f_s N^\Lambda P^\Lambda \bar{f}_t = Y_{ss}^\Lambda f_s \bar{f}_t = f_s \bar{f}_t Y_{tt}^\Lambda$$

with similar expressions for the other three types. Some pertinent facts:

1. The operators $Y_{\alpha\beta}^\Lambda$ Λ ranging, α, β range over *standard* YT constitute a basis for the group algebra of S_N.
2. The operators

$$\{Y_{\alpha\beta}^\Lambda \mid \beta \text{ fixed}, \alpha = 1, 2 \ldots |\Lambda|\}$$

span a minimal left ideal

$$g Y_{\alpha'\beta}^\Lambda = \sum_{\alpha''} \Lambda_{\alpha'\alpha''}^L(g) Y_{\alpha''\beta}^\Lambda$$

with Λ^L the left-natural representation.

3. The operators

$$\{Y_{\alpha\beta}^\Lambda \mid \alpha \text{ fixed}, \beta = 1, 2 \ldots |\Lambda|\}$$

span a minimal right ideal

$$Y_{\alpha\beta'}^\Lambda g = \sum_{\beta''} Y_{\alpha\beta''}^\Lambda \Lambda_{\beta''\beta'}^R(g)$$

with Λ^R the right-natural representation. With

$$Y_{\alpha\beta}^\Lambda = \sum_{g \in G} Y_{\alpha\beta}^\Lambda(g)\, g$$

where $Y_{\alpha\beta}^\Lambda(g)$ denotes the coefficient with which g occurs in $Y_{\alpha\beta}^\Lambda$ it follows easily that

$$\Lambda^L(g) = Y^\Lambda(g)\, \bar{Y}^\Lambda(e) \qquad \Lambda^R(g) = \bar{Y}^\Lambda(e)\, Y^\Lambda(g).$$

The matrix $Y^\Lambda(e)$ is called the NP structure matrix.

4. Multiplication rule

$$Y_{\alpha\beta}^\Lambda Y_{\mu\nu}^\Delta = \frac{|G|}{|\Lambda|} \delta(\Lambda\Delta)\, Y_{\mu\beta}^\Lambda(e)\, Y_{\alpha\nu}^\Lambda.$$

5. With

$$g = \sum_\Lambda \sum_{\mu\nu} Y_{\mu\nu}^\Lambda C_{\mu\nu}^\Lambda(g),$$

the expansion coefficients are given by

$$C^{\Lambda}_{\mu\nu}(g) = \frac{|\Lambda|}{|G|} \{\bar{Y}^{\Lambda}(e)\, Y^{\Lambda}(\bar{g})\, \bar{Y}^{\Lambda}(e)\}_{\nu\mu}.$$

6. Introduction of $(Y^{\Lambda}_{\alpha\beta})^{+}$ and

$$|C^{\Lambda}|\, \sigma^{\Lambda}_{\alpha\beta} = (Y^{\Lambda}_{\gamma\alpha})^{+} (Y^{\Lambda}_{\gamma\beta}), \text{ any } \gamma, \quad |R^{\Lambda}|\, \tau^{\Lambda}_{\alpha\beta} = (Y^{\Lambda}_{\alpha\gamma})(Y^{\Lambda}_{\beta\gamma})^{+},$$

yields the connection with Ch. 3.2.

Invariance groups for tableaux operators (See also App. E)

A right- (left-) invariance group of a tableau operator is a group of elements which, applied from the right (left) reproduce the operator apart from a possible change of sign. If the distinction left (right) is omitted it is understood that the invariance holds both from the left and the right. Then:

The groups $f_{\alpha} R \bar{f}_{\alpha}$ and $f_{\beta} C \bar{f}_{\beta}$ are right- and left-invariance groups of the operator $Y_{\alpha\beta}$.

In general however they are by no means the largest invariance groups. *The* invariance groups for $Y_{\alpha\beta}$ are given by the normalizers $N(f_{\alpha} R \bar{f}_{\alpha})$ and $N(f_{\beta} C \bar{f}_{\beta})$ respectively.

The normalizer N(H) of $H \subset G$ is defined by $N(H) = \{g \in G / gH\bar{g} = H\}$, i.e., N(H) is the largest subgroup of G in which H is a normal subgroup.

The group $N(R^{\Lambda})$ is the invariance group for the operator σ^{Λ}.
The group $N(C^{\Lambda})$ is the invariance group for the operator τ^{Λ}.

The invariance of the operator σ^{Λ} and τ^{Λ} with respect to the groups $N(R^{\Lambda})$ and $N(C^{\Lambda})$, respectively, is reflected by the following important theorem: $\tau^{\Lambda}(g)$, the coefficient of g in τ^{Λ} is a constant (up to sign) for all elements which belong to a given DC of G with respect to $N(C^{\Lambda})$

$$|\tau^{\Lambda}(g)| = |\tau^{\Lambda}(V)| \quad \forall g \in N(C^{\Lambda})\, V N(C^{\Lambda}), \qquad V \in G \text{ fixed}$$

and similarly,

$$|\sigma^{\Lambda}(g)| = |\sigma^{\Lambda}(W)| \quad \forall g \in N(R^{\Lambda})\, W N(R^{\Lambda}), \qquad W \in G \text{ fixed.}$$

In Ch. 3.3 it is shown that the algebraic numbers which must be calculated in the evaluation of matrix elements of a spin-free observable over antisymmetrized NP-spin-adapted N-electron functions are essentially the numbers $\sigma^{\Lambda}(g)$, or $\tau^{\Lambda}(g)$. The above theorem implies that the algebraic part of the problem can be solved in essence by a consideration of the DC decomposition of S_N with respect to normalizers of row-groups (column groups) of standard YT (App. E).

In order to show the practical relevance of the statement we compare the number of DC in the decomposition of G with respect to R^{Λ}, and $N(R^{\Lambda})$, respectively, for $(\Lambda) = [2^4 1^8]$. This is 93176758080 vs 69. Do you care?

2.3 Matrix elements and their evaluation

2.3.1 Introduction

In this section we consider matrix elements of a totally symmetric spin free operator Ω over antisymmetrized spin-adapted N-electron functions, i.e.,

$$(Q^{\rho\sigma}(1^N)\,[\phi'\Xi(N;W',S',M')]\,|\Omega|\,Q^{\rho\sigma}(1^N)\,[\phi''\Xi(N;W'',S'',M'')]). \tag{1}$$

From the discussion in (I App. A) the following correspondences and consequences hold:

$S \Leftrightarrow \Lambda_S$; Ω spin free $\Big\} \Rightarrow$ The matrix element factorizes according to $\delta(S'S'')$

$W \Leftrightarrow J$ — a row label for the IR Λ_S.

M refers to the multiplicity with which Λ_s occurs in the decomposition of spin space V^σ. On account of the orthonormality of 1-electron spin we have also a factorization according to $\delta(M'M'')$. Thus:

$$\Xi(N;W,S,M) = Q^{\Lambda_S}_{WM}\,\theta$$

$\theta \in V^\sigma$, an ordered N-spin product with eigenvalue M for S_Z in order to have a non-vanishing result.

Non vanishing matrix elements are then given by

$$(Q^{\rho\sigma}(1^N)\,[\phi' Q^{\Lambda_S}_{W'M}\,\theta]\,|\Omega|\,Q^{\rho\sigma}(1^N)\,[\phi'' Q^{\Lambda_S}_{W''M}\,\theta]). \tag{2}$$

In a conventional (i.e. spin-included) calculation one proceeds by evaluating the antisymmetrizers. This yields

$$\sum_{g\in G} (-)^g\,(\phi'|\Omega g|\phi'')\,(Q^{\Lambda_S}_{W'M}\,\theta\,|g|\,Q^{\Lambda_S}_{W''M}\,\theta) \tag{3}$$

after which the spin-matrix elements are to be calculated according to

$$(Q^{\Lambda_S}_{W'M}\,\theta\,|g|\,Q^{\Lambda_S}_{W''M}\,\theta) = \frac{|G|}{|\Lambda|}\,\{Q^\Lambda(\text{\textcircled{g}})\,\bar{Q}^\Lambda(e)\,\tau^\Lambda(e)\}_{W''W'}\quad(\theta\,|\sigma^\Lambda_{MM}|\,\theta)$$

where we used eqs. 7, 15, 17, 22 and 5 of Section 2, in that order.

Whatever the outcome of the spin matrix element on the right-hand side, it is a constant for ranging W' and W'' and therefore it can be omitted. Thus we get

$$\sum_{g\in G} (-)^g(\phi'|\Omega g|\phi'')\,\{Q^\Lambda(\text{\textcircled{g}})\,\bar{Q}^\Lambda(e)\,\tau^\Lambda(e)\}_{W''W'}. \tag{4}$$

In the spin-free evaluation one starts from (2) and applies (2.43). This yields

$$(\phi'|\Omega\,\{\tau^\Lambda(e)^t C^\Lambda P^{\tilde{\Lambda}} \bar{C}^\Lambda\}_{W'W''}|\phi''). \tag{5}$$

An example of (4) to be treated in the next section:

Q^Λ in the Young-Yamanouchi orthogonal matric basis.

Then (4) reads

$$\sum_{g\in G} (-)^{g} (\phi'|\Omega g|\phi'') \langle g|W'\Lambda W''\rangle \tag{6}$$

where $\langle g|W'\Lambda W''\rangle$ is the W', W'' matrix element of the matrix which represents g in Young-Yamanouchi's orthogonal representation.

An example of (5) is to be treated subsequently: Q^{Λ} in the Young-unit type NP matric basis. Then (5) reads

$$\sum_{g\in G} (\phi'|\Omega g|\phi'') (-)^{f_{W'}} \sigma^{\tilde{\Lambda}}_{\tilde{W}'\tilde{W}''}(g) (-)^{f_{W''}}. \tag{7}$$

It is seen that the two ways, spin-included and spin-free, of looking at the evaluation of matrix elements are equivalent. In both cases we have the set of orbital matrix elements $(\varphi'|\Omega g|\phi'')$, $g \in G$, and we have an algebraic problem, the determination of a set of (group theoretical) algebraic coefficients.

2.3.2 Matrix elements over Young-Yamanouchi N-electron spin (S, M) eigenfunctions

Young-Yamanouchi (YY) spin functions arise from consecutive coupling, 1 electron spin at a time.

This coupling procedure can be pictured in a branching diagram.

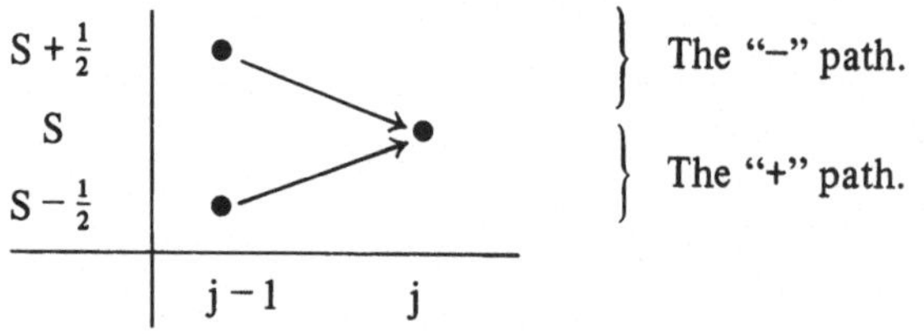

A j-electron spin (S, M) eigenfunction arises:

1. from the coupling of (j − 1)-electron spin $(S + \frac{1}{2})$ eigenfunctions with 1-electron spin $(\frac{1}{2})$ eigenfunctions

$$\begin{aligned} |j; S, M\rangle^{(-)} &= |j-1; S+\tfrac{1}{2}, M-\tfrac{1}{2}\rangle\, |1; \tfrac{1}{2}, \tfrac{1}{2}\rangle\, C^{-}_{+}(S, M) + \\ &+ |j-1; S+\tfrac{1}{2}, M+\tfrac{1}{2}\rangle\, |1; \tfrac{1}{2}, -\tfrac{1}{2}\rangle\, C^{-}_{-}(S, M) \end{aligned} \tag{8}$$

2. from the coupling of (j − 1)-electronspin $(S - \frac{1}{2})$ eigenfunctions with 1-electron spin $(\frac{1}{2})$ eigenfunctions

$$\begin{aligned} |j; S, M\rangle^{(+)} &= |j-1; S-\tfrac{1}{2}, M-\tfrac{1}{2}\rangle\, |1; \tfrac{1}{2}, \tfrac{1}{2}\rangle\, C^{+}_{+}(S, M) + \\ &+ |j-1; S-\tfrac{1}{2}, M+\tfrac{1}{2}\rangle\, |1; \tfrac{1}{2}, -\tfrac{1}{2}\rangle\, C^{+}_{-}(S, M). \end{aligned} \tag{9}$$

One easily shows that the coupling coefficients are given by

$$C^{+}_{\pm}(S, M) = +\left(\frac{S \pm M}{2S}\right)^{1/2} \qquad C^{-}_{\pm}(S, M) = \mp\left(\frac{S \pm M + 1}{2S + 2}\right)^{1/2}. \tag{10}$$

Note: the only zero coefficients are $C_+^+(S, -S) = 0 = C_-^+(S, S)$

the "–" sign only occurs for $C_+^-(S, M)$.

Any N-electronspin (SM) eigenfunction $\Xi(N; W, S, M)$ can be written as a linear combination

$$\Xi(N; W, S, M) = \sum_J \theta_J^M C_J^W(S, M)$$

where θ_J^M ranges over all N-spin products with fixed eigenvalue M for S_Z.
The label J can thus be specified by

a series $\{m(1), m(2), \ldots m(N)\}$ or by
a series $\{M_1, M_2, \ldots M_N\}$.

E.g.,

$$\theta_J^{M=1} = \beta\alpha\alpha\beta\alpha\alpha \Rightarrow \{-\tfrac{1}{2}, +\tfrac{1}{2}, +\tfrac{1}{2}, -\tfrac{1}{2}, +\tfrac{1}{2}, +\tfrac{1}{2}\}$$
and $\{(-\tfrac{1}{2})_1, (0)_2, (\tfrac{1}{2})_3, (0)_4, (\tfrac{1}{2})_5, (1)_6\}$ respectively.

The YY functions can be characterized by a series of consecutive S-values $S_1, S_2, \ldots S_N$, by means of which a certain path W is specified or alternatively by a series of + and – specifying the path. Thus

$$C_J^W(S, M) = \prod_{k=1}^{N} C_{\psi_k(J)}^{\phi_k(W)}(S_k^W, M_k^J) \tag{11}$$

where

$\phi_k(W) = \pm$ according to the path W followed
$\psi_k(W) = \pm$ according to the occurrence of α/β in θ_J
S_k^W is the eigenvalue of S^2 in the K-th step of path W
M_k^J is the eigenvalue of S_Z in the K-th step of θ_J.

Example:

$$C_{+-++}^{++-+}(1, 1) = C_+^+(\tfrac{1}{2}, \tfrac{1}{2})\, C_-^+(1, 0)\, C_+^-(\tfrac{1}{2}, \tfrac{1}{2})\, C_+^+(1, 1) =$$
$$= +\sqrt{1} \cdot +\sqrt{\tfrac{1}{2}} \cdot -\sqrt{\tfrac{1}{3}} \cdot +\sqrt{1} \quad \equiv -\tfrac{1}{6}\sqrt{6}$$

is the coefficient of $\alpha\beta\alpha\alpha$ in $\Xi(4; (++-+), 1, 1)$.

The matrix elements.in (6) become:

$$\langle \Xi(N; W', S, M) | g | \Xi(N; W'', S, M) \rangle \equiv \langle g | W' \Lambda_S W'' \rangle =$$
$$= \sum_{J'} \sum_{J''} C_{J'}^{W'}(S, M) \langle \theta_{J'}^M | g | \theta_{J''}^M \rangle C_{J''}^{W''}(S, M). \tag{12}$$

It is easily seen that

$$\langle \theta_{J'}^{M} | g | \theta_{J''}^{M} \rangle = D_{J'J''}(g) = \delta[J', g(J'')]$$

is a permutation matrix representation, so that

$$\langle g | W' \Lambda_S W'' \rangle = \sum_{J} C_{g(J)}^{W'}(S, M)\, C_J^{W''}(S, M) \tag{13}$$

which, of course does not depend on M.
The final result is given by

$$\sum_{g \in G} (-)^g (\phi' | \Omega g | \phi'') \sum_{J} C_{g(J)}^{W'}(S, M)\, C_J^{W''}(S, M). \tag{14}$$

One then proceeds with the simplification of the orbital matrix elements. This will be the subject of the following section not starting from (14) however. We choose the spin-free approach and show how to simplify matrix elements built from bases which are SA as discussed in (2.27).
The notation $\tilde{\Lambda}$ will not be used any longer. The tilde will be omitted and Λ then refers to the associates of IR that occur in the decomposition of spin space.

2.3.3 Matrix elements of a spin-free observable over sequence-symmetry-adapted bases

Here we consider the SA functions defined by (2.27).

$$Q(J \Lambda_k \lambda_t)\, |\phi; \lambda_t\rangle = |\phi(k\lambda_t); \Lambda J\rangle. \tag{15}$$

We recall: $|\phi; \lambda_t\rangle$ denotes an N-particle product function which transforms according to the trivial IR λ_t of its invariance group H^λ. The invariance group is defined as the direct product of symmetric groups, the degrees of which are given by the occupation numbers of the different 1-particle orbitals in the product ϕ.

$$Q(J \Lambda k \lambda_t) = \frac{|\Lambda|}{|G|} \sum_{g \in G} \langle g | J \Lambda k \lambda_t \rangle\, g \tag{16}$$

is an element of an orthogonal matric basis which is sequence adapted to $H^\lambda \subset G$ on the left-hand side.
The label Λ denotes an IR of G as usual, $J = 1, 2, \ldots |\Lambda|$ and $k = 1, 2, \ldots \langle \Lambda | \lambda_t \uparrow \rangle$, the number of times the IR Λ occurs in the induced representation $\lambda_t \uparrow$. This number is equal to the number of linearly independent functions of definite symmetry ΛJ which arise from $|\phi; \lambda_t\rangle$ by projection. The matrix elements we have to calculate in the induced bases are given by $(\phi'(k'\lambda_t'); \Lambda' J' | \Omega | \phi''(k''\lambda_t''); \Lambda'' J'')$ (17)

in which Ω is a spin-free operator, totally symmetric in the orbital coordinates of the N particles.
According to the Wigner-Eckart theorem the representation is block-diagonalized with respect to the symmetry of the group G so that (17) goes over into

$$(\phi'(k'\lambda_t'); \Lambda' \| \Omega \| \phi''(k''\lambda_t''); \Lambda')\, \delta(\Lambda' \Lambda'')\, \delta(J' J'') \tag{18}$$

implying that we only need to consider one particular symmetry, say ΛJ, at a time.

By definition of a reduced matrix element,

$$(\phi'(k'\lambda_t'); \Lambda \| \Omega \| \phi''(k''\lambda_t''); \Lambda) = \frac{1}{|\Lambda|} \sum_J^{|\Lambda|} (\phi'(k'\lambda_t'); \Lambda J | \Omega | \phi''(k''\lambda_t''); \Lambda J). \quad (19)$$

As a simple example we consider Ω_0, the identity operator in order to calculate overlap and norm of the induced basis functions. Thus,

$$(\phi'(k'\lambda_t'); \Lambda \| \phi''(k''\lambda_t''); \Lambda) = \frac{1}{|\Lambda|} \sum_J^{|\Lambda|} (\phi'(k'\lambda_t'); \Lambda J | \phi''(k''\lambda_t''); \Lambda J) =$$

"turn over"

$$= \frac{1}{|\Lambda|} \sum_J^{|\Lambda|} (\phi'; \lambda_t' | Q(k'\lambda_t' \Lambda J)\, Q(J \Lambda k''\lambda_t'') | \phi''; \lambda_t'') =$$

"contract"

$$= \frac{1}{|\Lambda|} \sum_J^{|\Lambda|} (\phi'; \lambda_t' | Q(k'\lambda_t' \Lambda k''\lambda_t'') | \phi''; \lambda_t'') = (\phi'; \lambda_t' | Q(k'\lambda_t' \Lambda k''\lambda_t'' | \phi''; \lambda_t'') =$$

"decompose"

$$= \frac{|\Lambda|}{|G|} \sum_Z l_Z \langle Z | k'\lambda_t' \Lambda k''\lambda_t'' \rangle (\phi'; \lambda_t' | Z | \phi''; \lambda_t'')$$

where "turn over" means $(Q\phi'| = (\phi'|Q^+$, "contract" means the application of the basic multiplication rule for matric basis elements, and "decompose" means that a *Double Coset* (DC) decomposition of the operator Q is performed (App. B).
Hence: Overlap and Norm are given by

$$(\phi'(k'\lambda_t'); \Lambda \| \phi''(k''\lambda_t''); \Lambda) = \frac{|\Lambda|}{|G|} \sum_Z l_Z \langle Z | k'\lambda_t' \Lambda k''\lambda_t'' \rangle (\phi'; \lambda_t' | Z | \phi''; \lambda_t''), \quad (20)$$

in which

a the sum ranges over all DC in the decomposition of the group $G = S_N$ with respect to the invariance groups H' of ϕ' and H'' of ϕ'' on the left- and right-hand side, respectively.

b each particular term is the product of three factors:

b1 l_Z the length of the DC which can be calculated from the associated double coset symbol (DCS),

b2 $\langle Z | k'\lambda_t' \Lambda k''\lambda_t'' \rangle$ the matrix element of the DC generator Z in the IR sequence-adapted with respect to $H' \subset G$ and $H'' \subset G$ on the left and right, respectively.

b3 $(\phi'; \lambda_t' | Z | \phi''; \lambda_t'')$ an integral, which can be written as an N-th degree monomial in 1-particle overlap, also to be determined from the DCS pertaining to the generator Z.

We will now turn our attention to the R-particle operator Ω_K. For the present discussion we omit all labels in (19) which do not explicitly refer to the symmetry of the group G, i.e., we omit all labels preceding the semicolon.

Then,

$$(\Lambda \| \Omega \| \Lambda) = \frac{1}{|\Lambda|} \sum_{J}^{|\Lambda|} (\Lambda J | \Omega | \Lambda J). \tag{21}$$

According to (2.1.1),

$$\Omega = \sum_{R=0} \Omega_R \tag{22}$$

and

$$\Omega_R = \binom{N}{R} \frac{1}{|G|} \sum_{g \in G} g \omega_R \bar{g} \tag{23}$$

in which the sum extends over all $g \in G$ and the operator ω_R denotes a prototype R-particle operator.

The latter operator is invariant under a subgroup $K_R = S_{N-R} \times S_R \subset S_N$. Its order is given by $|K_R| = (N-R)!\, R!$. Thus,

$$\Omega_R = \frac{1}{|K_R|} \sum_{g \in G} g \omega_R \bar{g}. \tag{24}$$

Using (24),

$$(\Lambda J | \Omega_R | \Lambda J) = \frac{1}{|K_R|} \sum_{g \in G} (\Lambda J | g \omega_R \bar{g} | \Lambda J) = \frac{|G|}{|\Lambda|\,|K_R|} \sum_{M}^{|\Lambda|} (\Lambda M | \omega_R | \Lambda M). \tag{25}$$

Substitution of (22, 25) into (21) and performing the sum over J, yields

$$(\Lambda \| \Omega \| \Lambda) = \frac{|G|}{|\Lambda|} \sum_{R=0}^{N} \frac{1}{|K_R|} \sum_{M}^{|\Lambda|} (\Lambda M | \omega_R | \Lambda M). \tag{26}$$

Consider the matrix elements $(\Lambda M | \omega_R | \Lambda M)$.

In order to take advantage of the fact that K_R is the invariance group of the operator ω_R we must use sequence-adapted labels with respect to $K_R \subset G$. Since the range over M is complete we have

$$\sum_{M}^{|\Lambda|} (\Lambda M | \omega_R | \Lambda M) = \sum_{n\mu m} (\Lambda n\mu m | \omega_R | \Lambda n\mu m) = \sum_{n\mu} (\Lambda n\mu \| \omega_R \| \Lambda n\mu)\, |\mu|$$

which, upon substitution into (26) yields:

The first intermediate result

$$(\Lambda \| \Omega \| \Lambda) = \frac{|G|}{|\Lambda|} \sum_{R=0}^{N} \sum_{n\mu} \frac{|\mu|}{|K_R|} (\Lambda n\mu \| \omega_R \| \Lambda n\mu) \tag{27}$$

where μ ranges over the IR of K_R and $n = 1, 2, \ldots \langle \Lambda | \mu \nearrow \rangle$.

For the moment we concentrate on a particular reduced matrix element

$$(\Lambda n\mu \| \omega_R \| \Lambda n\mu),$$

which, at this stage will again be written out explicitly. Thus,

$$(\phi'(k'\lambda_t'); \Lambda n\mu \| \omega_R \| \phi''(k''\lambda_t''); \Lambda n\mu) =$$

$$= \frac{1}{|\mu|} \sum_m^{|\mu|} (\phi'(k'\lambda_t'); \Lambda n\mu m | \omega_R | \phi''(k''\lambda_t''); \Lambda n\mu m) =$$

"turn over"

$$= \frac{1}{|\mu|} \sum_m^{|\mu|} (\phi'; \lambda_t' | Q(k'\lambda_t' \Lambda n\mu m)\, \omega_R\, Q(n\mu m \Lambda k''\lambda_t'') | \phi''; \lambda_t''). \tag{28}$$

Apply a DC decomposition twice, i.e., the operator $Q(k'\lambda_t'\Lambda n\mu m)$ is decomposed with respect to H' and K_R,

$$Q(k'\lambda_t' \Lambda n\mu m) = \frac{|\Lambda|}{|G||\mu|} \sum_{Z'} l_{Z'} \sum_s^{|\mu|} \langle Z' | k'\lambda_t' \Lambda n\mu s\rangle\, q(\lambda_t')\, Z' q(s\mu m) \tag{29}$$

the operator $Q(n\mu m\Lambda k''\lambda_t'')$ is decomposed with respect to K_R and H'',

$$Q(n\mu m \Lambda k''\lambda_t'') = \frac{|\Lambda|}{|G||\mu|} \sum_{Z''} l_{Z''} \sum_t^{|\mu|} \langle Z'' | n\mu t \Lambda k''\lambda_t''\rangle\, q(m\mu t)\, Z'' q(\lambda_t'') \tag{30}$$

and (15, 16) are substituted into (14).

This yields:

$$\frac{1}{|\mu|} \sum_m^{|\mu|} \left(\frac{|\Lambda|}{|G||\mu|}\right)^2 \sum_{Z'} \sum_{Z''} l_{Z'} l_{Z''} \sum_s^{|\mu|} \sum_t^{|\mu|} \langle Z' | k'\lambda_t' \Lambda n\mu s\rangle \langle Z'' | n\mu t \Lambda k''\lambda_t''\rangle \cdot$$

$$\cdot (\phi'; \lambda_t' | q(\lambda_t')\, Z' q(s\mu m)\, \omega_R\, q(m\mu t)\, Z'' q(\lambda_t'') | \phi''; \lambda_t''). \tag{31}$$

The operators $q(\lambda_t')$ and $q(\lambda_t'')$ act trivially to the left and right whereas the operators $q(s\mu m)$ and $q(n\mu t)$ commute with ω_R so that after contraction and performing the sum over m we get:

The second intermediate result

$$(\phi'(k'\lambda_t'); \Lambda n\mu \| \omega_R \| \phi''(k''\lambda_t''); \Lambda n\mu) =$$

$$= \left(\frac{|\Lambda|}{|G||\mu|}\right)^2 \sum_{Z'} \sum_{Z''} l_{Z'} l_{Z''} \sum_s^{|\mu|} \sum_t^{|\mu|} \langle Z' | k'\lambda_t' \Lambda n\mu s\rangle \langle Z'' | n\mu t \Lambda k''\lambda_t''\rangle \cdot$$

$$\cdot (\phi'; \lambda_t' | Z' \omega_R\, q(s\mu t)\, Z'' | \phi''; \lambda_t''). \tag{32}$$

In this expression there is a complete summation over labels s and t as yet unspecified. The operator $q(s\mu t)$ pertaining to the group K_R still contains permutations which act trivially to the left and to the right.

For a *fixed pair* of generators Z' and Z'' we therefore decompose this operator with respect to the groups

$$\bar{Z}'H'Z' \cap K_R \text{ on the left, and } Z''H''\bar{Z}'' \cap K_R \text{ on the right.}$$

In order to take advantage of this decomposition we replace the label s by a sequence-adapted label for $(\bar{Z}'H'Z' \cap K_R) \subset K_R$, say $(a'\chi'b')_{\bar{Z}'}$ and the label t by a sequence-adapted label for $(Z''H''Z'' \cap K_R) \subset K_R$, say $(a''\chi''b'')_{Z''}$. Subscripts $\bar{Z}'$ and Z'' will be omitted as long as no confusion arises. The DC decomposition reads

$$q((a'\chi'b')\,\mu\,(a''\chi''b'')) = \\ = \frac{|\mu|}{|\chi'||K_R||\chi''|} \sum_Z l_Z \sum_{c'}^{|\chi'|} \sum_{c''}^{|\chi''|} \langle Z|(a'\chi'c')\,\mu\,(a''\chi''c'')\rangle \cdot q(b'\chi'c')\,Z q(c''b''\chi''). \tag{33}$$

Substitution into (32) yields the following matrix elements to be considered

$$(\phi;\lambda_t'|Z'\omega_R q(b'\chi'c')\,Z q(c''\chi''b'')\,Z''|\phi'';\lambda_t'') = \\ = \delta(\chi'\chi_t')\,\delta(\chi''\chi_t'')\,(\phi';\lambda_t'|Z'\omega_R Z Z''|\phi'';\lambda_t'') \tag{34}$$

in which the second line follows from the fact that ket $Z''|\phi'';\lambda_t'')$ and bra $(\phi';\lambda_t'|Z'$ transform as the trivial IR of the groups $Z''H''\bar{Z}''$ and $\bar{Z}'H'Z'$ respectively whereas the operators $q(c''\chi''b'')$ and $q(b'\chi'c')$ pertain to subgroups $(Z''H''\bar{Z}'' \cap K_R)$ and $(\bar{Z}'H'Z$ $(\bar{Z}'H'Z' \cap K_R)$ of $Z''H''\bar{Z}''$ and $\bar{Z}'H'Z'$.

Thus if (33) is substituted into (32) and (34) is applied the sums over c', c'', χ' and χ'' drop out and the labels b' and b'' become superfluous.

Therewith we get:

The third intermediate result

$$\sum_{a'\chi'b'} \sum_{a''\chi''b''} \langle Z'|k'\lambda_t'\Lambda n\mu(a'\chi'b')\rangle \langle Z''|n\mu(a''\chi''b'')\,\Lambda k''\lambda_t''\rangle \cdot \\ \cdot (\phi';\lambda_t'|Z'\omega_R q((a'\chi'b')\,\mu\,(a''\chi''b''))\,Z''|\phi'';\lambda_t''\rangle = \\ = \frac{|\mu|}{|K_R|} \sum_{a'} \sum_{a''} \langle Z'|k'\lambda_t'\Lambda n\mu(a'\chi_t')\rangle \langle Z''|n\mu(a''\chi_t'')\,\Lambda k''\lambda_t''\rangle \cdot \\ \cdot \sum_Z l_Z \langle Z|(a'\chi_t')\,\mu\,(a''\chi_t'')\rangle (\phi';\lambda_t'|Z'\omega_R Z Z''|\phi'';\lambda_t''). \tag{35}$$

Substitution of (35) into (32) and the resulting expression into (27) yields: The integral

$$(\phi';\lambda_t'|Z'\omega_R Z Z''|\phi'';\lambda_t'')$$

occurs in

$$(\phi'(k'\lambda_t');\Lambda \,\|\,\Omega\,\|\, \phi''(k''\lambda_t'');\Lambda) \tag{36}$$

with coefficient

$$\frac{|\Lambda|}{|G|}\frac{1}{|K_R|^2}\, l_{Z'}\, l_Z\, l_{Z''} \sum_{n\mu}\ \sum_{(a')_{\overline{Z}'}}\ \sum_{(a'')_{Z''}} \langle Z'|k'\lambda'_t \Lambda n\mu (a'\chi'_t)_{\overline{Z}'}\rangle \cdot$$
$$\cdot \langle Z|(a'\chi'_t)_{\overline{Z}'}\, \mu (a''\chi''_t)_{Z''}\rangle \langle Z''|n\mu(a''\chi''_t)_{Z''}\, \Lambda k''\lambda''_t\rangle.$$

Here subscripts $\overline{Z}'$ and Z'' are reintroduced in order to show explicitly to which subgroups the sequence adaptation refers.

This result is slightly disturbing from an esthetical point of view. However, careful consideration shows that

$$\sum_{n\mu}\ \sum_{(a')_{\overline{Z}'}}\ \sum_{(a'')_{Z''}} \langle Z'|k'\lambda'_t \Lambda n\mu (a'\chi'_t)_{\overline{Z}'}\rangle \langle Z|a'\chi'_t)_{\overline{Z}'}\, \mu (a''\chi''_t)_{Z''}\rangle \cdot$$
$$\cdot \langle Z''|n\mu(a''\chi''_t)_{Z''}\Lambda k''\lambda''_t\rangle = \langle Z'ZZ''|k'\lambda'_t \Lambda k''\lambda''_t\rangle. \tag{37}$$

At last we consider

$$\frac{1}{|K_R|^2}\, l_{Z'}\, l_Z\, l_{Z''}$$

which contains the lengths of three DC. This factor can be rewritten, and yields

$$\frac{|H'||H''|}{d_Z} \tag{38}$$

with $d_Z = |(\overline{Z}'H'Z' \cap K_R) \cap Z(K_R \cap Z''H''\overline{Z}'')\,\overline{Z}|$, the frequency of the DC generated by Z.

Therewith: The integral

$$(\phi'; \lambda'_t | Z'\omega_R ZZ''|\phi''; \lambda''_t) \tag{39}$$

occurs in

$$(\phi'(k'\lambda'_t); \Lambda \,\|\Omega\|\, \phi''(k''\lambda''_t); \Lambda)$$

with coefficient

$$\frac{|\Lambda|}{|G|}\frac{|H'||H''|}{d_Z} \langle Z'ZZ''|k'\lambda'_t \Lambda k''\lambda''_t\rangle.$$

Hence:

$$(\phi'(k'\lambda'_t); \Lambda \,\|\Omega\|\, \phi''(k''\lambda''_t); \Lambda) =$$
$$= \frac{|\Lambda|}{|G|}\, |H'||H''| \sum_{R=0}^{N} \left\{ \sum_{Z'} \sum_{Z''} \left[\sum_{Z} \bar{d}_Z \langle Z'ZZ''|k'\lambda'_t \Lambda k''\lambda''_t\rangle \cdot \right.\right. \tag{40}$$
$$\cdot (\phi'; \lambda'_t | Z'\omega_R ZZ''|\phi''; \lambda''_t).$$

which expression must be applied as follows:

For given R,

a) decompose G with respect to H' and K_R; this yields a set of DC generators $\{Z'\}$,

b) decompose G with respect to K_R and H''; this yields a set of DC generators $\{Z''\}$,

c) decompose K_R with respect to $\bar{Z}'H'Z' \cap K_R$ and $Z''H''\bar{Z}'' \cap K_R$ for any pair Z', Z''; this yields a set of DC generators $\{Z\}_{Z',Z''}$.

Thus actually Z' and Z'' should carry a sub-index R whereas Z should be labelled by Z'_R, Z''_R.

Note that calculation of the reduced matrix elements

$$(\phi'(k'\lambda'_t); \Lambda \,\|\, \Omega_R \,\|\, \phi''(k''\lambda''_t); \Lambda)$$

k', k'' and Λ ranging, has been replaced by the calculation of

a) a set of integrals

$$(\phi'; \lambda'_t \,|\, Z'\omega_R Z Z'' \,|\, \phi''; \lambda''_t)$$

Z', Z and Z'' ranging, independent of k', k'' and Λ, and

b) sets of algebraic coefficients

$$\langle Z'ZZ'' \,|\, k'\lambda'_t \Lambda k''\lambda''_t \rangle$$

Z', Z and Z'' ranging, each set labelled by a triplet k', k'', Λ.

Without derivation we state: The integrals

$$(\phi'; \lambda'_t \,|\, Z'\omega_R Z Z'' \,|\, \phi''; \lambda''_t) \tag{41}$$

are unique, in that

a) they are all different, and

b) they cannot be simplified by symmetry arguments.

They will be called Basic Exchange Integrals (BEI).

Two important observations:

A. Whatever matric bases we use for the construction of SA N-particle functions we always end with the set of BEI defined above if the existing invariances H' for ϕ', H'' for ϕ'' and K_R for ω_R are exhaustingly used. Different matric bases will yield different sets of algebraic coefficients.

B. The actual use of (40) requires such an enormous number of DC decompositions that one may doubt its practicality.

The disadvantage can by partially cured group-theoretically by the introduction of *the normalizer of an invariance group.*

$$N(H') = \{g \in G \,|\, gH'\bar{g} = H'\} \tag{42}$$

i.e., the normalizer of $H' \subset G$ is the largest subgroup of G which contains H' as a normal subgroup.

Let the DC decomposition of G with respect to $N(H')$ and K_R be given by the generators $\{W'\}$, called a canonical set of generators.
Then for all $Z' \in N(H')\,W'K_R$, W' fixed one has $\overline{Z}'H'Z' \cap K_R = \overline{W}'H'W' \cap K_R$.

Similarly, let the DC decomposition of G with respect to K_R and $N(H'')$ be given by the generators $\{W''\}$.
Then for all $Z'' \in K_R W'' N(H'')$, W'' fixed one has $Z''H''\overline{Z}'' \cap K_R = W''H''\overline{W}'' \cap K_R$.

Thus the number of DC decompositions to be performed in (40^c) is reduced considerably.

In order to get an idea: $(\lambda') = (2^4, 1^2), (\lambda'') = (2^2, 1^6)$,
$R = 1$: step (40^c) requires 48 against 4, and
$R = 2$: step (40^c) requires 570 against 16 DC decompositions.

A second aspect of the use of the normalizer concerns equalities among the algebraic coefficients $\langle Z'ZZ'' | k'\lambda_t' \Lambda k''\lambda_t''\rangle$. This aspect will be not be treated here.

On account of observation A we always end with the set of BEI defined in (41). Let us first consider the simplest case, $R = 0$.
In that case $K_R = S_N$ so that only $Z' = e = Z''$ occurs. The BEI are

$$(\phi'; \lambda_t' | Z | \phi''; \lambda_t''). \tag{43}$$

The generators Z are associated with DCS, the rows and columns of which are labelled by the factors of the invariance groups H' of ϕ' and H'' of ϕ'' respectively. Thus they can also be labelled by the 1-particle orbitals a_j in ϕ' and b_j in ϕ''.

We recall from App. B and D:
A DCS D(Z) is a set of numbers d_{jk} $\quad j = 1, 2, \ldots s;\ k = 1, 2, \ldots t$

which satisfy $d_{jk} \geqslant 0$ integer and

$$\sum_j^s d_{jk} = d_{0k}, \quad \sum_k^t d_{jk} = d_{j0}, \quad \sum_j^s d_{j0} = N = \sum_k^t d_{0k}.$$

Their significance stems from the unique correspondence between D(Z) and $H^{(x)}\,Z\,H^{(y)}$, a DC of S_N, with $H^{(x)}$ and $H^{(y)}$ subgroups of S_N defined by the partitions $\{d_{j0} | j\}$ and $\{d_{0k} | k\}$, respectively.

$$H^{(x)} = \mathop{\times}_j^s S_{d_{j0}} \quad \text{and} \quad H^{(y)} = \mathop{\times}_k^t S_{d_{0k}}.$$

With

d_{j0} the occupation number of orbital a_j in ϕ',
d_{0k} the occupation number of orbital b_k in ϕ'',

and

$S_{jk} = (a_j | b_k)$ 1-particle overlap integrals,

it can be shown that with $\phi' = \ldots a_j^{d_{jo}} \ldots$ and $\phi'' = \ldots b_k^{d_{ok}} \ldots$, the BEI reads

$$(\phi'; \lambda_t' | Z | \phi''; \lambda_t'') = \prod_{jk} S_{jk}^{d_{jk}(Z)} \tag{44}$$

an N-th degree monomial in 1-particle overlap integrals.

Thus:

Characterization and evaluation of R = 0 BEI can be performed by means of DCS.

Similarly:

Characterization and evaluation of R ≠ 0 BEI can be performed by means of triple double coset symbols (TDCS).

A TDCS is defined as the following generalization of a DCS: (App. D)

A TDCS $D(Z; Z', Z'')$ is a set of numbers d_{jkl} $j = 1, 2, \ldots, s$
which satisfy $k = 1, 2, \ldots, t$
$d_{jkl} \geq 0$ integer $l = 1, 2, \ldots, v$

and

$$\sum_j^s d_{jkl} = d_{okl}, \quad \sum_k^t d_{jkl} = d_{jol}, \quad \sum_l^v d_{jkl} = d_{jko} \tag{45}$$

$$\sum_j^s d_{jol} = d_{ool} = \sum_k^t d_{okl}, \quad \sum_k^t d_{jko} = d_{joo} = \sum_l^v d_{jol}$$

$$\sum_l^v d_{okl} = d_{oko} = \sum_j^s d_{jko}, \quad \sum_j^s d_{joo} = \sum_k^t d_{oko} = \sum_l^v d_{ool} = N.$$

Then the three partitions define three subgroups of S_N,

$$\{d_{joo} | j\} \Leftrightarrow H^{(x)}, \quad \{d_{oko} | k\} \Leftrightarrow H^{(y)}, \quad \{d_{ool} | l\} \Leftrightarrow K.$$

The set

$$\{d_{jol} | j, l\}$$

defines a DCS, say $D(Z')$ with

$$D(Z') \Leftrightarrow H^{(x)} Z' K.$$

The set

$$\{d_{okl} | k, l\}$$

defines a DCS, say $D(Z'')$ with

$$D(Z'') \Leftrightarrow K Z'' H^{(y)}.$$

Without proof we state: (46)

The TDCS $D(Z;Z',Z'')$, Z' and Z'' fixed, Z ranging, correspond uniquely to the DC of the decomposition of K with respect to $(\bar{Z}'H^{(x)}Z' \cap K)$ on the left and $(K \cap Z''H^{(y)}\bar{Z}'')$ on the right.

If we now consider (40^{abc}) and we identify:

$H^{(x)} = H'$ the invariance group of the primitive bra ϕ',

$H^{(y)} = H''$ the invariance group of the primitive ket ϕ'',

$K = K_R$ the invariance group of the operator ω_R

we obtain the result in (45).

Thus:

$$(\phi';\lambda_t'|Z'\omega_R ZZ''|\phi'';\lambda_t'') \Leftrightarrow D(Z;Z',Z'')$$

i.e.

BEI $\Leftrightarrow$ TDCS.

The evaluation of the $R \neq 0$ BEI is nothing but the straightforward generalization of (44). If we label the numbers d_{joo} by orbitals a_j,

$\phi' = \ldots a_j^{d_{joo}} \ldots,$ the numbers d_{oko} by orbitals b_k,

$\phi'' = \ldots b_k^{d_{oko}} \ldots,$ then put $d_{oo1} = N - R$, $d_{oo2} = R$,

and introduce

1-particle charge distributions $\rho_{jk}^{(1)} = a_j(1)\, b_k(1)$

1-particle overlap integrals $S_{jk} = (a_j(1)|b_k(1))$

then, the BEI reads

$$(\phi';\lambda_t'|Z'\omega_R ZZ''|\phi'';\lambda_t'') = \prod_{jk} (\rho_{jk}^{d_{jk2}(Z;Z',Z'')}|\omega_R)\, S_{jk}^{d_{jk1}(Z;Z'Z'')}, \tag{47}$$

i.e. an R-particle integral times a monomial of degree $N - R$ in 1-particle overlap.

We note that: The factors $|H'|$ and $|H''|$ in (40) are given by

$$\prod_j d_{joo}! \quad \text{and} \quad \prod_k d_{oko}!$$

and the factor d_Z is given by

$$\prod_{jkl} d_{jkl}(Z;Z',Z'')! \tag{48}$$

where we recall that Z was defined for fixed pairs Z', Z''.

From (40^c) it follows that $Z \in K_R$ for all pairs Z', Z''. Thus we can write $Z = Y'Y''$ with $Y' \in S_{N-R}$ and $Y'' \in S_R$.

We then rewrite (47) as

$$(\phi'; \lambda_t' | Z'Y'\omega_R Y''Z'' | \phi''; \lambda_t'') = J^{(R)}(Y'') \, S(Y')$$

and (48) as

$$d_Z = d_{Y'} d_{Y''}, \quad d_{Y'} = \prod_{jk} d_{jk1}(Y'), \quad d_{Y''} = \prod_{jk} d_{jk2}(Y'').$$

The integrals $J^{(R)}(Y'')$ will be called *primitive R-particle* integrals. We can show that there exists a unique correspondence between "top sheets" of TDCS and primitive integrals.

Since we only have two sheets in our TDCS, it follows that for given

ϕ'	(specified by $\{d_{jo0} \| j\}$)
ϕ''	(specified by $\{d_{oko} \| k\}$)
$J^{(R)}(Y'')$	(specified by $\{d_{jk2}(Y'') \| j, k\}$)

the entities Z'' (specified by $\{d_{jol} | l = 1, 2; j\}$)
Z'' (specified by $\{d_{okl} | l = 1, 2; k\}$)

are fixed. Therewith the range of Y' is given by all possible $Z = Y'Y''$ that can exist for given Y''.

Hence: The primitive integral $J^{(R)}(Y'')$ occurs in the original matrix element (40) with a coefficient given by

$$\frac{|\Lambda|}{|G|} |H'||H''| \bar{d}_{Y''} \sum_{Y'} \bar{d}_{Y'} \langle Z'Y'Y''Z'' | k'\lambda_t' \Lambda k''\lambda_t'' \rangle \, S(Y').$$

With Z' and Z'' fixed by Y'', and Y' ranging over all possibilities which are compatible with the chosen Y''.

The compatibility condition is easily given; namely $d_{jk2}(Y'') + d_{jk1}(Y') = d_{jko}$ and the set of all numbers d_{jko} must represent a DCS pertaining to the decomposition of G with respect to H' and H'', respectively.

The element Z (for given Z', Z'') represents a DC of the type (see 40^c)

$$(\bar{Z}'H'Z' \cap K_R) \, Z \, (Z''H''\bar{Z}'' \cap K_R)$$

and since $(\bar{Z}'H'Z' \cap K_R) \subset \bar{Z}'H'Z'$, $(Z''H''\bar{Z}'' \cap K_R) \subset Z''H''\bar{Z}''$ this DC is a subset in the DC given by

$$(\bar{Z}'H'Z') \, Z \, (Z''H''\bar{Z}'').$$

Then, the DCS of Z with respect to $\bar{Z}'H'Z'$ and $Z''H''\bar{Z}''$ is equal to the DCS of $Z'ZZ''$ with respect to H' and H'', can easily be shown.

Since Y' and Y'' with $Y'Y'' = Z$ must "sum up" to a DCS defined with respect to H', H'' we may replace $Z'ZZ''$ by elements W which decompose G with respect to H' and H''.

In the expansion given above we therefore replace the sum over Y′ by a sum over W such that W is compatible with Y″, i.e., explicitly

$$d_{jko}(W) \geqslant d_{jk2}(Y'') \; \forall j, k.$$

If we introduce

$$S(W) = \prod_{jk} (S_{jk})^{d_{jko}(W)} \quad \text{and} \quad S(Y'') = \prod_{jk} (S_{jk})^{d_{jk2}(Y'')}, \tag{49}$$

then

$$S(Y') = S(W)/S(Y'').$$

With $d_W = \prod_{jk} d_{jko}(W)!$ it follows that

$$\bar{d}_{Y''}\bar{d}_{Y'} = \bar{d}_W \prod_{jk} \binom{d_{jko}(W)}{d_{jk2}(Y'')} \equiv C(W, Y'')$$

where

$$C(W, Y'') \neq 0 \quad \text{if} \quad d_{jko}(W) \geqslant d_{jk2}(Y'') \; \forall j, k,$$
$$= 0 \quad \text{otherwise.}$$

The expansion of (40) into primitive integrals then reads:

$$\frac{|G|}{|H'||\Lambda||H''|} (\phi'(k'\lambda_t'); \Lambda \,\|\Omega\|\, \phi''(k''\lambda_t''); \Lambda =$$

$$= \sum_W S(W) \langle W | k'\lambda_t' \Lambda k''\lambda_t'' \rangle \sum_{R=0}^{N} \sum_{Y''} \frac{J^{(R)}(Y'')}{S(Y'')} C(W, Y'') \tag{50}$$

where W ranges over the DC in the decomposition of G with respect to the invariance groups H′ of ϕ' and H″ of ϕ'', respectively.

The most interesting aspect of this expression is that we need no longer doubt the practicality of expansion (40). It is seen that (independent of R) only *one* DC decomposition must be performed, namely G with respect to H′ and H″. Even this one DC decomposition contains a large number of terms in general. Actually the number of DC in G with respect to any two subgroups H^α and H^β is given by

$$N^{\alpha\beta}(G) = \sum_\Lambda \langle \Lambda | \lambda_t^\alpha \nearrow \rangle \langle \Lambda | \lambda_t^\beta \nearrow \rangle \equiv \langle \lambda_t^\alpha \nearrow | \lambda_t^\beta \nearrow \rangle = \langle \lambda_t^\alpha \nearrow {}^G \swarrow_\beta | \lambda_t^\beta \rangle$$

i.e. the number of times the trivial IR λ_t^β of H^β occurs in the representation of H^β which is obtained by first inducing λ_t^α of H^α into G followed by subduction to H^β (App. D).

The number of TDCS of G with respect to H^α, H^β and H^γ is given by (App. D)

$$N^{\alpha\beta\gamma}(G) = \langle \lambda_t^\beta \nearrow {}^G \swarrow_\alpha | \lambda_t^\gamma \nearrow {}^G \swarrow_\alpha \rangle = \langle \lambda_t^\beta \nearrow {}^G \swarrow_\alpha \nearrow^G | \lambda_t^\gamma \nearrow^G \rangle = \langle \lambda_t^\beta \nearrow {}^G \swarrow_\alpha \nearrow^G \swarrow_\gamma | \lambda_t^\gamma \rangle.$$

Again the normalizers of the invariance groups H′ and H″ may be of great help in systematizing expansion (50) (App. E).

What remains in this scheme is the calculation of the algebraic coefficient $\langle W | k'\lambda_t' \Lambda k''\lambda_t'' \rangle$, i.e., matrix elements of the generators W in a sequence-adapted IR. This is a non-trivial, actually a complicated, problem.

In the next section we therefore discuss a similar scheme in which SA is not performed by means of the unitary matric bases $Q(J\Lambda k\lambda_t)$ but by the Young unit type NP operators.

2.3.4 Matrix elements of the spin-free N-electron Hamilton operator over Young unit-type Y^{Λ}-projected N-electron bases. Pauling numbers

In this section we start from (7) which, apart from minor changes in notation, reads

$$(\phi'; \lambda_t' | \Omega\, \sigma^{\Lambda}_{\beta'\beta''} | \phi''; \lambda_t''). \tag{51}$$

First we compare this matrix element with (17), the starting point of the previous section

$$(\phi'; \lambda_t' | \Omega\, Q(k'\lambda_t' \Lambda k''\lambda_t'') | \phi''; \lambda_t'').$$

We then see that the labels β' and β'' should play the role of k′ and k″ i.e., they should refer to linearly independent projections that occur in the induction from λ_t' and λ_t'' to Λ, respectively.

In this section we use projections of the NP-type, $Y^{\Lambda}_{\alpha\beta} | \phi; \lambda_t \rangle$, with α fixed and β ranging over standard Young tableaux (see App. C).

A linearly independent set of such functions pertaining to a given N-particle orbital product has been defined (2.2.4) by means of standard index tableaux. For the electron case where Λ represents a 2-columned IR we define a slightly modified set:

$$\{Y^{\Lambda}_{\alpha\beta} | \phi; \lambda_t) / \alpha \text{ fixed}, f_\beta \in C(H_\phi) \text{ ranging}\} \tag{52}$$

is a complete and linearly independent set of projections pertaining to the primitive product ϕ. Here H_ϕ denotes the invariance group of ϕ, $C(H_\phi)$ denotes the centralizer of H_ϕ, i.e.,

$$C(H_\phi) = \{g \in G / g h \bar{g} = h \;\; \forall h \in H_\phi\}$$

and f_β ranges over standard Young tableaux (YT).

On account of (2.2.4) it follows that the invariance group H_ϕ is of the type

$$H_\phi = \underbrace{S_2 \times S_2 \times \ldots S_2}_{q} \times \underbrace{S_1 \times S_1 \times \ldots \times S_1}_{N-2q} = S_2^q \times S_1^{N-2q} \quad \text{times.}$$

Then

$$\lambda_t = [2]^q [1]^{N-2q} \equiv \lambda_t^{(q)}. \tag{53}$$

On account of (2.2.4) it also follows that for $\Lambda = [2^p 1^{N-2p}] \equiv \Lambda^{(p)}$

$$\langle \Lambda^{(p)} | \lambda_t^{(q)} \nearrow \rangle = 0 \text{ for } p < q. \tag{54}$$

Moreover

$$\langle \Lambda^{(p)} | \lambda_t^{(q)} \nearrow \rangle = \binom{N-2q}{p-q} - \binom{N-2q}{p-q-1} \quad \text{for } p \geqslant q.$$

can be derived.

We note that for $p \geqslant q$ this quantity if equal to the dimension of the IR $[N-p-q, p-q]$ of the symmetric group of degree $N-2q$, i.e., the number of standard YT for the 2-columned diagramm $(N-p-q, p-q)$ pertaining to $N-2q$. It then follows:

For given ϕ, $H_\phi \subseteq R^\Lambda \ \forall \Lambda$ that occur in $\lambda_t \nearrow$, where λ_t denotes the trivial IR of the invariance group H_ϕ of ϕ.

With $H_\phi = S_2^q \times S_2^{N-2q}$, the centralizer of H_ϕ is given by $C(H_\phi) = H_\phi \times S_{N-2q}$. Thus the $f_\beta \in C(H_\phi)$ are elements of S_{N-2q}. Actually they are the permutations which define the standard YT for the diagramm $[N-p-q, p-q]$ in $S_{N-2q}(2q+1, 2q+2 \dots N)$. Their number is given by (54) and equals $\langle \Lambda | \lambda_t \nearrow \rangle$.

Thus we consider (51), where: (55)

$$\sigma^\Lambda_{\beta'\beta''} = f_{\beta'} (PNP)^\Lambda \bar{f}_{\beta''}, \quad f_{\beta'} \in C(H'), \ H' \text{ the invariance group of } \phi'$$
$$f_{\beta''} \in C(H''), \ H'' \text{ the invariance group of } \phi''.$$

Without derivation we state:

Expansion (40) holds if we replace $\langle Z'ZZ'' | k'\lambda_t' \Lambda k''\lambda_t'' \rangle$ by $\sigma^\Lambda_{\beta'\beta''}(Z'ZZ'')$.

Expansion (50) holds with the same replacement, i.e., apart from trivial constants,

$$(\phi'; \lambda_t' | \Omega\, \sigma^\Lambda_{\beta'\beta''} | \phi''; \lambda_t'') = \sum_W S(W)\, \sigma^\Lambda_{\beta'\beta''}(W) \sum_{R=0} \sum_{Y''} \frac{J^{(R)}(Y'')}{S(Y'')}\, C(W, Y''). \quad (56)$$

The proof of the above statement goes as follows:

a) $\sigma^\Lambda_{\beta'\beta''} \sim Q(\lambda_t^{\beta'} \Lambda \lambda_t^{\beta''})$

i.e. the PNP operator $\sigma^\Lambda_{\beta'\beta''}$ is proportional to an element of a unitary matric basis, namely $Q(\lambda_t^{\beta'} \Lambda \lambda_t^{\beta''})$, where

$\lambda_t^{\beta'}$ denotes the trivial IR of $f_{\beta'} R^\Lambda \bar{f}_{\beta'}$,

$\lambda_t^{\beta''}$ denotes the trivial IR of $f_{\beta''} R^\Lambda \bar{f}_{\beta''}$, with R^Λ the row group of the first standard YT T^Λ.

b) The operators $Q(\lambda_t^{\beta'} \Lambda \lambda_t^{\beta''})$ can be expanded in terms of the operators $Q(k'\lambda_t' \Lambda k''\lambda_t'')$ defined in the previous section.

One advantage of the use of $\sigma^\Lambda_{\beta'\beta''}$ over $Q(k'\lambda_t' \Lambda k''\lambda_t'')$ can be indicated immediately:

The coefficients $\langle W | k'\lambda_t' \Lambda k''\lambda_t'' \rangle$ depend on k' and k'', the multiplicity labels referring to the induction of λ_t' of H' (the invariance group of ϕ') and λ_t'' of H'', into Λ.

Such matrix elements are defined up to a unitary transformation on the left and the right. In order to calculate them one therefore has to work with irreducible representation matrices for the groups H′, H″ and G which are explicitly defined and moreover one has to specify the labels k′ and k″, e.g. as a series of IR of groups intermediate between $H' \subset \ldots \subset G$ and $H'' \subset \ldots \subset G$, respectively. This is a problem of sequential coupling or multiple-sequence-adaptation.

The coefficients $\sigma^{\Lambda}_{\beta'\beta''}(W)$ however are equal to the coefficients with which elements $\bar{f}_{\beta'} W f_{\beta''}$ occur in $(\bar{P}NP)^{\Lambda}$, or what amounts to the same

$$\sigma^{\Lambda}_{\beta'\beta''}(W') \sim \langle \bar{f}_{\beta'} W f_{\beta''} | \lambda_t \Lambda \lambda_t \rangle. \tag{57}$$

Since λ_t is the trivial IR of the rowgroup R^{Λ}, λ_t and Λ are defined with respect to one and the same partition of N, there does not exist a multiplicity problem.

The coefficients $\sigma^{\Lambda}_{\beta'\beta''}(W') = \sigma^{\Lambda}_{11}(\bar{f}_{\beta'} W f_{\beta''})$ are known as Pauling numbers.

In the previous section we characterized and evaluated the BEI by means of TDCS. Reconsideration shows that in the expansions given by (40) and (50) we have used the following parts of the symbol:

a) Z′ (set $\{d_{jol}\}$) and Z″ (set $\{d_{okl}\}$), the "first"- and "second projection" of the TDCS,

b) Y′ (set $\{d_{jk1}\}$) and Y″ (set $\{d_{jk2}\}$), the "bottom"- and "top sheet" of the TDCS, $Y'Y'' = Z$.

   This characterizes basic exchange integrals and primitive integrals.

c) sets $\{d_{joo}\}$ and $\{d_{oko}\}$ the "first"- and "second projection" on the axes. Therewith we calculate the coefficients $|H'|$ and $|H''|$.

d) the set $\{d_{jkl}\}$ itself, used to calculate d_Z.

It is tempting to investigate the possibility whether or not the remaining group-theoretical coefficients can be calculated from the "third projection", i.e., the set $\{d_{ool}\}$.

We cannot go into details here but we only state the result in the form of a prescription:

Let D (W) denote the "third projection". Then D (W) is a DCS with rows/columns labelled by the factors in the invariance groups $H' = S_2^{q'} \times S_1^{N-2q'}$ and $H'' = S_2^{q''} \times S_1^{N-2q''}$, respectively.

We use the notation $D^{q',q''}(W)$ and we define a "contracted" DCS $D^{p,p}(W)$ by addition of the

$$\begin{array}{ll} \text{rows} & q'+1 \text{ and } q'+2, q'+3 \text{ and } q'+4, \ldots, 2p-q'-1 \text{ and } 2p-q' \\ \text{columns} & q''+1 \text{ and } q''+2, q''+3 \text{ and } q''+4, \ldots, 2p-q''-1 \text{ and } 2p-q'' \end{array}$$

in $D^{q'q''}(W)$ (See also App. D).

The symbol $D^{p,p}(W)$ represents a DC $R^{\Lambda} W R^{\Lambda}$ with R^{Λ} the rowgroup of the first standard tableau $\mathcal{T}^{\Lambda}$, $\Lambda = [2^p 1^{N-2p}]$.

Let the rows (and columns) in $D^{p,p}(W)$ be labelled as follows

$$\left.\begin{array}{l} j, j', j'' \\ k, k', k'' \end{array}\right\} \in \{1, 2, \ldots, p\} \text{ and } \left.\begin{array}{l} s, s', s'' \\ t, t', t'' \end{array}\right\} \in \{p+1, p+2, \ldots, N-p\}. \tag{58}$$

Then our prescription reads:

Given $D^{q',q''}(W)$.

a) Permute the rows according to f'_β, the columns according to f''_β $\Rightarrow D^{q',q''}(\overline{f}_{\beta'} W f_{\beta''})$.

b) Determine the contracted symbol $\Rightarrow D^{p,p}(\overline{f}_{\beta'} W f_{\beta''})$.

c) Determine:

1. α_W, the number of series of entries in $D^{p,p}(\overline{f}_{\beta'} W f_{\beta''})$ of type $d_{jk} = d_{jk'} = d_{j'k'} = d_{j'k''} = \ldots = d_{j''k} = d_{jk}$ (= 1) including the special case $d_{jk} = 2$.
2. β_W, the number of series of entries in $D^{p,p}(\overline{f}_{\beta'} W f_{\beta''})$ of type $d_{sj} = d_{kj} = d_{kj'} = d_{k'j'} = \ldots = d_{j''t}$ (= 1).
3. γ_W, the number of series of entries in $D^{p,p}(\overline{f}_{\beta'} W f_{\beta''})$ of type $d_{sj} = d_{kj} = d_{kj'} = \ldots = d_{s'j''}$ (= 1) including the special case $d_{s'j} = d_{s''j}$ (= 1) and $d_{jt} = d_{jk} = d_{j'k} = \ldots = d_{j''t'}$ (= 1) including the special case $d_{jt'} = d_{jt''}$ (= 1).

d) Calculate

$$\epsilon_W = \sum_{s's''} d_{s's''} \sum_{t'<s'} \sum_{t''>s''} d_{t't''}. \tag{59}$$

Then

$$\sigma^{\Lambda}_{\beta'\beta''}(W) = (-)^{\epsilon_W}\, 2^{\alpha_W}\, 1^{\beta_W}\, 0^{\gamma_W} \quad (0^0 \equiv 1). \tag{60}$$

We remark that this is not an optimal prescription. In particular we should use the fact stated in App. C that the Pauling number is a constant (up to sign) for all $\overline{f}_{\beta'} W f_{\beta''}$ which belong to a DC defined with respect to the normalizer $N(R^{\Lambda})$.

2.3.5 Summary and Discussion

We presented group theoretical aspects pertaining to the quantum mechanical treatment of physical systems which contain N identical particles. Starting from a model Hamilton operator and a model space in which this operator acts the treatment was based on the concepts: Symmetry adaptation, antisymmetrization and double coset decomposition. The role which is played by group theory has been outlined. Symmetry adaptation involves the decomposition of a space into carrier spaces for irreducible representations of a group G and the construction of bases for the irreducible subspaces. Symmetry adaptation of $V_n \otimes^N$ to S_N has been achieved by means of the matric bases of the group algebra of S_N.

The determination of the antisymmetric component of the product of orbital and spin space has been discussed as a group theoretical coupling procedure. Matrix elements of a spin-free operator over antisymmetric spin-adapted N-particle functions have been evaluated in a spin-free scheme in the language of the symmetric group S_N. The double coset decompositions of S_N arise as a natural tool by means of which maximal use can be made of the invariance groups pertaining to the matrix element under consideration.

The evaluation of matrix elements has been treated as a three-step scheme. First, the transformation to basic exchange integrals, second, the transformation to primitive integrals and third, the calculation of the transformation coefficients.

It has been shown that the basic exchange intergrals can be uniquely characterized by triple double coset symbols and that these symbols contain all information needed in the evaluation of the integrals. The expansions of the original matrix elements in terms of basic exchange integrals, of basic exchange integrals in terms of primitive integrals and of the original matrix elements in terms of primitive integrals, have been obtained explicitly in a form well suited for actual calculations.

Of some importance are those physical systems which allow a multiplicity-free description the meaning of which will be briefly outlined.

Let F (= A, B, C, ...) denote (sub)systems (atoms, molecules, etc.) which contain N_F electrons. Let H_F denote the Hamilton operator for the F-th subsystem and $|\phi_F; (\lambda j)_F)$ the solutions of H_F.

We only allow "weak" interactions to take place between these subsystems and therefore define

$$H^{(0)} = \sum_F H_F$$

the "zero-order" Hamilton operator with eigenfunctions given by the products of the eigenfunctions of the separate systems, say $|\phi; \lambda^0 j^0)$.

Here $\lambda^0 j^0$ is the product of the separate symmetries $(\lambda j)_F$, i.e. the symmetry designation with respect to the zero-order symmetric group $G^0 = \bigtimes_F S_{N_F}$. The induced functions of final symmetry ΛJ are denoted by

$$|\phi(k^0\lambda^0); \Lambda J), \quad k^0 = 1, 2 \ldots, \langle \Lambda | \lambda^0 \nearrow \rangle.$$

A multiplicity-free description is at hand if we are only interested in those symmetries λ^0 and Λ for which $k^0 = 1$.

All applications discussed in the third part satisfy this condition. In the majority of these applications it turns out that we can focus our interest on the "first-order" energies, i.e. the expectation values of H, the complete Hamilton operator, with respect to the set

$$|\phi(\lambda^0); \Lambda J) \quad \text{fixed } \phi, \lambda^0, \text{and ranging } \Lambda.$$

We then have to calculate

$$(\phi(\lambda^0); \Lambda \| \Omega \| \phi(\lambda^0); \Lambda), \quad \phi(\lambda^0) \text{ fixed, } \Lambda \text{ ranging.}$$

The difference between this quantity and

$$(\phi;\lambda^0 \| \Omega \| \phi;\lambda^0)$$

is known as the "first-order exchange energy" for the state specified by Λ. The matrix elements can be calculated as outlined in the previous section. Here we will briefly describe an alternative formulation, often used in solid state physics, particularly in theories of magnetism.

From 2.3.1 it follows that the matrix elements defined above can also be obtained from the operator

$$(\phi;\lambda^0 j^0 | Q^{\rho\sigma}(1^N)\,\Omega | \phi;\lambda^0 j^0) = \frac{1}{|G|}\sum_g (-)^g (\phi;\lambda^0 j^0 | \Omega g | \phi;\lambda^0 j^0)\, g^\sigma$$

acting in a spin space the basis of which is sequence-adapted to $G^0 \subset G$.

If we decompose $Q^{\rho\sigma}(1^N)$ with respect to $G^0 \subset G$, i.e.,

$$Q^{\rho\sigma}(1^N) = \frac{1}{|G|}\sum_Z l_Z\, q^{\rho\sigma}(1^N)\, Z^\rho Z^\sigma q^{\rho\sigma}(1^N)$$

where Z ranges over all DC in the decomposition of G with respect to G^0 and $q^{\rho\sigma}(1^N)$ is the zero-order antisymmetrizer,

$$q^{\rho\sigma}(1^N) = q^{\rho\sigma}(1^{N_A})\, q^{\rho\sigma}(1^{N_B}) \ldots$$

and then expand $q^{\rho\sigma}(1^N)$ in terms of matric bases pertaining to the group G^0, we arrive at

$$(\phi;\lambda^0 j^0 | Q^{\rho\sigma}(1^N)\,\Omega | \phi;\lambda^0 j^0) =$$

$$= \frac{|G^0|}{|\lambda^0||G||\lambda^0|}\sum_Z (-)^Z l_Z \sum_{j'j''} (-)^{\phi_{j'}} (\phi;\lambda^0 j'^0 | \Omega Z | \phi;\lambda^0 j''^0)(-)^{\phi_{j''}} \cdot$$

$$\{q(\tilde{j}^0 \tilde{\lambda}^0 \tilde{j}'^0)\, Z (q(\tilde{j}''^0 \tilde{\lambda}^0 \tilde{j}^0)\}^\sigma.$$

This operator has expectation values in the sequence-adapted – $G^0 \subset G$ – spin space given by

$$\sum_Z J_Z \langle Z\rangle_\Lambda \quad \text{with} \quad \langle Z\rangle_\Lambda = \langle \tilde{\Lambda}\tilde{\lambda}^0\tilde{j}^0 | Z | \tilde{\Lambda}\tilde{\lambda}^0\tilde{j}^0\rangle$$

$$\text{and} \quad J_Z = (-)^Z l_Z (\phi;\lambda^0 j^0 | \Omega Z | \phi^0;\lambda^0 j^0).$$

The operator will be written $H_{eff} = \sum_Z J_Z Z^\sigma$ (eff for "effective").

The spin permutations Z^σ are replaced by spin operators by means of the Dirac identity in the following way:

$$p_{12}^\sigma = \frac{1}{2}(e + s_1 \cdot s_2)$$

$$p_{123} = p_{12}^\sigma p_{23}^\sigma = \frac{1}{4} e + s_1 \cdot s_2 + s_2 \cdot s_3 + (s_1 \cdot s_2)(s_2 \cdot s_3) \quad \text{etc.}$$

Then with p^σ_{12} $1 \in A$ and $2 \in B$ the same DC is generated if we use p^σ_{jk} $j \in A$ and $k \in B$. Therefore we replace

$$p^\sigma_{12} \quad \text{by} \quad \frac{1}{N_A N_B} \sum_{j \in A} \sum_{k \in B} \frac{1}{2}(e + 2s_j \cdot s_k) = \frac{1}{2} + \frac{S_A \cdot S_B}{N_A N_B} \quad \text{etc.}$$

Substitution yields a "spin hamiltonian" of the following type

$$H_{eff} = \sum_{k_{AB}, k_{AC}, \dots k_{BC}} (S_A \cdot S_B)^{k_{AB}} (S_A \cdot S_C)^{k_{AC}} \dots (S_B \cdot S_C)^{k_{BC}} \dots J(k_{AB}, k_{AC} \dots k_{BC} \dots)$$

where

$$J(k_{AB}, k_{AC} \dots k_{BC} \dots) = \sum_Z J_Z(k_{AB}, k_{AC} \dots k_{BC} \dots)$$

with $J_Z(k_{AB}, k_{AC}, \dots k_{BC}, \dots)$ a quantity which depends on the orbital matrix elements J_Z defined above and on the replacements we made to transform Z^σ into a linear combination of subsystem-spin-operators.

All exponents equal to zero, yields $H^{(0)}_{eff}$ which can be omitted in the calculation of the exchange energies.

All exponents equal to zero *but one* yields $H^{(1)}_{eff}$, the *Heisenberg hamiltonian.*

$$H^{(1)}_{eff} = -2 \sum_{A<B} J^{(1)}_{AB}(S_A \cdot S_B) \quad (\text{with } -2J^{(1)}_{AB} = J(1, 0, 0 \dots 0)$$

according to the usual definition).

$$H^{(2)}_{eff} = \sum_{A<B<C} J^{(2)}_{AB}(S_A \cdot S_B)^2 + J^{(2)}_{ABC}(S_A \cdot S_B)(S_B \cdot S_C) \quad \text{etc.}$$

The Heisenberg hamiltonian is a "first-order" operator, i.e. its expectation values in spin space (sequence-adapted to $G^0 \subset G$) are the *first-order exchange energies* according to

$$(\Xi^0; \Lambda_s \| H^{(1)}_{eff} \| \Xi^0; \Lambda_s)/(\Xi^0; \Lambda_s \| I^{(1)}_{eff} \| \Xi^0; \Lambda_s),$$

where $|\Xi^0; \Lambda_s)$ denotes a sequence-adapted spin function and $I^{(1)}_{eff}$ is the operator which is obtained if one replaces H by E, the identity, in $H^{(1)}_{eff}$. The constants $J^{(1)}_{AB}$ are called coupling constants. They can be calculated from the orbital matrix elements, and they can be compared with parameters obtained from experimental data.

As a last remark we note that the use of orthogonal 1-particle orbitals yields

$$J(k_{AB}, k_{AC}, \dots k_{BC} \dots) = 0,$$

if the sum of the entries $k_{AB}, k_{AB}, \dots$ is larger than one. On the basis of orthogonal 1-particle orbitals the Heisenberg hamiltonian is an exact spin hamiltonian.

Appendix D

The triple double coset symbol, double cosets, DC

Let $H^\alpha, H^\beta \subset G$ denote any two subgroups of a group G.

Definition: The DC $H^\alpha W_c^{\alpha\beta} H^\beta$ is the set of all different elements of G given by

$$\{g \in G/g = h^\alpha W_c^{\alpha\beta} h^\beta;\ h^\alpha \in H^\alpha,\ h^\beta \in H^\beta;\ W_c^{\alpha\beta} \in G \text{ fixed}\}. \tag{1}$$

The element $W_c^{\alpha\beta} \in G$ is called the generator of the DC.

Theorem: DC are equivalence classes, i.e.,

$$g \in H^\alpha W_c^{\alpha\beta} H^\beta \Longleftrightarrow g \sim W_c^{\alpha\beta}. \tag{2}$$

Corollaries: (3)

a) DC intersect trivially, i.e.,
   either $g \sim W_c^{\alpha\beta} \rightarrow H^\alpha g H^\beta = H^\alpha W_c^{\alpha\beta} H^\beta$
   or $g \not\sim W_c^{\alpha\beta} \rightarrow H^\alpha g H^\beta \cap H^\alpha W_c^{\alpha\beta} H^\beta = \phi$ in which case the DC generated by g and $W_c^{\alpha\beta}$ are disjoint. We write $g = W_{c'}^{\alpha\beta}$.

b) DC span the group, i.e.,
   there exists a DC decomposition G mod (H^α, H^β),

$$G = \bigoplus_c H^\alpha W_c^{\alpha\beta} H^\beta \quad \text{c ranging over disjoint DC.}$$

   Similarly, there exists an inverse decomposition, G mod (H^β, H^α), the generators of which can be chosen $W_{\bar{c}}^{\beta\alpha} = (W_c^{\alpha\beta})^{-1}$,

$$G = \bigoplus_c H^\beta W_{\bar{c}}^{\beta\alpha} H^\alpha.$$

c) The DC decomposition G mod (H^α, H^β) is unique up to order whereas the generators are unique up to equivalence.

A crucial concept in DC theory is the intersection group.

Definitions: (4)

a) The intersection group of the DC $H^\alpha W_c^{\alpha\beta} H^\beta$ is given by

$$H^\alpha \cap W_c^{\alpha\beta} H^\beta W_{\bar{c}}^{\beta\alpha} \equiv H_c^{\alpha\beta}.$$

b) The (repetition) frequency of the DC $H^\alpha W_c^{\alpha\beta} H^\beta$ is the number of times each element $g \in H^\alpha W_c^{\alpha\beta} H^\beta$ occurs if all multiplications $h^\alpha W_c^{\alpha\beta} h^\beta$, $h^\alpha \in H^\alpha$, $h^\beta \in H^\beta$ ranging, are performed. This frequency is denoted by $f_c^{\alpha\beta}$.

c) The length of the DC $H^\alpha W_c^{\alpha\beta} H^\beta$ is the number of different elements that occur among all $h^\alpha W_c^{\alpha\beta} h^\beta$, $\ell_c^{\alpha\beta}$.

Theorem: $f_c^{\alpha\beta} = |H_c^{\alpha\beta}| = |H_{\bar{c}}^{\beta\alpha}|$. (5)

Corollaries: $l_c^{\alpha\beta} f_c^{\alpha\beta} = |H^\alpha||H^\beta|$ (6)

and

$$\sum_c l_c^{\alpha\beta} = |G| \quad \sum_c \bar{f}_c^{\alpha\beta} = |G|(|H^\alpha||H^\beta|)^{-1}.$$

Special cases of DC arise if either H^α or H^β equals H^τ, the trivial subgroup of G which consists of the identity only.

The DC $H^\tau W_c^{\tau\beta} H^\beta = W_c^{\tau\beta} H^\beta$ are called left cosets (LC) of H^β.

The DC $H^\alpha W_c^{\alpha\tau} H^\tau = W_c^{\alpha\tau} H^\alpha$ are called right cosets (RC) of H^α.

The DC $H^\alpha W_c^{\alpha\beta} H^\beta$ can be decomposed according to LC of H^β,

$$H^\alpha W_c^{\alpha\beta} H^\beta = \bigoplus_n W_{c_n}^{\tau\beta} H^\beta \quad n = 1, 2 \ldots \frac{|H^\alpha|}{f_c^{\alpha\beta}} = \frac{l_c^{\alpha\beta}}{|H^\beta|},$$

RC of H^α,

$$H^\alpha W_c^{\alpha\beta} H^\beta = \bigoplus_m H^\alpha W_{c_m}^{\alpha\tau} \quad m = 1, 2 \ldots \frac{|H^\beta|}{f_c^{\alpha\beta}} = \frac{l_c^{\alpha\beta}}{|H^\alpha|}.$$

These decompositions are unique up to order. They can therefore be used to characterize DC.

Theorem [1]: The number of DC in the decomposition G mod (H^α, H^β) is given by

$$N_G^{\alpha\beta} = \frac{|G|}{|H^\alpha||H^\beta|} \sum_{C \in G} \frac{|H^\alpha \cap C||C \cap H^\beta|}{|C|} \tag{7}$$

in which the sum extends over all classes C of the group G.

Principle induced representation, PIR, and double coset matrices, DCM

Let $\vec{\omega}^H$ denote a 1-ROW matrix the entries of which ar the distinct LC of $H \subset G$, $\omega_j^H = g_j H$, $j = 1, 2, \ldots n^H = |G:H|$

Definition: The PIR of G relative to H is the representation generated by left-multiplication of $\vec{\omega}^H$ by $g \in G$,

$$g\,\vec{\omega}^H = \vec{\omega}^H P^H(g) \quad \text{i.e.,} \quad g\,\omega_{j'}^H = \sum_{j''} \omega_{j''}^H P_{j''j'}^H(g) \tag{8}$$

with

$$P_{j''j'}^H(g) = \delta[j'', g(j')] = 1 \quad \text{if } g_{j''}H = g g_{j'} H$$
$$= 0 \quad \text{otherwise.}$$

Theorem: (9)

a) The PIR P^H of G is the permutation matrix representation induced from the trivial 1-dimensional irreducible representation (IR) λ_t of $H \subset G$. Notation $P^H = \lambda_t^H \uparrow^G$. If no confusion arises we abbreviate $\lambda_t \uparrow$.

b) The character of the PIR is given by:

$$\langle g | \lambda_t \uparrow \rangle = \frac{|G|}{|C_g|} \cdot \frac{|C_g \cap H|}{|H|}$$

with C_g the class of g in G.

c) The decomposition of the PIR is given by:

$$\lambda_t \uparrow = \bigoplus_{\Lambda} \Lambda \langle \Lambda | \lambda_t \uparrow \rangle$$

where $\langle \Lambda | \lambda_t \uparrow \rangle$ denotes the frequency of induction

$$|H| \langle \Lambda | \lambda_t \uparrow \rangle = \langle H | \Lambda \rangle$$

and $\langle H | \Lambda \rangle$ denotes the sum of the characters of the elements $h \in H$ in the IR Λ of G.

We note that right multiplication of $\vec{\omega}^H$ by $g \in G$ makes no sense since the $\omega_j^H g = g_j H_g$ are not cosets. Right-multiplication of $\vec{\omega}^H$ by cosets ω_j^H however, yields linear combinations of cosets.

Let $\vec{\omega}^\alpha$ ($\vec{\omega}^\beta$) denote a 1-COLUMN matrix with entries given by the distinct LC

$$\omega_j^\alpha = g_j^\alpha H^\alpha,\ j = 1, 2 \dots n^\alpha = |G : H^\alpha|,\ (\omega_k^\beta = g_k^\beta H^\beta,\ k = 1, 2 \dots n^\beta = |G : H^\beta|).$$

Consider right-multiplication of $\vec{\omega}^\alpha$ by the DC $H^\alpha W_c^{\alpha\beta} H^\beta$ and express the result as a linear combination of LC of H^β:

$$\vec{\omega}^\alpha \cdot H^\alpha W_c^{\alpha\beta} H^\beta = A(W_c^{\alpha\beta})\, \vec{\omega}^\beta, \quad \text{i.e., explicitly}$$

i.e., explicitly

$$\omega_j^\alpha H^\alpha W_c^{\alpha\beta} H^\beta = \sum_k A_{jk}(W_c^{\alpha\beta})\, \omega_k^\beta.$$

Then

$$A_{jk}(W_c^{\alpha\beta}) = |H^\alpha| \quad \text{if} \quad g_k^\beta H^\beta \in g_j^\alpha H^\alpha W_c^{\alpha\beta} H^\beta, \quad \text{i.e.,} \quad \overline{g_j^\alpha}\, g_k^\beta \sim W_c^{\alpha\beta}$$
$$= 0 \quad \text{otherwise.}$$

Definition [2, 3]: The DCM $B(W_c^{\alpha\beta})$ pertaining to the decomposition G mod (H^α, H^β) are given by

$$B_{jk}(W_c^{\alpha\beta}) = \delta[\overline{g_j^\alpha}\, g_k^\beta, W_c^{\alpha\beta}] = 1 \quad \text{if } g_k^\beta H^\beta \in g_j^\alpha H^\alpha W_c^{\alpha\beta} H^\beta \tag{10}$$
$$= 0 \quad \text{otherwise,}$$

$j = 1, 2, \dots n^\alpha$; $k = 1, 2, \dots n^\beta$; c ranging.

Corollaries:

a) $$|H^\alpha| \sum_{j}^{n^\alpha} B_{jk}(W_c^{\alpha\beta}) = l_c^{\alpha\beta} = |H^\beta| \sum_{k}^{n^\beta} B_{jk}(W_c^{\alpha\beta}) \qquad (11)$$

and

$$f_c^{\alpha\beta} \sum_{j}^{n^\alpha} \sum_{k}^{n^\beta} B_{jk}(W_c^{\alpha\beta}) = |G|.$$

b) If $\alpha = \beta$, $|H^\alpha| S_p[B(W_c^{\alpha\alpha})] = \delta(c, 1)\,|G|$ and in general,

$$f_{c'}^{\alpha\beta} S_p[B(W_{c'}^{\alpha\beta})\,B(W_{\bar{c}''}^{\beta\alpha})] = \delta(c', c'')\,|G|.$$

Theorem [2, 3]:

$$B(W_c^{\alpha\beta}) \Longleftrightarrow H^\alpha W_c^{\alpha\beta} H^\beta, \quad \text{i.e.,} \qquad (12)$$

the correspondence between DC and DCM is unique.

Theorem [2, 3]:

$$P^\alpha(g)\,B(W_c^{\alpha\beta})\,P^\beta(\bar{g}) = B(W_c^{\alpha\beta}) \quad \forall g \in G, \ \forall c, \qquad (13)$$

i.e., DCM intertwine the PIR $\lambda_t^\alpha \nearrow$ and $\lambda_t^\beta \nearrow$.

Actually,

The DCM constitute a complete and linearly independent basis for the subalgebra of G which intertwines $\lambda_t^\alpha \nearrow$ and $\lambda_t^\beta \nearrow$.

Corollary [1, 2, 3]: The number of DC in the decomposition G mod (H^α, H^β) is equal to the intertwining number,

$$N_G^{\alpha\beta} = \langle \lambda_t^\alpha \nearrow | \lambda_t^\beta \nearrow \rangle = \sum_{\Lambda} \langle \Lambda | \lambda_t^\alpha \nearrow \rangle \langle \Lambda | \lambda_t^\beta \nearrow \rangle = \langle \lambda_t^\alpha \swarrow^G \nearrow_\beta | \lambda_t^\beta \rangle = \langle \lambda_t^\beta \nearrow^G \swarrow_\alpha | \lambda_t^\alpha \rangle. \qquad (14)$$

Concerning the DCM defined in (10) we state some more properties: (15)

a) $g \sim W_c^{\alpha\beta} \Longleftrightarrow B(g) = B(W_c^{\alpha\beta})$

b) $B_{jk}(W_c^{\alpha\beta}) = B_{g(j), g(k)}(W_c^{\alpha\beta}) \quad \forall g \in G$

c) $B_{jk}(W_c^{\alpha\beta}) = 1 \Rightarrow B_{jk}(g) = 0 \quad \forall g \not\sim W_c^{\alpha\beta}$

d) $\sum_{c} B(W_c^{\alpha\beta}) = \mathbb{B}, \ \mathbb{B}_{jk} = 1 \quad \forall j = 1, 2 \ldots n^\alpha, \ k = 1, 2 \ldots n^\beta.$

A case of special interest arises for $H^\alpha = H^\beta = H^\tau$.

Then: $P^\alpha = P^\beta = R$, the regular representation of G, and

$$R_{ab}(g) = \delta[g_a, g g_b] = \delta[g_a \bar{g}_b, g].$$

The matrices B, for this case denoted by S, satisfy

$$S_{jk}(g) = \delta[g_k, g_j g] = \delta[\bar{g}_j g_k, g]$$

and they also constitute a representation. R is generated by left-multiplication of the row $\vec{g}$ and S is obtained by right-multiplication of the column $\vec{g}$.
We note that the DCM defined in (10) can be obtained by contraction of the representation S, rows according to $g \in \omega_j^\alpha$, columns according to $g \in \omega_k^\beta$, i.e.,

$$\sum_{g' \in \omega_j^\alpha} \sum_{g'' \in \omega_k^\beta} S_{g'g''}(W_c) = f_c^{\alpha\beta} B_{jk}(W_c^{\alpha\beta}).$$

Double coset structure constants, DCSC.

Consider the following DC decompositions,

G mod (H^α, H^β) with generators $W_c^{\alpha\beta}$, G mod (H^β, H^γ) with generators $W_a^{\beta\gamma}$ and G mod (H^γ, H^α) with generators $W_b^{\gamma\alpha}$;

G mod (H^β, H^α) with generators $W_{\bar{c}}^{\beta\alpha}$, G mod (H^α, H^γ) with generators $W_{\bar{b}}^{\alpha\gamma}$ and G mod (H^γ, H^β) with generators $W_{\bar{a}}^{\gamma\beta}$.

Definition [4]: The DC multiplication constant $C_{ca\bar{b}}^{\alpha\beta\gamma}$ is the coefficient with which the DC $H^\alpha W_{\bar{b}} H^\gamma$ occurs in the product of DC, $H^\alpha W_c H^\beta$ and $H^\beta W_a H^\gamma$, i.e.,

$$H^\alpha W_c H^\beta \cdot H^\beta W_a H^\gamma = \sum_b H^\alpha W_{\bar{b}} H^\gamma C_{ca\bar{b}}^{\alpha\beta\gamma}. \tag{16}$$

The value of these coefficients is easily derived:

$$C_{ca\bar{b}}^{\alpha\beta\gamma} = \frac{|H^\beta|\, f_b^{\gamma\alpha}}{f_c^{\alpha\beta} f_a^{\beta\gamma}}\, |W_c^{\alpha\beta} H^\beta W_a^{\beta\gamma} \cap H^\alpha W_{\bar{b}}^{\alpha\gamma} H^\gamma|. \tag{17}$$

From this expression it is a simple task to find:

a)

$$C_{c'\bar{c}''1}^{\alpha\beta\alpha} = \delta(c', c'')\, l_{c'}^{\alpha\beta} \tag{18}$$

and

b)

$$C_{ca\bar{b}}^{\alpha\beta\gamma} = C_{\bar{a}\bar{c}b}^{\gamma\beta\alpha},\quad C_{ab\bar{c}}^{\beta\gamma\alpha} = C_{\bar{b}\bar{a}c}^{\alpha\gamma\beta},\quad C_{bc\bar{a}}^{\gamma\alpha\beta} = C_{\bar{c}\bar{b}a}^{\beta\gamma\alpha},$$

and

$$l_b^{\gamma\alpha} C_{ca\bar{b}}^{\alpha\beta\gamma} = l_c^{\alpha\beta} C_{ab\bar{c}}^{\beta\gamma\alpha} = l_a^{\beta\gamma} C_{bc\bar{a}}^{\gamma\alpha\beta}. \tag{19}$$

Definition: The quantity

$$\frac{l_b^{\gamma\alpha}}{|H^\alpha||H^\beta||H^\gamma|}\, C_{ca\bar{b}}^{\alpha\beta\gamma} = \langle W_{\bar{b}}^{\alpha\beta} | W_c^{\alpha\beta} W_a^{\beta\gamma} \rangle, \tag{20}$$

satisfies

$$\langle W_{\bar{b}}^{\alpha\gamma} | W_c^{\alpha\beta} W_a^{\beta\gamma} \rangle = \langle W_{\bar{c}}^{\beta\alpha} | W_a^{\beta\gamma} W_b^{\gamma\alpha} \rangle = \langle W_{\bar{a}}^{\gamma\beta} | W_b^{\gamma\alpha} W_c^{\alpha\beta} \rangle =$$

$$= \langle W_b^{\gamma\alpha} | W_{\bar{a}}^{\gamma\beta} W_{\bar{c}}^{\beta\alpha} \rangle = \langle W_c^{\alpha\beta} | W_{\bar{b}}^{\alpha\gamma} W_{\bar{a}}^{\gamma\beta} \rangle = \langle W_a^{\beta\gamma} | W_{\bar{c}}^{\gamma\alpha} W_{\bar{b}}^{\alpha\gamma} \rangle$$

and will be called a DCSC henceforth denoted by

$$\langle W_a^{\beta\gamma}, W_b^{\gamma\alpha}, W_c^{\alpha\beta} \rangle.$$

Corollaries:

a)

$$\sum_a \langle W_a^{\beta\gamma}, W_b^{\gamma\alpha}, W_c^{\alpha\beta} \rangle = \frac{l_b^{\gamma\alpha}\, l_c^{\alpha\beta}}{|H^\alpha||H^\beta||H^\gamma|} \tag{21}$$

and similar relations for "$\sum_b$" and "$\sum_c$".

b)

$$\sum_a \sum_b \langle W_a^{\beta\gamma}, W_b^{\gamma\alpha}, W_c^{\alpha\beta} \rangle = \frac{|G|\, l_c^{\alpha\beta}}{|H^\alpha||H^\beta||H^\gamma|}$$

and similar relations for "$\sum_a \sum_c$" and "$\sum_b \sum_c$".

c)

$$\sum_a \sum_b \sum_c \langle W_a^{\beta\gamma}, W_b^{\gamma\alpha}, W_c^{\alpha\beta} \rangle = \frac{|G|^2}{|H^\alpha||H^\beta||H^\gamma|}\,.$$

DCSC, DCM, and the REGULAR REPRESENTATION

Since DCSC arise from the multiplication of DC they will also appear in the multiplication of DCM on account of the correspondence (12).

The product rule for DCM reads:

$$|H^\beta|\, B(W_c^{\alpha\beta})\, B(W_a^{\beta\gamma}) = \sum_b B(W_{\bar{b}}^{\alpha\gamma})\, C_{ca\bar{b}}^{\alpha\beta\gamma}. \tag{22}$$

Multiplication from the left by $B(W_{b'}^{\gamma\alpha})$, calculation of spurs on the left and the right by means of (11^b), and introduction of (20) then yields

$$\langle W_a^{\beta\gamma}, W_b^{\gamma\alpha}, W_c^{\alpha\beta} \rangle = |G|^{-1}\, \mathrm{Sp}[B(W_a^{\beta\gamma})\, B(W_b^{\gamma\alpha})\, B(W_c^{\alpha\beta})], \tag{23}$$

which expresses DCSC as the trace of a triple product of DCM. This triple product can be written as a linear combination of DCM pertaining to the decomposition G mod (H^β, H^β), say $B(W_j^{\beta\beta})$. Since $B(W_1^{\beta\beta})$ is an n^β-square unit matrix, the diagonal elements of $B(W_j^{\beta\beta})$, $j \neq 1$, are all zero (15[c]) and we can write

$$S_p\,[B(W_a^{\beta\gamma})\,B(W_b^{\gamma\alpha})\,B(W_c^{\alpha\beta})] = \frac{|G|}{|H^\beta|}\,\{B(W_a^{\beta\gamma})\,B(W_b^{\gamma\alpha})\,B(W_c^{\alpha\beta})\}_{11}$$

where sub 11 denotes the (1, 1)-element of the product matrix.

Thus:

$$\begin{aligned}\langle W_a^{\beta\gamma}, W_b^{\gamma\alpha}, W_c^{\alpha\beta}\rangle &= |H^\beta|^{-1}\ \{B(W_a^{\beta\gamma})\,B(W_b^{\gamma\alpha})\,B(W_c^{\alpha\beta})\}_{11} = \\ &= |H^\gamma|^{-1}\ \{B(W_b^{\gamma\alpha})\,B(W_c^{\alpha\beta})\,B(W_a^{\beta\gamma})\}_{11} = \\ &= |H^\alpha|^{-1}\ \{B(W_c^{\alpha\beta})\,B(W_a^{\beta\gamma})\,B(W_b^{\gamma\alpha})\}_{11}.\end{aligned}$$

From this line of argument it is easily seen that DCSC are proportional to the frequency with which the identity occurs in the appropriate triple product of DC. Actually:

$$|H^\alpha|\,|H^\beta|\,|H^\gamma|\,\langle W_a^{\beta\gamma}, W_b^{\gamma\alpha}, W_c^{\alpha\beta}\rangle \tag{24}$$

equals the number of times the identity occurs in

$$\Omega(a, b, c) \equiv H^\beta W_a H^\gamma \cdot H^\gamma W_b H^\alpha \cdot H^\alpha W_c H^\beta,$$

and in

$$\Omega(b, c, a) \quad \text{and in} \quad \Omega(c, a, b).$$

Consider $\Omega(a, b, c)$ as an element of the GA represented by a matrix in the regular representation.

Then, since $\langle g|R\rangle$, the character of g in the representation R, is given by $|G|\,\delta(g, e)$, we have,

$$\langle \Omega(a, b, c)|R\rangle = |G|\,|H^\alpha|\,|H^\beta|\,|H^\gamma|\,\langle W_a^{\beta\gamma}, W_b^{\gamma\alpha}, W_c^{\alpha\beta}\rangle. \tag{25}$$

The left-hand side can be elaborated in 3 steps; first we decompose R according to

$$R = \bigoplus_{\Lambda} \Lambda\,|\Lambda|$$

so that

$$\langle \Omega(a, b, c)|R\rangle = \sum_{\Lambda} |\Lambda|\,\langle \Omega(a, b, c)|\Lambda\rangle; \tag{26}$$

secondly, the character $\langle \Omega(a, b, c)|\Lambda\rangle$ is evaluated as

$$\begin{aligned}&\sum_{(k\lambda j)^\beta}\ \sum_{(n\mu m)^\gamma}\ \sum_{(s\zeta t)^\alpha} \langle H^\beta W_a H^\gamma|(k\lambda j)^\beta\, \Lambda\,(n\mu m)^\gamma\rangle \cdot \\ &\cdot \langle H^\gamma W_b H^\alpha|(n\mu m)^\gamma\, \Lambda\,(s\zeta t)^\alpha\rangle \cdot \langle H^\alpha W_c H^\beta|(s\zeta t)^\alpha\, \Lambda\,(k\lambda j)^\beta\rangle\end{aligned} \tag{27}$$

where we introduced sequence-adapted labels for the rows and columns of the IR Λ; thirdly, we evaluate each of the factors as follows,

$$\langle H^\beta W_a H^\gamma | (k\lambda j)^\beta \Lambda (n\mu m)^\gamma \rangle = \frac{1}{f_a^{\beta\gamma}} \sum_{(k'\lambda' j')^\beta} \sum_{(n'\mu' m')^\gamma} \langle H^\beta | (k\lambda j)^\beta \Lambda (k'\lambda' j')^\beta \rangle \cdot$$

$$\cdot \langle W_a | (k'\lambda' j')^\beta \Lambda (n'\mu' m')^\gamma \rangle \cdot \langle H^\gamma | (n'\mu' m')^\gamma \Lambda (n\mu m)^\gamma \rangle$$

where

$$\begin{aligned} \langle H^\beta | (k\lambda j)^\beta \Lambda (k'\lambda' j')^\beta \rangle &= \delta(kk')\,\delta(\lambda\lambda')\,\langle H^\beta | j\lambda j' \rangle = \\ &= \delta(kk')\,\delta(\lambda\lambda')\,\delta(jj')\,\delta(\lambda\lambda_t^\beta)\ |H^\beta|, \end{aligned} \tag{28}$$

and similarly,

$$\langle H^\gamma | (n'\mu' m')^\gamma \Lambda (n\mu m)^\gamma \rangle = \delta(n'n)\,\delta(\mu'\mu)\,\delta(m'm)\,\delta(\mu\lambda_t^\gamma)\,|H^\gamma|,$$

so that

$$\langle H^\beta W_a H^\gamma | (k\lambda j)^\beta \Lambda (n\mu m)^\gamma \rangle = l_a^{\beta\gamma}\,\delta(\lambda^\beta \lambda_t^\beta)\,\delta(\mu^\gamma \lambda_t^\gamma)\,\langle W_a | k\lambda_t^\alpha \Lambda n \lambda_t^\gamma \rangle \tag{29}$$

which, upon substitution of (29) into (27) and the result into (26), finally yields:

$$\begin{aligned} |G| \langle W_a^{\beta\gamma}, W_b^{\beta\alpha}, W_c^{\alpha\beta} \rangle &= \frac{|H^\alpha||H^\beta||H^\gamma|}{f_a^{\beta\gamma} f_b^{\gamma\alpha} f_c^{\alpha\beta}} \cdot \\ &\cdot \sum_\Lambda |\Lambda| \sum_k \sum_n \sum_s \langle W_a | k\lambda_t^\beta \Lambda n \lambda_t^\gamma \rangle \langle W_b | n\lambda_t^\gamma \Lambda s \lambda_t^\alpha \rangle \langle W_c | s\lambda_t^\alpha \Lambda k \lambda_t^\beta \rangle . \end{aligned} \tag{30}$$

The sum over Λ extends over all IR of G, whereas $k = 1, 2, \ldots \langle \Lambda | \lambda_t^\beta \nearrow \rangle$, $n = 1, 2, \ldots$ $\ldots \langle \Lambda | \lambda_t^\gamma \nearrow \rangle$ and $s = 1, 2, \ldots \langle \Lambda | \lambda_t^\alpha \nearrow \rangle$.

Eq. (30) expresses DCSC in terms of the basic DC constants [3]

$$\langle W_a | k\lambda_t^\beta \Lambda n \lambda_t^\gamma \rangle = \langle g | k\lambda_t^\beta \Lambda n \lambda_t^\gamma \rangle \quad \forall g \in H^\beta W_a H^\gamma$$

$$\langle W_b | n\lambda_t^\gamma \Lambda s \lambda_t^\alpha \rangle = \langle g | n\lambda_t^\gamma \Lambda s \lambda_t^\alpha \rangle \quad \forall g \in H^\gamma W_b H^\alpha$$

and $\langle W_c | s\lambda_t^\alpha \Lambda k \lambda_t^\beta \rangle = \langle g | s\lambda_t^\alpha \Lambda k \lambda_t^\beta \rangle \quad \forall g \in H^\alpha W_c H^\beta$.

A large number of interesting relations can be derived from (30). Some of them are treated at the end of this appendix.

DC and DC SYMBOLS, DCS

The general theory will be applied to the following special case:
$G = S_N$, the symmetric group acting on the numbers $1, 2, \ldots N$.

$$\text{Let } V = \{1, 2, \ldots, N\}. \tag{31}$$

The subgroups H^α and H^β are defined by partitions of N, i.e.,

$$\text{let} \quad \lambda^\alpha = \{d_{jo} | j = 1, 2, \ldots s\} \quad \text{with} \quad \sum_j^s d_{jo} = d_{oo} \equiv N$$

$$\lambda^\beta = \{d_{ok} | k = 1, 2, \ldots t\} \qquad \sum_k d_{ok} = d_{oo} \equiv N,$$

denote any two partitions of N then we define

$$H^\alpha = \mathop{\times}_j^s S_{d_{jo}} \equiv \mathop{\times}_j^s H_j^\alpha \quad \text{and} \quad H^\beta = \mathop{\times}_k^t S_{d_{ok}} \equiv \mathop{\times}_k^t H_k^\beta. \tag{32}$$

Thus the subgroups we discuss in this chapter are direct products of symmetric groups. We complete the definition of $H^\alpha (H^\beta)$ by a specification of the order of factors H_j^α (e.g., $d_{10} \geqslant d_{20} \ldots \geqslant d_{so}$) and by assigning the numbers 1, 2, … N in natural order from the left to the right.

We introduce index sets $V_j^\alpha (V_k^\beta)$ which contain the numbers on which the symmetric groups $H_j^\alpha (H_k^\beta)$ act.

The intersection group $H_c^{\alpha\beta}$ of the DC $H^\alpha W_c^{\alpha\beta} H^\beta$ is given by

$$H_c^{\alpha\beta} = H \cap W_c^{\alpha\beta} H^\beta W_{\bar{c}}^{\beta\alpha} = \mathop{\times}_j^s \mathop{\times}_k^t (H_j^\alpha \cap W_c^{\alpha\beta} H_k^\beta W_{\bar{c}}^{\beta\alpha}). \tag{33}$$

The jk-factor

$$H_j^\alpha \cap W_c^{\alpha\beta} H_k^\beta W_{\bar{c}}^{\beta\alpha}$$

is a symmetric group which acts on the numbers common to the sets V_j^α and $W_c^{\alpha\beta} V_k^\beta$, i.e., the set $V_j^\alpha \cap W_c^{\alpha\beta} V_k^\beta$.

Definition [5, 6]: The DCS $D(W_c^{\alpha\beta})$ pertaining to the decomposition $G\mathrm{mod}(H^\alpha, H^\beta)$ are given by

$$D_{jk}(W_c^{\alpha\beta}) = |V_j^\alpha \cap W_c^{\alpha\beta} V_k^\beta|. \tag{34}$$

Corollaries:

a) $$\sum_j^s D_{jk}(W_c^{\alpha\beta}) = d_{ok} \quad \text{and} \quad \sum_k^t D_{jk}(W_c^{\alpha\beta}) = d_{jo} \quad \forall c, \tag{35}$$

and

$$\sum_j^s \sum_k^t D_{jk}(W_c^{\alpha\beta}) = d_{oo}.$$

b) $$f_c^{\alpha\beta} = \prod_j^s \prod_k^t \{D_{jk}(W_c^{\alpha\beta})\}!$$

Theorem [5, 6]:

$$D(W_c^{\alpha\beta}) \Leftrightarrow H^\alpha W_c^{\alpha\beta} H^\beta \quad \text{i.e.,} \tag{36}$$

the correspondence between DC and DCS is unique.

Corollary:

$$\sum_c \left\{ \prod_j^s \prod_k^t \frac{d_{jo}!\, d_{ok}!}{D_{jk}(W_c^{\alpha\beta})!} \right\} = d_{oo}! \tag{37}$$

where the sum extends over all distinct DC.

Definition: With $\vec{\omega} = (1, 2, \dots N)$ a 1-row matrix we define a permutation matrix representation Q of $G = S_N$ by left-multiplication

$$g\vec{\omega} = \vec{\omega}\, Q(g) \quad \text{with} \quad Q_{st}(g) = \delta[s, g(t)]. \tag{38}$$

The representation Q is the PIR of S_N relative to S_{N-1}.

Corollary: The DCS defined in (34) are the contraction of the representation Q,

rows according to the index sets V_j^α which define H^α, and
columns according to the index sets V_k^β which define H^β. (39)

Thus,

$$Q_{jk}^{(c)}(g) = \sum_{s \in V_j^\alpha} \sum_{t \in V_k^\beta} \delta[s, g(t)] = |V_j^\alpha \cap g V_k^\beta|$$

and

$$Q^{(c)}(g) = D(W_c^{\alpha\beta}) \quad \forall g \in H^\alpha W_c^{\alpha\beta} H^\beta.$$

The converse of contraction is expansion, a process which will be defined for future reference.

Definition: The "first" right coset symbol RCS pertaining to $D(W_c^{\alpha\beta})$ is obtained by expansion of columns according to

$$\begin{pmatrix} D_{1k} \\ D_{2k} \\ \vdots \\ D_{sk} \end{pmatrix} \Rightarrow \begin{pmatrix} \overbrace{1\,1\dots 1}^{} & & & \\ & 1\,1\dots 1 & & \\ & & \ddots & \\ & & & 1\,1\dots 1 \end{pmatrix} \quad k = 1, 2 \dots t. \tag{40}$$

$$\underbrace{}_{D_{1k}} \; \underbrace{}_{D_{2k}} \dots \underbrace{}_{D_{sk}} \quad \text{ones.}$$

This RCS represents the RC $H^\alpha W_{c_1}^{\alpha\tau} \in H^\alpha W_c^{\alpha\beta} H^\beta$.

All other RC $H^\alpha W_{c_\mu}^{\alpha\tau} \in H^\alpha W_c^{\alpha\beta} H^\beta$ can be obtained by permuting the d_{ok} columns in $D(W_{c_1}^{\alpha\tau})$ originating from the k-th column in $D(W_{\alpha\beta}^c)$, $k = 1, 2, \dots t$.

The range over μ is easily calculated

$$\prod_{k}^{t} \frac{\left(\sum_{j}^{s} D_{jk}\right)!}{\prod_{j} D_{jk}!} = \prod_{j}^{s} \prod_{k}^{t} \frac{d_{0k}!}{D_{jk}!} = \frac{|H^{\beta}|}{f_c^{\alpha\beta}} = \frac{l_c^{\alpha\beta}}{|H^{\alpha}|}.$$

Similarly, expansion of rows can be defined. The canonical expansion defined above yields $D(W_{c_1}^{\tau\beta})$ a LCS which represents $W_{c_1}^{\tau\beta}H^{\beta} \in H^{\alpha}W_c^{\alpha\beta}H^{\beta}$. All other LC $W_{c_\nu}^{\tau\beta}H^{\beta}$ are obtained by performing the appropriate permutations.

So far we have two characterizations for DC in S_N, namely by means of DCS and DCM. Some analogies have been pointed out e.g., both DCM and DCS are contracted PIR of certain type.

Crucial differences exist however: DCM span an algebra, they behave like matrices, i.e., multiplication, addition, spur, etc., have definite meaning and significance. DCS do not share these features.

Here we will derive a new property of DCS:

The multiplication rule for DCS

$$l_c^{\alpha\beta} l_a^{\beta\gamma} D(W_c^{\alpha\beta}) \bar{D}(W_1^{\beta\beta}) D(W_a^{\beta\gamma}) = |H^{\alpha}||H^{\beta}||H^{\gamma}| \sum_{b} D(W_{\bar{b}}^{\alpha\gamma}) \langle W_a^{\beta\gamma}, W_b^{\gamma\alpha}, W_c^{\alpha\beta} \rangle \tag{41}$$

where $\bar{D}(W_1^{\beta\beta})$ denotes the inverse of $D(W_1^{\beta\beta})$.

Consider

$$H^{\alpha}W_c^{\alpha\beta}H^{\beta} \cdot H^{\beta}W_a^{\beta\gamma}H^{\gamma} = \sum_{b} H^{\alpha}W_{\bar{b}}^{\alpha\gamma}H^{\gamma} \, C_{ca\bar{b}}^{\alpha\beta\gamma}$$

in the representation Q defined above and contract the left- and right-hand side of this expression with respect to H^{α} and H^{γ} on the left and right respectively.

This yields

$$|H^{\alpha}||H^{\gamma}| \sum_{\mu} \sum_{\nu} D(W_{c_\mu}^{\alpha\tau}) D(W_{a_\nu}^{\tau\gamma}) = \sum_{b} l_b^{\gamma\alpha} C_{ca\bar{b}}^{\alpha\beta\gamma} D(W_{\bar{b}}^{\alpha\gamma}).$$

The left-hand side can be evaluated with some effort and this yields

$$\frac{|H^{\alpha}||H^{\beta}|^2 |H^{\gamma}|}{f_c^{\alpha\beta} f_a^{\beta\gamma}} D(W_c^{\alpha\beta}) \bar{D}(W_1^{\beta\beta}) D(W_a^{\beta\gamma}) = \sum_{b} \frac{|H^{\alpha}||H^{\gamma}|}{f_{\bar{b}}^{\alpha\gamma}} D(W_{\bar{b}}^{\alpha\gamma}) C_{ca\bar{b}}^{\alpha\beta\gamma}.$$

Substitution of (20) yields the desired result. **

Compare (22),

$$B(W_c^{\alpha\beta})\,B(W_a^{\beta\gamma}) = \sum_b B(W_{\bar{b}}^{\alpha\gamma})\,f_b^{\gamma\alpha}\,\langle W_a^{\beta\gamma}, W_b^{\gamma\alpha}, W_c^{\alpha\beta}\rangle$$

with (41)

$$|H^\beta|\,\frac{D(W_c^{\alpha\beta})}{f_c^{\alpha\beta}}\,\bar{D}(W_1^{\beta\beta})\,\frac{D(W_a^{\beta\gamma})}{f_a^{\beta\gamma}} = \sum_b D(W_{\bar{b}}^{\alpha\gamma})\,\langle W_a^{\beta\gamma}, W_b^{\gamma\alpha}, W_c^{\alpha\beta}\rangle.$$

Then one easily sees that the modified DCS

$$E(W_c^{\alpha\beta}) = D^{-1/2}(W_1^{\alpha\alpha})\,\frac{D(W_c^{\alpha\beta})}{f_c^{\alpha\beta}}\,D^{-1/2}(W_1^{\beta\beta}) \tag{42}$$

multiply in exactly the same way as the modified DCM

$$B'(W_c^{\alpha\beta}) = |H^\alpha|^{-1/2}\,B(W_c^{\alpha\beta})\,|H^\beta|^{-1/2} \tag{43}$$

whereas the symbols $l_c^{\alpha\beta} D(W_c^{\alpha\beta})\,\bar{D}(W_1^{\beta\beta})$ multiply in exactly the same way as the DC $H^\alpha W_c^{\alpha\beta} H^\beta$ themselves.

Corollaries: (44)

a) $D^{\alpha\beta}(g'g'') = D^{\alpha\tau}(g')\,D^{\tau\beta}(g'')$

b) $D^{\alpha\beta}(g'gg'') = D^{\alpha\tau}(g')\,D^{\tau\tau}(g)\,D^{\tau\beta}(g'')$
where $D^{\tau\tau}(g) = Q(g)$ defined by (8).

c) $\langle W_a^{\beta\gamma}, W_b^{\gamma\alpha}, W_c^{\alpha\beta}\rangle = |H^\beta|^{-1} \sum_{\mu\nu} \delta\,[D(W_{c_\nu}^{\alpha\tau})\,D(W_{a_\mu}^{\tau\gamma}), D(W_{\bar{b}}^{\alpha\gamma})]$

and similar expressions by cyclically permuting α, β and γ.

The triple double coset symbol, TDCS

Given any positive integer $N = d_{ooo}$ and three of its partitions $\lambda^\alpha, \lambda^\beta$ and λ^γ,
$\lambda^\alpha = \{d_{joo}/j = 1, 2 \ldots s\}$, $\lambda^\beta = \{d_{oko}/k = 1, 2 \ldots t\}$, $\lambda^\gamma = \{d_{ool}/l = 1, 2 \ldots v\}$.

We define:
A TDCS $D(W; W_a^{\beta\gamma}, W_b^{\gamma\alpha}, W_c^{\alpha\beta})$ a "3-dim. matrix" with entries

$D_{jkl}(W; W_a^{\beta\gamma}, W_b^{\gamma\alpha}, W_c^{\alpha\beta})$ which satisfy: (45)

a) $D_{jkl}(W; W_a^{\beta\gamma}, W_b^{\gamma\alpha}, W_c^{\alpha\beta}) \geqslant 0$ integer,

b) $\sum_j^s D_{jkl}(W; W_a^{\beta\gamma}, W_b^{\gamma\alpha}, W_c^{\alpha\beta}) = D_{okl}(W_a^{\beta\gamma})$

$$\sum_{k}^{t} D_{jkl}(W; W_a^{\beta\gamma}, W_b^{\gamma\alpha}, W_c^{\alpha\beta}) = D_{jol}(W_{\bar{b}}^{\alpha\gamma})$$

$$\sum_{l}^{v} D_{jkl}(W; W_a^{\beta\gamma}, W_b^{\gamma\alpha}, W_c^{\alpha\beta}) = D_{jko}(W_c^{\alpha\beta})$$

such that,

c) $$\sum_{k}^{t} D_{jko}(W_c^{\alpha\beta}) = d_{joo} = \sum_{l}^{v} D_{jol}(W_{\bar{b}}^{\alpha\gamma})$$

$$\sum_{j}^{s} D_{jko}(W_c^{\alpha\beta}) = d_{oko} = \sum_{l}^{v} D_{okl}(W_a^{\beta\gamma})$$

$$\sum_{j}^{s} D_{jol}(W_{\bar{b}}^{\alpha\gamma}) = d_{ool} = \sum_{k}^{t} D_{okl}(W_a^{\beta\gamma})$$

and

d) $$\sum_{j}^{s} d_{joo} = \sum_{k}^{t} d_{oko} = \sum_{l}^{v} d_{ool} = d_{ooo} \equiv N.$$

The DCS $D(W_a^{\beta\gamma})$, $D(W_b^{\gamma\alpha})$ and $D(W_c^{\alpha\beta})$ will be called the projections of the TDCS. The label W distinguishes different TDCS for three fixed projections.

Theorem: (46)
Given any two fixed projections, say $D(W_b^{\gamma\alpha})$ and $D(W_c^{\alpha\beta})$, the TDCS $D(W; W_a^{\beta\gamma}, W_b^{\gamma\alpha}, W_c^{\alpha\beta})$, ranging W and $W_a^{\beta\gamma}$, correspond uniquely with the DC in the decomposition $H^\alpha \bmod (H_c^{\alpha\beta}, H_{\bar{b}}^{\alpha\gamma})$.

We will prove the theorem in three steps:

a) any element $h^\alpha \in H^\alpha$ yields a pair $(W, W_a^{\beta\gamma})$

b) elements $Z^\alpha \in H^\alpha$ which generate different DC in the decomposition $H \bmod (H_c^{\alpha\beta}, H_{\bar{b}}^{\alpha\gamma})$ yield different TDCS,

c) different TDCS correspond to different DC.

a) Any $h^\alpha \in H^\alpha$ is factorized $h^\alpha = h_1^\alpha \cdot h_2^\alpha \dots h_j^\alpha \dots h_s^\alpha \in H^\alpha = H_1^\alpha \times H_2^\alpha \times \dots \times H_s^\alpha$. Then h_j^α is represented by a set of numbers $D_{jkl}(h^\alpha; W_b^{\gamma\alpha}, W_c^{\alpha\beta})$ such that

$$\left.\begin{aligned} &\sum_{k}^{t} D_{jkl}(h^\alpha; W_b^{\gamma\alpha}, W_c^{\alpha\beta}) = D_{jol}(W_{\bar{b}}^{\alpha\gamma}) \quad \text{any } l \\ &\sum_{l}^{v} D_{jkl}(h^\alpha; W_b^{\gamma\alpha}, W_c^{\alpha\beta}) = D_{jko}(W_c^{\alpha\beta}) \quad \text{any } k \end{aligned}\right\} \quad j = 1, 2, \dots s.$$

Now, let $\sum_{j}^{s} D_{jkl}(h^\alpha; W_b^{\gamma\alpha}, W_c^{\alpha\beta}) = D_{okl}(h^\alpha)$,

then

$$\sum_{k}^{t} D_{okl}(h^\alpha) = \sum_{k}^{t} \sum_{j}^{s} D_{jkl}(h^\alpha; W_b^{\gamma\alpha}, W_c^{\alpha\beta}) = \sum_{j}^{s} D_{jol}(W_{\bar{b}}^{\alpha\gamma}) = d_{ool} \quad \text{for any } l$$

and

$$\sum_{l}^{v} D_{okl}(h^\alpha) = \sum_{l}^{v} \sum_{j}^{s} D_{jkl}(h^\alpha; W_b^{\gamma\alpha}, W_c^{\alpha\beta}) = \sum_{j}^{s} D_{jko}(W_c^{\alpha\beta}) = d_{oko} \text{ for any k.}$$

Thus, the third projection of $D(h^\alpha; W_b^{\gamma\alpha}, W_c^{\alpha\beta})$ satisfies the sum conditions implying that any $D(h^\alpha; W_b^{\gamma\alpha}, W_c^{\alpha\beta})$ can be identified with some $D(W; W_a^{\beta\gamma}, W_b^{\gamma\alpha}, W_c^{\alpha\beta})$. In particular this will be true for the generators $Z^\alpha \in H^\alpha$ pertaining to the decomposition $H^\alpha \bmod (H_c^{\alpha\beta}, H_{\bar{b}}^{\alpha\gamma})$.

b) Any DC $H_c^{\alpha\beta} Z^\alpha H_{\bar{b}}^{\alpha\gamma}$ is factorized (since H^α is factorized) as

$$\mathop{\times}_{j}^{s} [(H_j^\alpha \cap W_c^{\alpha\beta} H^\beta W_{\bar{c}}^{\beta\alpha}) Z_j^\alpha (H_j^\alpha \cap W_{\bar{b}}^{\alpha\gamma} H^\gamma W_b^{\gamma\alpha})], \; Z_j^\alpha \in H_j^\alpha. \tag{47}$$

Different DC thus differ in a number of factors, say $Z_j'^\alpha$ vs. $Z_j''^\alpha$, which give rise to different "j-sheets" in the TDCS.

c) Obvious.

The frequency of the DC $H_c^{\alpha\beta} Z^\alpha H_{\bar{b}}^{\alpha\gamma}$ corresponding with the TDCS $D(Z^\alpha; W_b^{\gamma\alpha}, W_c^{\alpha\beta})$ is given by

$$\prod_{j}^{s} \prod_{k}^{t} \prod_{l}^{v} \{D_{jkl}(Z^\alpha; W_b^{\gamma\alpha}, W_c^{\alpha\beta})\}! \tag{48}$$

The DC is factorized according to (47). Its frequency is the product of the frequencies for each factor Z_j^α j = 1, 2, ... s. The frequency of the j-th factor is given by the factorial product of the entries of the "j-th sheet" of the TDCS.

Corollary:

$$\sum_{Z^\alpha} \left\{\prod_{j}^{s} \prod_{k}^{t} \prod_{l}^{v} D_{jkl}(Z^\alpha; W_b^{\gamma\alpha} \cdot W_c^{\alpha\beta})!\right\}^{-1} = \frac{|H^\alpha|}{f_c^{\alpha\beta} f_b^{\gamma\alpha}}. \tag{49}$$

Definition:

$$F(W_a^{\beta\gamma}, W_b^{\gamma\alpha}, W_c^{\alpha\beta}) = \sum_{W} \left\{\prod_{j}^{s} \prod_{k}^{t} \prod_{l}^{v} D_{jkl}(W; W_a^{\beta\gamma}, W_b^{\gamma\alpha}, W_c^{\alpha\beta})!\right\}^{-1}. \tag{50}$$

Corollaries:

$$\text{a)}\ \sum_{a} F(W_a^{\beta\gamma}, W_b^{\gamma\alpha}, W_c^{\alpha\beta}) = \frac{l_b^{\gamma\alpha}\, l_c^{\alpha\beta}}{|H^\alpha||H^\beta||H^\gamma|}, \tag{51}$$

$$\text{b)}\ \sum_{a}\sum_{b} F(W_a^{\beta\gamma}, W_b^{\gamma\alpha}, W_c^{\alpha\beta}) = \frac{|G|\, l_c^{\alpha\beta}}{|H^\alpha||H^\beta||H^\gamma|},$$

$$\text{c)}\ \sum_{a}\sum_{b}\sum_{c} F(W_a^{\beta\gamma}, W_b^{\gamma\alpha}, W_c^{\alpha\beta}) = \frac{|G|^2}{|H^\alpha||H^\beta||H^\gamma|}.$$

On account (46) we may replace Z^α in (49) by $(W, W_a^{\beta\gamma})$. Therewith (49) goes over into (51^a).

Compare (51) with (21), not for interest's sake only, since we will prove the following

Theorem:

$$F(W_a^{\beta\gamma}, W_b^{\gamma\alpha}, W_c^{\alpha\beta}) = \langle W_a^{\beta\gamma}, W_b^{\gamma\alpha}, W_c^{\alpha\beta}\rangle. \tag{52}$$

i.e., the value of a DCSC equals the sum of the inverse factorial product of the entries of the TDCS which exist for three fixed projections (on account of (50)).

$$(H^\alpha \cap W_c^{\alpha\beta} H^\beta W_{\bar{c}}^{\beta\alpha})\, Z^\alpha\, (H^\alpha \cap W_{\bar{b}}^{\alpha\gamma} H^\gamma W_b^{\gamma\alpha}) \subset (W_c^{\alpha\beta} H^\beta W_{\bar{c}}^{\beta\alpha})\, Z^\alpha\, (W_{\bar{b}}^{\alpha\gamma} H^\gamma W_b^{\gamma\alpha}).$$

The DCS for Z^α with respect to $W_c^{\alpha\beta} H^\beta W_{\bar{c}}^{\beta\alpha}$ and $W_{\bar{b}}^{\alpha\gamma} H^\gamma W_b^{\gamma\alpha}$ is equal to the DCS for $W_{\bar{c}}^{\beta\alpha} Z^\alpha W_{\bar{b}}^{\alpha\gamma}$ with respect to H^β and H^γ.
Therefore $W_{\bar{c}}^{\beta\alpha} Z^\alpha W_{\bar{b}}^{\alpha\gamma}$ can be identified with some $W_a^{\beta\gamma}$, i.e.,

$$Z^\alpha = W_c^{\alpha\beta} W_a^{\beta\gamma} W_b^{\gamma\alpha}. \tag{53}$$

Then, for fixed $W_a^{\beta\gamma}$, the DC in H^α mod $(H_c^{\alpha\beta}, H_{\bar{b}}^{\alpha\gamma})$ for which the corresponding TDCS have one and the same third projection $D(W_a^{\beta\gamma})$, span $H^\alpha \cap W_c^{\alpha\beta} H^\beta W_a^{\beta\gamma} H^\gamma W_b^{\gamma\alpha}$. Summing their lengths yields

$$f_c^{\alpha\beta} f_b^{\gamma\alpha}\, F(W_a^{\beta\gamma}, W_b^{\gamma\alpha}, W_c^{\alpha\beta}) = |H^\alpha \cap W_c^{\alpha\beta} H^\beta W_a^{\beta\gamma} H^\gamma W_b^{\gamma\alpha}|.$$

Application of (17, 20) yields the desired result.

The number of TDCS

On account of (46): Given two fixed projections $W_c^{\alpha\beta}$, $W_b^{\gamma\alpha}$, the number of TDCS is equal to the number of DC in the decomposition H^α mod $(H_c^{\alpha\beta}, H_{\bar{b}}^{\alpha\gamma})$. This number is given by (14) as

$$\langle \lambda_t(H_c^{\alpha\beta}) \nearrow H^\alpha | \lambda_t(H_{\bar{b}}^{\alpha\gamma}) \nearrow H^\alpha\rangle = \langle \lambda_t(H_c^{\alpha\beta}) \nearrow H^\alpha \swarrow H_{\bar{b}}^{\alpha\gamma} | \lambda_t(H_{\bar{b}}^{\alpha\gamma})\rangle. \tag{54}$$

Summing over b yields the number of TDCS *for one fixed projection* $W_c^{\alpha\beta}$. Since

$$\sum_b \lambda_t(H_{\bar{b}}^{\alpha\gamma})\uparrow H^\alpha = \lambda_t(H^\gamma)\uparrow G\downarrow H^\alpha, \quad [7],$$

we get

$$\langle\lambda_t(H_c^{\alpha\beta})\uparrow H^\alpha|\lambda_t(H^\gamma)\uparrow G\downarrow H^\alpha\rangle = \langle\lambda_t(H_c^{\alpha\beta})\uparrow H^\alpha\uparrow G|\lambda_t(H^\gamma)\uparrow G\rangle =$$
$$= \langle\lambda_t(H_c^{\alpha\beta})\uparrow G|\lambda_t(H^\gamma)\uparrow G\rangle = \langle\lambda_t(H_c^{\alpha\beta})\uparrow G\downarrow H^\gamma|\lambda_t(H^\gamma)\rangle.$$

Summing (54) over b and c yields *the number of TDCS*, $N_G^{\alpha\beta\gamma}$.

$$\sum_{bc}\langle\lambda_t(H_c^{\alpha\beta})\uparrow H^\alpha|\lambda_t(H_{\bar{b}}^{\alpha\gamma})\uparrow H^\alpha\rangle = \langle\lambda_t(H^\beta)\uparrow G\downarrow H^\alpha|\lambda_t(H^\gamma)\uparrow G\downarrow H^\alpha\rangle =$$
$$= \langle\lambda_t(H^\beta)\uparrow G\downarrow H^\alpha\uparrow G|\lambda_t(H^\gamma)\uparrow G\rangle = \tag{55}$$
$$= \langle\lambda_t(H^\beta)\uparrow G\downarrow H^\alpha\uparrow G\downarrow H^\gamma|\lambda_t(H^\gamma)\rangle = N_G^{\alpha\beta\gamma}$$

and similar expressions by permuting α, β and γ.

TDCS and its application in quantum mechanics

The TDCS depends essentially on three partitions of N. We may therefore expect possible applications in the evaluation of quantum mechanical quantities which depend on three partitions of N where N is the number of identical particles of the system.

First we consider the DCS. Let

$$\phi^\alpha = \ldots a_j^{d_{jo}} \ldots \quad \text{and} \quad \phi^\beta = \ldots b_k^{d_{ok}} \ldots \tag{56}$$

denote N-particle product functions. The notation $a_j^{d_{jo}}$ means that the 1-particle orbital a_j is occupied by d_{jo} particles. Similarly b_k is occupied by d_{ok} particles. The occupation numbers are partitions of N,

$$\sum_j^s d_{jo} = N = \sum_k^t d_{ok}.$$

These partitions define the invariance groups H^α and H^β of the product functions ϕ^α and ϕ^β respectively.

This means

$$H^\alpha = \mathop{\times}_j^s S_{d_{jo}} \quad \text{and} \quad h^\alpha\phi^\alpha = \phi^\alpha \quad \forall h^\alpha \in H^\alpha,$$
$$H^\beta = \mathop{\times}_k^t S_{d_{ok}} \quad \text{and} \quad h^\beta\phi^\beta = \phi^\beta \quad \forall h^\beta \in H^\beta. \tag{57}$$

Then consider overlap matrix elements of the following type

$$(\phi^\alpha|g|\phi^\beta), \quad g \in G = S_N.$$

Then, obviously the DC in the decomposition $G \bmod (H^\alpha, H^\beta)$ characterize overlap integrals uniquely, i. e.,

$$(\phi^\alpha | g | \phi^\beta) = (\phi^\alpha | W | \phi^\beta) \quad \forall g \in H^\alpha W H^\beta. \tag{58}$$

By means of the corresponding DCS these overlap integrals can be evaluated as follows: Let

$$H^\alpha W H^\beta \Leftrightarrow D(W).$$

Introduce 1-particle overlap integrals $(a_j | b_k) = S_{jk}$. Then

$$(\phi^\alpha | g | \phi^\beta) = \prod_j^s \prod_k^t S_{jk}^{D_{jk}(W)} \quad \forall g \sim W, \tag{59}$$

where ϕ^α and ϕ^β are given by (56).

Applications for the TDCS may then be found in the evaluation of matrix elements which contain R-particle operators (R = 1, 2, ...), i.e., $(\phi^\alpha; \Lambda \| \Omega \| \phi^\beta; \Lambda)$ where Ω_R is the totally symmetric R-particle operator, ϕ^α and ϕ^β denote the product functions defined above and $|\phi^\alpha; \Lambda)$ and $|\phi^\beta; \Lambda)$ respectively denote induced N-particle functions of final symmetry Λ.

Such matrix elements can be transformed by first generating Ω_R from a prototype R-particle operator ω_R, defined with respect to particles 1, 2, ... R.

The invariance group of ω_R is $H^\gamma = S_R \times S_{N-R}$.

One arrives at reduced matrix elements of the type $(\phi^\alpha; \Lambda\lambda \| \omega_R \| \phi^\beta; \Lambda\lambda)$ where λ denotes an IR of H^γ.

In the evaluation of this type of matrix element three DC decompositions are performed, [6, 3]

a) $G \bmod (H^\alpha, H^\gamma)$ with generators $W_{\bar{b}}^{\alpha\gamma}$

b) $G \bmod (H^\gamma, H^\beta)$ with generators $W_{\bar{a}}^{\gamma\beta}$, and

c) for each pair $W_{\bar{b}}^{\alpha\gamma}$, $W_{\bar{a}}^{\gamma\beta}$ one has to calculate

$$H^\gamma \bmod (H_b^{\gamma\alpha}, H_{\bar{a}}^{\gamma\beta}).$$

The connection with the TDCS is therewith established.

The resulting integrals

$$(\phi^\alpha | W_{\bar{b}}^{\alpha\gamma} Z_\mu^\gamma \omega_R W_{\bar{a}}^{\gamma\beta} | \phi^\beta) \tag{60}$$

called Basic Exchange Integrals [6] are uniquely characterized by TDCS $D(Z_\mu^\gamma; W_a^{\beta\gamma}, W_b^{\gamma\alpha})$. (This is the theorem on page 169). These integrals can also be evaluated by means of the TDCS which characterize them.

With $Z^\gamma \in H^\gamma = S_R \times S_{N-R}$ we can write $Z_\mu^\gamma = Z_{\mu_R}^\gamma Z_{\mu_{N-R}}^\gamma$.

Introduction of 1-particle overlap $(a_j | b_k) = S_{jk}$

and 1-particle charge distributions $a_j b_k = \rho_{jk}$

then yields explicitly:

$$(\phi^{\alpha}|W_{\bar{b}}^{\alpha\gamma} Z_{\mu}^{\gamma} \omega_R W_{\bar{a}}^{\gamma\beta}|\phi^{\beta}) = \left(\prod_{jk}^{st} \rho_{jk}^{D_{jk1}(Z_{\mu R}^{\gamma})}|\omega_R\right) \prod_{jk}^{st} S_{jk}^{D_{jk2}(Z_{\mu N-R}^{\gamma})}. \tag{61}$$

These examples show the significance of the TDCS (See (2.3.3)).

Consequences of (30)

We derived

$$\frac{f_a^{\beta\gamma} f_b^{\gamma\alpha} f_c^{\alpha\beta}}{|H^{\alpha}||H^{\beta}||H^{\gamma}|} |G| \langle W_a^{\beta\gamma}, W_b^{\gamma\alpha}, W_c^{\alpha\beta}\rangle =$$
$$= \sum_{\Lambda} |\Lambda| \sum_k \sum_n \sum_s \langle W_a^{\beta\gamma}|k\lambda_t^{\beta}\Lambda n\lambda_t^{\gamma}\rangle \cdot \langle W_b^{\gamma\alpha}|n\lambda_t^{\gamma}\Lambda s\lambda_t^{\alpha}\rangle \cdot \langle W_c^{\alpha\beta}|s\lambda_t^{\alpha}\Lambda k\lambda_t^{\beta}\rangle \tag{30}$$

where Λ ranges over the IR of G and k, n and s range over the frequencies of induction, $\langle\Lambda|\lambda_t^{\beta}\uparrow\rangle, \langle\Lambda|\lambda_t^{\gamma}\uparrow\rangle$ and $\langle\Lambda|\lambda_t^{\alpha}\uparrow\rangle$, respectively.

We consider the special case $\beta = \gamma$ and $W_a^{\beta\gamma} = W_1^{\beta\beta}$. After some algebra we obtain:

$$l_{c'}^{\alpha\beta} \sum_{\Lambda} |\Lambda| \sum_s \sum_k \langle W_{c'}^{\alpha\beta}|s\lambda_t^{\alpha}\Lambda k\lambda_t^{\beta}\rangle \langle W_{c''}^{\alpha\beta}|s\lambda_t^{\alpha}\Lambda k\lambda_t^{\beta}\rangle^* = |G|\delta(c', c''). \tag{62}$$

This implies that the matrix with rows labelled by $W_c^{\alpha\beta}$ ranging c, columns labelled $s\lambda_t^{\alpha}\Lambda k\lambda_t^{\beta}$, $s\Lambda k$ ranging, and entries given by

$$\left(\frac{|\Lambda|}{|G|} l_c^{\alpha\beta}\right)^{1/2} \langle W_c^{\alpha\beta}|s\lambda_t^{\alpha}\Lambda k\lambda_t^{\beta}\rangle \tag{63}$$

has orthonormal rows.

The matrix is square (the range of c is given by $N_G^{\alpha\beta}$ and the range of $s\Lambda k$ is given by $\sum_{\Lambda}\langle\Lambda|\lambda_t^{\alpha}\uparrow\rangle\langle\Lambda|\lambda_t^{\beta}\uparrow\rangle$) on account of (14).

Therefore the columns of this matrix are orthonormal as well, i.e.,

$$|\Lambda| \sum_c l_c^{\alpha\beta} \langle W_c^{\alpha\beta}|s'\lambda_t^{\alpha}\Lambda' k'\lambda_t^{\beta}\rangle \langle W_c^{\alpha\beta}|s''\lambda_t^{\alpha}\Lambda'' k''\lambda_t^{\beta}\rangle^* = |G|\delta(s's'')\,\delta(\Lambda'\Lambda'')\,\delta(k'k''). \tag{64}$$

Equations (62) and (64) express orthonormality theorems pertaining to the basic DC constants, [3], (DCC).

With $\alpha = \beta = \tau$ we obtain the well-known orthogonality relations for matrix elements of IR.

Repetitive application of (61) to (30) yields a number of expressions which relate DCSC and DCC.

Thus: Multiply (30) with $\langle W_c^{\alpha\beta} | s'\lambda_t^\alpha \Lambda k'\lambda_t^\beta \rangle^*$, sum over c and apply (64). This yields:

$$f_a^{\beta\gamma} f_b^{\gamma\alpha} \sum_c \langle W_a^{\beta\gamma}, W_b^{\gamma\alpha}, W_c^{\alpha\beta} \rangle \langle W_{\bar{c}}^{\beta\alpha} | k\lambda_t^\beta \Lambda s\lambda_t^\alpha \rangle =$$

$$= |H^\gamma| \sum_n \langle W_a^{\beta\gamma} | k\lambda_t^\beta \Lambda n\lambda_t^\gamma \rangle \langle W_b^{\gamma\alpha} | n\lambda_t^\gamma \Lambda s\lambda_t^\alpha \rangle = \langle W_a^{\beta\gamma} H^\gamma W_b^{\gamma\alpha} | k\lambda_t^\beta \Lambda n\lambda_t^\gamma \rangle. \tag{65}$$

Multiply (65) with $\frac{|H^\alpha|}{f_b^{\gamma\alpha}} \langle W_b^{\gamma\alpha} | n'\lambda_t^\gamma \Lambda s'\lambda_t^\alpha \rangle^*$, sum over b and apply (64). This yields:

$$\begin{aligned} &|H^\alpha| f_a^{\beta\gamma} \sum_b \sum_c \langle W_a^{\beta\gamma}, W_b^{\gamma\alpha}, W_c^{\alpha\beta} \rangle \langle W_{\bar{c}}^{\beta\alpha} | k\lambda_t^\beta \Lambda' s'\lambda_t^\alpha \rangle \langle W_{\bar{b}}^{\alpha\gamma} | s''\lambda_t^\alpha \Lambda'' n\lambda_t^\gamma \rangle = \\ &= \frac{|G|}{|\Lambda|} \langle W_a^{\beta\gamma} | k\lambda_t^\beta \Lambda' n\lambda_t^\gamma \rangle \, \delta(s's'') \, \delta(\Lambda'\Lambda''). \end{aligned} \tag{66}$$

Multiply (66) with $l_a^{\beta\gamma} \langle W_a^{\beta\gamma} | k'\lambda_t^\beta \Lambda'' n'\lambda_t^\gamma \rangle^*$, sum over a and apply (64). This yields

$$\begin{aligned} &|H^\alpha||H^\beta||H^\gamma| \sum_a \sum_b \sum_c \langle W_a^{\beta\gamma}, W_b^{\gamma\alpha}, W_c^{\alpha\beta} \rangle \langle W_{\bar{c}}^{\beta\alpha} | k\lambda_t^\beta \Lambda s\lambda_t^\alpha \rangle \cdot \\ &\cdot \langle W_{\bar{b}}^{\alpha\gamma} | s'\lambda_t^\alpha \Lambda' n\lambda_t^\gamma \rangle \langle W_{\bar{a}}^{\gamma\beta} | n'\lambda_t^\gamma \Lambda'' k'\lambda_t^\beta \rangle = \left(\frac{|G|}{|\Lambda|} \right)^2 \delta(\Lambda\Lambda') \, \delta(\Lambda'\Lambda'') \, \delta(ss') \, \delta(nn') \, \delta(kk') \end{aligned} \tag{67}$$

and in particular:

$$\begin{aligned} &\sum_{abc} \langle W_a^{\beta\gamma}, W_b^{\gamma\alpha}, W_c^{\alpha\beta} \rangle \langle W_{\bar{c}}^{\beta\alpha} | k\lambda_t^\beta \Lambda s\lambda_t^\alpha \rangle \langle W_{\bar{b}}^{\alpha\gamma} | s\lambda_t^\alpha \Lambda n\lambda_t^\gamma \rangle \langle W_{\bar{a}}^{\gamma\beta} | n\lambda_t^\gamma \Lambda k\lambda_t^\beta \rangle = \\ &= \left(\frac{|G|}{|\Lambda|} \right)^2 (|H^\alpha||H^\beta||H^\gamma|)^{-1}. \end{aligned} \tag{68}$$

At last we multiply (68) with $|\Lambda|$ and perform a sum over Λ, k, n and s and apply (30):

$$\begin{aligned} &\sum_{abc} \langle W_a^{\beta\gamma}, W_b^{\gamma\alpha}, W_c^{\alpha\beta} \rangle f_a^{\beta\gamma} f_b^{\gamma\alpha} f_c^{\alpha\beta} \langle W_{\bar{c}}^{\beta\alpha}, W_{\bar{b}}^{\alpha\gamma}, W_{\bar{a}}^{\gamma\beta} \rangle = \\ &= |G|^2 \sum_\Lambda \frac{\langle \Lambda | \lambda_t^\alpha \nearrow \rangle \langle \Lambda | \lambda_t^\beta \nearrow \rangle \langle \Lambda | \lambda_t^\gamma \nearrow \rangle}{|\Lambda|}. \end{aligned} \tag{69}$$

Then we consider (66) with $s' = s''$ and $\Lambda' = \Lambda''$, perform a sum over s and apply (65). This yields

$$\sum_{a'} M_{a'a''} C_{a''} = C_{a'}, \tag{70}$$

where

$$C_a = \langle W_a^{\beta\gamma} | k\lambda_t^\beta \Lambda n\lambda_t^\gamma \rangle$$

and

$$M_{a'a''} = \frac{|\Lambda|}{|G|} \langle W^{\beta\gamma}_{a'}, W^{\gamma\alpha}_{b}, W^{\alpha\beta}_{c} \rangle \frac{f^{\beta\gamma}_{a'} f^{\gamma\alpha}_{b} f^{\alpha\beta}_{c}}{\langle \Lambda | \lambda^{\alpha}_{t} \nearrow \rangle} \langle W^{\beta\alpha}_{\bar{c}}, W^{\alpha\gamma}_{\bar{b}}, W^{\gamma\beta}_{\bar{a}''} \rangle . \tag{71}$$

Thus the vector $\mathbb{C}$ with elements C_a is an eigenvector of the matrix $\mathbb{M}$ with eigenvalue 1 for all k and n.

References

[1] *Ruch, E., Hässelbarth, W., Richter, B.*, Theor. Chim. Acta, **19** (1972) 288.
[2] *Frame, J. S.*, Bull. Amer. Math. Soc. **49** (1943) 81.
[3] *Roël, R. W. J.*, Thesis, Amsterdam, 1976.
[4] *Frame, J. S.*, Bull. Amer. Math. Soc. **54** (1948) 740.
[5] *Coleman, A. J.*, "Induced representations with applications to S_N and GL (n)", Queen's Papers in Pure and Applied Mathematics, no. 4, Queen's University, Kingston, Ontario, 1966.
[6] *Kramer, P., Seligman, T. H.*, Nucl. Phys. **A186** (1972) 49.
[7] *Mackey, G. W.*, Anuals of Math. **55** (1952) 101; **58** (1952) 193.

Appendix E

The canonical double coset symbol

Let T^{λ} denote the first standard Young tableau defined with respect to the partition (λ) of N,

$$(\lambda) = 1^{\lambda_1} 2^{\lambda_2} \dots j^{\lambda_j} \dots, \quad \sum_j{}' j\lambda_j = N, \quad \sum_j \lambda_j = N(\lambda). \tag{1}$$

Let T^{λ}_{J} denote the part of T^{λ} which consists of the λ_j rows of length j. Let the rows of T^{λ}_{J} be labelled $\lambda, \lambda = 1, 2, \dots \lambda_j$ and let $V(J_\lambda)$ denote the set of numbers in the λ-row.

By $R^{(\lambda)}_j$ we denote the symmetric group acting on the numbers in $V(J_\lambda)$; $R^{(\lambda)}_j \cong S_j$ $\forall \lambda = 1, 2, \dots \lambda_j$.

Let the columns of T^{λ}_{J} be labelled $t, t = 1, 2, \dots j$ and let $V(J_t)$ denote the set of numbers in the t-column.

By $C^{(t)}_{\lambda_j}$ we denote the symmetric group acting on the numbers in $V(J_t)$; $C^{(t)}_{\lambda_j} \cong S_{\lambda_j}$ $\forall t = 1, 2, \dots j$.

We define:

$$R(J^{\otimes}) = \mathop{\times}_{\lambda=1}^{\lambda_j} R^{(\lambda)}_j, \quad C(J^{\otimes}) = \mathop{\times}_{t=1}^{j} C^{(t)}_{\lambda_j}, \tag{2}$$

and

$$C(J^{\boxtimes}) \subset C(J^{\otimes}), \qquad C(J^{\boxtimes}) \cong S_{\lambda_j} \tag{3}$$

where $C(J^{\boxtimes})$ denotes the diagonal subgroup of $C(J^{\otimes})$. This group can be considered as the group of permutations on the λ_j factors in $R(J^{\otimes})$. At last:

$$R^{\lambda} = \underset{J}{\times} R(J^{\otimes}) \qquad \text{the rowgroup of } T^{\lambda} \tag{4}$$

$$C^{\lambda} = \underset{J}{\times} C(J^{\boxtimes}) \qquad \text{the (sub)group (of the columngroup of } T^{\lambda}\text{) which permutes}$$

rows of equal length in T^{λ}. (5)

Note that the semidirect product $R(J^{\otimes}) \wedge C(J^{\boxtimes})$ is isomorphic with the wreath product $S_j \sim S_{\lambda_j}$ [2].

Definitions:

Double Coset (DC): (6)

With $H', H'' \subset G$ any two subgroups, the DC $H'ZH''$ is the set of different elements of G given by:

$$\{g \in G / g = h'Zh''; h' \in H', h'' \in H''; Z \in G \text{ fixed}\}.$$

The element Z is called the generator of the DC.

Intersection group: (7)

This group is defined as $H' \cap ZH''\overline{Z}$.

(Repetition) frequency: (8)

The number of times each element $g \in H'ZH''$ occurs among all products $h'Zh''$, $h' \in H''$, $h'' \in H''$ ranging, is called the frequency of the DC.

The frequency is denoted by $f_Z : f_Z = |H' \cap ZH''\overline{Z}|$, i.e. the frequency equals the order of the intersection group.

Length: (9)

The number of different elements that occur among all products $h'Zh''$, $h' \in H'$, $h'' \in H''$ ranging, is called the length l_Z of the DC. Obviously

$$l_Z f_Z = |H'||H''|.$$

Normalizer (10)

The normalizer $N(H)$ of $H \subset G$ is the set of elements

$$\{g \in G / gH\overline{g} = H\}.$$

$N(H)$ is the largest subgroup of G in which H is normal,

$$H \leqslant N(H) \subseteq G.$$

Permutation matrix representation, PMR: (11)

By $Q^{(N)}$ we denote the PMR of the symmetric group S_N defined on the set of numbers 1, 2, ... N by

$$Q^{(N)}_{st}(g) = \delta[s, g(t)] \quad \forall g \in S_N, \; s, t = 1, 2 \dots N.$$

The (λ', λ'')*-contraction of* $Q^{(N)}$: With (12)

$$(\lambda') = \dots j'^{\lambda'_{j'}} \dots, \; (\lambda'') = \dots j''^{\lambda''_{j''}} \dots$$

any two partitions of N for which number sets $V(J'_{\lambda'})$ and $V(J''_{\lambda''})$ have been introduced above, we define.

The (λ', λ'')-contraction of $Q^{(N)}$ is the set of matrices, the elements of which are given by

$$Q^{\lambda'\lambda''}_{J'_{\lambda'}J''_{\lambda''}}(g) = \sum_{s \in V(J'_{\lambda'})} \; \sum_{t \in V(J''_{\lambda''})} Q^{(N)}_{st}(g) = |V(J'_{\lambda'}) \cap g V(J''_{\lambda''})|,$$

These matrices have $N(\lambda')$ rows and $N(\lambda'')$ columns.

The $J'_{\lambda'}J''_{\lambda''}$ element of the matrix indicates how many numbers from the set $V(J''_{\lambda''})$ have been mapped into the set $V(J'_{\lambda'})$ under the action of $g \in S_N$. These contracted matrices are known as *double coset symbols, DCS.*

Theorems:

A DC are equivalence classes. As such, (13)

A1 DC intersect trivially, i.e., two DC are either identical or they are disjoint.

A2 DC span the group, i.e., the group is a union of DC.

B With subgroups R^{λ} defined in (4), [1], (14)

$$R^{\lambda'} Z R^{\lambda''} \Leftrightarrow Q^{\lambda'\lambda''}(Z).$$

Calling these DC, R-type DC, this theorem states that R-type DC and DCS are 1-1-corresponding.

Corollary:

A $$f^{\lambda'\lambda''}_Z = \prod_{J'_{\lambda'}} \prod_{J''_{\lambda''}} Q^{\lambda'\lambda''}_{J'_{\lambda'}J''_{\lambda''}}(Z)! \quad (15)$$

i.e., the frequency of an R-type DC is equal to the factorial product of the entries of the corresponding DCS.

B $$\sum_{J'_{\lambda'}} Q^{\lambda'\lambda''}_{J'_{\lambda'}J''_{\lambda''}}(Z) = j'' \quad \forall \lambda'' = 1, 2 \dots \lambda''_{j''}; \; j'' = 1, 2 \dots; \; \forall Z$$

(16)

$$\sum_{J''_{\lambda''}} Q^{\lambda'\lambda''}_{J'_{\lambda'}J''_{\lambda''}}(Z) = j' \quad \forall \lambda' = 1, 2 \dots \lambda'_{j'}; \quad j' = 1, 2 \dots; \; \forall Z.$$

These expressions will be called the sum conditions (of the DCS).

They imply: Any DCS $Q^{\lambda'\lambda''}(Z)$ contains:

$N(\lambda')$ rows; these rows are partitioned into subsets of rows labelled, $J'_{\lambda'}$, $\lambda' = 1, 2, \ldots \lambda'_{j'}$ and each of these rows has the sum of its entries equal to j'. The $\lambda'_{j'}$ rows pertaining to j' ($j' = 1, 2, \ldots$) will be called permutable rows, for reasons soon to become clear.

$N(\lambda'')$ columns; these columns are partioned into subsets of $\lambda''_{j''}$ columns each and any column in such a subset has the sum of its entries equal to j''. The columns $J''_1, J''_2, \ldots J''_{\lambda''_{j''}}$ will be called permutable columns.

Let, according to (11), $Q^{(\lambda_j)}$ denote the PMR of $S_{\lambda_j} \cong C(J^{\boxtimes})$. Then, the direct sum

$$Q^{(\lambda)} = \bigoplus_j Q^{(\lambda_j)}$$

is the PMR of the group C^λ. With $c \in C^\lambda$ the elements of C^λ, we use $Q^\lambda(c)$ to denote the matrices which represent c in the PMR Q^λ.

Theorem: The normalizer N^λ of R^λ in S_N is given by

$$N^\lambda = R^\lambda \wedge C^\lambda = \mathop{\times}_J (R(J^{\otimes}) \wedge C(J^{\boxtimes})). \qquad (17)$$

We note that these normalizers are direct products of wreath products: With $(\lambda) = \ldots j^{\lambda_j} \ldots$, one has

$$N^\lambda \cong \mathop{\times}_j (S_j \sim S_{\lambda_j})$$

Theorem [3]:

$$Q^{\lambda'}(c')\, Q^{\lambda'\lambda''}(Z)\, Q^{\lambda''}(c'') = Q^{\lambda'\lambda''}(c'Zc'') \qquad (18)$$

and

$$Q^{\lambda'\lambda''}_{J'_{\lambda'}J''_{\lambda''}}(c'Zc'') = Q^{\lambda'\lambda''}_{J'_{\bar{c}'(\lambda')}J''_{c''(\lambda'')}}(Z)$$

where $\bar{c}$ denotes the inverse of the element c.

This theorem states that the DCS $Q^{\lambda'\lambda''}(\bar{c}'Zc'')$ where $c' \in N^{\lambda'}/R^{\lambda'}$ and $c'' \in N^{\lambda''}/R^{\lambda''}$ can be obtained from the DCS $Q^{\lambda'\lambda''}(Z)$ by permuting its rows/columns according to the action of c'/c'' respectively.

Since

$$N^\lambda/R^\lambda = C^\lambda = \mathop{\times}_J C(J^{\boxtimes})$$

is a direct product, each factor of which pertains to the λ_j rows of equal length j in the previously defined Young tableau, it follows that the elements $c \in C^\lambda$ are factorized accordingly and, as such, they permute rows (columns) with equal sum condition only. Therefore these rows (columns) were called permutable.

The above theorem (18) defines an equivalence relation on the set of DCS, i.e.,

$$Q^{\lambda'\lambda''}(Z') \sim Q^{\lambda'\lambda''}(Z'') \Leftrightarrow (Z'' \in C^{\lambda'}Z'C^{\lambda''}) \\ \Leftrightarrow Z'' \in N^{\lambda'}Z'N^{\lambda''}. \tag{19}$$

Therewith, the set of DCS is partitioned into disjoint equivalence classes, called DCS-classes:

$$\mathbb{Q}^{\lambda'\lambda''}(Z) = \{Q^{\lambda'\lambda''}(c'Zc'')/c' \in C^{\lambda'}, c'' \in C^{\lambda''}\}. \tag{20}$$

Corollary:

$$\mathbb{Q}^{\lambda'\lambda''}(Z) \Leftrightarrow N^{\lambda'}ZN^{\lambda''} \tag{21}$$

i.e., N-type DC and DCS-classes are 1-1-corresponding.

For the DCS-classes we can define a frequency and a length in exactly the same way as was done for a DC. Thus:
The frequency of the DCS-class $\mathbb{Q}^{\lambda'\lambda''}(Z)$ is the number of times each DCS occurs if all multiplications $Q^{\lambda'\lambda''}(c'Zc'')$, $c' \in C^{\lambda'}$, $c \in C^{\lambda''}$ ranging, are performed. The length of the DCS-class is the number of different DCS that occur among all $Q^{\lambda'\lambda''}(c'Zc'')$ c', c'' ranging.

Instead of an intersection group we define

The invariance group of a DCS: (22)

With $C^{\lambda'} \times C^{\lambda''}$ the abstract direct product of the groups $C^{\lambda'}$ and $C^{\lambda''}$, i.e. a group with elements (c', c'') and multiplication

$$(c_j', c_k'')(c_m', c_m'') = (c_j'c_m', c_k''c_m''),$$

the invariance group

$$C_Z^{\lambda'\lambda''} \quad \text{of the DCS} \quad Q^{\lambda'\lambda''}(Z)$$

is the subgroup of $C^{\lambda'} \times C^{\lambda''}$ the elements of which satisfy

$$Q^{\lambda'}(c')\,Q^{\lambda'\lambda''}(Z)\,Q^{\lambda''}(\overline{c}'') = Q^{\lambda'\lambda''}(Z),$$

or alternatively

$$Q^{\lambda'\lambda''}_{J'_{c'(\lambda')}J''_{c''(\lambda'')}}(Z) = Q^{\lambda'\lambda''}_{J'_{\lambda'}J''_{\lambda''}}(Z) \;\; \forall J'_{\lambda'}, J''_{\lambda''}.$$

Obviously: The frequency of a DCS-class equals the order of the invariance group of its generator. (23)

Moreover: The frequency of an R-type DC is a DCS-class constant, i.e.,

$$f_{Z'}^{\lambda'\lambda''} = f_Z^{\lambda'\lambda''} \;\; \forall Z' \quad \text{for which} \quad Q^{\lambda'\lambda''}(Z') \in \mathbb{Q}^{\lambda'\lambda''}(Z). \tag{24}$$

Hence: The frequency of the N-type DC $N^{\lambda'}ZN^{\lambda''}$ is equal to

$$f_Z^{\lambda'\lambda''}|C_Z^{\lambda'\lambda''}| = F_Z^{\lambda'\lambda''}. \tag{25}$$

The length of the N-type DC $N^{\lambda'}ZN^{\lambda''}$ is equal to

$$\frac{|N^{\lambda'}||N^{\lambda''}|}{F_Z^{\lambda'\lambda''}} = \frac{|R^{\lambda'}||R^{\lambda''}|}{f_Z^{\lambda'\lambda''}} \cdot \frac{|C^{\lambda'}||C^{\lambda''}|}{|C_Z^{\lambda'\lambda''}|},$$

i.e., the length of the corresponding R-type DC times the length of the corresponding DCS-class.

Our aim is three-fold:

1. The definition of a canonical DCS, i.e. the characterization of N-type DC by *one* uniquely defined (and, if possible, easily generated) member of the corresponding DCS-class.
2. The determination of the invariance group of the canonical DCS.
3. The determination of the frequency of the N-type DC.

Our interest is completely general, but here we confine ourselves to the electron case. In this case the relevant partitions are $(\lambda) = [1^{N-2P}2^P] \equiv (P)$.

The canonical DCS in the electron case

From (18):
If $Q^{\lambda'\lambda''}(g') \sim Q^{\lambda'\lambda''}(g'')$, and if for chosen $c' \in C^{\lambda'}, c'' \in C^{\lambda''}$,

$$Q^{\lambda'}(c')\, Q^{\lambda'\lambda''}(g')\, Q^{\lambda''}(g'') = Q^{\lambda'\lambda''}(g),$$

then there always exist $d' \in C^{\lambda'}$, $d'' \in C^{\lambda''}$ such that

$$Q^{\lambda'}(d')\, Q^{\lambda'\lambda''}(g'')\, Q^{\lambda''}(d'') = Q^{\lambda'\lambda''}(g).$$

This is a trivial statement but nevertheless it will be the basis by means of which the canonical DCS will be defined. (We omit (λ', λ''), in this case (p', p'') from now on).
Let Q(g) have the following appearance (Fig. 2.1).

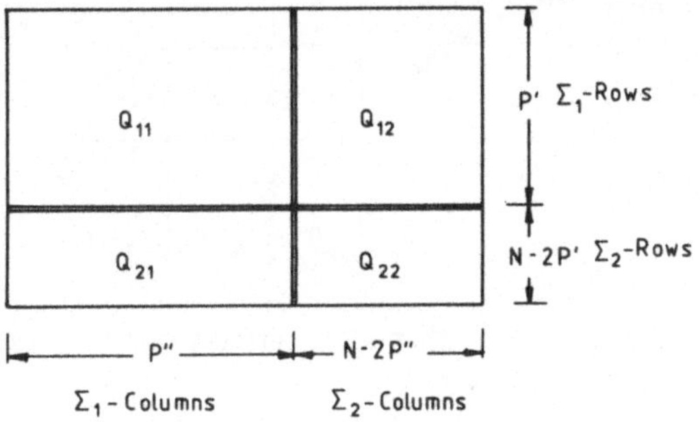

Fig. 2.1 DCS

Let a Σ_j-row/column denote a row/column for which the sum of entries equals j. Then we distinguish the rows as follows:

type Σ_1 with an entry 1 in Q_{11}, say N_1 of them,
type Σ_1 with no entry 1 in Q_{11}, $N - 2P' - N_1$,
type Σ_2 with two entries 1 in Q_{21}, say A_1 of them,
type Σ_2 with one entry 1 in Q_{21}, say A_1' of them,
type Σ_2 with no entry 1 in Q_{21}, $P' - A_1 - A_1' = A_1''$.

The columns are distinguished similarly:

type Σ_1 with an entry 1 in Q_{11}, N_1,
type Σ_1 with no entry 1 in Q_{11}, $N - 2P'' - N_1$,
type Σ_2 with two entries 1 in Q_{12}, say B_1 of them,
type Σ_2 with one entry 1 in Q_{12}, say B_1' of them,
type Σ_2 with no entry 1 in Q_{12}, $P'' - B_1 - B_1' = B_1''$.

Necessarily, we have the relations

$$N_1 + 2A_1 + A_1' = N - 2P''$$
$$N_1 + 2B_1 + B_1' = N - 2P'.$$

Obviously there exist permutations among the permutable rows and columns such that the following partially ordered symbol results (Fig. 2.2):

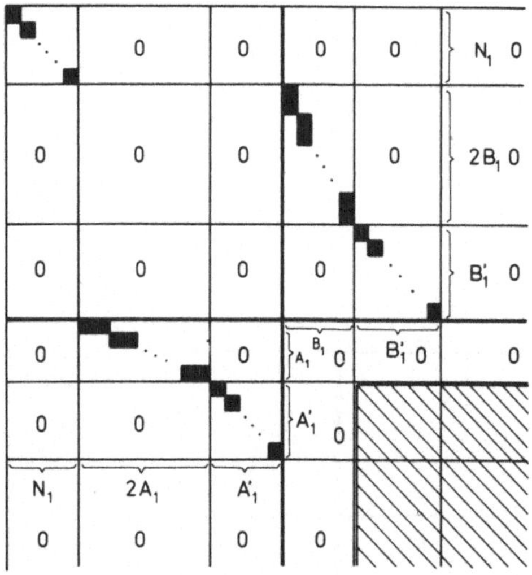

Fig. 2.2
Partially ordered DCS

If $Q(g') \sim Q(g'')$ then after partially ordering the two symbols become equal, apart possibly from the shaded area.

Now consider the A_1'-square unit matrix: for any permutation on the A_1' rows the inverse permutation on the A_1' columns restores the diagonal order and does not affect the symbol. This holds for the B_1' square as well.

Now, permute the A_1' rows and the B_1' columns such that the entries 1 in the $A_1' \times B_1'$ rectangular matrix are put in the upper left diagonal positions. Generally, the A_1'-square and B_1'-square are not longer diagonal but according to the above argument we can restore diagonal order without affecting the symbol.

Thus, we can proceed with the ordering procedure without worrying about the part already ordered. This means that the partially ordered symbol can as well be represented in a contracted way.

We contract as follows:

a) Rows $N_1 + 2B_1 + B_1'$ Columns $N_1 + 2A_1 + A_1'$

b) Rows A_1 Columns B_1.

This yields (Fig. 2.3):

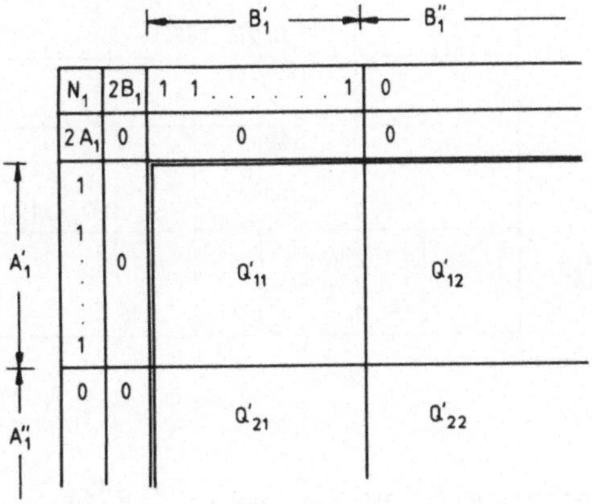

Fig. 2.3
Partially ordered contracted DCS

The ordering – followed by contraction – procedure is repeated for the Q′-part of the symbol (Figs. 2.4 and 2.5).

The procedure becomes repetitive once $N_k = A_k = A_k' = B_k = B_k' = 0$ in which case $A_k'' = B_k'' = R_k \geqslant 0$.

In the k + 1-step one then has to order the DCS pertaining to the decomposition $S_{2R_k} \bmod (S_2^{R_k}, S_2^{R_k})$ which will be discussed later.

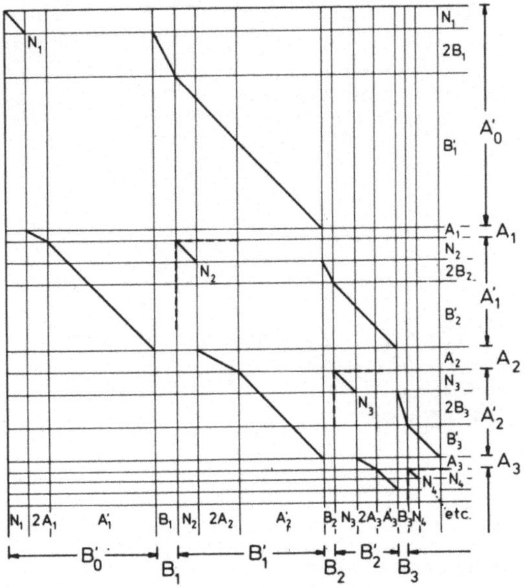

Fig. 2.4
Ordering-procedure

The symbols obtained in this way, called *contracted canonical DCS*, denoted by q(Z), have the following appearance (Fig. 2.5):

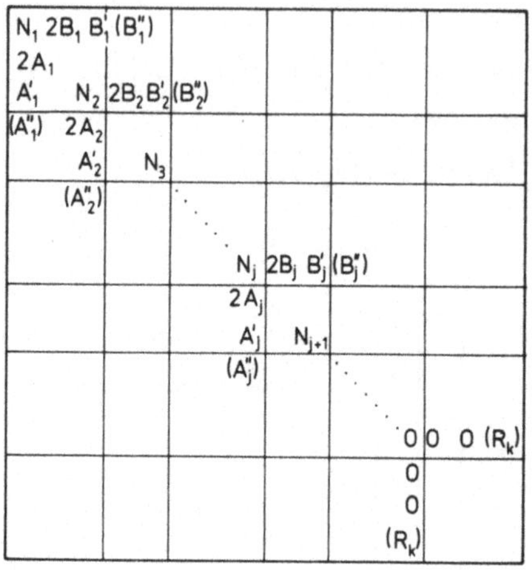

Fig. 2.5
Contraction procedure

With $(\phi) = 1^{\phi_1} 2^{\phi_2} \dots t^{\phi_t} \dots, \quad \sum_t t\phi_t = R_k$

$$F = \prod_{j=1}^{k} N_j!\, A_j!\, B_j!\, 2^{A_j + B_j} \times \prod_{t=1}^{R_k} (2t)^{\phi_t} \phi_t!$$

$$f = 2^{\phi_1}$$

The shaded area represents an ordered DCS for the special case $S_{2R_k} \bmod (S_2^{R_k}, S_2^{R_k})$.
The following lemma is easily proved:

a) $$Q^{p'p''}(Z') \sim Q^{p'p''}(Z'') \Rightarrow q^{p'p''}(Z') = q^{p'p''}(Z'') \equiv q^{p'p''}(Z_c) \qquad (26)$$
(c for canonical)

b) $$Q^{p'p''}(Z') \nsim Q^{p'p''}(Z'') \Rightarrow q^{p'p''}(Z') \neq q^{p'p''}(Z''). \qquad (27)$$

Theorem:

$$q^{p'p''}(Z_c) \Longleftrightarrow N^{p'} Z_c N^{p''} \qquad (28)$$

i.e., contracted canonical DCS and N-type DC are 1-1-corresponding.

Suppose $q^{p'p''}(Z) = q^{p'p''}(Z_c)$.
Then:

either $Q^{p'p''}(Z) \in \mathbb{Q}^{p'p''}(Z_c)$
or $Q^{p'p''}(Z) \notin \mathbb{Q}^{p'p''}(Z_c)$.

In the second case, lemma (b) states that $q^{p'p''}(Z) \neq q^{p'p''}(Z_c)$ which contradicts the assumption. Thus the first case holds, i.e.,

$$q^{p'p''}(Z) = q^{p'p''}(Z_c) \Rightarrow Q^{p'p''}(Z) \in \mathbb{Q}^{p'p''}(Z_c)$$

From lemma (a)

$$Q^{p'p''}(Z) \in \mathbb{Q}^{p'p''}(Z_c) \Rightarrow q^{p'p''}(Z) = q^{p'p''}(Z_c).$$

Since, from (21), $\mathbb{Q}^{p'p''}(Z) \Longleftrightarrow N^{p'} Z N^{p''}$, the theorem is proved.

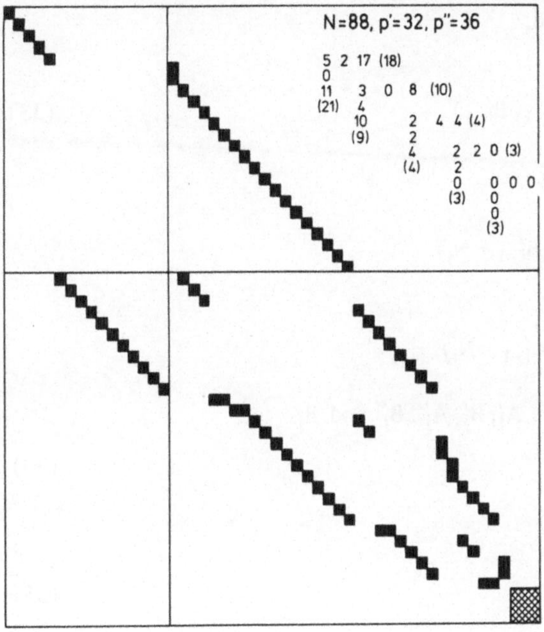

Fig. 2.6
Example of canonical DCS

The way in which the symbols $q^{p'p''}(Z_c)$ have been constructed leads to an obvious and unique expansion procedure, (see Fig. 2.6)

$$q^{p'p''}(Z_c) \Rightarrow Q^{p'p''}(Z_c) \tag{29}$$

the canonical DCS (Figs. 2.4 and 2.5). Hence:

$$Q^{p'p''}(Z_c) \Leftrightarrow N^{p'} Z_c N^{p''}. \tag{30}$$

The contracted canonical DCS, $q^{p'p''}(Z_c)$.

From the ordering-contraction-procedure it follows that the entries of the symbols satisfy certain sum conditions, namely

$$\begin{aligned} &N_j + 2B_j + B_j' = A_{j-1}' && A_0' = N - 2P' \\ &B_j + B_j' + B_j'' = B_{j-1}'' && B_0'' = P'' \\ &N_j + 2A_j + A_j' = B_{j-1}' && B_0' = N - 2P'' \\ &A_j + A_j' + A_j'' = A_{j-1}'' && A_0'' = P'. \end{aligned} \tag{31}$$

From these sum conditions one can prove the following relations:

$$\begin{aligned} &B_j' + 2B_j'' = A_j' + 2A_j'' \; (\equiv \alpha_j) && j \geqslant 0 \\ &B_j' - 2A_j'' = A_j' - 2B_j'' \; (\equiv \beta_j) && j \geqslant 0 \\ &N_j = \alpha_j + \beta_{j-1} && j \geqslant 1. \end{aligned} \tag{32}$$

The following algorithm can be devised.

a) Choose N_j such that

$$\mathrm{Max}\,(0, \beta_{j-1}) \leqslant N_j \leqslant \mathrm{Min}\,(A'_{j-1}, B'_{j-1}) \tag{33}$$

and calculate α_j.

b) Choose A_j such that

$$\mathrm{Max}\,(0, A''_{j-1} - \alpha_j) \leqslant A_j \leqslant \frac{1}{2}(B'_{j-1} - N_j).$$

Choose B_j such that

$$\mathrm{Max}\,(0, B''_{j-1} - \alpha_j) \leqslant B_j \leqslant \frac{1}{2}(A'_{j-1} - N_j).$$

c) For any choice (N_j, A_j, B_j) calculate A'_j, B'_j, A''_j, B''_j and B_j.

Start: $\beta_0 = N - 2P' - 2P''$ (34)

$A'_0 = N - 2P', \quad B'_0 = N - 2P''$

$A''_0 = P' \qquad B''_0 = P''.$

Finish: $N_k = A_k = A'_k = B_k = B'_k = 0$ (35)

$A''_k = B''_k = R_k \geqslant 0.$

The canonical DCS, $Q^{pp}(Z_c)$.

In the algorithm defined above one ends with $A''_k = B''_k = R_k \geqslant 0$ and if $R_k > 0$ one has to determine the ordered DCS for the case $S_{2p} \bmod (S_2^p, S_2^p)$, $p = R_k$.

Theorem [4, 5]:

$$N^p Z_c N^p \Leftrightarrow \text{partitions of p.} \tag{36}$$

With $S_p^2 = S_p \otimes S_p = S_p(1, 3, \ldots 2p-1)\, S_p(2, 4, \ldots 2p)$

and $S_2^p = S_2(1, 2)\, S_2(3, 4) \quad \ldots \quad S_2(2p-1, 2p)$

one has

$$S_{2p} = S_2^p S_p^2 S_2^p.$$

Since $S_p^2 = S_p \otimes S_p = S_p \boxtimes S_p S_p S_p \boxtimes S_p$

one has $S_{2p} = (S_2^p S_p \boxtimes S_p)\, S_p (S_p \boxtimes S_p S_2^p) = N^p S_p N^p$.

With $g \in S_p$, C_g the class of g in S_p

one has $N^p g N^p \cap S_p = C_g$.

Hence

$$N^p g N^p \Leftrightarrow C_g \Leftrightarrow \text{partitions of p.}$$

The canonical DCS will be denoted $Q^{pp}(Z_\pi)$ with $(\pi) = 1^{\pi_1} 2^{\pi_2} \ldots t^{\pi_t} \ldots$ a partition of p.
We define Q_t^{pp} as in Fig. 2.7, and $Q^{pp}(Z_\pi)$ by (37)

$$Q^{pp}(Z_\pi) = \bigoplus_t \bigoplus_{\pi_t} Q_t^{pp}$$

i.e., a matrix in which Q_t^{pp} occurs π_t times along the diagonal. Note that 1^{π_1} is represented by π_1 diagonal entries equal to 2.

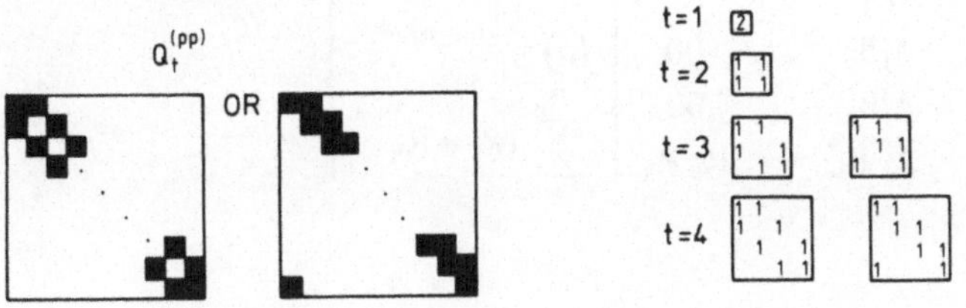

Fig. 2.7 $Q_t^{(pp)}$

The invariance of the canonical DCS [6]

We only present results insofar the order of the invariance group, which determines the frequency of the N-type DC, is concerned.
First, the $R_k > 0$ part.

Consider $Q_t^{pp}, t \geq 1$: invariance group of order $2t$

Then

$\bigoplus_{\pi_t} Q_t^{pp}, \quad t \geq 1$: invariance group of order $(2t)^{\pi_t}\, \pi_t!$

Further Q_1^{pp}: invariance group of order 1

so that

$\bigoplus_{\pi_1} Q_1^{pp}$: invariance group of order $\pi_1!$

From (25) it follows that we also need the frequency of the corresponding R-type DC. This frequency is given by 2^{π_1}.

Thus the contribution of the $R_k > 0$ part to the frequency of the N-type DC is given by

$$\prod_{t=1} (2t)^{\pi_t}\, \pi_t! \quad \text{where} \quad \sum_t t\pi_t = R_k. \tag{38}$$

Next we consider the remaining part of the symbol.

With respect to N_j there exists and invariance group of order N_j.

A_j	$2^{A_j} A_j!$
B_j	$2^{B_j} B_j!$
A_j', B_j'	1.

This gives rise to a contribution

$$\prod_{j=1}^{k} N_j!\, A_j!\, B_j!\, 2^{A_j + B_j}. \tag{39}$$

Theorem: The frequency of the N-type DC $N^{p'}Z_cN^{p''}$ can be determined from the corresponding contracted canonical DCS, $q^{p'p''}(Z_c)$, here represented as (40)

$$q^{p'p''}(Z_c) = \left\{ \begin{array}{ccccc|c} N_1 & & N_j & & 0 & \\ A_1B_1 & \dots & A_jB_j & \dots & 00 & (\pi) = \dots t^{\pi_t} \dots \\ A_1'B_1' & & A_j'B_j' & \dots & 00 & \overset{R_k}{} \\ A_1''B_1'' & & A_j''B_j'' & & R_kR_k & \sum_t t\pi_t = R_k \end{array} \right\}$$

by means of

$$F^{p'p''}(Z_c) = \prod_{j=1}^{k} N_j!\, A_j!\, B_j!\, 2^{A_j + B_j} \prod_{t=1}^{R_k} (2t)^{\pi_t}\, \pi_t!$$

The factor 2^{π_1} is the frequency of the corresponding R-type DC.

Example: $N = 8$, $p' = 3$, $p'' = 2$: The complete set of canonical DCS (and their frequencies) denoted by $\{Q_8^{3,2}(Z_c)\}$ is given in Fig. 2.8.

The corresponding contracted symbols from which they are constructed are e.g.,

no. 3   no. 6   no. 9

$$\left\{ \begin{array}{cc|c} 0 & 0 & \\ 2,1 & 0,0 & (1) \\ 0,0 & 0,0 & \\ 1,1 & 1,1 & \end{array} \right\}, \left\{ \begin{array}{cccc|c} 1 & 0 & 1 & 0 & \\ 1,0 & 0,0 & 0,0 & 0,0 & (0) \\ 1,1 & 1,1 & 0,0 & 0,0 & \\ 1,1 & 0,0 & 0,0 & 0,0 & \end{array} \right\}, \left\{ \begin{array}{cc|c} 2 & 0 & \\ 1,0 & 0,0 & (1^2) \\ 0,0 & 0,0 & \\ 2,2 & 2,2 & \end{array} \right\}$$

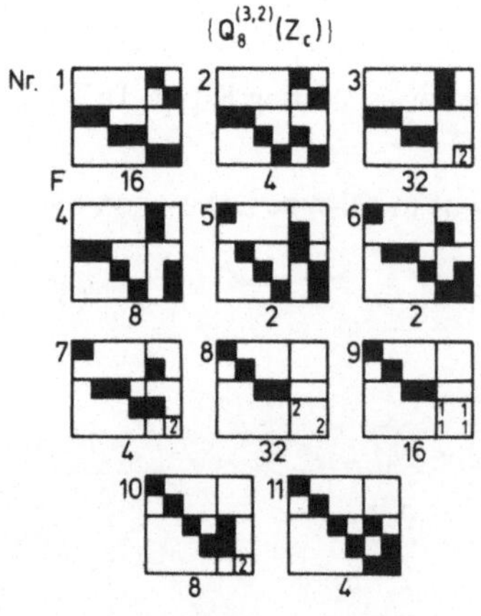

Fig. 2.8 $\{Q_8^{(3,2)}(z_c)\}$

The significance and importance of the canonical DCS

DC decomposition of a group G with respect to subgroups H^α, H^β turns out to be an extremely useful tool for the characterization, classification and evaluation of matrix elements.

Definition:

A DC constant is any function over the group, $\psi(g)$, which satisfies

$$\psi(g) = \psi(Z) \quad \forall g \in H^\alpha Z H^\beta.$$

Sometimes a slightly more general definition is more convenient

$$|\psi(g)| = |\psi(Z)| \quad \forall g \in H^\alpha Z H^\beta,$$

where the vertical bars denote the absolute value, i.e.,

$\psi(g)$ is a constant up to sign.

Examples:

a) Suppose $\phi^\alpha = \ldots (\phi_j^\alpha)^{d_j^\alpha}$

$\phi^\beta = \ldots (\phi_k^\beta)^{d_k^\beta}$

are N-particle product functions where d_j^α (d_k^β) particles occupy the 1-particle orbital ϕ_j^α (ϕ_k^β).

Then with $H^\alpha = \underset{j}{\times} S_{d_j^\alpha}$, a direct product of symmetric groups of degree d_j^α, one has

$$h^\alpha \phi^\alpha = \phi^\alpha \quad \forall h^\alpha \in H^\alpha,$$

and the group H^α is called the invariance group of ϕ^α. Similarly we define H^β with respect to ϕ^β. Obviously:

$$(\phi^\alpha|g|\phi^\beta) = (\phi^\alpha|Z|\phi^\beta) \quad \forall g \in H^\alpha Z H^\beta.$$

Thus, such matrix elements are DC constants. These matrix elements can be classified according to the DC in the decomposition G mod (H^α, H^β).

Since the DC of this type correspond uniquely to DCS, these matrix elements can be characterized by DCS, Q(Z).

If we label the rows of Q(Z), corresponding to the occupation numbers d_j^α, by ϕ_j^α, the columns by ϕ_k^β, and if we introduce 1-particle overlap integrals $S_{jk} = (\phi_j^\alpha|\phi_k^\beta)$, then

$$(\phi^\alpha|Z|\phi^\beta) = \prod_{jk} S_{jk}^{Q_{jk}(Z)}.$$

Thus, the overlap matrix elements can be evaluated by means of the DCS which characterize them.

Actually one only needs the canonical Z_c since with $Z = \bar{c}^{\alpha} Z_c c^{\beta}$, $c^{\alpha} \in N^{\alpha}/H^{\alpha}$ and $c^{\beta} \in N^{\beta}/H^{\beta}$ one has

$$(\phi^{\alpha} | Z | \phi^{\beta}) = \prod_{jk} S^{Q_{jk}(Z_c)}_{c^{\alpha}(j) c^{\beta}(k)}.$$

b) Suppose $\Lambda(g)$ is a unitary matrix which represents $g \in G$ in the IR (irreducible representation) Λ of G.

Let $\Lambda_{jk}(g) \equiv \langle g | J \Lambda k \rangle = \langle \bar{g} | k \Lambda J \rangle^*$ denote its matrix elements. Subducing $\Lambda \swarrow H^{\alpha}$ yields

$$\Lambda \swarrow H^{\alpha} = \bigoplus_{\lambda^{\alpha}} \lambda^{\alpha} \langle \lambda^{\alpha} | \Lambda \swarrow \rangle$$

where $\langle \lambda^{\alpha} | \Lambda \swarrow \rangle$ denotes the frequency with which λ^{α} occurs in $\Lambda \swarrow$. Then there exists a unitary matrix U^{α} such that the matrices $U^{\alpha} \Lambda(h^{\alpha}) U^{\alpha \dagger}$ are block diagonalized $\forall h^{\alpha} \in H^{\alpha}$,

$$U^{\alpha} \Lambda(h^{\alpha}) U^{\alpha \dagger} = \bigoplus_{\lambda^{\alpha}} \lambda^{\alpha}(h^{\alpha}) \langle \lambda^{\alpha} | \Lambda \swarrow \rangle.$$

Then $U^{\alpha} \Lambda(g) U^{\alpha \dagger}$ is called a sequence – $(H^{\alpha} \subset G)$ – adapted IR. [7] (App. B). The elements are given by

$$\langle g | k \lambda^{\alpha}_j \Lambda n \mu^{\alpha} m \rangle, \quad k = 1, 2, \ldots \langle \lambda^{\alpha} | \Lambda \swarrow \rangle, \; n = 1, 2, \ldots \langle \mu^{\alpha} | \Lambda \swarrow \rangle$$

λ^{α} and μ^{α} range over the IR of H^{α},
$j = 1, 2, \ldots |\lambda^{\alpha}|$, $m = 1, 2, \ldots |\mu^{\alpha}|$, and

$$\langle h^{\alpha} | k \lambda^{\alpha}_j \Lambda n \mu^{\alpha} m \rangle = \delta(kn)\, \delta(\lambda \mu) \langle h^{\alpha} | j \mu^{\alpha} m \rangle \quad \forall h^{\alpha} \in H^{\alpha}.$$

Similarly we can decompose Λ with respect to $H^{\beta} \subset G$.

Then generally $U^{\alpha} \Lambda(g) U^{\beta^*}$ is a matrix with elements $\langle g | k \lambda^{\alpha}_j \Lambda n \mu^{\beta} m \rangle$. Again, let $H(H^{\alpha}, H^{\beta})$ denote a direct product of symmetric groups and let λ denote any of its 1-dimensional IR, i.e., λ is a product of 1-dimensional IR, one from each of the factors of H, i.e., either $[d_j]$ or $[1^{d_j}]$ for each factor S_{d_j}.

Then

$\langle g | k \lambda^{\alpha} \Lambda n \lambda^{\beta} \rangle$ equals $\langle Z | k \lambda^{\alpha} \Lambda n \lambda^{\beta} \rangle$ up to the sign,
$\forall g \in H^{\alpha} Z H^{\beta}$, $\forall$ 1-dimensional IR λ^{α} of H^{α} and λ^{β} of H^{β},

the elementary DC constants.

Since N(H) is a group between H and G, its IR can be used for a partial or complete specification of the frequencey label k in $\Lambda k \lambda$. In particular, if the specification is complete, i.e., $H < N(H) \subset G$ is a frequency free sequence for the IR $\lambda(H)$ and $\Lambda(G)$ of interest then the group N(H) is extremely useful, and the DC decomposition $G \bmod (N^{\alpha}, N^{\beta})$ is a prerequisite.

Here we consider an important and simple application [3, 4, 8]: Symmetry-adaptation of orbital space $V_n \otimes^N$ to S_N can be performed e.g. by means of tableau-type operators

$\sigma^{\Lambda}_{\alpha\beta} = \bar{f}_{\alpha} (PNP)^{\Lambda} f_{\beta}$ where

P denotes the row symmetrizer, and
N denotes the column antisymmetrizer of the first standard Young tableau pertaining to the IR Λ of S_N, and
f_α and f_β are permutations which refer to the α- and β-standard Young tableaux.
Consider $(PNP)^\Lambda = \sigma_{11}^\Lambda$.

Theorem: The invariance group of σ_{11}^Λ, both from the left and the right is the normalizer of the rowgroup, $N(R^\Lambda)$.

Corollary:

With

$$\sigma_{11}^\Lambda = \sum_{g \in G} \sigma_{11}^\Lambda(g)\, g,$$

$|\sigma_{11}(g)|$ is a constant $\forall g \in N^\Lambda Z_c N^\Lambda$.

More general:

$$\sigma_{\alpha\beta}^\Lambda(g) = \sigma_{11}^\Lambda(f_\alpha g \bar{f}_\beta) = \sigma_{11}^\Lambda(Z_c)(-)^{\phi(f_\alpha g \bar{f}_\beta, Z_c)}$$

where $\phi(f_\alpha g \bar{f}_\beta, Z_c)$ is a phase which depends on the R-type DC to which $f_\alpha g \bar{f}_\beta$ and Z_c respectively, belong.
The absolute value of the "Pauling number" $\sigma_{\alpha\beta}^\Lambda(g)$ however only depends on the N-type DC to which $f_\alpha g \bar{f}_\beta$ belongs.

References

[1] *Kramer, P., Seligman, T. H.*, Nucl. Phys. **A186** (1972) 49.

[2] *Huppert, B.*, "Endliche Gruppen I" (page 94), Springer-Verlag, Berlin, Heidelberg, New York, 1967.

[3] *Seligman, T. H.*, Hab. Schrift, Tübingen, 1974.
Klein, D. J., Junker, B. R., J. Chem. Phys. **54** (1971) 4290.
Junker, B. R., Klein, D. J., J. Chem. Phys. **54** (1971) 5532.

[4] *Roël, R. W. J.*, Thesis, Amsterdam, 1976.

[5] *Siems, W. F., Poshusta, R. D.*, J. Chem. Phys. **64** (1976) 20.

[6] *Roël, R. W. J.*, unpublished.

[7] *Klein, D. J., Carlisle, C. H., Matsen, F. A.*, Adv. Qu. Chem. V, Ed. P.-O. Löwdin, Ac. Press, New York, London, 1970.

[8] *Roël, R. W. J.*, Lecture Notes in Physics 50, Group Theoretical Methods in Physics, V-th Int. Coll., Nijmegen, 1975, Springer-Verlag, Berlin, Heidelberg, New York, 1976.

Appendix F

General References

1:`

[1] *Jansen, L., Boon, M.*, "Theory of Finite Groups. Applications in Physics", N.-H. Publ. Cy., Amsterdam, 1967.

[2] *Wigner, E. P.*, "Group Theory and its Applications to the Quantum Mechanics of Atomic Spectra", Ac. Press, New York, 1959.

[3] *H. Weyl*, "The Theory of Groups and Quantum Mechanics", Dover Publ. Inc., (S. Hirzel Verlag, Leipzig, 1931).

[4] *Weyl, H.*, "The Classical Groups", Princeton University Press, 1946.

2:

[5] *Löwdin, P.-O.*, "Algebra, Convolution Algebra, Applications to Quantum Mechanics", Rev. Mod. Phys., **39** (1967) 259.

[6] *Löwdin, P.-O., Goscinski, O.*, "The exchange phenomenon, the Symmetric Group, the spin-degeneracy problem", Int. J. Qu. Chem., **3S** (1970) 533.

[7] *Matsen, F. A.*, "Spin-free Quantum Chemistry", in Adv. in Qu. Chem. , Ed. Löwdin, P.-O., Vol. 1, Ac. Press, New York, 1964.

[8] *Klein, D. J., Carlisle, C. H., Matsen, F. A.*, "Symmetry adaptation to Sequences of Finite Groups", in Adv. in Qu. Chem. , Ed. Löwdin, P.-O., Vol. 5, Ac. Press, New York, 1970.

[9] *Klein, D. J.*, "Finite Groups and Semi simple Algebras in Quantum Mechanics", in Group Theory and its Applications , Ed. Loebl, E. M., Vol. 3, Ac. Press, New York, 1975.

[10] *Boerner, H.*, "Representations of Groups", N.-H. Publ. Cy., Amsterdam, 1970.

[11] *Hamermesh, M.*, "Group Theory", Addison-Wesley Publ. Cy., Reading, Massachusetts, 1962.

[12] *Rutherford, D. E.*, "Substitutional Analysis", Edinburgh University Press, Edinburgh, 1948.

3.

No general references available.

Appendix G

Special References.

2.2.3; 2.2.4; 2.2.5; Appendix A

[13] *Wormer, P.*, "Intermolecular Forces and the Group Theory of Many-body Systems", Thesis, University of Nijmegen, Nijmegen, 1975.

2.2.2; Appendix B; 2.3.3; 2.3.4

[14] *Roël, R. W. J.*, "Perturbation Theory for Interacting Systems, The Quantum Mechanical Eigenvalue Problem and Double Coset Decompositions of the Symmetric Group", Thesis, University of Amsterdam, Amsterdam, 1976.

2.3.2

[15] *Grabenstetter, J. E., Tseng, T. J., Grein, F.*, Int. J. of Qu. Chem., **10** (1976) 143.

2.3.3

[16] *Kramer, P., Seligman, T. H.*, Nucl. Phys., **A123** (1969) 161, **A136** (1969) 545, **A186** (1972) 49.

[17] *Seligman, T. H.*, Rev. Mex. Fis., **22** (1973) 151.

2.3.4

[18] *Roël, R. W. J.*, "Invariance Groups of Young Operators, Pauling Numbers", in "Group Theoretical Methods in Physics", Ed. *Janner, A., Janssen, T.* and *Boon, M.*, Springer Verlag, Berlin, Heidelberg, New York, 1976.

2.3.5

[19] *Matsen, F. A., Klein, D. J., Foyt, D. C.*, "Spin-free Quantum Chemistry X The Effective Spin Hamiltonian", J. Phys. Chem., **75** (1971) 1866.

[20] *Block, R.*, "Physical Aspects of Direct and Indirect Exchange Interactions in Molecular Aggregates and in Non-Conducting Solids", Thesis, University of Amsterdam, Amsterdam, 1974.

Results of 2.2.2; 2.3.3; and 2.3.4 have not yet been published.

3 The effective electron model: Applications

3.1 Introduction

We will now go over to the analysis of a number of specific physical problems, using some of the group-theoretical concepts developed in Part 2. As with any physical theory, the aim is to produce numbers which can be compared with experimental results.

Before doing so, however, we must develop *models* for the physical systems under consideration. This necessity of simplification follows from the complexity of the problems selected: it is simply impossible to carry out first-principle calculations to solve the Schrödinger equation. A *model* constitutes an *a priori* reduction of the physical complexity of a system. Once these simplifications have been made, we carry out an as-rigorous-as-possible analysis on the basis of the model.

The phenomena to be analysed are:

a) exchange interactions in non-conducting solids of composition MX with paramagnetic 3d-cations, such as solid MnS, NiO, etc.;

b) stability and conformation of rare-gas halides such as XeF_2, XeF_4 and KrF_2;

c) rotational barriers in simple molecules;

d) low-temperature spin-structures of the Mn-pyrites MnS_2, $MnSe_2$ and $MnTe_2$.

We intend to show that the observed phenomena can be described on the basis of one-an-the-same model, and that they can all be accounted for in terms of exchange interactions. It should be kept in mind that the applications of the group-theoretical machinery developed in Part 2 extend, of course, far beyond the elemental physical systems to be discussed in this Part. All we will need, in fact, is the spin-free evaluation of matrix elements occurring in symmetry-adapted perturbation theory through the method of symmetric double-coset decompositions of permutation groups.

3.2 Effective-electron model for weak chemical bonding

It is obvious that an *a priori* evaluation of exchange interactions for a general system of, say, two paramagnetic cations and one diamagnetic anion, generally lies beyond the capabilities of present first-principle theory. Therefore, a model based on a number of drastic approximations must be adopted; the validity of these assumptions is partly hypothetical, partly empirical and must be ultimately established by comparison with experimental facts. We adopt the following simplifications:

A) *Weak* exchange interactions may be evaluated by using perturbation methods, starting from free-ion wavefunctions in zeroth order of approximation;

B) *Permutation symmetry* of the Hamiltonian determines the essential characteristics of (weak) exchange interactions, i.e. in the symmetry group G_H we may replace the group F_H, containing all symmetries except permutations, by the *identity* element;

C) Properties of (weak) exchange interactions are not *primarily* dependent upon the number of unpaired electrons present. As a consequence, we replace, on each paramagnetic cation, its unpaired electrons by one "*effective*" electron. The number of unpaired electrons is reflected indirectly by the "*magnetic size*" of the orbital of the effective electron.

Assumption (A) implies that the experimental facts to be explained must be reproduced already in the *lowest* orders of the perturbation series which yield a non-zero contribution to the exchange. We choose a Rayleigh-Schrödinger type of exchange perturbation theory [1] and limit ourselves to effects of first and second orders. The adoption of *free-ion* wavefunctions in zeroth order implies automatically that those at different centers are *not* orthogonal. Assumption (B) implies, first of all, that translational symmetry of the crystal is but of secondary importance for the phenomena of weak coupling in insulating solids. In view of the short-range character of exchange interactions, this *local* aspect of the model seems reasonable. Further, given validity of (A), we are here assuming that the effect of the ligands (the diamagnetic anions) on the coupling between unpaired electrons of two cations may be described in terms of a valence-bond *cluster expansion* (first one, then two, etc. anions and two cations). The rest-symmetry group F_H for a system of two cations and an anion placed in an arbitrary position consists of the identity only. At the same time, the full permutation group S_N reduces to the subgroup referring to the subsystem we are considering.

To place this procedure into proper perspective, it is instructive to make a brief comparison with a quite different, very ingenious, approach to the problem of magnetic interactions in solids developed by Anderson [2]. His evaluation of exchange interactions proceeds in two steps: *first*, Anderson considers *one* magnetic electron in the field of all the anions (ligands) and of all the other cations in the solid, while leaving out exchange between the unpaired electron considered and those of the other cations, in a Hartree-Fock scheme, i.e. in a *one-electron approximation.* If we start from a state in which the spins of all unpaired electrons are parallel (ferromagnetic case), then the Hartree-Fock operator for the electron considered has the periodicity of the (cation-) lattice, so that the solutions are running Bloch waves. These can be transformed to Wannier functions, localized around the cation positions; by their definition, Wannier functions centered at different lattice points are *orthogonal* to one another. In the *second* step, Anderson considers exchange interactions between two electrons on different Wannier orbitals, taking those orbitals as zeroth-order wavefunctions. Because of orthogonality, first-order exchange between two electrons on Wannier functions favors parallel alignment of their spins (so-called "potential exchange"). In second order of perturbation, with a "delocalization part" of the Hartree-Fock operator as the perturbation (it mixes into a given Wannier function those located at other lattice points), a gain in energy is obtained only if the spins of the two electrons are *anti*parallel ("kinetic exchange").

The Anderson method has the advantage that, in principle, it can lead to either parallel or antiparallel alignment of the spins on different cations, dependent upon whether "potential" of "kinetic" exchange, respectively, is more important. It is, however, clear that an actual calculation on this basis is hardly feasible, whereas simplified versions of Anderson's method are hopelessly confusing regarding validity of the approximations made. Furthermore, in Anderson's method the effect of the ligands is replaced by a smeared-out Coulomb and exchange potential. This tends to over-emphasize delocalization of the magnetic orbitals and to under-emphasize the rôle of the ligands. We note that, on the basis of assumption (B), cation- and anion-electrons explicitly play the same rôle.

Finally, we consider assumption (C) of the model, introducing one "effective electron" per cation in replacement of the unpaired electrons (five per Mn^{2+}-, seven per Eu^{2+}-cation, etc.). This assumption may be made plausible by observing that the stable magnetic patterns of solids with 3d-cations are *all* antiferromagnetic; moreover, they are all of the so-called *second kind* (except for CrN). In addition, the orders of magnitude of their Néel temperatures are the same. This indicates that the essential aspects of magnetic coupling may be described by means of one magnetic electron per cation and that the number of unpaired electrons may be incorporated as a "size"-parameter of the effective-electron orbital (the larger this number, the larger is the orbital). In accordance with this description, we replace the electrons of a closed-shell anion by *two*, spin-paired, effective electrons. A system of two cations and one anion is then described by three centers and four effective electrons, the so-called "*three-center, four-electron model*", proposed by Kramers [3] more than forty years ago.

3.3 Applications of the model

3.3.1 Indirect exchange interactions in magnetic solids

Since the model adopted for weak chemical bonding has been devised to describe magnetic coupling in insulating solids, its first application must obviously deal with this class of phenomena. Consider, as the first term in a ligand-cluster expansion, two paramagnetic cations A and B, each with one effective electron, and one diamagnetic anion, C, with two spin-paired effective electrons. We start from an assignment of electrons 1, 2, 3 and 4 in which electron 1 is on cation A, 2 on cation B, and the electrons 3, 4 on anion C. The wavefunctions for the (effective) electrons are denoted by ϕ_A, ϕ_B and ϕ_C, respectively. As stated before, we do not assume orthogonality between the orbitals on different centers.

In neglecting the effect of the net charges of cations and anion, the three centers will further be called "atoms". In terms of atomic wavefunctions, the correct zeroth order function describing the singlet and triplet groundstates of the system can be written as

$$\psi_\sigma^{ABC} = A\,\phi_A(1)\,\phi_B(2)\,\phi_C(3)\,\phi_C(4)\,\sigma \equiv A\,\phi\sigma;$$

A is the antisymmetrizer for the total system and σ denotes the triplet or singlet spin-eigenfunction of the squared total-spin operator.
In good approximation [1], the associated first-order interaction energy $E_\sigma^{(1)}$ (only terms linear in the perturbation Hamiltonian) is given by

$$E_\sigma^{(1)} = (\phi\sigma, VA\,\phi\sigma)/(\phi\sigma, A\,\phi\sigma),$$

where V is the perturbation Hamiltonian for the system ABC in which the labels of the electrons on the individual atoms are fixed.
In view of the fact that there are only two unpaired electrons, the first-order interaction energy can always be expressed as the expectation value of a purely bilinear spin-Hamiltonian of the form

$$E_\sigma^{(1)} = M + N\,\langle S_1 \cdot S_2\rangle_\sigma, \qquad (1)$$

where M and N consist of contributions due to the pure *interatomic* permutations of electrons (M contains as well the identity permutation.) Employing the method of double-coset decomposition of the antisymmetrizer A, we have evaluated $E_\sigma^{(1)}$ adopting Slater functions $r^{n-1}\exp(-pr)$ for each of the effective-electron orbitals ϕ_A, ϕ_B and ϕ_C. Actually, it turns out that the results are very similar [4] for n = 1, 2, 3 and 4, so that we limit ourselves to 1s-functions (n = 1). Note that spherical symmetry of the unperturbed wavefunctions is consistent with the assumption that no spatial symmetries must be taken into account. Furthermore, we only consider those geometric configurations (cation-anion-cation) in which the two cation-anion distances are equal. The opening angle θ, at the site of the anion C, is varied from 40° to 180°. The cations A and B, with fixed distance $R_{AB} = 7$ au, are assumed to be of the same kind. The choice of the orbital parameters $\underline{p}$ was $p_{cation} = 1\ \text{au}^{-1}$; $p_{anion} = 0.75\ \text{au}^{-1}$. These values roughly correspond to those for a unit Ni-Cl-Ni in an application of the effective-electron model to the analysis of 180°-superexchange [5].
In Figure 3.1 we present the contributions [6] to the coefficient N in eq. (1) due to the interatomic permutations P_{12} (cation-cation exchange only), P_{123} and P_{132} (give the same contribution for the isosceles configurations considered) and $P_{13,24}$. A positive contribution implies that the *singlet* state of the coupled system is favored, whereas a negative contribution implies *triplet* coupling. The unit of energy is 10^{-3} au.

From the figure we conclude that first-order exchange favors antiparallel coupling of the cation spins, except for a region of weak ferromagnetic alignment around $\theta = 80°$. The permutations $P_{13,24}$ and P_{12} always lead to antiparallel coupling, whereas P_{123} and P_{132} always favor a parallel alignment. *Direct* cation-cation exchange is very small. (This contribution varies somewhat with θ because of dependence of the energy denominator on the cation-anion overlap.) For large openings θ the coupling is strongly antiferromagnetic, in agreement with experiments [6].

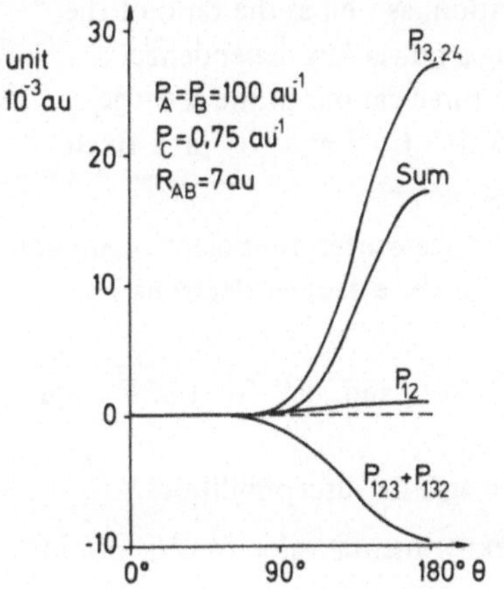

Fig. 3.1
Contributions to singlet-triplet splitting in the three-center, four-electron model, from permutations involving both electrons with unpaired spin (1,2), as well as their sum, as a function of the opening angle θ at the anion. Parameters p_A, p_B (cations) and p_C (anion) of the 1s-Slater orbitals are given. The unit of energy is 10^{-3} au. Positive values imply singlet, negative values triplet stability.

Of particular interest for comparison with experiment is the narrow θ-range of weak *ferro*-magnetic coupling. Considering, in particular, the results for $\theta = 85°$ ($-78 \cdot 10^{-6}$ au) and for $\theta = 95°$ ($+178 \cdot 10^{-6}$ au) and translating the energy of interaction in terms of a Heisenberg Hamiltonian $C - 2J_{AB}S_A \cdot S_B$, where S_A and S_B are the *total*-spin operators for the cations, we find for a unit Ni-Cl-Ni a variation in J_{AB}/k between + 3 and − 7 K. Experimentally, it is known that the sign of J_{AB} does indeed sensitively depend on the opening θ *just in this region* [5].

The accuracy with which a simple model of three centers and four electrons, in an exchange-perturbation approach, reproduces the experimental findings, is indeed surprising. It is further important to note that the *angle dependence* of the exchange is here due *solely* to permutation symmetry of the zeroth-order function in the non-orthogonal basis of atomic wavefunctions. If non-orthogonality is neglected then the coupling appears only in higher order of the perturbation series.

The fact that the non-orthogonal basis gives rise, in a Heitler-London formalism, to a θ-range of triplet stability between $\theta = 180°$ (singlet) and $\theta = 0°$ (singlet), can be easily explained. We may always (Schmidt-) orthogonalize the cation wavefunctions to that of the anion, since this leaves the total (determinantal) wavefunction unchanged. If we decrease the opening angle θ, starting from $\theta = 180°$ (singlet stability), there will be a value of θ at which the Schmidt-orbitals for the cations are mutually orthogonal (the square of the cation-anion overlap is then equal to the cation-cation overlap). We have in this case *three orthogonal orbitals* for the system; the only first-order exchange splitting is due to direct exchange between electrons on these Schmidt-orbitals, thus favoring *parallel* assignment of their spins. Upon decreasing θ further, there finally remains only the direct exchange between electrons on the non-orthogonal cation orbitals, again favoring the *singlet* state.

On the basis of the above simple model, 180°-indirect exchange occurring in solids of composition XMF_3 and X_2MF_4 (X = K, Rb, Tl; M = Mn, Co, Ni) has been evaluated and compared with experimental values [5]. The available experimental data on the exchange constants J/k and on the cation nearest-neighbor distance R enable one to give an accurate estimate of the dependence of J/k on R for the same cation, as well as the ratio of the J/k-values for Mn^{2+}, Co^{2+} and Ni^{2+} at *fixed* R. Assuming a power-law dependence, $J/k \sim R^{-n}$, the analysis of the data yields $n \sim 12$ for the three cations in the R-range covered experimentally. The ratio of the J/k-values for Mn^{2+}, Co^{2+} and Ni^{2+} at a fixed R (= 4.074 au) is found to be 1 :3.6 :7.7.

These results are compared with values predicted by the three-center, four-electron model. The task of determining the Slater orbital parameters $\underline{p}$ for the effective electrons was accomplished as follows:

a) estimate the ratios of the cation parameters for Mn^{2+}, Co^{2+} and Ni^{2+} from SCF-calculations of the $\langle r^2 \rangle$-values of the cation electrons;

b) estimate the ratio $p_{Mn^{2+}}/p_{F^-}$ from experimental diamagnetic susceptibilities;

c) assign an absolute value to p_{F^-} through a fit on the experimental value for J/k of Mn^{2+}. In this way, all other parameters are determined.

Proceeding in this manner, the ratio of the J/k-values is predicted to be 1 :3.3 :7.6, *in close agreement* with the ratio derived from experiments (1 :3.6 :7.7). Similarly good agreement with experiment is found for the R-dependence of J/k for fixed cation. Although the model actually predicts an exponential R-dependence, the theoretical results for the limited R-region covered by experiments can *indeed* be closely approximated by a R^{-n}-dependence with $n \sim 12$. For further details we refer to the literature [5].

The main conclusion to be drawn from these results is that the effective-electron model, coupled with first-order exchange perturbation analysis and spherically symmetric zeroth-order wavefunctions, is capable of reproducing the principal features of magnetic ordering in the solids considered. Quantitatively, the calculated exchange constants depend, of course, strongly on the choice of the orbital parameters. Nevertheless, it is intriguing to observe that a set of parameters can be found which *simultaneously* yields the correct ratio of the J/k-values for the three different cations, the correct R-dependence of J/k in the experimental range as well as exchange constants themselves which are in close agreement with experiment. Additional results (as yet unpublished) of calculations on 180°-superexchange paths with other types of cation and anion strongly support this conclusion.

An application of the three-center, four-electron model to magnetic ordering in solids with superexchange paths involving also angles $\theta < 180°$ had already previously been undertaken [7] for the case of solid MnS. This compound is of particular importance since it crystallizes in three different structures: Bl (NaCl), B3 (sphalerite) and B4 (wurtzite). It is often supposed in the literature that directed, covalent, cation-anion bonding plays an important rôle in phenomena of magnetic ordering, as, for example, in the Goodenough-Kanamori rules [8]. If this should be the case, then we must expect that the experimental data on MnS in its three different modifications *cannot* be explained on the basis of a simple model of cation-anion-cation exchange interactions between effective electrons on spherically-symmetric zeroth-order orbitals. The observed differences in magnetic ordering (antiferromagnetic of second kind in B1, of third kind in the B3 –, (B4 –) modification) should then be ascribed to *different covalent character* of the Mn-S bonding in an octahedral (B1) and a tetrahedral (B3, B4) environment.

We note that in B3 and B4 superexchange paths with $\theta = 180°$ are completely absent. The analysis [7] (we used simple Gaussian orbitals for the effective electrons for programme-technical reasons) shows that the stable magnetic structure in the B1-configuration is indeed antiferromagnetic of the second kind, but that this spin arrangement is ruled out if the structure is B3 (B4) just because of the lack of strongly anti-ferromagnetic 180°-superexchange in the latter case. For tetrahedral environment the predicted structure is antiferromagnetic of the third kind, again in agreement with experiment. (In fact, an antiferromagnetic spin arrangement of the first kind is almost equally stable.) These results at least indicate that *covalent cation-anion bonding need not be invoked* in the case of MnS; instead, three-ion exchange satisfactorily accounts for the experimental data.

3.3.2 Stability of rare-gas halides: A case of selective valence

The above considerations were based on properties of solid-state systems. However, more essential results bearing on the concept of "valence" may be expected from applications of the model to bonding in (inorganic) molecules. As such, we select the rare-gas halides as prime example, because their very existence seems to invalidate Lewis' octet rule and, thus, it attacks the traditional interpretation of "valence" at one of its principal roots.

Since 1962 it has been known [9] that XeF_2, XeF_4 and XeF_6, as well as KrF_2 and KrF_4, are stable molecules; their binding energy (per bond) is about 30 kcal/mole for the Xe-, about 15 kcal/mole for the Kr-compounds. Extensive theoretical attempts have been undertaken to explain the observed stability, geometric configurations (XeF_2 is linear, XeF_4 is square-planar and *not* tetrahedral) as well as the observed selectivity (only Xe and Kr form molecules, and only with fluorine), both on a valence-bond and a molecular-orbital basis, as well as with other types of approach (hybridization, resonance and spin correlations), which we will discard here. The binding is much too large to be explained by Van der Waals forces, so that the concept of valence must be invoked for an explanation. For a review of these analyses we refer to the literature [10].

Typical differences between the VB- and MO-methods come to the fore in these analyses. First of all, an electron-pair description of bonding is practically hopeless from the outset: with increasing coordination in the xenon fluorides, more and more electrons must be promoted from inner shells of the xenon atom to construct its appropriate valence state. Although the VB-method is highly stereospecific in that fitting hybridization schemes can always be found, this approach offers no chance of quantitative success. The MO-method, on the other hand, has a decisive advantage with respect to the energy of molecule formation. However, in accordance with what was said in the beginning, its stereospecificity and selectivity are much too low to serve as a satisfactory starting point.

The three-atom exchange model used for the analysis of magnetic ordering in solids is also applicable to these molecules. A halogen atom has an open p-shell of electrons and spin 1/2, represented in the model by one effective electron. The rare-gas atom has closed shells, replaced by two spin-paired effective electrons. A complex XRX, with X a halogen atom, R a rare-gas atom, is then represented by the *same* three-center, four-electron model as we have already applied in the discussion of magnetic ordering. The extension to molecules RX_n with $n > 2$, is straightforward. For the orbital parts of the effective-electron wavefunctions we chose a simple Gaussian form [11]

$$\phi(r) = (\beta/\pi^{1/2})^{3/2} \exp(-\beta^2 r^2/2),$$

where $\underline{r}$ is the distance from the effective electron to its nucleus and where β is a parameter characteristic for the size of the atom. If we denote the parameter for the rare-gas atom by β, that for the halogen atom by β', then $\gamma = (\beta'/\beta)^2$ is a measure for their relative size; for $\gamma < 1$ the halogen atom is larger than the rare-gas atom, whereas for $\gamma > 1$ the situation is reversed. The interaction energy for a complex XRX is evaluated in *first and second orders* of a Rayleigh-Schrödinger type of exchange perturbation theory [1]. In the following Figure 3.2 we wish to demonstrate *selectivity* of these three-atom exchange interactions. In the figure, the calculated binding energy (per bond) for linear symmetric rare-gas halides in the singlet state is plotted as a function of the relative-size parameter γ, for a value of the dimensionless parameter $\beta R_0 = 1.1$, with R_0 the rare-gas-halogen atom distance. Estimated values [12] for γ are ~ 2.5 for XeF_2, ~ 1.8 for KrF_2 and ~ 1 for $XeCl_2$. For other halides the γ-values are still lower.

The curve *strikingly* agrees with experimental findings in predicting XeF_2 and KrF_2 to be stable, $XeCl_2$ unstable, whereas other halides cannot exist. Also, the binding energy in KrF_2 is considerably lower than in XeF_2, again in agreement with experiment. The binding is very stereospecific in that a XeF_2-molecule with a 90°-opening at the xenon atom exhibits no bonding at all. Dihalides in the triplet state are all unstable.

Concerning stereospecificity of molecules with coordination higher than two, we carried out the calculations [13] for XeF_4 in both a square-planar and a tetrahedral configuration. Here, we have different states of total spin to consider; these are indexed in Figures 3.3 and 3.4 by different values of total spin *S*. Figure 3.3 gives the result for the square-planar, Figure 3.4 for the tetrahedral arrangement; the βR_0-value lies here at approximately 1.15.

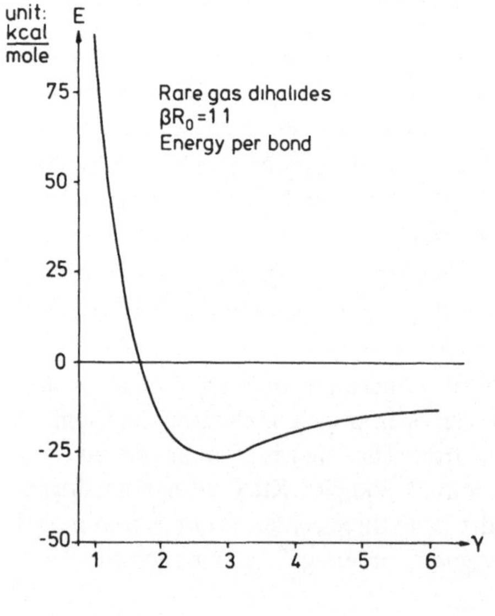

Fig. 3.2
Sum of first- and second-order interaction energies, in kcal/mole, for xenon dihalides, as a function of the relative size γ of the xenon and the halogen atoms.

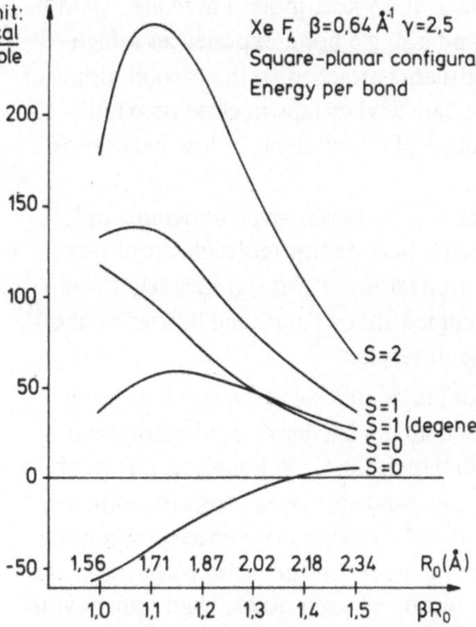

Fig. 3.3 Calculated energy per bond, in kcal/mole, of the square-planar configuration for the molecule XeF_4, as a function of the dimensionless parameter βR_0, where R_0 is the Xe-F distance.

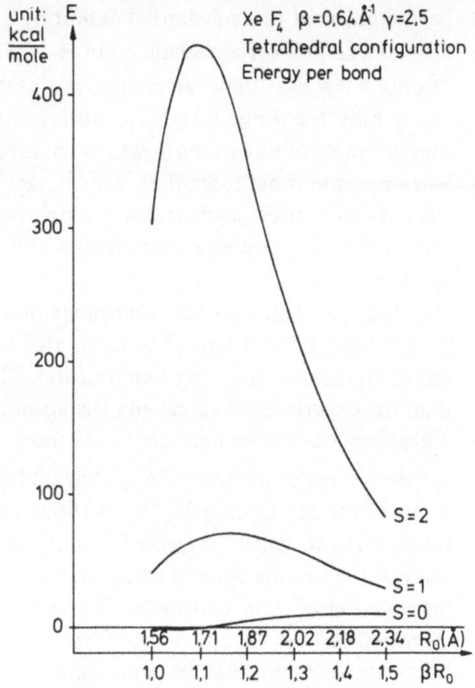

Fig. 3.4 Calculated energy per bond, in kcal/mole, of the tetrahedral configuration for the molecule XeF_4, as a function of the dimensionless parameter βR_0, where R_0 is the Xe-F distance.

The results are seen to be in very good agreement with experiment, both with regard to stereospecificity and to binding energy (exp. ~ 30 kcal/mole and per bond). The introduction of a *relative-size* parameter as an important quantity in determining the possibility of chemical bonding is, of course, an old idea. Stability of ionic crystals in the Goldschmidt-model (1927) depends sensitively on the ratio r_+/r_- of cation- and anion-radii, the ions being considered as rigid spheres. This size concept was also used extensively by Van Arkel and collaborators in an analysis of properties of inorganic compounds and solids [14]. In a sense, therefore, we seem, through the introduction of three-atom exchange interactions, to re-establish the significance of some old ideas regarding chemical bonding and valence.

3.3.3 Rotational barriers in simple molecules

A last type of application of many-atom exchange interactions in the above model formalism concerns the explanation and evaluation of *rotational barriers* in simple organic and inorganic molecules. The subject of rotational barriers dates back to the early days of stereochemistry and to Van't Hoff (1874); he postulated that rotation about a *single* C–C bond, as in ethane C_2H_6, should be free, that around a *double* bond as in ethylene C_2H_4, should be restricted. Now, 100 years later, it can only be stated that these early views have been confirmed to a substantial extent by a great many experiments. Specifically, rotation around a carbon-carbon single bond is, while not completely free, hampered by a barrier of only a few kcal/mole, an energy very small on the chemical scale. The rotational barrier in e.g. ethylene is much higher, namely, of the order of 65 kcal/mole. There are "pathological" cases of molecules in which rotation around a single bond experiences a high barrier, but in those cases there are "bulky" substituents attached to the carbon atoms of the axis of rotation, such as, for example, nitro or carboxyl groups in close proximity to each other. Here, we limit ourselves, within the scope of our subject, to low barriers of rotation.

Conclusive evidence for thc occurrence of a barrier in ethane was given by Kemp and K. S. Pitzer [15] in 1936. They calculated the specific heat of this molecule from statistical thermodynamics, assuming free rotation. The theoretical result was significantly lower than the experimental value, and the authors concluded that a rotational barrier of about 3 kcal/mole would be necessary to account for the difference.

A first successful theoretical explanation for the origin of this barrier was achieved by R. M. Pitzer and Lipscomb [16] in 1963. They calculated a barrier of 3.3 kcal/mole in favor of the so-called "staggered" configuration, in which the C–H bonds on different sides of the axis are rotated away from each other, in good agreement with the experimental value of 2.88 kcal/mole. The barrier measures the energy difference between the staggered and the "eclipsed" conformations of the molecule; in the latter form the C–H bonds on different sides lie in projection pairwise on top of each other. Uncertainty with respect to the reliability of the (variational) calculation arose, however, since it appeared that the energies of the two conformations *themselves* were in error by as much as 500 kcal/mole, i.e. 160 times the barrier!

During the past 10 years, many detailed analyses have been undertaken on a number of molecules, aided by ever-more-powerful computing facilities. Of these, it is appropriate in our context to single out those on ethane and methanol by Kern, (R. M.) Pitzer and collaborators [17], carried out on the basis of the so-called "bond orbital" concept. The wavefunction of a C–H bond was here constructed from a sp^3-hybrid at the C-atom and a ls-function on hydrogen, the latter multiplied by a polarity parameter, λ. The bond orbitals were used as building blocks for a wavefunction of the whole molecule, in two versions: a *simple* product of bond orbitals on one hand, and an *anti-symmetrized* combination on the other hand.

The results are extremely interesting and illuminating. First, it is found that, on the basis of a *simple* product of bond-orbital functions, i.e. neglecting exchange interactions (the Pauli principle), the barrier has the *wrong sign* (i.e., is in favor of the eclipsed conformation) in the region of acceptable λ-values ($\lambda > 0.5$). Second, the authors find that *with* exchange the barrier (now of the *correct* sign) varies only very weakly with the polarity (λ) of the bond. This implies that magnitude and sign of the barrier depend only slightly upon the detailed representation of the C–H bond as long as the wavefunction satisfies the Pauli principle. Very similar results were obtained for the methanol molecule.

We have carried out [18] a model calculation of rotational barriers in about 40 simple molecules including ethane, its halogen derivatives as well as inorganic molecules such as

B_2Cl_4, Si_2H_6, etc., on the basis of first-order exchange perturbation theory. The influence of the atoms on the axis of rotation was neglected. The electrons of each other atom were replaced by one effective electron with a Gaussian distribution of charge; the spins of these electrons are taken as parallel, to avoid chemical bonding. The results are in general in quite good agreement with experiment. They will, however, not be reported here, since we have to do in this case with so-called "*repulsive dominant*" barriers, i.e. the stable configuration of the molecule is that of minimal *repulsive* interactions between atoms on different sides of the axis of rotation.

Very different and intriguing are the phenomena involving rotational barriers in molecules which contain a carbon-oxygen double bond (C=O). The simplest molecule of this category, closely related to ethane, is acetaldehyde, CH_3COH. In the analysis we have also included the halogen derivatives of acetaldehyde of the types CH_3COX and CH_2XCOY, with X and Y halogen atoms.

With acetaldehyde and its derivatives the situation is indeed quite different from that observed for molecules without a double bond. The stable conformation for molecules CH_3COX is *always* the one where the C=O double bond and one of the C–H bonds of the CH_3-group are in *cis* position. As a consequence, C–X, C–H lie on different sides of the C=O bond and are coplanar with it (see sketch). This preference cannot possibly be explained on the basis of repulsive interactions between non-bonded atoms only. Thus, the barriers in these molecules may be called "*attractive dominant*". In molecules of the type CH_2XCOY we always find C–X and C–Y in *trans* position, i.e., on different sides of, and coplanar with, the C=O double bond (see sketch). The same applies for molecules of composition CY_3COY. The experimental values of the barriers are low and amount to 1–2 kcal/mole.

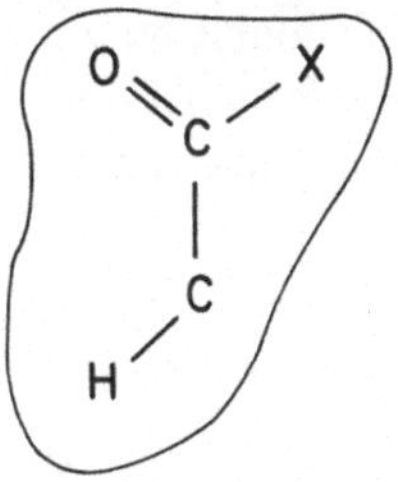

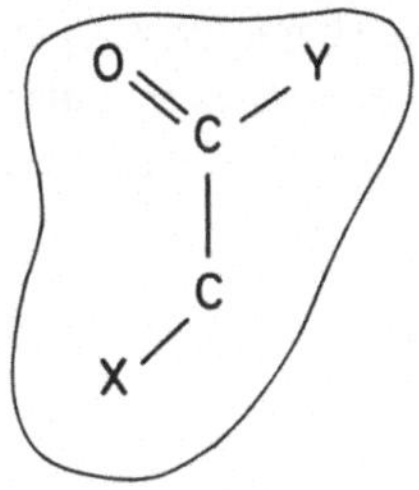

For the analysis of these experimental observations we adopt again an effective-electron model. *Attractive* many-body exchange interactions, which are apparently needed to explain the observed particular stability of the complexes XCOH and YOCX, can be taken over from the phenomenon of "superexchange" in insulating solids containing paramagnetic 3d-cations, discussed in Section I.

We apply the same model to the calculation of rotational barriers in the above molecules; the C=O bond replaces the diamagnetic anion; X and H, or Y and X replace the two cations. Together with the two remaining atoms of the methyl group one then has a five-center, six-electron system, which is again treated on the basis of exchange perturbation theory, with simple Gaussian orbitals for all six electrons. Values for the parameter β of the atoms are taken over from earlier work, whereas the value of β_{db} for the double bond is left as a parameter for the time being. We find [19], first, that the barrier is indeed sensitive to variations in β_{db}, and that it has the correct sign. Second, the "best" value of β_{db} varies little from molecule to molecule and, as in the case of rotation about single bonds, quite good agreement with experiment is obtained for series of related molecules.

Through a *cluster decomposition* of the barrier we will now investigate whether *indeed* the proposed superexchange mechanism is essentially responsible for the barrier of rotation in molecules of the types considered here. This decomposition was carried through for acetaldehyde; the components of the (first-order) perturbation energy are denoted by E_2, ΔE_3, ΔE_4, and ΔE_5, where E_2 is the sum of pair interactions (a pair consisting of two H-atoms or a H-atom and the double bond), ΔE_3 denotes the total simultaneous exchange involving three centers, etc. In Figure 3.5 the results are given as a function of the angle of rotation θ ($\theta = 0°$ implies *cis* position of C=O and C–H. Note the energy scale.).

From the figure we conclude that the two-particle energy, E_2, varies only very weakly with θ and that the stability of the *cis* formation results principally from the three-particle exchange energy, ΔE_3 (and, surprisingly enough, also from ΔE_4). Further, we find that indeed ΔE_3 is a result of super-exchange in the unit H(C=O)H. We have also carried out a *second*-order perturbation calculation for acetaldehyde. The result is that the energies of first and second order vary in a very similar manner with the angle θ. This similarity renders it possible to include second-order effects in the first-order results through a parametric adaptation of β_{db} for the C=O group. It must be expected that those adapted values will be somewhat different from molecule to molecule. For molecules of the type CH_3COX these variations in β_{db} are found to be only small.

The model has been applied also to molecules of the type CH_2XCOY, with X and Y again halogen atoms. Although the model is still found to be qualitatively correct, the height of the barriers depends more sensitively on the characteristic parameter for the double bond. The quantitative accuracy of the model here requires that also the *two-particle* interaction is reproduced with precision. However, the model was developed with the principal aim

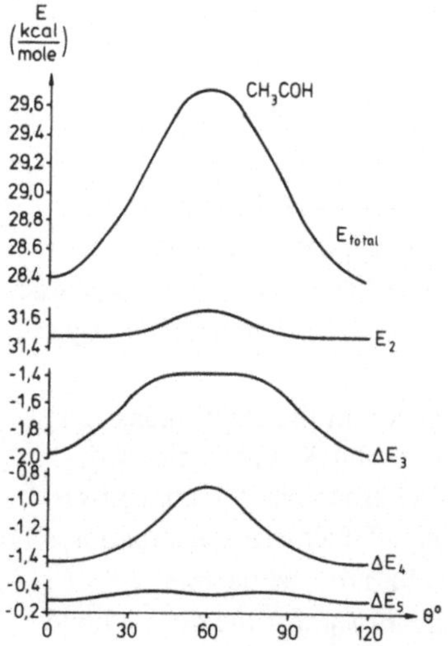

Fig. 3.5
Cluster decomposition of the first-order energy for the molecule CH_3COH in terms of two-atom interactions (E_2), simultaneous interactions between three atoms (ΔE_3), between four (ΔE_4) and between five atoms (ΔE_5), as a function of the dihedral angle θ (at $\theta = 0°$ the C = O double bond eclipses one of the C–H bonds of the methyl group). All energies in kcal/mole. Note the change of scale for each curve.

of providing a good description for *many*-body interactions. It is, therefore, not surprising that the inner consistency is less satisfactory in the exceptional case of molecules CH_2XCOY.

3.3.4 Magnetic structures of the manganese pyrites [20, 21]

In the application of the effective-electron model to super-exchange phenomena in insulating solids, discussed at the beginning of this Chapter, the "exchange unit" contained only one anion. It should be mentioned that this is not always a satisfactory approximation: sometimes two anions must be included in the exchange unit.

A very interesting example is presented by the manganese pyrites MnS_2, $MnSe_2$ and $MnTe_2$. Their solids are semiconductors, magnetically ordered at low temperatures. However, the magnetic structure is *not* the expected antiferromagnetic ordering of the second type (AF2), but the third type (AF3) in MnS_2, the first type (AF1) in $MnTe_2$ and a mixture of these two types in $MnSe_2$. This implies, compared to AF2, a larger number of spins of nearest-neighbour cations antiparallel to that of the central one (eight instead of six), and a larger number of spins parallel to the central one for second neighbours (four in AF3, all six in AF1, compared to zero in AF2). In addition, we note an increasing tendency towards ferromagnetism going from MnS_2 to $MnTe_2$. No theory has as yet accounted for this deviating behaviour of the Mn-pyrites.

We have undertaken a first analysis of this problem assuming the complex $Mn^{2+}-X_2^{2-}-Mn^{2+}$ to be the exchange unit; X = S, Se, Te. The anion here consists of two centers and two spin-paired effective electrons; the latter are placed on a molecular orbital $x \pm y$. The plus sign refers to a bonding, the minus sign to an antibonding orbital; x and y are again 1s-Slater orbitals with variable parameter λ.

The analysis is formally the same as for the solids discussed in the first application. The expression for the first-order energy can easily be written as a sum of contributions from different interatomic permutations by making use of a symmetric double-coset (SDC) decomposition of the antisymmetrizer A, with respect to the invariance group K of the product function $\phi = \phi_A(1)\,\phi_B(2)\,\phi_C(3)\,\phi_C(4)$. Here A and B denote the two cations, C denotes the "molecular anion" X_2^{2-}; 1–4 denote the electrons. In our case, K is simply the unit operator and the permutation P_{34} interchanges the electron labels 3 and 4.

Once the first-order energy has been evaluated, the "magnetic energies" of the three antiferromagnetic structures AF1, AF2 and AF3 (and the energy of the ferromagnetic configuration) at 0 K can be readily determined and compared. The result is unambiguous, but *wrong*: all Mn-pyrites should have the "normal" magnetic structure AF2, in contradiction with experiment. The preference for AF2 over AF1 and AF3 is very pronounced if the molecular orbital on X_2^{2-} is of the *bonding* type, less pronounced if it is of the *antibonding* type. In the latter case, it is even possible to stabilize MnS_2 in AF3, but for a very narrow range of Slater-orbital parameters. In addition, the difference in magnetic energy between AF3 and AF2 is extremely small, and the calculated Néel temperature (of transition between ordered and disordered states) is much lower than the observed value for MnS_2.

It is thus clear that the problem cannot be solved in this way. From previous calculations on $(CuCl_4)$-ferromagnets [22] we had concluded that in *certain* cases the inclusion of a *second* anion in the exchange unit gives an appreciable contribution to the magnetic energy in the form of a ferromagnetic shift in the coupling constant J_2. Since the preference for the AF2 structure in the model is caused by the fact that J_2 is too antiferromagnetic (too negative), and since the AF1 ordering ($MnTe_2$) even necessitates a *ferromagnetic* $J_2 (> 0)$, we were led to including a second X_2^{2-}-anion in the exchange unit.

The extended model thus comprises six electrons on six centers. Formally, the analysis proceeds as before, with obvious modifications. Remarkably enough, the results are now in striking agreement with experiment (the molecular orbitals on the two X_2^{2-}-anions must here be of the *bonding* type, which is physically more satisfactory) in that MnS_2 is definitely more stable in the magnetic structure AF3, MnTe2 in AF1 ($J_2 > 0$), whereas $MnSe_2$ has $J_2 \sim 0$ and assumes a mixed AF3/AF1 configuration. Moreover, the Néel temperatures are in reasonable agreement with experiment.

References

[1] *Jansen, L.*, Phys. Rev., **162** (1967) 63; *Beyers Brown, W.*, Chem. Phys. Lett., **2** (1968) 105; *Farberov, D. S., Mitrofanov, V. Ya., Men, A. N.*, Int. J. Qu. Chem., **6** (1972) 1057.

[2] *Anderson, P. W.*, in "Solid State Physics", Ed. *Seitz, F., Turnbull, D.*, Vol. 14, Ac. Press, New York; 1963; *Anderson, P. W.*, Phys. Rev., **115** (1959) 2.

[3] *Kramers, H. A.*, Physica **1** (1934) 182.

[4] *Lombardi, E., Tarantini, G., Block, R., Roël, R. W. J., Maten, G. ter., Jansen, L., Ritter, R.*, Chem. Phys. Lett., **12** (1972) 534.

[5] *de Jongh, L. J., Block, R.*, Physica **79B** (1975) 568.

[6] *Block, R., Jansen, L.*, in "Quantum Science", Plenum Publ. Corp., New York, 1976.

[7] *Jansen, L., Ritter, R., Lombardi, E.*, Physica **71** (1974) 425.

[8] References 11, 12 and 13, of Section 1.

[9] *Claasen, H. H., Selig, H., Malm, J. G.*, J. Am. Chem. Soc., **84** (1962) 3593; *Hoppe, R., Dähne, W., Mattauch, H., Rödder, K. M.*, Angew. Chem., **74** (1962) 903.

[10] *Coulson, C. A.*, J. Chem. Soc., **A1964**, 1442; *Malm, J. G., Selig, H., Jortner, J., Rice, S. A.*, Chem. Rev., **65** (1965) 199; *Selig, H.*, in "Halogen Chemistry", Ed. *Gutmann, V.*, Vol. 1, Ac. Press, New York, 1967.

[11] Calculations were carried out in Geneva with Gaussian functions. Later we used Slater functions and they did not show qualitative differences.

[12] *Lombardi, E., Ritter, R., Jansen, L.*, Int. J. Qu. Chem., **7** (1973) 155.

[13] *Lombardi, E., Pirola, L., Tarantini, G., Jansen, L., Ritter, R.*, Int. J. Qu. Chem., **8** (1974) 335.

[14] See e.g., *Van Arkel, A. E., de Boer, J. H.*, "Chemische Bindung als elektrostatische Erscheinung", Hirzel Verlag, Leipzig, 1931.

[15] *Kemp, J. D., Pitzer, K. S.*, J. Chem. Phys., **4** (1936) 749; J. Am. Chem. Soc., **57** (1937) 276.

[16] *Pitzer, R. M., Lipscomb, W. N.*, J. Chem. Phys., **39** (1963) 1995.

[17] *Sovers, O. J., Kern, C. W., Pitzer, R. M., Karplus, M.*, J. Chem. Phys., **49** (1968) 2592; *Kern, C. W., Pitzer, R. M., Sovers, O. J.*, J. Chem. Phys., **60** (1974) 3583.

[18] *Lombardi, E., Tarantini, G., Pirola, L., Jansen, L., Ritter, R.*, J. Chem. Phys., **61** (1974) 894.

[19] *Lombardi, E., Tarantini, G., Pirola, L., Ritter, R.*, J. Chem. Phys., **63** (1974) 2553.

[20] *Van Kalkeren, G., Block, R., Jansen, L.*, Physica **85B** (1977) 259.

[21] *Van Kalkeren, G., Block, R., Jansen, L.*, Physica **93B** (1978) 195.

[22] *Jansen, L., Block, R.*, Physica **86–88B** (1977) 1012.

Chapter IV
Groups and Semigroups for Composite Nucleon Systems

P. Kramer

1 Introduction

The detailed exploration of nuclear structure and reaction data has led to a number of refined theoretical developments in nuclear theory [1]. The theory of nuclear structure has developed from simple shell models towards sophisticated many-body theories. The theory of nuclear reactions started from the simple R-matrix theory which treats the separate nuclear fragments as point particles outside the reaction region. More recent many-body reaction theories elaborate in a systematic fashion the boundary conditions and the effects of the Pauli principle. The necessary reduction of the degrees of freedom is achieved through variational procedures. For a recent review of the corresponding point of view we refer to Tang and Wildermuth [2]. The resonating-group method [2] and the generator coordinate method [3] are theories of nuclear composite particles and their interaction.

In what follows we shall describe some features of composite particle theories in nuclear physics which are in a natural correspondence to concepts of group theory. We shall explain the underlying ideas and the way in which groups and representations enter, but we shall not attempt to give the proofs of the fundamental assertions. This seems justified since a much more detailed account of this approach is to be published [4].

2 Exchange and double cosets of the symmetric group

We shall see through examples that the physical concept of exchange in n-body systems leads in a natural way to a concept in the theory of the symmetric group.

Consider first two nuclei or atoms located at different center positions 1, 2. For simplicity, we consider two s-states φ_1 and φ_2 of a single particle located at the two centers. We do not assume that these states are normalized and introduce the overlap matrix

$$\epsilon = \begin{bmatrix} \epsilon_{11} & \epsilon_{12} \\ \epsilon_{21} & \epsilon_{22} \end{bmatrix}$$

where

$$\epsilon_{il} = (\varphi_i | \varphi_l).$$

Now we build up an n-body state $\psi_{w_1 w_2}$ by letting w_1 and w_2 particles respectively occupy the two states φ_1 and φ_2. The matrix element of an orbital permutation operator U(p) must be a product of powers of the overlaps ϵ_{il} of the form

$$(\psi_{w_1 w_2} | U(p) | \psi_{w_1 w_2}) = \epsilon_{11}^{k_{11}} \epsilon_{12}^{k_{12}} \epsilon_{21}^{k_{21}} \epsilon_{22}^{k_{22}} .$$

The numbers k_{il} are not arbitrary since we fixed the occupation numbers w_1, w_2. The conditions are best seen from the symbol

$$\left\{ \begin{matrix} k_{11} & k_{12} \\ k_{21} & k_{22} \end{matrix} \right\}.$$

The first row and column sum must yield w_1 and the second row and column sum w_2. These conditions were precisely the ones found in the description of double cosets for the symmetric group in section 2.1 of chapter I. The left and right subgroups are just the stability groups of the bra and ket states (which in our example coincide), and clearly the matrix elements of U(p) are functions on the double cosets which we call *basic exchange integrals.*

A second example is provided by the nuclear reaction

$$^4\mathrm{He} + \mathrm{d} \rightarrow {}^3\mathrm{He} + {}^3\mathrm{H}.$$

The configuration in the entrance channel may be described as $s^4 + s^2$ while in the exit channel, we have the configuration $s^3 + s^3$. We characterize the channels by *weights* $\tilde{w} = (3\ \ 3)$ and $\tilde{w} = (4\ \ 2)$ whose components are the occupation numbers. The matrix elements of permutations between these two configurations are described by the numbers of particles transferred between the fragments in the two channels. The 2×2 symbol k takes the possible values

$$\left\{ \begin{matrix} 3 & 0 \\ 1 & 2 \end{matrix} \right\}, \quad \left\{ \begin{matrix} 2 & 1 \\ 2 & 1 \end{matrix} \right\}, \quad \left\{ \begin{matrix} 1 & 2 \\ 3 & 0 \end{matrix} \right\}$$

where the first and second column sum must yield 4 and 2 respectively while the first and second row sum must yield 3 and 3 respectively. We conclude that there are three basic types of exchange in the system.

The examples demonstrate the following interpretation of exchange phenomena: Given a bra and a ket configuration characterized by two weights $\tilde{w} = (\tilde{w}_1 \tilde{w}_2 \ldots \tilde{w}_j)$, $w = (w_1 w_2 \ldots w_j)$, these two weights determine two subgroups of the symmetric group which are the stability groups of the two weights. Then, the basic and different exchange integrals or terms correspond to the double cosets of the symmetric group with respect to these stability groups and are identified by a matrix $k = \{k_{il}\}$ subject to sum rules for the columns and rows. The sums are fixed by the components of the weights $\tilde{w}$ and w. For configurations based on single-particle states, any basic exchange integral has the form

$$(\psi_{\tilde{w}} | U(p) | \psi_w) = \prod_{i,l} (\epsilon_{il})^{k_{il}} = \epsilon^k .$$

For the relation between exchange properties and double cosets we refer to [4] and [5].

3 Orbital symmetry and the representation of the symmetric and general linear groups

We shall now describe how the physical states of an n-body system of nucleons may be characterized by quantum numbers arising from the symmetric group. Our analysis will be focussed on the orbital states and their transformation properties. The correct fermion property is provided by coupling these orbital states with appropriate spin-isospin states. The coupling to antisymmetric states is possible if and only if the orbital partition f and the spin-isospin partition are obtained from one another by interchange of rows with columns in the Young diagram. Since a single nucleon has four spin-isospin states, the spin-isospin diagrams are restricted to at most four rows. Hence, the orbital partitions can have at most four columns.

The orbital state will be assumed to arise from the occupation of j in general non-orthogonal single-particle states. We denote the set of occupation numbers by a weight $w = (w_1 w_2 \dots w_j)$. The orbital partition f is established through application of a Young operator. The second index of the Young operator is a Gelfand pattern for the weight w. Instead of the Young operators given in chapter I section 5.4 we introduce the normalized operators

$$c(rfq) = \left[\frac{n!}{|f|\,w!}\right]^{1/2} c^f_{rq}, \quad w! = \prod_i (w_i!)$$

and write the orbital state as

$$|\alpha^n frq) = c(rfq)|\psi_w).$$

The possible Gelfand patterns for given weight w and partition f exhaust the linearly independent states. The properties of the Young operator guarantee that the states transform according to the representation d^f under S(n). Moreover we know that the states transform irreducibly according to the representation D^f of the general linear group GL(j, C) whose elements effect linear transformations of the j-dimensional single-particle state space. This correlation of representations appears for example in the value of the scalar product between two n-body states. The methods of the symmetric group lead to the multiplication of the Young operators and their double coset decomposition according to

$$c(\tilde{q}fr)\,c(rfq) = \frac{n!}{|f|}\,[\tilde{w}!\,w!]^{-1/2}\,c^f_{\tilde{q}q}$$

$$(\psi_{\tilde{w}}|c^f_{\tilde{q}q}|\psi_w) = \frac{|f|}{n!}\,\tilde{w}!\,w!\sum_k \{d^f_{\tilde{q}q}(z_k)\,[k!]^{-1}\,(\psi_{\tilde{w}}|U(z_k)|\psi_w)\}.$$

These equations yield the scalar product of the states as

$$(\alpha^n f r\tilde{q}|\alpha^n frq) = [\tilde{w}!\,w!]^{1/2}\sum_k d^f_{\tilde{q}q}(z_k)\,[k!]^{-1}\,\epsilon^k, \quad k! = \prod_{i,l}(k_{il}!),$$

and we refer to this sum as the *exchange decomposition* of the scalar product. Alternatively, the same expression may be written as

$$(\alpha^n f\tilde{q} \mid \alpha^n f q) = D^f_{\tilde{q}q}(\epsilon).$$

This form stresses the relation of this scalar product to the general linear group $GL(j, C)$. The representations of the latter group are available from other sources. We give here some results based on the work of Nagel and Moshinsky [6] and Louck [7] on the unitary group. Define the expression

$$\Delta^{i_1 i_2 \dots i_l}_{j_1 j_2 \dots j_l}$$

as the determinant of a $l \times l$ submatrix of ϵ with the elements from the rows $j_1 j_2 \dots j_l$ and the columns $i_1 i_2 \dots i_l$. Define the highest weight as the one whose components coincide with the components of the partition f. There is a single Gelfand pattern associated with this highest weight which we call q_{max}. Then

$$D^f_{q_{max} q_{max}}(\epsilon) = (\Delta^1_1)^{f_1 - f_2} (\Delta^{12}_{12})^{f_2 - f_3} \dots (\Delta^{12\dots j-1}_{12\dots j-1})^{f_{j-1} - f_j} (\Delta^{12\dots j}_{12\dots j})^{f_j}.$$

Example: The system $^4He + {}^4He$.

The orbital state of the nucleus ^4He is to a good approximation stable under the group $S(4)$. Hence, for the system $^4He + {}^4He$, the weights may be taken as $\tilde{w} = w = (4\ \ 4)$. The only admissible orbital partition is $f = [4\ \ 4]$ and the generalized Young tableau and Gelfand pattern become

1	1	1	1
2	2	2	2

$$\begin{matrix} f \\ q \end{matrix} = \begin{matrix} 4\ 4 \\ 4 \end{matrix}$$

The scalar product in a single-particle basis becomes

$$(\alpha^8 [44]\, r q_{max} \mid \alpha^8 [44]\, r q_{max}) = D^{[44]}_{q_{max} q_{max}}(\epsilon) = (\Delta^{12}_{12})^4 = (\epsilon_{11}\epsilon_{22} - \epsilon_{12}\epsilon_{21})^4.$$

The methods of the symmetric and linear group may be extended to the analysis of the interaction as explained in detail in [4].

4 Weyl operators, linear canonical transformations and Bargmann Hilbert space

In the Schrödinger representation of the canonical commutation relations for m degrees of freedom one uses a Hilbert space of square integrable functions $\varphi(\xi)$ of m real variables $\xi = (\xi_1 \xi_2 \dots \xi_m)$. Following the notation of Bargmann [8], translations α' of position and α'' of momentum in phase space are unitarily and irreducibly represented by

$$\alpha = (\alpha' \alpha'') \to T_\alpha, \quad T_\alpha \text{ being the } \textit{Weyl operator}$$

$$T_\alpha : \varphi \to T_\alpha \varphi, (T_\alpha \varphi)(\xi) = \exp\left[-i\alpha'' \cdot \left(\xi - \frac{1}{2}\alpha'\right)\right] \varphi(\xi - \alpha').$$

The Weyl operators multiply according to the *Weyl relations*

$$T_\alpha \circ T_\beta = \exp[i\{\alpha,\beta\}]\, T_\beta \circ T_\alpha = \exp \frac{i}{2}\{\alpha,\beta\}\ T_{\alpha+\beta}$$

where $\{\alpha,\beta\}$ is the real symplectic form

$$\{\alpha,\beta\} = \alpha \cdot K\beta, \quad K = \begin{bmatrix} 0 & 1_m \\ -1_m & 0 \end{bmatrix}$$

which characterizes the geometry of classical phase space. The real symplectic form $\{,\}$ is preserved by real symplectic transformations $g \in Sp(2m, \mathbb{R})$. It follows that the operators $T_{g\alpha}, T_{g\beta}$ provide a pair obeying the Weyl relations with the same factor as T_α, T_β. Now a theorem of von Neumann states that two irreducible unitary sets of operators obeying the Weyl relations are equivalent and related by a unitary operator S determined up to a phase factor. In the present case, $S = S_g$ must then obey

$$S_g^{-1} \circ T_{g\alpha} \circ S_g = T_\alpha$$

and the operators S_g form a ray representation of $Sp(2m, \mathbb{R})$.
We investigate now the geometrical interpretation of the operators T_α and S_g. For this purpose, we introduce the position operator Ξ_i and the momentum operator Π_l in the Schrödinger representation by the usual prescription

$$\Xi_i \colon \varphi(\xi) \to \xi_i \varphi(\xi)$$
$$\Pi_l \colon \varphi(\xi) \to \frac{1}{i}\frac{\partial}{\partial \xi_l}\varphi(\xi).$$

Now it is very simple to write down the commutators of these operators with the Weyl operators with the result

$$T_\alpha \circ \Xi_i \circ T_\alpha^+ = \Xi_i - \alpha_i' I$$
$$T_\alpha \circ \Pi_l \circ T_\alpha^+ = \Pi_l + \alpha_l'' I.$$

Hence the operators of position and momentum are translated by the Weyl operators as expected. Moreover, these operators actually generate the Weyl translations. A direct computation shows that the operator L_α defined by

$$(L_\alpha \varphi)(\xi) = \lim_{\tau \to 0} (i\tau^{-1}[T_{\tau\alpha} - I]\varphi)(\xi)$$

is given as

$$L_\alpha = \sum_j (\alpha_j' \Pi_j + \alpha_j'' \Xi_j).$$

If we consider a family of Weyl operators depending on the real parameter τ and write

$$S_g^{-1} \circ T_{\tau\alpha} \circ S_g = T_{g^{-1}\tau\alpha},$$

differentiation at the point $\tau = 0$ yields

$$S_g^{-1} \circ L_\alpha \circ S_g = L_{g^{-1}\alpha}.$$

We write in a short-hand notation

$$L_\alpha = -(\alpha', \alpha'')\, K \begin{bmatrix} \Xi \\ -\Pi \end{bmatrix}$$

and use

$$^t g^{-1} K = K g$$

to derive the transformation

$$S_g^{-1} \circ \begin{bmatrix} \Xi \\ -\Pi \end{bmatrix} \circ S_g = g \begin{bmatrix} \Xi \\ -\Pi \end{bmatrix}.$$

Hence, the operators S_g acting on the position and momentum operators yield a linear symplectic transformation belonging to the group $Sp(2m, \mathbb{R})$. In conclusion we have seen that the operators T_α and S_g represent in quantum mechanics the translations and linear canonical transformations of classical phase space.

The explicit form of the operators S_g has been derived from the canonical commutation relations by Moshinsky and Quesne [9]. Bargmann [8] constructed a Hilbert space F_m of entire functions $f(z)$ of m complex variables $z = (z_1 z_2 \dots z_m)$ which yields the Fock representation of the canonical commutation relation obeyed by oscillator annihilation and creation operators. The scalar product in F_m is defined as

$$(f, g) = \int \overline{f(z)}\, g(z)\, d\mu(z)$$

$$d\mu(z) = \pi^{-m} \exp[-z \cdot \bar{z}] \prod_{i=1}^{m} d\mathrm{Re}(z_i)\, d\mathrm{Im}(z_i).$$

Moreover he constructed a unitary mapping between H_m and F_m by the integral transform

$$f(z) = \int A(z, \xi)\, \varphi(\xi)\, d\xi$$

$$\varphi(\xi) = \int \overline{A(z, \xi)}\, f(z)\, d\mu(z)$$

where

$$A(z, \xi) = \pi^{-1/4\, m} \exp\left[-\frac{1}{2}\xi \cdot \xi - \frac{1}{2} z \cdot z + \sqrt{2}\, \xi \cdot z\right].$$

The Weyl operators in Bargmann space F_m become operators W_{α_c} with the property

$$W_{\alpha_c}: f \to W_{\alpha_c} f, \quad (W_{\alpha_c} f)(z) = \exp\left[\bar{a} \cdot \left(z - \frac{1}{2} a\right)\right] f(z - a)$$

where $\alpha_c = \binom{a}{\bar{a}}$ is related to $\alpha = \binom{\alpha'}{\alpha''}$ by

$$\alpha_c = R\alpha, \quad R = \sqrt{\frac{1}{2}} \begin{bmatrix} 1_m & i\,1_m \\ 1_m & -i\,1_m \end{bmatrix}.$$

The Weyl relations become

$$W_{\alpha_c} \circ W_{\beta_c} = \exp[-\{\alpha_c, \beta_c\}]\, W_{\beta_c} \circ W_{\alpha_c}.$$

One derives from the properties of R that

$$\{\alpha, \beta\} = i\,\{\alpha_c, \beta_c\}$$
$$\{\bar{\alpha}, \beta\} = i(\alpha_c | \beta_c)$$

where

$$(\alpha_c | \beta_c) = \bar{\alpha}_c \cdot M\beta_c, \quad M = \begin{bmatrix} 1_m & 0 \\ 0 & -1_m \end{bmatrix}.$$

Conversely, if g_c preserves both $\{,\}$ and $(|)$, it belongs to the group $Sp(2m, \mathbb{C}) \cap U(m,m)$ and can be shown to be equivalent to an element g of $Sp(2m, \mathbb{R})$.

We shall require a factorization of Weyl operators acting on finite dimensional linear subspaces L^N of F_m. We denote by L^N the polynomial states of degree up to N or in other words, all linear combinations of oscillator states up to excitation N. In addition, we introduce the linear subspaces $W_{\alpha_c} L^N$, $W_{\beta_c} L^N$ as the images of L^N under W_{α_c} or W_{β_c}. Then we define [4]:

$$V_a : f \to V_a f, \quad (V_a f)(z) = f(z-a)$$
$$\Lambda_a : f \to \Lambda_a f, \quad (\Lambda_a f)(z) = (W_{\alpha_c} \circ V_{-a} f)(z) = \exp[z \cdot \bar{a}]\, f(z)$$

and claim the following properties for V_a, Λ_a:

V_a is a bounded operator on any domain $W_{\beta_c} L^N$ with range $W_{\beta_c} L^N$.

Two operators V_a, V_b multiply as $V_a \circ V_b = V_{a+b}$.

The inverse of V_a is $V_a^{-1} = V_{-a}$.

Λ_a is a bounded operator on the domain $W_{\beta_c} L^N$ with range $W_{\alpha_c + \beta_c} L^N$.

Two operators Λ_a, Λ_b multiply as $\Lambda_a \circ \Lambda_b = \Lambda_{a+b}$.

The inverse of Λ_a is $\Lambda_a^{-1} = \Lambda_{-a}$.

The next properties apply to combinations of Λ and V:

$$V_a \circ \Lambda_b = \Lambda_b \circ V_a \exp[-a \cdot \bar{b}]$$
$$W_{\alpha_c} = \Lambda_a \circ V_a \exp\left[-\frac{1}{2} a \cdot \bar{a}\right] = V_a \circ \Lambda_a \exp\left[\frac{1}{2} a \cdot \bar{a}\right]$$
$$\Lambda_a^+ = V_{-a}, \quad V_a^+ = \Lambda_{-a}$$
$$\Lambda_a \circ V_b \circ \Lambda_{a'} \circ V_{b'} = \Lambda_{a+a'} \circ V_{b+b'} \exp[-b \cdot \bar{a}'].$$

All these relations are to be understood with appropriate arrangements of range and domain. The proof [4] is essentially based on the fact that V_a applied to a polynomial cannot raise the degree.

Since in later applications we shall use Weyl operators up to a factor, we introduce them by the definition

$$\widehat{W}_{\alpha_c} = W_{\alpha_c} \exp\left[\frac{1}{2} a \cdot \bar{a}\right] = \Lambda_a \circ V_a.$$

By von Neumann's theorem, W_{α_c} and $W_{g_c \alpha_c}$ are related by

$$S_{g_c}^{-1} \circ W_{g_c \alpha_c} \circ S_{g_c} = W_{\alpha_c}$$

where the operators S_{g_c} are related to the operators S_g discussed earlier. In Bargmann space, operators may be written as integral operators with kernels $K(\tilde{z}, z)$. In particular, the identity operator is

$$S_e(\tilde{z}, z) = \exp[\tilde{z} \cdot \bar{z}].$$

Of particular interest are the coherent states

$$e_a(z) = \exp[z \cdot \bar{a}]$$

which upon transforming back to H_m become Gaussian states whose average position and momentum are determined by the real and imaginary part of a. Noting the relation of the coherent states to the identity operator, one easily finds that the kernel of any operator K may be obtained as a scalar product

$$K(\tilde{z}, z) = (e_{\tilde{z}}, K e_z).$$

The kernels of the Weyl operators are derived by Bargmann and read

$$W_{\alpha_c}(\tilde{z}, z) = \exp\left[\tilde{z} \cdot \bar{a} + \tilde{z} \cdot \bar{z} - a \cdot \bar{z} - \frac{1}{2} a \cdot \bar{a}\right].$$

Finally, the operators S_{g_c} have the kernel

$$S_{g_c}(\tilde{z}, z) = [\det C^{-1}]^{1/2} \exp\left[\frac{1}{2} \tilde{z} \cdot A \tilde{z} + \tilde{z} \cdot C \bar{z} + \frac{1}{2} \bar{z} \cdot B \bar{z}\right]$$

where

$$g_c = \begin{bmatrix} \lambda & \mu \\ \nu & \rho \end{bmatrix} = \begin{bmatrix} 1_m & 0 \\ A & 1_m \end{bmatrix} \begin{bmatrix} {}^tC^{-1} & 0 \\ 0 & C \end{bmatrix} \begin{bmatrix} 1_m & -B \\ 0 & 1_m \end{bmatrix}$$

and ${}^tA = A$, ${}^tB = B$ guarantees that all factors are complex symplectic (but in general do not belong to $U(m, m)$). For the geometry and representation of linear canonical transformations, we refer to Grossmann [10] and Sternberg [11].

5 Canonical transformations for interacting n-body systems

As a first step towards a theory of composite particles we shall introduce appropriate coordinates which will always be taken by linear orthogonal transformations from the single-particle coordinates ξ. The same orthogonal transformation applies to the momenta and the corresponding $2n \times 2n$ matrix in phase space is

$$g(B) = \begin{bmatrix} B & 0 \\ 0 & B \end{bmatrix}, \quad {}^tBB = 1_n.$$

For complex coordinates in Bargmann space this matrix becomes

$$g_c(B) = R\,g(B)\,R^{-1} = g(B) = \begin{bmatrix} B & 0 \\ 0 & B \end{bmatrix}$$

and hence (for orthogonal B!) the complex coordinates transform as the real ones. As an example we shall consider a splitting of the n particles into two sets of n_1 and n_2 particles. If we introduce standard orthogonal Jacobi coordinates for both sets, we may permute them by a matrix Q to get the two center coordinates

$$z_1 = [n_1]^{-1/2} \sum_{1}^{n_1} x_i \quad z_2 = [n_2]^{-1/2} \sum_{n_1+1}^{n} x_i$$

plus $n-2$ additional relative coordinates y. We write the orthogonal transformation as

$$x = JQ\begin{pmatrix} z \\ y \end{pmatrix}$$

From these new coordinates we may pass to relative or cluster coordinates s between the centers and maintain the coordinates y,

$$\begin{pmatrix} z \\ y \end{pmatrix} = W\begin{pmatrix} s \\ y \end{pmatrix}.$$

For two centers the relevant part of W is described by

$$s_1 = \left[\frac{n_2}{n}\right]^{1/2} z_1 - \left[\frac{n_1}{n}\right]^{1/2} z_2$$

$$s_2 = \left[\frac{n_1}{n}\right]^{1/2} z_1 + \left[\frac{n_2}{n}\right]^{1/2} z_2$$

and s_2 is the overall c.m. vector. To transform states from single-particle to center or cluster plus internal coordinates, we employ the operator S_{g_c} to get

$$l(x) = (S_{g_c^{-1}(JQ)}\, l)(zy) = (S_{g_c^{-1}(JQW)}\, l)(sy)$$

where for orthogonal B the kernel of $S_{g_c(B)}$ is found to be

$$S_{g_c(B)}(\tilde{z}, z) = \exp[\tilde{z}_i \cdot B_{ij}\bar{z}_j].$$

To project states of permutational symmetry we employ Young operators $c(rfq)$. Since in single-particle coordinates a permutation is represented by an orthogonal permutation matrix P, we get the Young operators in the form

$$c(rfq) = \left[\frac{|f|}{n!\,w!}\right]^{1/2} \sum_{p \in S(n)} d^f_{rq}(p)\, S_{g_c(P)}.$$

Let us analyze in one dimension the complex symplectic 2 X 2 matrix

$$h = \begin{bmatrix} 1 & 0 \\ -iq & 1 \end{bmatrix}, \quad q > 0.$$

By constructing formally the corresponding operator in H_1, we obtain the bounded Gaussian operator

$$S_h(\xi, \xi') = \exp\left[-\frac{1}{2} q\xi^2\right] \delta(\xi - \xi').$$

This allows us to formally describe a Gaussian interaction between two particles by a symplectic matrix. The matrix h may be transformed by the matrix R into a matrix h_c which is symplectic but does not belong to U(n, n). Decomposing h_c in a similar fashion as g_c allows one to define three blocks A, B, C and to formally write down a kernel in Bargmann space. The question how these manipulations may be justified was studied in [12, 13, 14] and we briefly mention the main results. On a 2n dimensional linear complex space with the scalar product (|) mentioned in section 4, one may introduce the length-increasing semigroups $U^{>}(n, n)$ and $U^{\geq}(n, n)$ respectively by the condition that the elements h and h′ of these semigroups obey respectively the inequalities

$$(hv|hv) - (v|v) > 0$$
$$(h'v|h'v) - (v|v) \geq 0$$

for any element v = 0 of the linear space. The geometry and factorization of such semigroups is developed in [13].
If now one defines the semigroups

$$U^{>}(n, n) \cap Sp(2n, \mathbb{C}) \quad \text{and} \quad U^{\geq}(n, n) \cap Sp(2n, \mathbb{C})$$

it is shown in [14] that the kernels corresponding to the first semigroup form a representation by Hilbert-Schmidt operators, while the kernels corresponding to the second semigroup form a representation by bounded operators.
These results have an immediate application to the many-body system since they allow the interpretation of Gaussian interactions as representations of the second semigroup. Combining all the steps we may write the operator of a Gaussian two-body interaction between projected n-body states depending on center coordinates in the form

$$S_{g_c^{-1}(\tilde{J}\tilde{Q})} \circ c(\tilde{q}\tilde{f}r) \circ S_{h_c} \circ c(rfq) \circ S_{g_c(JQ)}.$$

This expression may be expanded into a sum of terms each of which is the representation of a semigroup element.

6 Interaction of composite particles

The methods developed in the last section allow us to express general two-body interactions in an n-body system in terms of operators acting on functions of the coordinates (z, y) or, with simple modifications, of (s, y). The variational derivation of the dynamics of composite particles is based on the assumption that a reasonable but fixed choice of the state dependence on the internal coordinates y be made. A possible choice of these internal states would be eigenstates of the corresponding internal Hamiltonian. We shall work with the much simpler assumption that the internal states are described by harmonic oscillator configurations. This assumption may be justified by the observation that these configurations describe indeed a reasonable fraction of a nuclear state when effective nucleon-nucleon forces are employed [15]. It offers the great advantage that one obtains closed expressions for the composite particle interactions as we shall see in this and the following sections.

The dynamics of j composite particles will be written in a Hilbert space F_j. The projection from F_n to F_j will be obtained by a variational principle from variational states which in the zy-coordinates are assumed in the form

$$(S_{g_c^{-1}(JQ)}\, l)\,(zy) = u(z)\, v(y)$$

subject to the restricted variation

$$(\delta\, S_{g_c^{-1}(JQ)}\, l)\,(zy) = (\delta u)\,(z)\, v(y).$$

For simplicity, we consider the case j = 2 of two composite particles. The variational principle for stationary states leads to an integral equation for u(z) of the form

$$\int [H(\tilde{z}, z) - E\, N(\tilde{z}, z)]\, u(z)\, d\mu(z) = 0.$$

The kernels $H(\tilde{z}, z)$ and $N(\tilde{z}, z)$ act in F_2 but may be obtained, comparing the remarks of section 3, as scalar products in F_n upon replacing

$u(z')\, v(y)$ by $e_z(z')\, v(y)$ in the form

$$H(\tilde{z}, z) = (c(r\tilde{f}\tilde{q}) \circ S_{g_c(\tilde{J}\tilde{Q})}\,(e_{\tilde{z}} v),\ H \circ c(rfq) \circ S_{g_c(JQ)}\,(e_z v))$$

$$N(\tilde{z}, z) = (c(r f\tilde{q}) \circ S_{g_c(\tilde{J}\tilde{Q})}\,(e_{\tilde{z}} v),\ c(rfq) \circ S_{g_c(JQ)}\,(e_z v)).$$

Note that these operators incorporate both interactions and orbital projections. We call H the interaction and N the normalization operator.

We now claim that the normalization and interaction operators may be evaluated in a way involving only states built from single-particle configurations. This result is not obvious since the dynamics of composite particles was defined in terms of relative coordinates. Consider two configurations 1, 2 of n_1 and n_2 particles in harmonic oscillator states which are non-spurious, that is, carry no center-of-mass excitation. Apply to the total state a Weyl operator $\Lambda_t \circ V_t$ such that the n components of t are

$$t = (\underbrace{t_1 t_1 \ldots t_1}_{n_1}\ \underbrace{t_2 t_2 \ldots t_2}_{n_2}).$$

A similar operator $\Lambda_{\tilde{t}} \circ V_{\tilde{t}}$ may be applied to a bra configuration of $n = \tilde{n}_1 + \tilde{n}_2$ particles. We now claim that the interaction and normalization operators are given by

$$H(\tilde{z}, z) = (c(r\tilde{f}\tilde{q}) \circ \Lambda_{\tilde{t}} \circ V_{\tilde{t}}\tilde{l}, H \circ c(rfq) \circ \Lambda_t \circ V_t l)$$

$$N(\tilde{z}, z) = (c(rf\tilde{q}) \circ \Lambda_{\tilde{t}} \circ V_{\tilde{t}}\, \tilde{l}, c(rfq) \circ \Lambda_t \circ V_t\, l)$$

where

$$\sqrt{\tilde{n}_i}\,\tilde{t}_i = \tilde{z}_i \quad \sqrt{n_i}\,t_i = z_i, \qquad i = 1, 2$$

and the state $\tilde{l}, l$ describe the configurations mentioned above. We shall not give the proof of these equations which stems from a straight-forward analysis of the transformation of the kernels under change of coordinates [4]. Rather we shall elaborate some applications in the following section. The important point to be noticed is that now the kernels have been cast into the framework of single-particle configurations. These single-particle configurations consist of oscillator shell configurations which have been shifted by Weyl operators to different center positions.

7 Configurations of simple composite particles

In Bargmann space, the single-particle harmonic oscillator states take the form

$$l_N(x) = [N!]^{-1/2}\, x^N$$

or, for particles in three dimensions

$$l_{NLM}(x) = P^N_{LM}(x) = A_{NL}\,(x \cdot x)^{1/2\,(N-L)}\, Y_{LM}(x)$$

where the constants A_{NL} are given in [16].
We shall discuss a configuration where at center i there are w_i nucleons in the unexcited level $N = 0$,

$$l_{000}(x) = 1.$$

Application of the Weyl operator to this state yields

$$(\Lambda_{t_i} \circ V_{t_i} l_{000})(x) = \exp[x \cdot \bar{t}_i]$$

and the overlaps become

$$\epsilon_{il} = (\Lambda_{\tilde{t}_i} \circ V_{\tilde{t}_i} l_{000}, \Lambda_{t_l} \circ V_{t_l} l_{000}) = \exp[\bar{\tilde{t}}_i \cdot \bar{t}_l] = \exp[[\tilde{n}_i n_l]^{-1/2}\, \bar{\tilde{z}}_i \cdot z_l].$$

The technique of GL (j, C) developed in section 3 gives for the n-particle state

$$N(\tilde{z}, z) = D^f_{\tilde{q}q}(\epsilon) = [\tilde{w}!\, w!]^{1/2} \sum_k d^f_{\tilde{q}q}(z_k)\, [k!]^{-1}\, \epsilon^k$$

$$= [\tilde{w}!\, w!]^{1/2} \sum_{k_{il}} d^f_{\tilde{q}q}(z_k)\, [\Pi\, k_{il}!]^{-1} \exp\left[\sum_{i,l} \bar{\tilde{z}}_i \frac{k_{il}}{[\tilde{n}_i n_l]^{1/2}} \cdot z_l\right].$$

This is an operator acting in a Bargmann space F_j and may be interpreted as a linear combination of operators which describe complex extensions of linear canonical transformations. Moreover, $N(\tilde{z}, z)$ commutes with the oscillator hamiltonian of j particles and its eigenstates must belong to fixed total excitations.

Example: The system ^{4}He + ^{4}He

The general form of the scalar product of orbital states was determined in section 3 as

$$D^{[44]}_{4\ 4}(\epsilon) = (\epsilon_{11}\epsilon_{22} - \epsilon_{12}\epsilon_{21})^4.$$

Insertion of the values ϵ_{il} yields

$$N(\tilde{z}_1\tilde{z}_2, z_1 z_2) = \sum_{\alpha=0} (-1)^\alpha \binom{4}{\alpha} \exp\left[\frac{4-\alpha}{4}\tilde{z}_1\cdot\bar{z}_1 + \frac{\alpha}{4}\tilde{z}_1\cdot\bar{z}_2 + \frac{\alpha}{4}\tilde{z}_2\cdot\bar{z}_1 + \frac{4-\alpha}{4}\tilde{z}_2\cdot\bar{z}_2\right]$$

One recognizes the exchange decomposition according to the double cosets

$$\begin{Bmatrix} 4-\alpha & \alpha \\ \alpha & 4-\alpha \end{Bmatrix}.$$

Upon introduction of relative and overall c.m. coordinates one finds

$$N(\tilde{s}_1\tilde{s}_2, s_1 s_2) = \exp[\tilde{s}_2\cdot\bar{s}_2] \sum_{\alpha=0}^{4} (-1)^\alpha \binom{4}{\alpha} \exp\left[\frac{4-2\alpha}{4}\tilde{s}_1\cdot\bar{s}_1\right].$$

The first factor represents the unit operator with respect to the overall c.m. state, the second factor reflects the effect of the Pauli principle on the state of relative motion.

With Gaussian interactions one still gets representations of linear canonical transformations. States of up to three such composite particles have been applied to the structure of light nuclei up to n = 10 [15, 4]. One obtains very good agreement with experimentally known data and predictions of new resonances in these nuclei.

8 Composite particles with a closed-shell configuration

The computations for complex clusters are greatly simplified if the Weyl operators are modified. With respect to oscillator states we found the factorization of Weyl operators (see section 4):

$$\hat{W}_{\tau_c} = \Lambda_t \circ V_t.$$

If for the scalar products ϵ_{il} we modify this operator according to

$$\Lambda_{t_l} \circ V_{t_l} \to \Lambda_{t_l} \circ V_{\tilde{t}_l}$$

for the ket states, and according to

$$\Lambda_{\tilde{t}_i} \circ V_{\tilde{t}_i} \to \Lambda_{\tilde{t}_i} \circ V_{t_i}$$

for bra states, then this does not affect the n-body matrix element after orbital projection. But since

$$[\Lambda_{\tilde{t}_i} \circ V_{t_i}]^+ \circ \Lambda_{t_i} \circ V_{\tilde{t}_i} = \Lambda_{-t_i} \circ V_{-\tilde{t}_i} \circ \Lambda_{t_i} \circ V_{\tilde{t}_i} = I \cdot \exp[\tilde{t}_i \cdot \bar{t}_i]$$

this modification yields a biorthogonal basis at each pair of centers [4].

A second useful transformation is obtained if the projection operator P_1 is introduced which projects on all states occupied at the center 1 (undisplaced). Then the states at center 2 may be modified according to

$$(\Lambda_{t_2} \circ V_{\tilde{t}_2}\, l)(x) \to ([1 - \Lambda_{t_1} \circ V_{\tilde{t}_1} \circ P_1 \circ \Lambda_{-t_1} \circ V_{-\tilde{t}_1} \exp[-\tilde{t}_1 \cdot \bar{t}_1]] \circ \Lambda_{t_2} \circ V_{\tilde{t}_2}\, l)(x)$$
$$= (\Lambda_{t_1} \circ V_{\tilde{t}_1} \circ [1 - P_1] \circ \Lambda_{t_2 - t_1} \circ V_{\tilde{t}_2 - \tilde{t}_1}\, l)(x) \exp[-\tilde{t}_1 \cdot \bar{t}_1 + \tilde{t}_1 \cdot \bar{t}_2].$$

These states are biorthogonal to all states at center one while their scalar product with similar bra states at center two is determined by the matrix element of the operator

$$\Lambda_t \circ V_{\tilde{t}} \circ [1 - P_1] \circ \Lambda_{-t} \circ V_{-\tilde{t}} \exp[\tilde{t}_2 \cdot \bar{t}_2 - \tilde{t} \cdot \bar{t}],$$

where $\tilde{t} = \tilde{t}_1 - \tilde{t}_2$, $t = t_1 - t_2$. We note the following properties of the transformation to relative coordinates s_1, s_2:

$$n_1 \tilde{t}_1 \cdot t_1 + n_2 \tilde{t}_2 \cdot t_2 = \tilde{s}_1 \cdot \bar{s}_1 + \tilde{s}_2 \cdot \bar{s}_2$$
$$s_1 = \left[\frac{n_1 n_2}{n}\right]^{1/2} t.$$

The second transformation amounts to a transformation of the overlap matrix ϵ by triangular matrices such that the determinant is unchanged.

We examine first a single nucleon interacting with a closed oscillator shell of N_1 quanta of excitation at center 1 corresponding to ^4He, ^{16}O or ^{40}Ca. The number of single-particle states at center 1 we call $j - 1$. With the modified states, the matrix ϵ is now diagonal with entries

$$\epsilon_{il} = \delta_{il} \exp[\tilde{t}_i \cdot \bar{t}_l] \quad i, l = 1, 2 \ldots j - 1,$$

while for $i = l = j$ we get

$$\epsilon_{jj} = (l_{000}, \Lambda_t \circ V_{\tilde{t}} \circ [1 - P_1] \circ \Lambda_{-t} \circ V_{-\tilde{t}}\, l_{000}) \exp[\tilde{t}_2 \cdot \bar{t}_2 - \tilde{t} \cdot \bar{t}]$$
$$= q_{N_1+1}(\tilde{t} \cdot \bar{t}) \exp[\tilde{t}_2 \cdot \bar{t}_2 - \tilde{t} \cdot \bar{t}],$$

where we introduced the function

$$q_m(\zeta) = \sum_{l=m}^{\infty} (l!)^{-1} \zeta^l.$$

The only permutational symmetry of physical interest is $f = [4^{j-1}\, 1]$ with the representation

$$D^{[4^{j-1}\, 1]}_{q_{max} q_{max}}(\epsilon) = (\Delta^{1 \ldots j-1}_{1 \ldots j-1})^3 (\Delta^{1 \ldots j}_{1 \ldots j})^1$$

and where the choice $q = q_{max}$ means that we have a single nucleon interacting with a closed oscillator shell. Now

$$\Delta^{1\dots j-1}_{1\dots j-1} = \exp[(j-1)\tilde{t}_1 \cdot \bar{t}_1]$$

and

$$\Delta^{1\dots j}_{1\dots j} = \exp[(j-1)\tilde{t}_1 \cdot \bar{t}_1 + \tilde{t}_2 \cdot \bar{t}_2 - \tilde{t} \cdot \bar{t}]\, q_{N_1+1}(\tilde{t} \cdot \bar{t})$$

which gives

$$N(\tilde{t}_1 \tilde{t}_2, t_1 t_2) = \exp[n_1 \tilde{t}_1 \cdot \bar{t}_1 + \tilde{t}_2 \bar{t}_2 - \tilde{t} \cdot \bar{t}]\, q_{N_1+1}(\tilde{t} \cdot \bar{t})$$

or, after transforming to cluster coordinates,

$$N(\tilde{s}_1 \tilde{s}_2, s_1 s_2) = \exp\left[\tilde{s}_2 \cdot \bar{s}_2 + \tilde{s}_1 \cdot \bar{s}_1 - \frac{n_1+1}{n_1} \tilde{s}_1 \cdot \bar{s}_1\right] q_{N_1+1}\left(\frac{n_1+1}{n_1} \tilde{s}_1 \cdot \bar{s}_1\right).$$

This expression yields the normalization kernel in explicit form. The dependence on the overall c.m. vector s_2 has the form of a reproducing kernel describing the unit operator. The dependence on $\tilde{s}_1$, s_1 is only through $\tilde{s}_1 \cdot \bar{s}_1$ which means that the eigenstates of N are the oscillator states with respect to the vector s_1. The eigenvalues may be explicitly obtained upon expanding N in the form

$$N(\tilde{s}_1 \tilde{s}_2, s_1 s_2) = \exp[\tilde{s}_2 \cdot s_2] \sum_{N=N_1+1}^{\infty} \eta_N (N!)^{-1} (\tilde{s}_1 \cdot \bar{s}_1)^N$$

since clearly η_N is the eigenvalue for an oscillator state of excitation N. The values N below $N_1 + 1$ are forbidden by the Pauli principle. The eigenvalue for $N_1 + 1$ is clearly

$$\eta_{N_1+1} = \left[\frac{n_1+1}{n_1}\right]^{N_1+1}$$

which for not too small values of n_1 is close to unity. For very high excitations the kernel N with respect to s_1 behaves like a unit operator. An estimate of the lowest possible value N_{min} is obtained as follows: If all nucleons are put into a single center, one obtains a minimum total excitation N_{12}. If we subtract from this number the excitations N_1 and N_2 at the separated centers, we get the estimate

$$N_{min} = N_{12} - N_1 - N_2 .$$

The second example which we shall analyze is a closed oscillator shell at the first center and a simple composite particle with $n_2 = 2, 3, 4$ nucleons at the second center. The normalization kernel is now given by

$$D^{[4^{j-1} n_2]}_{q_{max}\, q_{max}}(\epsilon) = (\Delta^{1\dots j-1}_{1\dots j-1})^{4-n_2} (\Delta^{1\dots j}_{1\dots j})^{n_2}$$

which yields

$$N(\tilde{t}_1 \tilde{t}_2, t_1 t_2) = \exp[n_1 \tilde{t}_1 \cdot \bar{t}_1 + n_2 \tilde{t}_2 \bar{t}_2 - n_2 \tilde{t} \cdot \bar{t}]\, [q_{N_1+1}(\tilde{t} \cdot \bar{t})]^{n_2}.$$

Upon transforming to cluster coordinates one finds

$$N(\tilde{s}_1\tilde{s}_2, s_1 s_2) = \exp\left[\tilde{s}_2 \cdot s_2 + \tilde{s}_1 \cdot \bar{s}_1 - n_2 \frac{n}{n_1 n_2} \tilde{s}_1 \cdot \bar{s}_1\right] \left[q_{N_1+1}\left(\frac{n}{n_1 n_2} \tilde{s}_1 \cdot \bar{s}_1\right)\right]^{n_2}.$$

The general properties of this kernel are similar to the ones for $n_2 = 1$. The lowest non-vanishing eigenvalue of this kernel is given by

$$\eta_{n_2(N_1+1)} = \left[\frac{n}{n_1 n_2}\right]^{n_2(N_1+1)} \frac{[n_2(N_1+1)]!}{[(N_1+1)!]^{n_2}}.$$

This number is in many cases much smaller than 1. Obviously, the effect of the Pauli principle extends to higher excitations.

The last example to be considered is the system $^{16}O + {}^{16}O$. The number of single-particle states at each center is $\nu = 4$ corresponding to the configuration sp^3. The overlap matrix is an 8×8 matrix and the orbital symmetry is $f = [4^8]$. We have

$$D^{[4^8]}_{q_{max} q_{max}}(\epsilon) = (\Delta^{1\ldots8}_{1\ldots8})^4$$

and the computation yields

$$\Delta^{1\ldots8}_{1\ldots8} = \exp[4\tilde{t}_1 \cdot \bar{t}_1 + 4\tilde{t}_2 \cdot \bar{t}_2 - 2\tilde{t} \cdot \bar{t}]$$

$$[q_6(2\tilde{t} \cdot \bar{t}) + q_6(-2\tilde{t} \cdot \bar{t}) - 4q_6(\tilde{t} \cdot \bar{t}) - 4q_6(-\tilde{t} \cdot \bar{t}) - (\tilde{t} \cdot \bar{t})^2 q_4(\tilde{t} \cdot \bar{t}) - (\tilde{t} \cdot \bar{t})^2 q_4(-\tilde{t} \cdot \bar{t})].$$

Without taking the fourth power, the determinant would describe the composite system $sp^3 + sp^3$. The combination of terms means that these composite particles behave like bosons under exchange. As a consequence, $\Delta^{1\ldots8}_{1\ldots8}$ starts with $N = 6$, not with $N_{min} = N_{12} - N_1 - N_2 = 11 - 3 - 3 = 5$. If we take the fourth power to describe $^{16}O + {}^{16}O$, $(\Delta^{1\ldots8}_{1\ldots8})^4$ starts with $N = 6 \cdot 4 = 24$ instead of $N_{min} = 4 \cdot 5 = 20$. This means that by a collision of $^{16}O + {}^{16}O$ nuclei one cannot reach the lower levels of the combined system and gets a rather selective excitation. Similar viewpoints were developed from a completely different angle by Harvey [17].

9 Conclusion

The interplay between nuclear composite particle dynamics and group theory may be summarized as follows: Semi-group extensions of these transformations have bounded operator representations in n-body Hilbert space. The n-body dynamics, including transformations to appropriate coordinates and application of permutations, is expressible in terms of linear combinations of these operators with coefficients from representations of the symmetric group. By employing Weyl translation operators one arrives at explicit expressions for the composite particle interaction in the corresponding reduced Hilbert space. The groups and semigroups are not symmetries of this interaction but could be called dynamical semigroups. Since the present approach provides explicit expressions for the composite particle dynamics, one is in a position to analyze the structure of these interactions. The fermion nature of the underlying n-body system is represented by the normalization operator. This operator prevents the composite particles from moving into forbidden states, modifies the interaction for the lowest allowed oscillator states and exhibits the boson (or fermion) nature of identical composite particles.

References

[1] *Bohr, A.*, and *Mottelson, B. R.*, Nuclear Structure, vol. 1 New York 1969, vol. 2 Reading 1975.

[2] *Wildermuth, K.* and *Tang, Y. C.*, A Unified Theory of the Nucleus, Braunschweig 1977.

[3] *Brink, D. M.* and *Weiguny*, Nucl. Phys. **A120**, 59 (1968).

[4] *Kramer, P., John, G.* and *Schenzle, D.*, Group Theory and the Interaction of Composite Nucleon Systems, Braunschweig 1980.

[5] *Kramer, P.* and *Seligman, T. H.*, Nucl. Phys. **A123**, 161 (1969); Nucl. Phys. A136, 545 (1969); Nucl. Phys. **A186**, 49 (1972).

[6] *Nagel, J. G.* and *Moshinsky, M.*, J. Math. Phys. **6**, 682 (1965).

[7] *Louck, J. D.*, Amer. J. Phys. **38**, 3 (1970).

[8] *Bargmann, V.*, Group Representations in Hilbert Spaces of Analytic Functions, in: Analytic Methods in Mathematical Physics, ed. by *Gilbert, R. P.* and *Newton, R. G.*, New York 1968.

[9] *Moshinsky, M.* and *Quesne, C.*, J. Math. Phys. **14**, 692 (1973).

[10] *Grossmann, A.*, Geometry of Real and Complex Canonical Transformations, Proc. VI. Int. Coll. Group Theoretical Methods in Physics, Tübingen 1977.

[11] *Sternberg, S.*, Metaplectic Structures, Proc. VI. Int. Coll. Group Theoretical Methods in Physics, Tübingen 1977.

[12] *Kramer, P. Moshinsky, M.* and *Seligman, T. H.*, Complex Extension of Canonical Transformations in Quantum Mechanics, in: Group Theory and its Applications III, ed. by *Loebl, E. M.*, New York 1975.

[13] *Kramer, P.* and *Brunet, M.*, to be published.

[14] *Kramer, P.* and *Brunet, M.*, Proc. VI. Int. Coll. Group Theoretical Methods in Physics, Nijmegen 1975.

[15] *Kramer, P.* and *Schenzle, D.*, Nucl. Phys. **A204**, 593 (1973); Proc. II. Int. Conf. Clustering Phenomena in Nuclei, College Park, Maryland 1975.

[16] *Kramer, P.* and *Schenzle, D.*, Revista Mexicana de Fisica **22**, 25 (1973).

[17] *Harvey, M.*, Proc. II. Int. Conf. Clustering Phenomena in Nuclei, College Park, Maryland 1975.

Chapter V
An Algebraic Approach for Spontaneous Symmetry Breaking in Quantum Statistical Mechanics

G. G. Emch

1 Introduction

The language of this first section is quite heuristic, as it only aims at a broad presentation of the problems we want to turn our attention to. A rigorous formulation of some of these questions, and of their answers, is reviewed in section 2. A few illustrative models are mentioned briefly in Section 3. Section 4 is entirely devoted to the question raised by the very existence of crystals in a world where the fundamental interactions are believed to be covariant at least under the group of rigid motions in $\mathbb{R}^3$.

Spontaneous symmetry breaking could roughly be characterized as a situation where the fundamental, or stable, solutions ψ_0 of a problem have a lower symmetry than that of the equation, say $H(\psi) = 0$, one starts from.

One of the earliest, and most entertaining, account of this phenomenon, cast in the language of the best mathematical physics, can be found in Poincaré's lectures on cosmology [1]. He there discusses a tentative explanation for the origin of the Earth-Moon system, or possibly of the double-stars, from the cooling down of an homogeneous, rotating and self-graviting "nebula". Radiation losses naturally cause the temperature to decrease, and thus the fluid to contract; its angular velocity thus increases and the following scenario unfolds. At small angular velocity ω, one finds the expected result, namely that the stable equilibrium configuration of the fluid is a rotationally symmetric ellipsoid, flattened in the direction of the rotation axis (see Fig. 1.a). These are the so-called MacLaurin ellipsoids, one for each value of ω. As the angular velocity reaches a first critical value ω_C these solutions become unstable, whereas new stable, but rotationally asymmetric solutions appear, namely elongated ellipsoids, with three unequal axis: the so-called Jacobi ellipsoids (see Fig. 1.b). Poincaré notices that a point ω_M is subsequentially reached where the Jacobi ellipsoids in turn become unstable; new equilibrium configurations do then appear, which have a pearlike form such as that suggested in Fig. 1.c. It is these "pears" which would ultimately develop

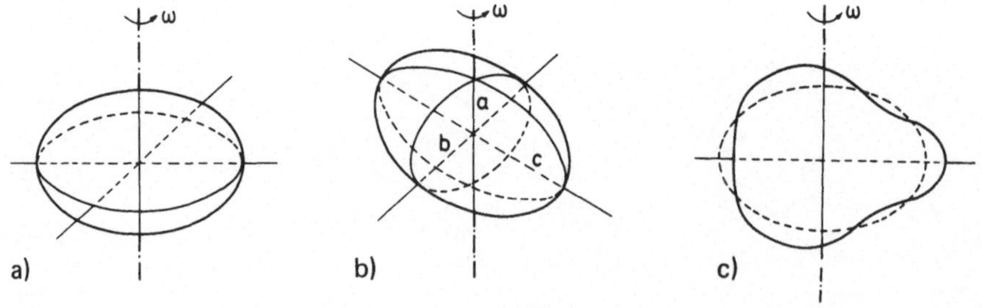

Fig. 1

into two distinct bodies of roughly comparable sizes: the double-stars, or the Earth-Moon system. From the point of view which will be taken in these lectures, it is interesting to comment that the rotational symmetry of the original problem is only lost when one looks at these equilibrium configurations separately, i.e. when the system is looked upon at an arbitrary, but fixed time. The original symmetry still manifests itself by the fact that any rotation will bring any non-rotationally symmetric solution onto another solution with the same control parameters ... and so it actually does in the course of a single revolution of the system.

In a less speculative vein, we certainly believe that the fundamental laws of physics governing our everyday experience are at least euclidian-covariant. Still we know of ferromagnets (breaking of internal rotational symmetry) and of crystals (breaking of both translational and rotational symmetries). We also have come to learn about internal symmetry breaking in elementary particle physics, and in constructive quantum field theory. In all of these cases, the heuristic description attempted in the second paragraph of this section seems to apply.

To exhibit explicitly the spontaneous symmetry breaking phenomenon in a mathematical model, one often looks at it from a point of view which makes it appear as a kind of structural instability. One indeed first breaks the symmetry by fiat, only to then let this perturbation formally fade away. This is for instance done by adding explicitly a perturbation h of lower symmetry than H, looking thus for the fundamental solution(s) of:

$$(H + \lambda h)(\psi_\lambda) = 0;$$

one then checks that

$$\psi_0 = \lim_{\lambda \to 0} \psi_\lambda$$

exists. When ψ_0 does exist and is not invariant under the full group of symmetries of H, one could claim to have obtained a model for spontaneous symmetry breaking. Another way to achieve this is to play with the boundary conditions for finite volume, and then to examine whether the infinite volume limit still depends on the boundary conditions chosen to reach it. One thus studies [2]:

$$H_\Lambda^{BC}(\psi_\Lambda) = 0 \quad \text{and} \quad \psi_0 = \lim_{\Lambda \to \infty} \psi_\Lambda.$$

When spontaneous symmetry breaking can indeed be so interpreted as a structural instability, one often tries to trace the latter to a collective behavior emerging from a large assembly of interacting particles.

This raises two types of questions. Firstly, how does one get *all* fundamental solutions, i.e. in the above two schemes, how does one guess all interesting h or BC? Even for the Ising model, this problem has only been recently discussed in a satisfactory manner [3]. Secondly, one might wonder whether this explicit breaking of the symmetry, only to "regret" it immediately thereafter, is really necessary; or whether one should not try to get more directly at the phenomenon. To be more precise, one would like in statistical mechanics to have a definition of equilibrium states which would allow to get at them in a covariant way, and nevertheless allow for a canonical decomposition into pure thermodynamical phases, in such a manner that only the latter may not inherit the full symmetry

of the problem. In that sense, one could really speak of a "spontaneous" symmetry breaking. To provide such a scheme is precisely the aim of the next section.
In that section, we will follow the so-called "*algebraic approach*" [4]. Some apology might be in order at this point. It should first be recognized that this is a conceptual "approach"; as such, it is not necessarily the most efficient way to explicitly solve intricate and specific models, although it has provided some useful tools [5]. It is elsewhere that the main strength of this approach is to be found. Indeed its strength lies principally in its convenience for *properly formulating problems* and in the *discussion of the possible relations between the different notions* which are called upon for the solution of these problems.

In the sequel, we shall mainly concentrate our attention on quantum statistical mechanics, thus only paying in passing our respects to classical statistical mechanics and to quantum field theory. For classical statistical mechanics, much of what the algebraic approach says can already be said in the context of measure theory. It is actually one of the benefits of the algebraic approach to generalize classical measure theory (and ergodic theory) in such a manner that it can be applied as well to quantum situations. Concerning constructive quantum field theory, with the particular problem of spontaneous symmetry breaking in view, we mostly refer to Fröhlich's lecture notes [7].

2 General theory

In this section we review the general line of argument leading to the C^*-algebraic description of the spontaneous symmetry breaking in quantum statistical mechanics. We shall assume the reader to be familiar with the basic notions of C^*-algebra, state, representation, and symmetry group as they have been presented by Dr. Rieckers earlier in this School [8].

2.1 Finite statistical mechanics

Consider a finite quantum mechanical system enclosed in a finite box Λ. As usual in ordinary quantum statistical mechanics, the Hamiltonian-operator H has three physical avatars: it is interpreted as an observable, the energy; it appears in the Gibbs canonical state, or density matrix; and it is the generator of the time evolution. The consistency of this threefold nature of the Hamiltonian is insured by the following mathematical result.

Proposition 1: *Let* H *be a self-adjoint operator, acting on a Hilbert space* H; *assume that* H *has discrete spectrum* $\{\lambda_i\}$ *such that* $\Sigma_i \exp(-\beta\lambda_i)$ *converges for all* $0 < \beta < \infty$. *Let further* ρ *be a trace-class, positive operator on* H *with* $\mathrm{Tr}\rho = 1$. *Then the following conditions on* ρ *are equivalent:*

(i) $S(\rho) \equiv -\mathrm{Tr}\rho \ln \rho$ *is maximal, subject to* $\mathrm{Tr}\rho \mathrm{H} = \mathrm{E}$;

(ii) $\rho = \exp(-\beta \mathrm{H})/\mathrm{Tr}\exp(-\beta \mathrm{H})$;

(iii) *For every pair* A, B *of bounded operators on H there exists a function* $F_{AB}(z)$ *analytic in, and continuous on the boundaries of, the strip* $\{0 < \mathrm{Im}(z) < \beta\}$ *such that:*

$$F_{AB}(t) = \mathrm{Tr}\rho\, A\alpha(t)[B] \quad \textit{and} \quad F_{AB}(t + i\beta) = \mathrm{Tr}\rho\, \alpha(t)[B]\, A$$

with $\alpha(t)[B] = \exp(-iHt)\, B \exp(iHt)$.

This result is well-known, and a few comments should suffice. Condition (i) is a variational principle: the entropy is maximized subject to the constraint that the expectation value of the energy be fixed. β thus appear in (ii) as a Lagrange multiplier. The equivalence of (i) and (ii) is proven in details in [9]. Condition (iii) is the KMS condition [10]. Checking that (ii) implies (iii) is trivial. To check that (iii) can only be satisfied by (ii), first write $A = I$ to prove that $\alpha(t)[\rho] = \rho$ for all t in $\mathbb{R}$, i.e. ρ commutes with H; reintroduce then A arbitrary in (iii), and use the fact that $B(H)$ is irreductible.

In connection with the main purpose of these lectures, we should notice that the equivalence of conditions (i) and (ii) in Proposition 1 rules out spontaneous symmetry breaking in finite quantum statistical mechanics. Indeed, if U is a symmetry for the Hamiltonian H, i.e. $[U, H] = 0$, we also have $U\rho U^* = \rho$, i.e. ρ inherits the full symmetry of H. Furthermore, if the KMS condition should be taken as a definition of an equilibrium state (as we shall subsequently argue it should), then the equivalence of conditions (iii) and (ii) is a uniqueness theorem which rules out the coexistence of several thermodynamical phases.

To escape from these two "no-go" consequences of Proposition 1, the line of least resistance seems, at this point, to take the thermodynamical limit, which we will analyze in the next subsection.

2.2 Thermodynamical limit

Besides the fact that important physical phenomena, such as spontaneous symmetry breaking and phase transitions, cannot be described consistently within the usual formalism of *finite* statistical mechanics, there are further reasons for considering the thermodynamical limit. Amongst these reasons, one should note that a finite system is a somewhat unwarranted physical idealization in that it assumes perfect walls. Rather than enclosing the system of interest with such "perfect walls", it seems more physical to consider it as a part of a larger system of the same kind, which is taken to be infinite in order to avoid irrelevant boundary effects.

Contrary to the statement sometimes encountered in the literature, the C*-algebraic approach does *not* start from the outset with infinite systems. Rather, it *first* gives a mathematically precise meaning to *the process of taking the thermodynamical limit,* starting from finite systems; it *then* studies the general properties of the objects obtained in this limit.

Assuming that we ultimately want to describe an infinite system with configuration space Ω (such as $\mathbb{Z}^\nu$, $\mathbb{R}^\nu$, or M^{s+1}), we suppose that we are given a family F of bounded (open) regions Λ in Ω satisfying the following properties.

(A) Under the partial ordering relation provided by the set-theoretic inclusion, we have:

(i) for any pair Λ_1, Λ_2 of elements of F there exists at least one element Λ_{12} of F such that $\Lambda_1 \subseteq \Lambda_{12}$ and $\Lambda_2 \subseteq \Lambda_{12}$;

(ii) every bounded region in Ω is contained in at least one element Λ of F.

(B) To every $\Lambda \in F$ is associated a C*-algebra $\mathbf{A}(\Lambda)$, called the algebra of observables for the region Λ. Except for an explicit mention to the contrary, we tacitly assume that all C*-algebras appearing in these lectures possess a unit (here I_Λ), and are equipped with their w*-topology. Physical consistency requires us to assume, for any pair Λ_1, Λ_2 of elements of F satisfying $\Lambda_1 \subseteq \Lambda_2$, the existence of an injective *-homomorphism $i_{21}: \mathbf{A}(\Lambda_1) \to \mathbf{A}(\Lambda_2)$ such that:

(i) $i_{21}(I_1) = I_2$

(ii) $\Lambda_1 \subseteq \Lambda_2 \subseteq \Lambda_3$ implies $i_{31} = i_{32} \circ i_{21}$.

In summary, we suppose that we are given an isotonic family $\{\mathbf{A}(\Lambda) \mid \Lambda \in F\}$ of local C*-algebras.

For any such family, there exists then [11] a C*-algebra $\mathbf{A}$ (unique up to isomorphism) and a family of injective *-homomorphisms $\{i_\Lambda \mid \Lambda \in F\}$ such that:

(i) $i_\Lambda: \mathbf{A}(\Lambda) \to \mathbf{A}$

(ii) $i(I_\Lambda) = I$

(iii) $\mathbf{A} = \overline{\cup_{\Lambda \in F} i_\Lambda(\mathbf{A}(\Lambda))}^{\text{norm}}$.

Mathematically, $\mathbf{A}$ is the C*-*inductive limit* of $\{\mathbf{A}(\Lambda) \mid \Lambda \in F\}$; we will omit in the sequel to write i_Λ, thus identifying $\mathbf{A}(\Lambda)$ with its image in $\mathbf{A}$. Physically, $\mathbf{A}$ is interpreted as the *algebra of quasi-local observables* for the infinite system to be considered.

The simplest example to be offered here is probably the one-dimensional Heisenberg chain. To every $n \in \mathbb{Z}$ we associate a copy $\mathbf{A}_n$ of the C*-algebra $M(2, \mathbb{C})$ of the 2-by-2 complex matrices; physically $\mathbf{A}_n$ is the algebra of observables for a spin-half particle pinned at the site n of the one-dimensional lattice $\mathbb{Z}$. For every finite region $\Lambda \subset \mathbb{Z}$, we define:

$$\mathbf{A}(\Lambda) = \otimes_{n \in \mathbb{Z}} \mathbf{A}_n \simeq M(2^{|\Lambda|}, \mathbb{C}).$$

For any pair $\Lambda_1 \subseteq \Lambda_2$ we write:

$$\Lambda_2 = \Lambda_1 \cup (\Lambda_2 \setminus \Lambda_1) \text{ and thus:}$$

$$\mathbf{A}(\Lambda_2) = \mathbf{A}(\Lambda_1) \otimes \mathbf{A}(\Lambda_2 \setminus \Lambda_1):$$

this suggests the natural embedding:

$$i_{21}: A \in \mathbf{A}(\Lambda_1) \mapsto A \otimes I_{\Lambda_2 \setminus \Lambda_1} \in \mathbf{A}(\Lambda_2).$$

Clearly $\{\mathbf{A}(\Lambda) \mid \Lambda \in F\}$ together with $\{i_{21} \mid \Lambda_1, \Lambda_2 \in F, \Lambda_1 \subseteq \Lambda_2\}$ satisfy the conditions (A) and (B) above. Takeda's theorem then assures the existence of the C*-algebra of quasi-local observables for the infinite chain $\mathbb{Z}$, as the C*-inductive limit $\mathbf{A}$ of $\{\mathbf{A}(\Lambda) \mid \Lambda \in F\}$:

$$\mathbf{A} \cong \otimes_{n \in \mathbb{Z}} \mathbf{A}_n.$$

Notice that each $\mathbf{A}(\Lambda)$ was defined, as is customary in finite quantum statistical mechanics, as an algebra of operators acting on a Hilbert space. However $\mathbf{A}$ only appears here as an abstract C*-algebra, without any reference to an underlying Hilbert space on which it should act. We will take advantage of this flexibility in the sequel.

We now return to the general case. The next step in the process of taking the thermodynamical limit is to further suppose that to every $\Lambda \in F$ is associated a state ϕ_Λ on $\mathbf{A}(\Lambda)$ in such a manner that, for every $\Lambda_0 \in F$ and every $A \in \mathbf{A}(\Lambda_0)$,

$$\lim_{\substack{\Lambda \to \Omega \\ (\Lambda \supseteq \Lambda_0)}} \langle \phi_\Lambda ; A \rangle = \langle \phi ; A \rangle$$

exists, and thus defines a positive linear functional

$$\phi : A \in \cup_{\Lambda \in F} \mathbf{A}(\Lambda) \mapsto \langle \phi ; A \rangle = \lim_{\Lambda \to \Omega} \langle \phi_\Lambda ; A \rangle$$

with $\langle \phi ; I \rangle = 1$. This extends by continuity to a state on the quasi-local algebra $\mathbf{A}$ itself; we again denote this state by ϕ.

In the case of the Heisenberg chain discussed above, the ϕ_Λ can be chosen to be:

$$\phi_\Lambda : A \in \mathbf{A}(\Lambda) \to \langle \phi_\Lambda ; A \rangle = \mathrm{Tr}\, \rho_\Lambda A \text{ with}$$

$$\rho_\Lambda = \exp(-\beta H_\Lambda)/\mathrm{Tr} \exp(-\beta H_\Lambda) \text{ and}$$

$$H_\Lambda = -B\, \Sigma_{n \in \Lambda}\, \sigma_n^z + J\, \Sigma_{n \in \Lambda}\, \vec{\sigma}_n \cdot \vec{\sigma}_{n+1}$$

(the latter being taken for instance with periodic boundary conditions.

The existence of the limiting Gibbs state ϕ has been proven in [6]; the proof is actually valid for interactions which are quite more general than the simple nearest neighbor interaction taken here.

Now that we have a state on our system of interest, we can construct a representation of $\mathbf{A}$ as an algebra of operators acting on a Hilbert space adapted to the situation at hand. This is the well-known GNS construction [12] which we only recall here for self-completeness.

Proposition 2: *Let ϕ be a state on a C*-algebra $\mathbf{A}$. Then there exists a triple $\{H_\phi, \Phi, \pi_\phi\}$, unique up to unitary equivalence, consisting of a Hilbert space H_ϕ, a normalized vector $\Phi \in H_\phi$, and a *-representation π_ϕ: $A \in \mathbf{A} \mapsto \pi_\phi(A) \in B(H_\phi)$ such that:*

(i) $\langle \phi ; A \rangle = (\pi_\phi(A)\, \Phi, \Phi)$ *for all* A *in* $\mathbf{A}$

(ii) $\pi_\phi(\mathbf{A})\, \Phi$ *is dense in* H_ϕ.

It is often useful to know whether H_ϕ is separable. This does occur in particular when ϕ is a locally normal [13] state on a quasi-local algebra, a condition often verified in physical applications.

The third, and last step in taking the thermodynamical limit is to suppose that for each $\Lambda \in F$ we are given a time evolution $\alpha_\Lambda : t \in \mathbb{R} \mapsto \alpha_\Lambda(t) \in \text{Aut}\,[\mathbf{A}(\Lambda)]$, such as that obtained for instance from an Hamiltonian H_Λ:

$$\alpha_\Lambda(t)\,[A] = \exp(-iH_\Lambda t)\, A \exp(iH_\Lambda t).$$

In the simplest cases one can again prove the existence of a group representation $\alpha: t \in \mathbb{R} \to \alpha(t) \in \text{Aut}\,\mathbf{A}$ such that for every $\Lambda_0 \in F$ and $A \in \mathbf{A}(\Lambda_0)$:

$$\lim_{\substack{\Lambda \to \Omega \\ (\Lambda \supseteq \Lambda_0)}} \|\alpha_\Lambda(t)\,[A] - \alpha(t)\,[A]\| = 0.$$

This is in particular the case for quantum lattice systems with sufficiently short-range interactions [14]. These however are rather exceptionally favorable circumstances [15]. It nevertheless seems [16] that one could reasonably expect the following to hold:

$$\lim_{\Lambda \to \Omega} \| \{\pi_\phi \circ \alpha_\Lambda(t)\,[A] - \exp(-iHt)\,\pi_\phi(A)\exp(iHt)\}\,\Psi\| = 0$$

with

$$\exp(-iHt)\,\pi_\phi(\mathbf{A})''\exp(iHt) = \pi_\phi(\mathbf{A})''$$

where

ϕ is the Gibbs state obtained in the second step of the process of taking the thermodynamical limit (see above);

π_ϕ is the GNS representation constructed from ϕ;

H is a (in general unbounded) self-adjoint operator acting on the Hilbert space H_ϕ;

Ψ is a vector running over H_ϕ.

It might be argued that it could be only a peculiarity of non-relativistic physics that the time evolution cannot in general be defined in a representation-independent manner, i.e. as a group of automorphisms of the quasi-local algebra $\mathbf{A}$. Indeed in these theories, either classical or quantum, it is very easy to exhibit catastrophic initial conditions such as for instance they would lead, in a finite time, to an infinite accumulation of particles, or of energy, in a finite volume. One could thus try to reason away the peculiarity that $\alpha(t)$ is only defined in a representation-dependent manner, by saying that the representation in its very construction only involves "smooth" states, such as those encountered when one modifies the Gibbs state only by a local perturbation. For lattice systems the same phenomena (namely that $\alpha(t)$ be representation-dependent) can occur due to very long range interactions which enhance, all the way up to the finite-time evolution, the collective effects of the whole system; this is the case in mean free field theories [17] where different automorphisms of $\mathbf{A}$ itself are produced for different pure thermodynamical phases.

Now, even in the cases where the time evolution has to be defined in a representation-dependent manner, first selecting the proper states to construct these, it seems that one most often keeps enough flexibility by assuming (as suggested above) that $\{\alpha(t)\,|\,t \in \mathbb{R}\}$ is an automorphism group of $\pi_\phi(\mathbf{A})''$. Since $\mathbf{N} = \pi_\phi(\mathbf{A})''$ is a von Neumann algebra, and thus a special kind of C^*-algebra, we can proceed further with our discussion.

In the favorable circumstances where the thermodynamical limit can be carried out as just discussed, we thus have:

$$\alpha(t)[A] = \exp(-iHt)\, A \exp(iHt)$$
$$\exp(iHt)\, \Phi = \Phi.$$

This operator H, which appears as the generator of the time evolution, should however not be interpreted too hastily as an ordinary Hamiltonian. In general, its spectrum is neither necessarily discrete, nor even bounded below. We therefore have definitely moved away from a situation which could be handled by the Gibbs prescription (see Proposition 1). It is however remarkable that the KMS condition (condition iii of Proposition 1) persists to make sense for the infinite systems considered in the present subsection. We indeed can rewrite this condition, independently of the setting of subsection 1 above, as follows.

Definition 1: *Let* **A** *be a* C^**-algebra;* α: $\mathbb{R} \to \mathrm{Aut}(\mathbf{A})$ *be a group homomorphism; and* ϕ *be a state on* **A**. ϕ *is said to be KMS on* **A** *with respect to* $\alpha(\mathbb{R})$ *if:*

(i) $\langle\phi; A\alpha(t)[B]\rangle$ *and* $\langle\phi; \alpha(t)[B]A\rangle$ *are continuous in* t *for all* A, B *in* **A**;

(ii) *For every* $A, B \in \mathbf{A}$, *there exists a function* $F_{AB}(z)$ *analytic in the strip* $\{0 < \mathrm{Im}(z) < 1\}$ *such that* $F_{AB}(t) = \langle\phi; A\alpha(t)[B]\rangle$ *and* $F_{AB}(t+i) = \langle\phi; \alpha(t)[B]A\rangle$.

In this definition, we wrote $\beta = 1$; the case $\beta \neq 1$ can be handled by a scale change in t. We should also note that we tacitly assume, as usual, that **A** has a unit; if this condition were not realized, we should further require in the above definition that $\phi \circ \alpha(t) = \phi$ for all t in $\mathbb{R}$.

If one assumes (as is customarily done in these lectures) that **A** posseses a unit, one checks easily the following result:

Scholium 3: *If* ϕ *is KMS on* **A** *with respect to* $\alpha(\mathbb{R})$, *then* $\phi \circ \alpha(t) = \phi$ *for all* t *in* $\mathbb{R}$.

This result is evidently a sine qua non prerequisite for the forthcoming argument that the KMS condition is an equilibrium condition. We should also notice the evidence provided by the fact that whenever the thermodynamical limit has been carried out one has found, in this limit, a Gibbs state to satisfy the KMS condition with respect to the proper time evolution.

2.3 KMS and thermodynamical stability

In the past few years several proposals have been made to rewrite the KMS condition in such a manner that it could be naturally justified from first principles. A small, but hopefully representative sample of these will be mentioned in the present subsection.

Several correlation-inequalities generalizing, strengthening or otherwise modifying the fundamental Bogoliubov inequality have been discussed in this context, and variously interpreted as physical variational principles [18]. Amongst these we mention here that

Sewell derived, from a local thermodynamical stability condition, the following inequality which is equivalent to the KMS condition.

Proposition 4: *Let* **A** *be a* C*-*algebra;* ϕ *be a state on* **A**; $\alpha(\mathrm{IR})$ *be a continuous group of automorphisms of* **A**; *and* δ *be the derivation obtained as the generator of* $\alpha(\mathrm{IR})$, *with domain* $D(\delta)$. *Then the following two conditions are equivalent:*

(i) ϕ *is KMS on* **A** *with respect to* $\alpha(\mathrm{IR})$;

(ii) $-i\langle\phi; A^*\delta(A)\rangle \geqslant S(\langle\phi; AA^*\rangle, \langle\phi; A^*A\rangle)$

for all A *in* $D(\delta)$; *and with:*

$$S(v,u) = \begin{cases} u\ln(u/v) & \text{if } u>0 \text{ and } v>0 \\ 0 & \text{if } u=0 \\ +\infty & \text{if } u>0 \text{ and } v=0. \end{cases}$$

The KMS condition has also recently been derived as a consequence of the second principle of thermodynamics [19]. Namely, with **A** a C*-algebra, and $\alpha(\mathrm{IR})$ a continuous group of automorphisms of **A**, a state ϕ on **A** is said to be *passive* if

$$L^h(\phi) = \int_0^T dt \left\langle \phi; \alpha_t^h\left[\frac{dh_t}{dt}\right]\right\rangle \geqslant 0$$

for all continuously differentiable h: $t \in \mathrm{IR} \to h_t \in \mathbf{A}_{sa}$ with $h_t = 0$ for $t \leqslant 0$ and $t \geqslant T$ for some $T < \infty$. In the above expression α_t^h is defined by the differential equation

$$\frac{d}{dt}\alpha_t^h[A] = \alpha_t^h[\delta_t(A)] \text{ for all A in } D(\delta),$$

with initial condition $\alpha_0^h = \mathrm{id}$;

δ is the derivation obtained as the generator of $\alpha(\mathrm{IR})$, and

$$\delta_t = \delta + i[h_t, \ ].$$

Note that $L^h(\phi)$ is interpreted as the energy transfered to the system during the finite time interval $[0, T]$. ϕ is further said to be *completely passive* if for every $N = 1, 2, \ldots$ the state $\otimes_N \phi$ on the C*-algebraic system $\{\otimes_N \mathbf{A}, \otimes_N \alpha(\mathrm{IR})\}$ is passive.

The connection between passivity and KMS is then expressed as the following result [19].

Proposition 5: (i) ϕ *is a KMS state (with* $\beta > 0$*) (resp. a ground state) on* $\{\mathbf{A}, \alpha(\mathrm{IR})\}$ *implies that* ϕ *is completely passive;* (ii) ϕ *completely passive (resp. passive and* η*-clustering* [20] *with respect to some locally compact amenable group of automorphisms of* **A** *commuting with* $\alpha(\mathrm{IR})$*) implies that* ϕ *is either a KMS state for some* $\beta \geqslant 0$ *or a ground state.*

Finally we recall the argument originally proposed in [21]. We again consider a C*-algebra **A**, a continuous group $\alpha^0(\mathbb{R})$ of automorphisms of **A**, and define for an arbitrary element $h = h^*$ in **A** the unitary cocycle $P^h : t \in \mathbb{R} \to P^h_t \in U(\mathbf{A})$ determined by

$$\frac{d}{dt} P^h_t = i P^h_t \alpha^0_t [h]; \quad P^h_0 = I.$$

As a consequence of

$$P^h_{t+s} = P^h_t \alpha^0_t [P^h_s],$$

$$\alpha^h_t \equiv \mathrm{ad}_{[P^h_t]} \circ \alpha^0_t$$

defines a group homomorphism

$$\alpha^h : t \in \mathbb{R} \mapsto \alpha^h_t \in \mathrm{Aut}(\mathbf{A})$$

which satisfies:

$$\frac{d}{dt} \alpha^h_t \Big|_{t=0} = \frac{d}{dt} \alpha^0_t \Big|_{t=0} + i[h, \ ].$$

This perturbed evolution $\{\alpha^h_t \mid t \in \mathbb{R}\} \subset \mathrm{Aut}(\mathbf{A})$ is now used to formulate the following stability concept. A state ϕ on **A**, with $\phi \circ \alpha^0_t = \phi$ for all t in $\mathbb{R}$, is said to be *locally stable* if there exist a neighborhood N_0 of $0 \in \mathbf{A}_{sa}$ and $\phi : h \in N_0 \mapsto \phi^h \in \mathbf{S}(\mathbf{A})$ such that:

(i) $\phi^h \circ \alpha^h_t = \phi^h$ for all t in $\mathbb{R}$;

(ii) $w^*\text{-}\lim_{\lambda \to 0} \phi^{\lambda h} = \phi$;

(iii) $w^*\text{-}\lim_{t \to \pm\infty} \phi^h \circ \alpha^0_t = \phi$.

$\{\mathbf{A}, \alpha^0(\mathbb{R})\}$ is said to be *L^1-asymptotically abelian* in time if there exists a norm-dense, $\alpha^0(\mathbb{R})$-stable, *-subalgebra $\mathbf{A}_0$ of **A** such that

$$\int_{-\infty}^{+\infty} dt \, \|[A, \alpha^0_t [B]]\| < \infty \quad \text{for all } A, B \in \mathbf{A}_0.$$

With $\mathbf{A}_0$ as above, a state ϕ on **A** is said to be *hyperclustering of order* k with respect to $\alpha^0(\mathbb{R})$ if for every $\{A_1, \ldots, A_p \mid p \leqslant k\} \subseteq \mathbf{A}_0$ there exist $C > 0$ and $\epsilon > 0$ such that

$$|\langle \phi^T_{(p)} ; \alpha^0_{t_1}[A_1] \ldots \alpha^0_{t_p}[A_p] \rangle| \leqslant C [1 + \operatorname{Sup}_{i \neq j} |t_i - t_j|^{-(1+\epsilon)}].$$

Equipped with these definitions we can now state the main result concerning the justification of the KMS condition, according to [21].

Proposition 6: *If $\{\mathbf{A}, \alpha^0(\mathbb{R})\}$ is L^1-asymptotically abelian, and if ϕ is a state on* **A** *which is: hyperclustering of order* 4, *$\alpha^0(\mathbb{R})$-invariant and locally stable, and is not a trace. Then either ϕ is a β-KMS with respect to $\alpha^0(\mathbb{R})$; or $\alpha^0(\mathbb{R})$ is unitarily implemented, in the GNS representation for ϕ, by a $U_\phi(\mathbb{R})$, the generator of which has one-sided spectrum.*

We should first remark that in the above proposition the condition of hyperclustering of order 4 can [21] be replaced by the weaker condition of η-clustering [20] if one imposes, in addition, that the composite system formed by the juxtaposition (i.e. tensor product) of several copies of $\{\mathbf{A}, \alpha^0(\mathrm{IR}), \phi\}$ is still locally stable.

The approach leading to this proposition presents two main rewards, and one drawback which will have to be elucidated later. Firstly, it succeeds in justifying the formulation of equilibrium statistical mechanics from ideas which involve some postulate to the effect that there is in nature a dominant tendency for equilibrium to establish itself from a non-equilibrium situation. Secondly, it generalizes to a theory where not only the temperature β, but also the chemical potential μ enter in a natural manner. On the other hand, one of the most delicate assumptions of the theory is that concerning asymptotic abelianness in time; we shall come back to it in Subsection 6 below. As we shall see there, some property of this kind, but probably weaker, should apparently hold if one insists that pure thermodynamical phases have certain ergodic properties. We must nevertheless point out here that we suffer from a bad paucity of Hamiltonian models which have actually been proven to enjoy a property of asymptotic abelianness in time.

Propositions 4–6, as well as the last remarks of the preceding subsection, seem thus to indicate that the KMS condition can be made, from first principles, into a physically reasonable requirement on the equilibrium states of a large quantum mechanical system. This claim evidently begs for a classical formulation of this condition, and for a comparison between the latter and the various known classical equilibrium conditions. This has fortunately been recently achieved [22].

Finally, it might be of interest to point out here that a rather tight connection has also recently been noticed [23] between the KMS condition and a "detailed balance" condition for a small quantum system weakly coupled with a large thermal bath.

2.4 Properties of KMS states

KMS states were defined at the end of Subsection 2. Some of the physical motivation behind this condition was given in Subsection 3.

The aim of the present subsection is now to collect some mathematical information on KMS states and the KMS condition. The mathematical theory is nowadays very rich, and a choice had to be made; the selection presented here is evidently biased by the physical problem at hand, and by the general perspective of these lectures.

Let $\mathbf{A}$ be a C*-algebra, ϕ be a state on $\mathbf{A}$, and π_ϕ be the GNS representation of $\mathbf{A}$ induced by ϕ; let further

$$\mathrm{Ker}\,\pi_\phi = \{A \in \mathbf{A} \mid \pi_\phi(A) = 0\}$$

$$\mathrm{Ker}\,\phi = \{A \in \mathbf{A} \mid \langle\phi; A^*A\rangle = 0\}.$$

In general we only have $\mathrm{Ker}\,\pi_\phi \subseteq \mathrm{Ker}\,\phi$. For KMS states however, the following result follows from the analyticity of the function $F_{X*Y*Y,X}(z)$ entering in Definition 1 (see Subsection 2).

Scholium 7: ϕ *KMS on* **A** *implies* Ker π_ϕ = Ker ϕ. *In particular every KMS state on a simple* C*-*algebra is faithful.*

We now carry over the KMS property from ϕ on **A** to the von Neumann algebra $\pi_\phi(\mathbf{A})''$ generated by the GNS representation for ϕ. Since $\phi \circ \alpha(t) = \phi$, $\alpha(t)$ is unitarily implemented by $U_\phi(t)$, defined as the continuous extension to H_ϕ of the map:

$$U_\phi(t): \pi_\phi(A)\,\Phi \in H_\phi \mapsto \pi_\phi(\alpha(t)[A])\,\Phi \in H_\phi.$$

One then defines

$$\tilde{\alpha}(t): N \in \pi_\phi(\mathbf{A})'' \mapsto U_\phi(t)\,N\,U_\phi(-t) \in \pi_\phi(\mathbf{A})'' \equiv \mathbf{N}$$

$$\tilde{\phi}: N \in \pi_\phi(\mathbf{A})'' \mapsto (N\Phi, \Phi) \in \mathbb{C}.$$

With this notation we have:

Scholium 8: ϕ *KMS on* **A** *with respect to* $\alpha(\mathrm{IR})$ *implies* $\tilde{\phi}$ *KMS on* **N** *with respect to* $\tilde{\alpha}(\mathrm{IR})$, *and* $\tilde{\phi}$ *faithful state on* **N**.

In particular, we thus not only have that Φ is cyclic in H_ϕ for $\pi_\phi(\mathbf{A})$, but also that Φ is cyclic in H_ϕ for $\pi_\phi(\mathbf{A})'$, i.e.

$$\overline{\pi_\phi(\mathbf{A})\,\Phi} = H_\phi = \overline{\pi_\phi(\mathbf{A})'\,\Phi}$$

which already indicates that $\pi_\phi(\mathbf{A})'$ is rather large. Indeed the following, most remarkable result has been proven by Takesaki [24].

Proposition 9: *Let* **N** *be a von Neumann algebra acting on a Hilbert space* H; *suppose that there exists* Φ *in* H, *both cyclic and separating for* **N**; *let* $\phi: N \in \mathbf{N} \to (N\Phi, \Phi) \in \mathbb{C}$. *Then there exist:*

> *an involutive, antiunitary operator* J *on* H, *and a continuous unitary representation* $U: \mathrm{IR} \to \mathcal{U}(H)$

such that:

(i) $J\mathbf{N}J = \mathbf{N}'$

(ii) $\sigma(t): N \in \mathbf{N} \to \sigma(t)[N] = U(t)\,N\,U(-t) \in \mathbf{N}$

(iii) ϕ *is KMS on* **N** *with respect to* $\sigma(\mathrm{IR})$.

Furthermore:

(iv) $\sigma(\mathrm{IR})$ *is the only continuous one-parameter group of automorphisms of* **N**, *with respect to which* ϕ *is KMS on* **N**;

(v) $\alpha \in \mathrm{Aut}(\mathbf{N})$, $\phi \circ \alpha = \phi$ *imply* α *commutes with* $\{\sigma(t) \mid t \in \mathrm{IR}\}$.

The proof of the existence of J and U(t), as given in [24], is constructive, and quite involved. It has been somewhat simplified in [25]; a simple characterization of the operators J and U(t) can be found in [26]. Let indeed $\mathbf{N}_{sa}$ be the self-adjoint part of **N**; H_{IR} be the real Hilbert space the elements of which are those of H, and with scalar product $\langle \ldots, \ldots \rangle =$ $= \mathrm{Re}(\ldots, \ldots)$; K be the closed, real linear span of $\mathbf{N}_{sa}\Phi$; and L be iK. Note that $K \cap L = 0$

and $K + L$ dense in $H_{\mathbb{R}}$. Let then P (resp. Q) be the projector from $H_{\mathbb{R}}$ onto K (resp. L). Form then $R = P + Q$, and the polar decomposition JT of (P–Q). Upon lifting these objects back to H, R becomes a linear, self-adjoint operator on H, with $0 \leqslant R \leqslant 2$, R and $(2 - R)$ both injective; J becomes an anti-unitary involutive operator on H. This J is the operator J appearing in Proposition 9, whereas $U(t) = (2 - R)^{it} R^{-it}$ implements $\sigma(t)$.

From the KMS condition, now extended to $\mathbf{N} = \pi_\phi(\mathbf{N})''$, one checks easily the following result.

Scholium 10: *Let ϕ be a KMS state on* $\mathbf{A}$ *with respect to* $\alpha(\mathbb{R})$*; and* $\tilde{\alpha}(\mathbb{R})$ *be the continuous extension of* $\alpha(\mathbb{R})$ *to* $\mathbf{N} = \pi_\phi(\mathbf{A})''$. *Then for every* $Z \in Z_\phi = \mathbf{N} \cap \mathbf{N}'$, $\tilde{\alpha}(t)[Z] = Z$ *for all* t *in* $\mathbb{R}$.

Coming back to Proposition 9, we should notice that the latter asserts that, given $\tilde{\phi}$ faithful on $\mathbf{N} = \pi_\phi(\mathbf{A})''$, there exists an *unique* automorphism group $\tilde{\alpha}(\mathbb{R})$ of $\mathbf{N}$ such that $\tilde{\phi}$ is KMS on $\mathbf{N}$ with respect to $\tilde{\alpha}(\mathbb{R})$; this claims transfers down to $\mathbf{A}$ itself. However, and this is *the main point in connection with the problem of spontaneous symmetry breaking*, the proposition does *not* assert the converse; namely, given $\alpha(\mathbb{R})$ and β (here $\beta = 1$ for convenience), nothing tells us that ϕ should be unique, subject to the KMS condition. Indeed, as a simple application of Scholium 10, we obtain the following statement, which is instrumental in the proof of the forthcoming Proposition 12.

Scholium 11: *Let ϕ be a KMS state on* $\mathbf{A}$ *with respect to* $\alpha(\mathbb{R})$. *Then, for every* $Z \in Z_\phi = \pi_\phi(\mathbf{A})'' \cap \pi_\phi(\mathbf{A})'$ *with* $\|Z\Phi\| = 1$,

$$\psi_Z : A \in \mathbf{A} \to (\pi_\phi(A) Z\Phi, Z\Phi)$$

is a KMS state on $\mathbf{A}$ *with respect to* $\alpha(\mathbb{R})$. *Moreover there exists* $\lambda \in \mathbb{R}^+$ *such that* $\langle \psi_Z ; A^*A \rangle \leqslant \lambda \langle \phi ; A^*A \rangle$ *for all* A *in* $\mathbf{A}$. *Conversely, if* ψ *is a KMS state on* $\mathbf{A}$ *with respect to* $\alpha(\mathbb{R})$, *such that* $\langle \psi ; A^*A \rangle \leqslant \lambda \langle \phi ; A^*A \rangle$ *for some* $\lambda \in \mathbb{R}^+$, *and all* A *in* $\mathbf{A}$, *then there exists an unique* $Z \in \pi_\phi(\mathbf{A})'' \cap \pi_\phi(\mathbf{A})'$ *such that:* $\|Z\Phi\| = 1$, $Z \geqslant 0$ *and*

$$\langle \psi ; A \rangle = (\pi_\phi(A) Z\Phi, Z\Phi).$$

A KMS state ϕ on $\mathbf{A}$ with respect to $\alpha(\mathbb{R})$ is said to be *extremal KMS* if and only if it cannot be decomposed into a convex sum of other KMS states with respect to the same $\alpha(\mathbb{R})$. It then follows immediately from Scholium 11 that extremal KMS states are characterized, amongst other KMS states, by the following criterion.

Proposition 12: *Let ϕ be a KMS state on* $\mathbf{A}$ *with respect to* $\alpha(\mathbb{R})$. *Then the following two conditions are equivalent*

(i) ϕ *is extremal KMS*

(ii) $\pi_\phi(\mathbf{A})'' \cap \pi_\phi(\mathbf{A})' = \mathbb{C}I$.

2.5 What is a pure thermodynamical phase?

If we were now to agree, in particular on the basis of the results of Subsection 3, that the KMS condition is *the* equilibrium condition, then physical consistency would immediately require that *pure thermodynamical phases* be identified as *extremal KMS states,* i.e. as equilibrium states which cannot be decomposed as mixtures of other equilibrium states.

Alternate definitions have however been proposed, based on some other desirable property that a pure thermodynamical phase should have. Amongst these definitions, we should mention here at least the following two. Firstly, pure thermodynamical phases could be identified as stationary states with some strong clustering properties; this condition emphasizes the link which seems to exists between long-range order and coexistence of several phases. Secondly, pure thermodynamical phases could be identified as KMS states which cannot be decomposed into other stationary states; this condition presents the peculiarity of precluding any such definition for a system which is finite in the sense of Subsection 1; this could nevertheless be a reasonable a priori candidate in the case of infinitely extended systems.
In line with the claim, made towards the end of Section 1, on the possible usefulness of the C*-algebraic approach, we shall now compare, and refine when needed, the above tentative definitions. In particular we shall weight the supportive evidence in favor of the definition of pure thermodynamical phases as extremal KMS states.

2.6 Cluster properties and ergodic theory

Recall that in the context of these lectures the C*-algebra $\mathbf{A}$ is to be physically interpreted as an algebra of quasi-local observables:

$$\mathbf{A} = \overline{\cup_{\Lambda \in F} \mathbf{A}(\Lambda)}^{\text{norm}} .$$

Clearly, when we know how to associate a local algebra $\mathbf{A}(\Lambda)$ to an open bounded region Λ of the configuration space Ω, we also know how to do it for the region $(\Lambda + a)$ for every translation a in S where

$$S = \begin{cases} \mathbb{Z}^\nu & \text{if} \quad \Omega = \mathbb{Z}^\nu \\ \mathbb{R}^\nu & \text{if} \quad \Omega = \mathbb{R}^\nu \\ \mathbb{R}^s & \text{if} \quad \Omega = \mathbf{M}^{s+1} \end{cases} .$$

We shall therefore suppose that F is closed under the operation $\Lambda \mapsto (\Lambda + a)$ for every a in S. Moreover, we assume that for every a in S, $\mathbf{A}(\Lambda)$ is isomorphic to $\mathbf{A}(\Lambda + a)$, and thus that we are given a representation of the translation group S:

$$\alpha\colon a \in S \mapsto \alpha(a) \in \operatorname{Aut}(\mathbf{A}).$$

The physical principle of locality implies that given any $\Lambda \in F$ there exists $a_0 > 0$ such that

$$[A, B] = 0 \text{ for all } \begin{cases} A & \text{in} \quad \mathbf{A}(\Lambda) \\ B & \text{in} \quad \mathbf{A}(\Lambda + a) \\ a & \text{in} \quad |a| > a_0 \end{cases} .$$

This is easily shown to imply that

$$\lim_{|a| \to \infty} \|[A, \alpha(a) B]\| = 0$$

for all A, B in $\mathbf{A} = \overline{\cup_{\Lambda \in F} \mathbf{A}(\Lambda)}^{\text{norm}}$.

This property is known as *norm-asymptotic abelianness.* For space translations this property is thus a universal consequence of locality.
It would be "nice" if this property were to hold as well for the time-evolution; this however is extremely hard to satisfy. It is nevertheless true on the even algebra of the x–y model [27], this result has been used in [28] to show, for this model, that the Gibbs state is dynamically stable under "even" local perturbations; see also [29]. We also saw that a strong form of norm-asymptotic abelianness in time is used in Proposition 6. Finally, a slightly weaker condition of norm-asymptotic abelianness, isolated in [30], has been proven to hold for reversible Generalized K-Flows [31], a structure where the time-evolution responsible for the approach to equilibrium is described in the interaction picture for the van Hove weak-coupling, long-time limit.
It should nevertheless be emphasized again that, contrary to the case of space translations, norm-asymptotic abelianness in time is so genuine a constraint to be imposed on an Hamiltonian dynamical system that one does not yet know how to either ensure it, or cure the extreme paucity of models for which it is proven to hold.

In view of the above remarks, it is interesting to search for conditions of "asymptotic abelianness" which are weaker than the condition of norm-asymptotic abelianness. This quest will also help in elucidating the mathematical structure of the theory. We examine here a few amongst these alternate forms of asymptotic abelianness; for a more complete review, and the proofs of the forthcoming statements, see [4, Chap. II and IV].

A locally compact group G is said to be *amenable* if the C*-algebra $C(G)$ of all complex-valued, continuous bounded functions on G admits at least one G-invariant state η. This state is then said to be an *invariant mean* on G. It is useful to know that all compact groups, all locally compact abelian groups, and all their semi-direct products are amenable. Hence, the time-evolution T, the space-translation group S, the rotation group O_3^+, the euclidian group E_3 are amenable. However, no non-compact, semi-simple Lie group is amenable; in particular the homogeneous and the inhomogeneous Lorentz groups are not amenable; for most applications however, it is sufficient that the translation part of the latter is amenable. Note finally that an amenable group may admit several invariant means; some of the forthcoming results nevertheless are independent of the mean used to formulate them.

Scholium 13: *Let G be an amenable symmetry group of a C*-algebra* **A** *(with unit!). Then* **A** *admits at least one G-invariant state* ϕ.

We now introduce the concepts of G-abelian and η-abelian states.
Let **A** be a C*-algebra; G be an amenable symmetry group of **A**; and η be an invariant mean on G. We denote the action of G on **A** by $\alpha: G \to \mathrm{Aut}(\mathbf{A})$. A state ϕ on **A**, with $\phi = \phi \circ \alpha(g)$ for all g in G, is said to be *G-abelian* if

$$\eta\langle\psi; [A, \alpha(g)[B]]\rangle = 0 \quad \text{for all } A, B \text{ in } \mathbf{A}$$

whenever

$$\psi: A \in \mathbf{A} \mapsto (\pi_\phi(A)\,\Psi, \Psi)$$

with

$$\Psi \in E_\phi \mathcal{H}_\phi$$

and

$$E_\phi = \{\Psi \in \mathcal{H}_\phi \mid U_\phi(g)\,\Psi = \Psi \quad \text{for all g in G}\}$$

where

$$U_\phi: g \in G \mapsto U(g) \in \mathcal{U}(\mathcal{H}_\phi)$$

is defined by

$$U_\phi(g)\,\pi_\phi(A)\,\Phi = \pi_\phi(\alpha(g)[A])\,\Phi$$

with g running over G and A running over **A**.

Note that the condition of G-abelianness on ϕ is actually independent of the mean η used to formulate it; it is indeed equivalent with the condition that $(E_\phi \pi_\phi(\mathbf{A})'' E_\phi)$ be abelian. It furthermore implies that $\pi_\phi(\mathbf{A})' \cap U_\phi(G)'$ is also abelian.
A state ϕ on **A**, with $\phi = \phi \circ \alpha(g)$ for all g in G, is said to be *η-abelian* if

$$\eta\langle\phi; C^*[A, \alpha(g)[B]]\,C\rangle = 0 \quad \text{for all } A, B, C \text{ in } \mathbf{A}.$$

Clearly, G-abelianness follows from η-abelianness of ϕ; on the other hand, if G acts in a norm-asymptotic abelian manner on **A**, then every G-invariant state ϕ is η-abelian. We have thus obtained two degrees of potential weakening for the condition of norm-asymptotic abelianness. The respective roles of these weaker forms of asymptotic abelianness will be discussed later on in this subsection; see in particular Proposition 15 and Scholium 17.

We need one more concept before being able to state our main ergodic theorem, namely the concept of the average of an observable with respect to the action of an amenable symmetry group. Since the average of an observable with respect to the translation group is evidently far from being a local observable, we should not expect this averaging procedure to be in general defined within **A** itself; on the contrary, we should anticipate that it is a representation-dependent concept; this is indeed what the next result shows.

Scholium 14: *Let* **A** *be a* C**-algebra;* G *be an amenable symmetry group of* **A***;* η *be an invariant mean on* G*; and* ϕ *be a* G*-invariant state on* **A***. Let* $U_\phi : G \to \mathcal{U}(H_\phi)$ *and* E_ϕ *be defined by:*

$$U_\phi(g)\,\pi_\phi(A)\,\Phi = \pi_\phi(\alpha(g)[A])\,\Phi \quad \textit{for all } A \textit{ in } \mathbf{A}$$
$$E_\phi = \{\Psi \in H_\phi \mid U_\phi(g)\,\Psi = \Psi \quad \textit{for all } g \textit{ in } G\}$$

Then there exists a mapping

$$\eta_\phi : \mathbf{A} \to \pi_\phi(\mathbf{A})'' \cap U_\phi(G)'$$

such that

(i) $\eta(\pi_\phi(\alpha(g)[A])\,\Psi_1, \Psi_2) = (\eta_\phi[A]\,\Psi_1, \Psi_2)$

for all Ψ_1, Ψ_2 *in* H_ϕ*, and all* A *in* **A**;

(ii) $\eta_\phi[A]\,E_\phi = E_\phi \eta_\phi[A] = E_\phi\,\pi_\phi(A)\,E_\phi$ *for all* A *in* **A**.

We are now ready to state the following ergodic theorem, which illustrates the respective roles of the conditions of asymptotic abelianness just introduced.

Proposition 15: *With the assumptions and notations of Scholium 14, the logical scheme:*

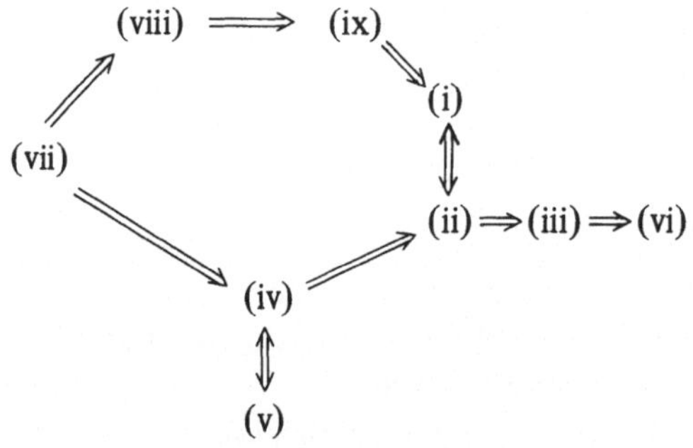

gives the relations between the following nine conditions.

(i) ϕ *is extremal* G*-invariant*

(ii) $\mathbf{M}_\phi' \equiv \pi_\phi(\mathbf{A})' \cap U_\phi(G)' = \mathbb{C}I$

(iii) $\mathbf{M}_\phi'' \cap \mathbf{M}_\phi' = \mathbb{C}I$

(iv) $\eta\langle\phi; \alpha(g)[A]\,B\rangle = \langle\phi; A\rangle\langle\phi; B\rangle$ *for all* A, B *in* **A**

(v) $\dim E_\phi H_\phi = 1$

(vi) $\pi_\phi(\mathbf{A})'' \cap \pi_\phi(\mathbf{A})' \cap U_\phi(G)' = \mathbb{C}I$

(vii) $\eta_\phi[A] = \langle\phi; A\rangle\,I$ *for all* A *in* **A**

(viii) ϕ *is the only normal* G*-invariant state on* $\pi_\phi(\mathbf{A})''$

(ix) ϕ *is the only* G*-invariant vector state on* $\pi_\phi(\mathbf{A})$.

If ϕ is G-abelian, then the first five of the above conditions are equivalent. If furthermore ϕ is η-abelian, then all nine conditions are equivalent. In particular, if ϕ is η-abelian and primary, it satisfies all nine conditions above.

A brief comment on the various conditions of this proposition is perhaps in order. Condition (i) is the non-commutative analog of metric indecomposability in classical ergodic theory, whereas conditions (viii) and (ix) are variations around it. Condition (v) is a well-known ergodicity condition in the classical theory, and it carries over directly to the present non-commutative scheme. Condition (ii) is particularly interesting in case ϕ is η-abelian; we have then indeed $\mathbf{M}'_\phi = \pi_\phi(\mathbf{A})'' \cap U_\phi(G)'$, so that the condition becomes to mean that there are no non-trivial G-invariant elements in the global observable-algebra $\pi_\phi(\mathbf{A})''$ associated to ϕ. Conditions (iii) and (vi) are weakenings of the original condition (ii), but again they are equivalent to the condition $\pi_\phi(\mathbf{A})'' \cap U_\phi(G)' = \mathbb{C}I$ when ϕ is η-abelian. Condition (iv) is the so-called η-clustering property. It indicates the absence of long-range order in the mean. Condition (vii) strengthens this condition by requiring that the mean over G of every observable be a c-number in the representation associated to ϕ. The last condition introduced in the proposition, namely that ϕ be primary, has to be looked upon in the perspective of Proposition 12, i.e. in connection with the identification of pure thermodynamical phases as extremal KMS states. Take indeed for G the group S of the translations; η-abelianness of ϕ follows then from norm-asymptotic abelianness, itself a consequence of locality. We thus have that every translation-invariant, extremal KMS state ϕ satisfies also conditions (i)–(ix) of the proposition, with respect to $G \equiv S$.

The Proposition admits a sort of converse, which we shall now state, and will use towards the end of this subsection.

Scholium 16: *Let* $\mathbf{A}$ *be a* C*-*algebra;* G *be an amenable symmetry group of* $\mathbf{A}$; η *be a invariant mean on* G; ϕ *be an extremal* G*-invariant state on* $\mathbf{A}$ *such that* Φ *is cyclic for* $\pi_\phi(\mathbf{A})'$. *Then* ϕ *is* η*-abelian.*

This result calls for several remarks. First of all, it is one of these results which is independent of the mean η used to formulate it; it thus holds for all means on G. Secondly, we should notice that a state satisfying the assumptions of this scholium, also satisfies all nine conditions of Proposition 15. Thirdly, we saw (Scholium 8) that ϕ KMS on $\mathbf{A}$ implies that $\tilde{\phi}$ is faithful on $\pi_\phi(\mathbf{A})''$, i.e. that Φ is separating for $\pi_\phi(\mathbf{A})''$ and thus cyclic for $\pi_\phi(\mathbf{A})'$. Hence the last assumption of the scholium is satisfied for every KMS state, not just for extremal KMS states. We should thus be open to the possibility that some states be extremal translation invariant, KMS but not extremal KMS; we shall later see that this situation can be sorted out through a careful study of the actual clustering properties of these states. Finally, one can replace in the Scholium the condition that ϕ be extremal G-invariant, by the conditions that it be G-invariant and such that $\pi_\phi(\mathbf{A})'' \cap U_\phi(G)' = \mathbb{C}I$, a condition which we already discussed in our commentary of Proposition 15.

In most of this subsection G was taken to be an arbitrary symmetry group; when we particularized it, we most often were thinking to the translation-group S, thus concentrating on the *spacial properties* of pure thermodynamical phases. We would now like to derive some further consequences of Proposition 15 when G is identified with the *time-evolution.*

Scholium 17: *Let ϕ be a KMS state on* **A** *with respect to* $\alpha(\mathbb{R})$. *Then the following three conditions*

(i) *ϕ is extremal KMS*

(ii) *ϕ is extremal time-invariant*

(iii) *ϕ is η-abelian with respect to $\alpha(\mathbb{R})$*

are in the logical relations:

$$\text{(iii)} \Rightarrow \{\text{(i)} \Longleftrightarrow \text{(ii)}\};$$
$$\text{(ii)} \Rightarrow \text{(i) } \textit{and } \text{(iii)}.$$

Notice again that this result is independent of the mean η (over $\mathbb{R}$) used to formulate it. Condition (iii) of this scholium is evidently a fortiori realized when the time-evolution is norm-asymptotic abelian; in that case, there is then no difference in requesting that a KMS state be extremal KMS or extremal time-invariant, or that it satisfy any of the eight other conditions of Proposition 15. When this very favorable circumstance is met, the proper definition of a pure thermodynamical phase is thus conceptually much easier to write down. On the other hand, the second part of the scholium gives some information on a definition of pure thermodynamical phases which would incorporate some ergodicity in time. Specifically, if one wishes to identify pure thermodynamical phases as (extremal) KMS states which are extremal time-invariant, then *some* asymptotic abelianness in time *must* be satisfied, namely these states must be η-abelian with respect to the time-evolution. Although η-abelianness on particular states is a weaker requirement than norm-asymptotic abelianness, we should recall here our discussion of asymptotic abelianness as given earlier in this subsection. This requirement that ϕ be η-abelian in time is evidently not necessary if one is satisfied with simply requiring that pure thermodynamical phases be identified as extremal KMS states. We recall that KMS extremality and space invariance were already sufficient to ensure quite a list of spacial properties that pure thermodynamical phases might be expected to enjoy.

As a final side-comment to this subsection, we might mention that several of the ergodic theorems mentioned here do hold as well if one replaces $\alpha(\mathbb{R}) \subseteq \text{Aut}(\mathbf{N}, \phi)$ by $\gamma(\mathbb{R})^+ \subseteq \text{CP}(\mathbf{N}, \phi)$, the semi-group of completely positive maps of **N** on itself for which $\phi \circ \gamma = \phi$ [32].

2.7 Uniform clustering

Whereas our main interpretation of Proposition 15 only uses a weak consequence of locality, namely η-abelianness with respect to space translations, one should expect stronger results when one takes into account the full strength of locality, namely norm-asymptotic abelianness. This is indeed the case. In particular, the last statement of Proposition 15 can be strengthened as follows.

Scholium 18: *Let ϕ be space-translation invariant and primary. Then*

$$w - \lim_{|a| \to \infty} \pi_\phi(\alpha(a)[A]) = \langle \phi; A \rangle I \textit{ for all } A \textit{ in } \mathbf{A}.$$

This statement thus replaces a result holding only in the mean, namely condition (vii) of Proposition 15, by an assertion on the existence of point-wise, weak operator limits. This results implies immediately the following strengthening of the η-clustering property.

Scholium 19: *Let ϕ be space-translation invariant and primary. Then*

$$\lim_{|a| \to \infty} \langle \phi; B\alpha(a)[A]C \rangle = \langle \phi; A \rangle \langle \phi; B\,C \rangle .$$

A well-known application of this scholium is the computation of the spontaneous magnetization in the two-dimensional Ising model. One indeed has (see for instance [33]):

$$\lim_{n \to \infty} \langle \sigma_j^z \sigma_{j+n}^z \rangle_\pm = \langle \sigma_j^z \rangle_\pm^2$$

for the two pure phases $\langle \ldots \rangle_+$ and $\langle \ldots \rangle_-$. On the other hand the flip-flop invariant state obtained in the thermodynamical limit is:

$$\langle \ldots \rangle = \tfrac{1}{2} (\langle \ldots \rangle_+ + \langle \ldots \rangle_-)$$

and thus

$$\langle \sigma_j^z \rangle = 0 \text{ but}$$

$$\lim_{n \to \infty} \langle \sigma_j^z \sigma_{j+n}^z \rangle = \langle \sigma_j^z \rangle_\pm^2 .$$

The left-hand side of the above expression is the quantity computed by Onsager [34]. The clustering property of Scholium 19 can actually be further strengthened. To explain this we need a few definitions which we shall now give.

Let ϕ be a state on the quasi-local algebra

$$\mathbf{A} = \overline{\cup_{\Lambda \in F} \mathbf{A}(\Lambda)}^{\text{norm}} .$$

We form:

$$\mathbf{B}_\phi(\Lambda) = \pi_\phi(\mathbf{A}^c(\Lambda))''$$

where

$$\mathbf{A}^c(\Lambda) = \overline{\cup_{M \in F(\Lambda)} \mathbf{A}(M)}^{\text{norm}}$$

with

$$F(\Lambda) = \begin{cases} \{M \in F \mid M \cap \Lambda = \emptyset\} \text{ if } \Omega = Z^\nu \text{ or } \mathbb{R}^\nu \\ \{M \in F \mid M \times \Lambda\} \text{ if } \Omega = M^{s+1} . \end{cases}$$

The von Neumann *algebra of observables at infinity* with respect to ϕ, is then defined as

$$\mathbf{B}_\phi = \cap_{\Lambda \in F} \mathbf{B}_\phi(\Lambda).$$

It is then easy to check that the following result is true.

Scholium 20: $\mathbf{B}_\phi \subset \mathbf{Z}_\phi \equiv \pi_\phi(\mathbf{A})'' \cap \pi_\phi(\mathbf{A})'$.

On the other hand, a state ϕ on the quasi-local algebra $\mathbf{A}$ is said to be *uniformly clustering* if, given $\epsilon > 0$ and A in $\mathbf{A}$, there exists $\Lambda \in F$ such that:

$$|\langle \phi; A\,B \rangle - \langle \phi; A \rangle \langle \phi, B \rangle| \leqslant \epsilon \,||\, B \,||$$

for all B in $\mathbf{A}^c(\Lambda)$.

Proposition 21: *A state ϕ on the quasi-local algebra* $\mathbf{A}$ *is uniformly clustering if and only if its algebra of observable at infinity is trivial, i.e.* $\mathbf{B}_\phi = \mathbb{C}\, I$.
From this Proposition and Scholium 20 the following result is trivially obtained.

Corollary 22: *Every primary state on the quasi-local algebra* $\mathbf{A}$ *is uniformly clustering.*
This is a clear strengthening of the previous versions of clustering, and of the general results obtained in Subsection 6. Physically, this result is important in that it asserts that every extremal KMS state on the quasi-local algebra is uniformly clustering in space (compare with Proposition 15).
It could be argued that a pure thermodynamical phase should be characterized by the property of uniform clustering, leaving alone any explicit reference to the KMS condition. This will not be the point of view adopted in these lectures, for reasons which should be clear by now. It could nevertheless be mentioned here that there are definite physical situations where not only (see Scholium 20) $\mathbf{B}_\phi \subseteq Z_\phi$ but actually $\mathbf{B}_\phi = Z_\phi$. This is in particular the case for quantum lattice systems, where we thus have that every uniformly clustering KMS state is extremal KMS.
These notions were originally considered in [35]; the discussion given here follows [4] where our assertions are all proven in details.
When one is interested in investigating whether a specific model exhibits or not a phase transition, the following technique is worth knowing.

Let P be the projection on $\mathbf{H}_\infty^\perp$ where $\mathbf{H}_\infty = \overline{\mathbf{B}_\phi \Phi}$, and define:

$$\langle A^*A \rangle^T = \langle A^*PA \rangle = (PA\,\Phi, A\,\Phi).$$

One then introduce the *long-range parameter*

$$\alpha_A = \langle A^*A \rangle - |\langle A \rangle|^2 - \langle A^*A \rangle^T .$$

One verifies immediately that if $\mathbf{B}_\phi = \mathbb{C}\,\mathrm{I}$ then $\alpha_A = 0$ for all A in **A**. To prove the existence of a phase transition (in the sense that $\mathbf{B}_\phi \neq \mathbb{C}\,\mathrm{I}$) it is therefore sufficient to find some A for which one can compute:

(a) an upper bound $\langle A^*A\rangle^T \leqslant C_1$

(b) a lower bound $\langle A^*A\rangle - |\langle A\rangle|^2 \geqslant C_2$

(c) $C_2 - C_1 > 0$.

To prove (a) it is sometimes convenient to consider the projector P_I on $\mathbf{H}_I^\perp$ where $\mathbf{H}_I$ is the subspace of all translation-invariant vectors in $\mathbf{H}_\phi$. Since $\mathbf{H}_I \subseteq \mathbf{H}_\infty$, $P_I \geqslant P$. Define then:

$$\langle A^*A\rangle^c = \langle A^* P_I A\rangle = (P_I A\Phi, A\Phi).$$

We thus have:

$$\langle A^*A\rangle^c \geqslant \langle A^*A\rangle^T .$$

It so happens that in some models one can lay ones hands on $\langle A^*A\rangle^c$, and thus obtain, via the above inequality, the upper bound (a) one is looking for. This alley has been successfully walked in [7].

2.8 Symmetry-breaking and decomposition theory

We start with the following consequences of the considerations of subsection 6 (see in particular Proposition 15).

Scholium 23: *Let ϕ be a state on the quasi-local algebra* **A**, *such that both of the following conditions are satisfied:*

(a) *ϕ is invariant under the translation group* S

(b) *ϕ is clustering.*

Suppose further that α is a symmetry of **A**, *commuting with* S. *Then: either $\phi \circ \alpha = \phi$; or $\phi \circ \alpha$ is not a vector-state in $\mathbf{H}_\phi$.*

The terms of the alternative in the above Scholium correspond to: either α *is*, or α *is not* unitarily implemented in the representation canonically associated to the state a. This is the basis for the assertion according to which symmetry breaking is a notion pertaining to the spacial (i.e. Hilbert space) properties of the system of interest, namely that the "internal symmetry" α is or is not unitarily implemented in the representation of interest. This is the point of view taken for instance in [7], and in the last reference of footnote [21]. We shall take here a slightly more general point of view, which makes it easier to incorporate in the proposed scheme such things as the phenomenon of crystallization; we will therefore formulate the theory in a way which does not preclude a priori a possible breaking of the translation-invariance. We shall in fact avoid for a while any reference to this particular symmetry.

We now set the stage for this final subsection of our account of the algebraic approach to spontaneous symmetry breaking in (quantum)-statistical mechanics.

In view of the evidence accumulated in the preceding subsections, we can now choose to call *pure thermodynamical phase* a state ψ which is extremal KMS. We also expect that, in a theory where the equilibrium state (or "Gibbs state") is computed in a completely covariant way, we will get in general a KMS state ϕ which is invariant under the original symmetry G of the theory. We however have no general constraint imposing that this Gibbs state should be extremal KMS, i.e. should be a pure thermodynamical phase. Our next task will thus be to find a (unique, or canonical) way to decompose this equilibrium state ϕ into its pure thermodynamical phase components. We should however be ready to face a situation where the resulting extremal KMS states do not inherit of the full symmetry G of ϕ. When indeed the symmetry H of the pure thermodynamical phase components of the Gibbs state is lower than the symmetry G of that state, we shall speak of a *spontaneous symmetry breaking.*

This will be our description of spontaneous symmetry breaking in (quantum) statistical mechanics. This description indicates a logical potentiality of the theory. We shall see in Section 3 that this potential is realized, specifically, we shall see that this scheme covers enough Hamiltonian models to be physically acceptable.

To introduce softly the general decomposition theory we should be looking for, we first consider the following particular case.

Let ϕ be a KMS state on $\mathbf{A}$, with $\mathbf{H}_\phi$ separable. Let $\mathcal{Z}_\phi = \pi_\phi(\mathbf{A})'' \cap \pi_\phi(\mathbf{A})'$; this is clearly an abelian von Neumann algebra acting on a separable Hilbert space; suppose further that $\mathcal{Z}_\phi$ is generated by an hermitian operator Z_ϕ which has discrete spectrum; this latter restriction is the only peculiarity of the present illustrative example, and it will be removed later on. There is therefore, in this example, a partition $\{P_i\}$ of the identity on $\mathbf{H}_\phi$ such that one has exactly:

$$\mathcal{Z}_\phi = \{\Sigma_i\, z_i P_i \,|\, z_i \in \mathbb{C}\}.$$

Define now

$$\psi_i : A \in \mathbf{A} \mapsto \langle \psi_i, A\rangle = \langle \tilde{\phi}; P_i A\rangle \,/\, \langle \tilde{\phi}; P_i\rangle \in \mathbb{C}\,;$$
$$\lambda_i = \langle \tilde{\phi}; P_i\rangle\,.$$

One first checks easily that ϕ KMS implies that all ψ_i are again KMS with respect to the same time-evolution. It is furthermore easy to see that the very construction of the ψ_i, using the minimal projectors of P_i, implies that the restriction of $\pi_\phi(\mathbf{A})$ to $P_i\mathbf{H}_\phi$ is primary, and coincides with the GNS representation canonically associated to ψ_i; hence the ψ_i are primary KMS states, i.e. they are extremal KMS. Finally, one checks by substitution that

$$\phi = \Sigma_i\, \lambda_i\, \psi_i\,.$$

We have thus obtained a decomposition of the KMS state ϕ into a convex combination of extremal KMS states ψ_i. A simple application of Scholium 11 shows that this decomposition is unique. We have thus achieved, for this particular case, the decomposition we were looking for, namely that of an equilibrium state into its pure thermodynamical phase components. A closer scrutiny will allow the general case to be infered from this particular situation. Before doing so however, let us take advantage of the simplicity of this particular case in order to illustrate another general phenomenon.

Suppose further that α is a symmetry of **H** such that $\phi \circ \alpha = \phi$. There exists then a unitary operator U_ϕ, acting on $\mathbf{H}_\phi$, and such that

$$\pi_\phi(\alpha[A]) = U_\phi \pi_\phi(A)\, U_\phi^{-1} \quad \text{for all } A \text{ in } \mathbf{A}\,,$$

and

$$U_\phi \phi = \phi\,.$$

Let then $\tilde{\alpha}$ denote the extension of α to $\mathbf{N} = \pi_\phi(\mathbf{A})''$ defined by:

$$\tilde{\alpha} : N \in \mathbf{N} \mapsto U_\phi N\, U_\phi^{-1} \in \mathbf{N}\,.$$

Clearly the center Z_ϕ of **N** is stable under the action of $\tilde{\alpha}$. We however have no reason to assume that Z_ϕ is point-wise invariant. Since Z_ϕ is discrete with atoms P_i, and since $\tilde{\alpha}$ is a symmetry of Z_ϕ, the image $\tilde{\alpha}[P_i]$ of an atom should again be an atom $P_{j(i)}$ of Z_ϕ. Since moreover $\phi \circ \alpha = \phi$, we must have $\lambda_{j(i)} = \lambda_i$. Hence, whenever $j(i) \neq i$ the symmetry α is *broken* in the pure thermodynamical phase ψ_i; the latter however is sent by the action of $\tilde{\alpha}$ to another pure thermodynamical phase $\psi_{j(i)}$ which occurs in the decomposition of ϕ with the same weight $\lambda_{j(i)}$ as that of ψ_i.

To step up to the general decomposition theory from this particular case, we should further notice that we have, still in this simple case, for every A in **A** and every $Z = \Sigma_i z_i P_i$ in Z_ϕ:

$$\langle \tilde{\phi}; Z\,A \rangle = \Sigma_i \lambda_i z_i \langle \psi_i; A \rangle\,.$$

The general form of this simply verifyable fact is now given.

Proposition 24: *For every state* ϕ *on a* C*-*algebra* **A** *there exists a unique measure* μ_ϕ *on the state space* **S** *of* **A** *such that there is a* σ*-continuous isomorphism:*

$$\varphi : Z \in Z_\phi \mapsto \varphi_Z \in L^\infty(\mathbf{S}, \mu_\phi)$$

with the property:

$$\langle \phi; Z\,A \rangle = \int d\mu_\phi(\psi)\, \varphi_Z(\psi) \langle \psi; A \rangle$$

for all A *in* **A** *and all* Z *in* Z_ϕ *where* $\tilde{\phi}$ *denotes the normal extension of* ϕ *from* **A** *to* $\pi_\phi(\mathbf{A})''$. *Furthermore* μ_ϕ *is concentrated in the Baire sense on the set* F, *the elements of which are the primary states on* **A**. *Moreover if* ϕ *is a (separable) KMS state on* **S** *with respect to some* $\alpha(\mathrm{IR})$, *then so is every state in the support of* μ_ϕ.

The measure μ_ϕ the existence of which is asserted in the above Proposition is called the *central measure* of ϕ. Notice that the Proposition does not require **A** to be separable. For a proof of this Proposition, see [36]. For the purpose of dealing specifically with equilibrium states and their decomposition into pure thermodynamical phases, the above Proposition can be rephrased as follows.

Scholium 25: *Let* A *be a* C*-*algebra and* $\alpha(\mathrm{IR})$ *be a one-parameter group of symmetries of* A. *Let further* $\mathbf{S}_\beta^\alpha$ *denote the convex set of all KMS states on* A *with respect to* $\alpha(\mathrm{IR})$, *and* ϵ_β^α *denote the set of all its extreme points. Then for every* ϕ *in* $\mathbf{S}_\beta^\alpha$ *there exists a unique (maximal) measure* μ_ϕ *concentrated on* ϵ_β^α *which decomposes* ϕ.

This fact is often alluded to by saying that $\mathbf{S}_\beta^\alpha$ is a *Choquet simplex.*

The fact that μ_ϕ is in general concentrated on F only in the *Baire sense* means that for every Baire set $T \subset \mathrm{S}$, we can only have $\mu_\phi(T) = 0$ when $T \cap F = \phi$. We recall that the Baire sets of a locally compact Hausdorff space Γ are the elements of the σ-ring generated by all compact G_δ of Γ, and that a subset S of Γ is said to be a G_δ if there exists a sequence $\{U_n\}$ of open sets in Γ such that $S = \cap_n U_n$. Dealing with Baire sets is a mathematical nuisance. Fortunately for most physical applications, the measure μ_ϕ is actually concentrated on F in the Borel sense [37].

Proposition 24 can be viewed as decomposition of ϕ with respect to the abelian von Neumann algebra Z_ϕ. This point of view allows to obtain other decompositions of a state ϕ. For instance, one could have wished to have the emphasis laying in the uniform clustering property, rather than on the KMS condition itself. In this case the decomposition would be with respect to the abelian von Neumann algebra of the observables at infinity associated to ϕ. Since in general $\mathbf{B}_\phi \subseteq Z_\phi$, the resulting decomposition would be coarser than the decomposition we obtained above, with the possible effect of a lesser breaking of the symmetry. It should nevertheless be remembered that in many cases $\mathbf{B}_\phi = Z_\phi$, so that the two decompositions then do coincide. Finally, suppose that ϕ is a G-invariant state. If it is furthermore G-abelian, then the von Neumann algebra $\mathbf{M}_\phi' = \pi_\phi(\mathrm{A})' \cap U_\phi(\mathrm{G})'$ is abelian. This fact allows again to use Proposition 24 to obtain a unique decomposition of ϕ into its extremal G-invariant components; see [4] for details and references. We shall use the latter fact in Section 4.

3 Exactly solvable models

One of the reasons one has for studying exactly solvable models is to male sure (hence the requirement "exactly solvable") that some broad understanding and/or speculations (such as those of Section 2) are indeed confirmed in controlable situations (or "models"). When the general theory will have passed that first test, it can be used as a guide in the exploration of new territories (see for instance Section 4) where it is hoped to be of some predictive value.

This section thus concentrates principally on models which exhibit a spontaneous symmetry breaking associated to a phase transition. The converse problem, namely whether there are models for which a phase transition is not accompanied by any symmetry breaking, is not touched upon here; see however [38]. We shall thus mainly review models for which the decomposition of the Gibbs state into its pure thermodynamical phase components produces extremal KMS states of a lower symmetry than that of the original equilibrium state, or Hamiltonian.

3.1 The BCS model

This model was first discussed at a semi-heuristic level. Its splendid success and scope [39] were however slightly shadowed by two puzzling features. The elementary excitations coming out as solutions of the model were not gauge-invariant, and they were characterized by a temperature-dependent energy-gap. It soon became apparent that those features were not artifacts to be simply blamed on an imperfect approximation scheme, since these excitations occured as exact solutions, in the thermodynamical limit, of a reliable diagonalization procedure for the Hamiltonian of the model. Some explanation was thus required to ensure the internal consistency of the description. This model actually is the first example for which a C*-algebraic approach was proven to be of some help. Indeed, the upshot of Haag's pionneering paper [40] was that the spontaneous symmetry breaking occuring in this model is linked to the fact that the representations of the CCR associated to the temperature states are not irreducible, not equivalent to the Fock representation, and not equivalent amongst themselves. The latter fact was soon confirmed on another model, the free Bose gas, by Araki and Woods [41]. Haag's paper on the BCS model was to be the germ of a thorough investigation [42] of the mathematical structure of this model. Today's consensus, which confirms Haag's views, seems indeed to be that the Gibbs state of this model can be written as a gauge-invariant state; and that the decomposition of this equilibrium state into its pure thermodynamical phase components produces primary states which are not themselves individually gauge-invariant (spontaneous symmetry breaking of Gauge-invariance of the 1st kind). The energy-gap appears furthermore as a representation-dependent, gauge invariant element of the center; for each pure thermodynamical phase, it is thus a (temperature-dependent!) c-number. So are also the gauge-dependent coefficients of the Bogoliubov transformation leading to the elementary excitations proper to each pure thermodynamical phase.

The mathematical analysis of the model is somewhat complicated by the fact that this is a "mean free field" model (albeit in momentum space) for which the time-evolution, in the thermodynamical limit, cannot be defined in a representation-independent manner. A simpler model of this kind will be described next.

3.2 The Weiss-Ising (anti-)ferromagnet

The model is defined by giving the local Hamiltonian

$$H(\Lambda) = -B \sum_{n\in\Lambda} \sigma_n^z - \frac{1}{2} \sum_{n,m\in\Lambda} J_{n,m}(\Lambda)\, \sigma_n^z \sigma_m^z$$

where Λ runs over the finite subsets of $\mathbf{Z}$ such that card $(\Lambda) \equiv |\Lambda| \in p\,\mathbf{Z}^+$ with p an arbitrary, but fixed, finite positive integer; the following conditions are imposed on the coupling constants $J_{n,m}(\Lambda)$.

(i) $J_{n,m}(\Lambda) = J_{m,n}(\Lambda)$ real, so that $H(\Lambda)$ be self-adjoint;

(ii) $J_{n,m}(\Lambda) = J_{|n-m|}(\Lambda)$, so that $H(\Lambda)$ be translation-invariant;

(iii) $J_n(\Lambda) = J_{n+p}(\Lambda)$, i.e. periodic interaction of period p; we assume cyclic boundary conditions on Λ;

(iv) $J_n(\Lambda) = p\,f(n) / |\Lambda|$, which is a stability condition.

For $p = 1$ and $f(n) = J > 0$ (resp. $p = 2$ and $f(n) = (-1)^n J$) this model reduces to the ordinary Weiss-Ising model for ferromagnetism (resp. anti-ferromagnetism). Since the formalism is the same throughout, we can as well discuss the general case described above. The model can be considered as an archetype for mean free field theories.

In the context of these lectures the interest of this model is that the KMS condition can be used to compute directly the quantities of interest. Specifically, one can show [43] that the extremal KMS states are characterized by the solutions of the classical [44] "self-consistency" equations:

$$\begin{cases} \langle M_n \rangle = \tanh \left[\beta \left(B + \sum_{m=0}^{p-1} f(n-m) \langle M_m \rangle\right) \right] \\ n = 0, 1, \ldots, p-1 \end{cases}$$

where

$$M_n = \text{w-op}\,\mathrm{l} \lim_{\substack{\Lambda\to\mathbf{Z} \\ |\Lambda|\in p\mathbf{Z}^+}} \frac{p}{|\Lambda|} \sum_{\substack{m\in\mathbf{Z} \\ mp+n\in\Lambda}} \sigma_{mp+n}^z$$

is the partial magnetization for the sublattice $(n + p\mathbf{Z})$.

Each of the pure thermodynamical phase so obtained occurs, with equal weight, in the central decomposition of any translation-invariant KMS state (and thus of the canonical equilibrium, or Gibbs state).

The model consequently satisfies formally the scheme presented in Section 2 for the phenomenon of spontaneous symmetry breaking in (quantum) statistical mechanics. Some care however should be exercised. In particular, the time-evolution of the infinite system cannot be defined in a representation-independent manner. This can fortunately be controlled [43]; moreover the physical explanation for the latter peculiarity can be understood naturally from the very nature of any mean free field model exhibiting phase transitions.

The techniques pertaining to the solution of this model, and some of its principal features, have recently been generalized to the realm of continuous models [45].

Finally the question of which extremal KMS state in the Weiss-Ising model is a *dynamically* stable pure thermodynamical phase has recently been discussed [46], in a manner which uses further the underlying quantum features of the model.

3.3 Ising and Heisenberg models

The mean free field models mentioned up to this point evidently suffer from the fact the cooperative effects are brutally forced upon them by the enormously long range of the interactions. Consequently, their value is mainly that of good illustrative examples to be only mentioned in the most primitive stages of the theory.

The next step is to consider short-range interactions. For this case, the prime model is the nearest-neighbor, two dimensional (square) Ising model. This model, as it involves in a crucial manner only interactions of the form $\sigma_n^z \sigma_m^z$, does really pertain to classical statistical mechanics. Moreover it suffers in turn from one main drawback, namely that the model is only exactly solvable in the absence of magnetic field. It has nevertheless one main virtue over all mean free field models, namely that it produces impressively much better critical coefficients, these however are not the subject of these lectures. We shall therefore not go into this model further than we already did earlier in these lectures (see Subsection 2.7 and the allusion to [3] in Section 1).

The following step in sophistication is then the isotropic Heisenberg model. Although we know [47] that this model does not allow for phase transitions when the supporting lattice is one- or two-dimensional, these cases still present some illustrative interest in connection with the formalism of Section 2.

It has indeed been possible [6] to solve exactly the one-dimensional Heisenberg model with finite-range interactions by an astute quantum, C*-algebraic generalization of the transfer-matrix formalism. In particular, this allows to prove that the Gibbs state for this model satisfies the KMS condition, is lattice-translation invariant, is invariant under the internal rotation symmetry, and is exponentially uniformly clustering in space; it is thus extremal KMS. This confirms the general scheme in the sense that, for a model which does not exhibit phase transitions, one should indeed expect that the Gibbs state inherits the full symmetry of the Hamiltonian and is a pure thermodynamical phase.

Whereas one expects a phase transition to occur when the model is spread over a three-dimensional regular lattice, this model has apparently still escaped a complete and rigorous solution [48]. The classical equivalent of the Heisenberg model has however recently been shown [38] to exhibit a phase transition with spontaneous breaking of the internal 0_3^+ symmetry.

3.4 Conclusion

On the basis of the three families of models briefly discussed in this section, it now appears that the general scheme of Section 2 has some real substance, it does indeed capture all the essential features of spontaneous symmetry breaking as it is known to occur in some specific, completely controlable situations. On the other hand, one could further add that the literature on spontaneous symmetry breaking does not seem to offer any glaring counterexample to that scheme.

We can thus be satisfied that the theory has passed its first test, as formulated in the beginning of this section.

We should nevertheless make it clear that the theory is not closed. First, we recall the problems mentioned in Subsection 2.2 in connection with the proper, i.e. most general, definition of the time-evolution for infinite, nonrelativistic systems. Second, although it goes outside the limits of our subject, it must be added that the general scheme has not yet been fructfully deployed in the study of critical phenomena [49].

However urgent these problems might be, they should not prevent us from looking towards some exploratory applications of the formalism. Section 4 is precisely devoted to such an application.

4 Existence of crystals

The aim of this section is to show that the general scheme of Section 2 can be used to guide a speculative investigation on the existence of both crystals and fluid phases in theories where the fundamental interactions have at least the symmetry of the Euclidian group in three dimensions. The investigation is speculative on several accounts, the most glaring of which being that is does not present a specific Hamiltonian model but rather is carried entirely in a general (some would say "abstract") framework.

We mainly follow here the line of arguments first presented in [50]. A comparative discussion of some of the related literature [53–56] will be attempted towards the end of this section.

4.1 A general theory

We denote by E^3 (resp. $\mathbb{R}^3$; and O_+^3) the group of all rigid motions (resp. all translations; and all rotations) in the euclidian three-dimensional space $\mathbb{R}^3$.

We first specify the scope of this study by assuming that the equilibrium state we start from satisfies Properties P and S below.

Property P: (1) ϕ *is a KMS state (which is denoted* $\phi \in S_\beta$); (2) ϕ *is extremal* E^3*-invariant* ($\phi \in \epsilon(E^3)$); ϕ *is locally normal* ($\phi \in N$).

Property T: *There exists a least one $\psi \in \text{supp}\, \mu_\phi$ such that $\mu_\phi(O_\psi(E^3)) = 1$, where $O_\psi(E^3)$ denotes the orbit of ψ under the action of* E^3.

Property S. *ϕ satisfies Property* T, *and there exists at least one $\psi \in \text{supp}\, \mu_\phi$ such that the orbit $O_\psi(\mathrm{IR}^3)$ of ψ under IR^3 is closed.*

We denote by P (resp. T and S) the subset of all states which satisfy Property P (resp. T and S).

Let us briefly discuss the reasons for these assumptions.

First of all, (P.1) should be clear in the light of Section 2.

Concerning (P.2), we certainly want to assume that ϕ is E^3-invariant, i.e. that the equilibrium (or "Gibbs") state ϕ we start from inherits the full symmetry of the fundamental interactions; this again is in line with the arguments presented in Section 2. The present assumption requires more. The reason for it, however, is simple. If ϕ were only E^3-invariant, but not necessarily extremal E^3-invariant, we would first decompose ϕ into its extremal E^3-invariant components, which is possible without loosing the other assumed properties of ϕ. We have indeed $E^3 \supset \mathrm{IR}^3$ with the latter acting in an asymptotically abelian manner. We thus have that $Z_\phi(\mathbf{A})$ contains $\pi_\phi(\mathbf{A})' \cap U_\phi(E^3)'$; this is precisely the abelian von Neumann algebra with respect to which ϕ is decomposed into its extremal E^3-invariant components. We might thus as well start our analysis with those states, which is the actual content of (P.2).

As for (P.3), this assumption requires the prior introduction of more local structure than we have cared to discuss up to this point. We now suppose that the algebra **A** of quasi-local observables has been obtianed in the following manner. To every cube $\Lambda \subset \mathrm{IR}^3$ we suppose we can associate a Hilbert space $H(\Lambda)$ (e.g. the Fock space corresponding to Λ) in such a manner that $\mathbf{A}(\Lambda) = \mathbf{B}(H(\Lambda))$. Let $\mathbf{C}(\Lambda)$ be the algebra of compact operators on $H(\Lambda)$. We now define **A** as the usual C*-inductive limit of this family $\{\mathbf{A}(\Lambda)\}$. We say that a state ϕ on **A** is *locally normal* (or $\phi \in N$) if its restriction to each $\mathbf{A}(\Lambda)$ is a normal state on this von Neumann algebra. Recall that $\phi \mid \mathbf{A}(\Lambda)$ normal means that any, and therefore all, of the following equivalent conditions is satisfied:

(i) $\phi \upharpoonright \mathbf{A}(\Lambda)$ is ultraweakly continuous;

(ii) $\| \phi \upharpoonright \mathbf{C}(\Lambda) \| = 1$;

(iii) there exists a positive, self-adjoint operator $\rho(\Lambda)$ acting on $H(\Lambda)$ such that $\langle \phi; A \rangle = \mathrm{Tr}\{\rho(\Lambda) A\}$ for all A in $\mathbf{A}(\Lambda)$.

It might be interesting to note at this point that this structure for **A** can certainly be generalized without our loosing any of the essential results of the theory; moreover this can even be done [51] in such a manner that (P.3) is not independent of our other assumptions, in particular (P.1); we nevertheless listed (P.3) as a separate condition to emphasize its physical importance in the present scheme: only finitely many particles are allowed to coexist is any finite region Λ of the infinite space IR^3.

Property T is refered to by saying that ϕ is "E^3-transitive" with respect to its central measure. We should notice for its defense that, given Property P, ϕ is either E^3-transitive or is such that $\mu_\phi(O_\psi(E^3)) = 0$ for all $\psi \in \mathbf{S}_\beta$.

Property S requires in addition the existence of at least one ψ "in" the central decomposition of ϕ such that the orbit of ψ under translations is closed. This mathematical restriction guarantees, as we shall see below, that the pure thermodynamical components of ϕ enjoy the physical property that they keep some translation invariance in three non-coplanar directions. The following result is interesting in this connection.

Scholium 26: *Let $\phi \in P$. Then the following two conditions on ϕ are equivalent:*

(i) $\phi \in S$;

(ii) $\mathbb{R}^3/H_\psi(\mathbb{R}^3)$ *compact for all* $\psi \in \operatorname{supp}\mu_\phi$ *where* $H_\psi(\mathbb{R}^3) = \{a \in \mathbb{R}^3 \mid \psi \circ \alpha(a) = \psi\}$.

We are now going to see that Properties S and P together imply that μ_ϕ is concentrated on ϵ_β in a good sense (and not simply in the Baire sense guaranteed from the general theory of Section 2). The discussion of this fact is better brought in focus if we first notice the following intermediary result.

Scholium 27: $\phi \in \mathbf{S}_\beta \cap N \cap T$ *implies* $\operatorname{supp}\mu_\phi \subseteq \epsilon_\beta \cap N$.

Note that $\phi \in \mathbf{S}_\beta \cap N$ already implies that μ_ϕ is concentrated on $\epsilon_\beta \cap N$ in the Borel sense. It is transitivity which then makes the containment to be as strict as asserted in the above Scholium. It is therefore interesting to know when this condition can be realized, under the general constraints imposed in the beginning of this subsection. The following result goes in that direction.

Scholium 28: *Let $\phi \in P$. Then the following conditions on ϕ are equivalent:*

(i) $\phi \in T$;

(ii) $E^3/H_\psi(E^3)$ *compact for all* $\psi \in \operatorname{supp}\mu_\phi$ *where* $H_\psi(E^3) = \{g \in E^3 \mid \psi \circ \alpha(g) = \psi\}$;

(iii) $O_\psi(E^3)$ *compact, and* $O_\psi(E^3)$ *isomorphic to* $E^3/H_\psi(E^3)$ *for all* $\psi \in \operatorname{supp}\mu_\phi$.

The above three Scholiums can now be used directly to exhibit the strength of our initial restriction that Properties P and S be satisfied. The following Proposition can indeed easily be inferred from these results.

Proposition 29: *Let $\phi \in S \cap P$; μ_ϕ be its central measure; ψ and ψ' be any two states in* $\operatorname{supp}\mu_\phi$. *Then*

(i) $\operatorname{supp}\mu_\phi \subseteq \epsilon_\beta \cap N$;

(ii) $\mu_\phi(O_\psi(E^3)) = 1$;

(iii) $E^3/H_\psi(E^3) \simeq O_\psi(E^3)$ *compact,*

(iv) $\mathbb{R}^3/H_\psi(\mathbb{R}^3)$ *compact;*

(v) *there exists an element* g *in* E^3 *such that* $\psi' = \psi \circ \alpha(g)$;

(vi) $H_\psi(E^3)$ *and* $H_{\psi'}(E^3)$ *belong to the same conjugacy class of* E^3.

We should remark that (iv) means that every pure thermodynamical phase appearing in the decomposition of ϕ keeps the following subtranslation invariance: $H_\psi(\mathbb{R}^3)$ contains at least three non-coplanar translations. Furthermore (vi) tells us that the conjugacy class of $H_\psi(E^3)$ in E^3 is uniquely determined by ϕ. We shall therefore call it the *intrisic symmetry* of ϕ, and denote it by $H_\phi(E^3)$ (a notation the slight ambiguity of which should not trouble us in the sequel).
We are now going to classify the states ϕ in $S \cap P$ by their intrinsic symmetry with respect to $\mathbb{R}^3$. First of all, note that the following mutually exclusive classes exhaust $S \cap P$:

SP_1 : $\phi \in \epsilon_\beta$ (and thus $H_\phi(E^3) = E^3$);
SP_2 . $H_\phi(\mathbb{R}^3) = \mathbb{R}^3$, but $H_\phi(E^3) \neq E^3$;
SP_3 : $H_\phi(\mathbb{R}^3)$ discrete in one or two directions;
SP_4 : $H_\phi(\mathbb{R}^3)$ generated by three non-coplanar translations.

Our aim will now be to give some alternate characterization of these classes so that the occurence of a spontaneous symmetry breaking can be identified better than by a mere tautology. We will indeed classify the states in $S \cap P$ by either their spectral properties or by their clustering properties with respect to the translation group. We should however be aware of the following fact.

Scholium 30: *Let* ϕ *be any* E^3*-invariant state;* π_ϕ *be the GNS representation associated to* ϕ*; and* P_ϕ *be the generator of the unitary representation implementing* $\alpha(\mathbb{R}^3)$ *in* π_ϕ*. Then the discrete spectrum* $Sp_d(P_\phi)$ *of* P_ϕ *consists exactly of the point* 0.

As a set, the discrete spectrum of P_ϕ is thus trivially the same for all states in $S \cap P$. The only direct information we could hope to get from this discrete spectrum should thus come from its multiplicity. We have indeed the following result.

Proposition 31: *Let* $\phi \in S \cap P$*, and* $E_\phi(\mathbb{R}^3)$ *be the invariant subspace of the unitary representation implementing* $\alpha(\mathbb{R}^3)$ *in* π_ϕ*. Then the following conditions are equivalent:*

(i) ϕ *belongs to the class* SP_1;
(ii) $\dim E_\phi(\mathbb{R}^3) = 1$,
(iii) ϕ *is uniformly clustering.*

To proceed further we should take notice of the following facts.

Scholium 32: *Let* $\phi \in S \cap P$*;* μ_ϕ *be its central measure; and* ν_ϕ *be the measure associated to the decomposition of* ϕ *into its extremal* $\mathbb{R}^3$*-invariant components. Let further* η *be any invariant mean over* $\mathbb{R}^3$*. Then:*

(a) *For every* χ *in* $\operatorname{supp} \nu_\phi$ *there exists a state* ψ *in* $\operatorname{supp} \mu_\phi$ *such that* $\chi = \eta \circ \psi \circ \alpha(a)$;
(b) *For any* ψ *in* $\operatorname{supp} \mu_\phi$*, let* $\chi = \eta \circ \psi \circ \alpha(a)$*, then* χ *satisfies the following conditions:*
  (i) $\chi \in \epsilon(\mathbb{R}^3) \cap S_\beta$,
  (ii) $Sp_d(P_\chi) = H_\phi^*(\mathbb{R}^3)$;
  (iii) $Sp_d(P_\chi)$ *is simple.*

We thus notice that the intrinsic symmetry $H_\phi(\mathrm{IR}^3)$ of ϕ with respect to translations occurs in property (b ii) above through its reciprocal group:

$$H_\phi^*(\mathrm{IR}^3) = \{p \in \mathrm{IR}^3 \mid \exp(-i\,p\,a) = 1 \text{ for all } a \in H_\phi(\mathrm{IR}^3)\}.$$

This strongly suggests to use χ for our classification purpose, since the discrete spectrum of the corresponding P_χ tells us to which of the SP classes ϕ belongs. The physical content of this information will be discussed in Subsection 2 below. For the time-being, we would like to further investigate the mathematical properties of the states χ in supp ν_ϕ. In particular, property (bi) in Scholium 32 implies directly (see Section 2) that χ are η-clustering. The question now is whether these states might perhaps have stronger clustering properties than that, depending on the SP class to which ϕ belongs. A precise answer is provided by the next result.

Proposition 33: *Let $\phi \in S \cap P$ with ϕ not belonging to the class* SP_1. *Then:*

(a) *the following three conditions are equivalent*

(i) *ϕ belongs to the class* SP_2,

(ii) $Sp_d(P_\chi) = \{0\}$;

(iii) *χ is weak-mixing,* i.e.

$$\eta \mid \langle\chi; \alpha(a)[A]B\rangle - \langle\chi; A\rangle\langle\chi; B\rangle \mid = 0;$$

(b) *the following three conditions are equivalent*

(i) *ϕ belongs to the class* SP_3;

(ii) $Sp_d(P_\chi)$ *is generated by one (resp. two) non-coplanar vectors in* IR^3;

(iii) *χ is not weak-mixing, but is partially weak-mixing, i.e.*

$$\eta(\mathrm{IR}^d) \mid \eta(\mathrm{IR}^{3-d}) \langle\chi; \alpha(a)[A]B\rangle - \langle\chi; A\rangle\langle\chi; B\rangle \mid = 0.$$

with $d = 2$ (resp. 1);

(c) *the following three conditions are equivalent*

(i) *ϕ belongs to the class* SP_4;

(ii) $Sp_d(P_\chi)$ *is generated by three non-coplanar vector in* IR^3;

(iii) *χ is not partially weak-mixing, but is η-clustering (or "weak-clustering") i.e.*

$$\eta\,\langle\chi; \alpha(a)[A]B\rangle - \langle\chi; A\rangle\langle\chi; B\rangle = 0.$$

4.2 Physical interpretation

Proposition 31 characterizes the states belonging to the class SP_1. These states are interpreted as *fluid phases.* Indeed they are extremal KMS, uniformly clustering states, whose intrinsic symmetry is the unbroken euclidian group E^3. These states are moreover not only extremal E^3-invariant, but also extremal $\mathbb{R}^3$-invariant. The fact that for these states one has $Sp_d(P_\phi) = Sp_d(P_\chi) = \{0\}$ indicates that their X-ray diffraction pattern is without any structure.

At the opposite extreme one finds the states of the class SP_4, which are characterized by Proposition 33 c. The physical interpretation of these states as crystalline states is justified as follows. These states ϕ decompose into extremal KMS states ψ which are invariant under a proper subgroup $H_\psi(E^3)$ of E^3, characterized by the fact that the translation part $H_\phi(\mathbb{R}^3)$ of the intrinsic symmetry of ϕ is generated by three non-coplanar translations. As there is no "normality" restriction imposed on $H_\psi(E^3)$ we conclude from the above observation that $H_\phi(E^3)$ is a crystalline group. The fact that $Sp_d(P_\chi) = H_\phi^*(\mathbb{R}^3)$ indicates that the X-ray diffraction pattern obtained from such states is that which is characteristic of crystals. One might object at this point that the states χ are artificial constructs, that only the pure thermodynamical phases ψ make physical sense, and that "Nature does not make averages". All that is indeed true, except that in this case it is the measuring apparatus instead that makes the average!

Although it is evident from the definitions themselves, we want to emphasize here the hierarchical order of the clustering properties appearing in the above analysis. Each of the following properties is weaker than the one preceding it: uniform clustering, weak-mixing, partial weak-mixing, η-clustering. Hence fluid phases are characterized by the strongest clustering property, whereas crystalline phases (quite precisely each χ in supp ν_ϕ) enjoy only the weakest of all cluster properties.

This remark appears to provide a basis for Landau's argument [52] on the absence of critical point in the solid-liquid phase transition; this argument rests indeed on the assumption that solids and liquids have qualitatively different clustering properties in space. Propositions 31 and 33 c do establish such a link between the strength of the clustering properties and the intrinsic symmetry of ϕ : E^3 for a fluid, and $H_\phi(E^3) \neq E^3$ with $H_\phi(\mathbb{R}^3)$ generated by three non-coplanar translations, for crystals.

Whereas the classes SP_1 and SP_4 are those of interest for a theory of crystallization, one might still wonder at the possible physical content of the other two classes. For the case SP_2, only rotation symmetry is spontaneously broken, while the translation symmetry remains intact. This could perhaps occur in a homogeneous system exhibiting spontaneous magnetization. For the case SP_3, one might tentatively try to think about liquid crystals, although this might very well be a rather wild guess at this point, certainly not to be treated at the same level as the interpretation given above for the classes SP_1 (ordinary fluids) and SP_4 (ordinary crystals).

4.3 Related theories

The line of approach threaded here was initiated in [53]. A significantly expanded version of this early contribution was subsequently published [54]. The paper [50] which we followed in the present lecture appeared in the meantime, with a motivation to combine [53] and [10]. This paper [50] itself was a simplification of two earlier attempts [55] where η-asymptotic abelianness in time however played an unnecessary restrictive role. For other antecedents, see the bibliography of [50] and [54]. For this subsection, we found it useful to limit our discussion to a comparison of [50] and [54].

The *first difference* between these two papers lies in the motivation explicitly given for the identification of pure thermodynamical phases as primary states. In [50] pure thermodynamical phases are identified from the outset as extremal KMS states; it then follows directly from KMS theory that these states are primary. In contradistinction, the KMS condition itself is not invoked in [54]. Instead pure thermodynamical phases are identified as time-transitive, time-ergodic states which are A-stable; moreover the time evolution is assumed to act in an asymptotically abelian manner.

Specifically, the central assumptions of [54] are the following:

(i) ϕ is time transitive, i.e. $\mu_\phi(O_\phi(\mathrm{IR})) = 1$

(ii) ϕ is time-ergodic, i.e. $\pi_\phi(\mathbf{A})' \cap \pi_\phi(\mathbf{A})' \cap U_\phi(\mathrm{IR})' = \mathbb{C}\, I$

(iii) with $\phi_A : B \in \mathbf{A} \mapsto \langle\phi; A^*BA\rangle / \langle\phi; A^*A\rangle \in \mathbb{C}$, one has
$w^* - \lim_{t\to\infty} (\phi_A - \phi) \circ \alpha(t) = 0$ for all $A \in \mathbf{A}$

(iv) $\lim_{t\to\infty} \langle\psi; [A, \alpha(t)[B]]\rangle = 0$ for all ψ in $\mathbf{S}$, and all A, B in $\mathbf{A}$.

Condition (iii) is to be interpreted as a condition of return to equilibrium for local perturbations. Condition (iv) is their condition of asymptotic abelianness in time. We shall not come back here again on the discussion presented in Section 2 concerning this kind of condition. Let it suffice to say here that our reluctance to assume explicitly any asymptotic abelianness in time is at the core of our effort to pass to [50] from the previous papers [53] and [55].

The *second difference* between the approaches of [50] and [54] is that we restricted our attention to those states which satisfy both properties P *and* S. Condition (P.1) is the KMS condition; (P.2) is for convenience, as explained above; (P.3) is a physical condition of local normality which can actually be dispensed with in [50] as well as in [54]. It is therefore not in Property P that other essential differences have to be looked for. On the other hand P and S together had three important consequences which we pointed out. First, they imply transitivity under the action of E^3, a condition which is also assumed in [54]. Second, because of transitivity, we know where μ_ϕ is really concentrated. Third, given P, S is equivalent to $\mathbb{R}^3/H_\phi(\mathbb{R}^3)$ compact. No such condition is imposed in [54]. They however show a very interesting fact in this connection. In addition to the possibilities listed in Subsection 2, namely $H_\phi(\mathbb{R}^3) = \mathbb{R}^3$ (i.e. SP_1 and SP_2, or class 4 in [54]), $H_\phi(\mathbb{R}^3) = \mathbb{Z} + \mathbb{R}^2$ or $\mathbb{Z}^2 + \mathrm{IR}$ (i.e. SP_3, or classes 3 and 2 in [54]), and $H_\phi(\mathbb{R}^3) = \mathbb{Z}^3$ (i.e. SP_4, or class 1 in [54]), only one other case can occur, namely $H_\phi(\mathbb{R}^3) = \mathbb{R}^2$, while a less careful look at the effect of discarding Property S could have wrongly suggested the six following possibilities: $H_\phi(\mathbb{R}^3) = \{0\}, \mathbb{Z}, \mathbb{R}, \mathbb{Z}^2, \mathbb{Z} + \mathbb{R}$, and $\mathbb{R}^2$. In the latter case, the complete intrinsic symmetry groups of ϕ is the group generated by (i) the translations in the plane $\mathbb{R}^2$, (ii) an helicoidal transformation (a, r) along the axis u perpendicular to $\mathbb{R}^2$, with angle $\theta(r)$ irrational to π, and (iii) none, one, two, or three of the following generators: (α) a rotation around u of an angle rational to π, (β) a rotation of π around an axis in the plane $\mathbb{R}^2$ (γ) the symmetry through the plane $\mathbb{R}^2$ or through a plane containing u. This family, as pointed out in [54], is usually forgotten; they propose, as possible physical realization of it (as well as of their class 3) the cholesteric liquids, or matter in an helimagnetic state.

In addition to the above curiosity, one can find in [54] a complete list of the possible groups $H_\phi(E^3)$ corresponding to their five classes; this thus is an actual complement to the analysis presented in the main part of this lecture.

On the other hand [50] establishes the links, which we have described in Subsection 1, between: intrinsic symmetries, spectral properties, and clustering properties. These links were indispensible for much of the physical interpretations proposed in Subsection 2.

Finally, it should be mentioned that all the papers listed in the present subsection only present an abstract, or general, analysis of the ways in which the E^3-symmetry can be broken spontaneously. It is therefore interesting to note that some of the theoretical predictions so achieved are actually supported by a concrete, primitive, but exactly solvable model [56]. The hope naturally is that the assumptions needed here to arrive at these predictions will be met in a significantly wider class of models.

References

[1] *Poincaré, H.*, Leçons sur les Hypothèses Cosmogoniques, 2nd édition, Hermann, Paris, 1913; see in particular pp. LX–LXI, 53–58, and 184–189. See also *G. Bertin* and *L. A. Radicati*, Astrophys. Journ. **206** (1976) 815–821. Both of these works contain useful cross-references.

[2] We should emphasize here that in all this section the notation is very loose; H is not necessarily meant to denote an Hamiltonian, but is only to be taken as a symbol for a prescription which defines the model and the solutions of interest.

[3] *J. L. Lebowitz*, in Rome International Conference on the Mathematical Problems in Theoretical Physics, June 6–15, 1977.

[4] *Emch, G. G.*, Algebraic Methods in Statistical Mechanics and Quantum Field Theory, Wiley-Interscience, New York, 1972.

[5] See for instance the quantum generalization of the transfer-matrix idea as elaborated for the Heisenberg model by H. Araki (see [6]); other models will be mentioned as well in section III.

[6] *Araki, H.*, Commun. math. Phys. **14** (1969) 120–157.

[7] *Fröhlich, J., in* Elementary Particles and Mathematical Physics, P. Urban, Ed., Springer-Verlag, 1976, pp. 133–269.

[8] *Rieckers, A.*, Lecture Notes, *in* Tübingen International Summer School on Groups and Many-Body Physics, July 11–16, 1977; his list of references also provides a good background for the present lectures.

[9] *von Neumann, J.*, Grundlagen der Quantenmechanik, Springer, Berlin, 1932.

[10] *Kubo, R.*, J. Phys. Soc. Japan **12** (1957) 570–586; *Martin, P. C.* and *Schwinger, J.*, Phys. Rev. **115** (1959) 1342–1373; *Haag, R., Hugenholtz, N.* and *Winnink, M.*, Commun. math. Phys. **5** (1967) 215–236.

[11] *Takeda, Z.*, Tohoku Math. Journ. **7** (1955) 68–86.

[12] *Gelfand, I.* and *Naimark, M. A.*, Mat. Sborn, N.S. **12** 54 (1943) 197–217, and *Segal, I. E.*, Ann. Math. **48** (1947) 930–948. See [8] in this volume.

[13] For a discussion of this point, see [4] pp. 282–3 and references quoted therein; in particular: *Hugenholtz, N.* and *Wieringa, J. D.*, Commun. math. Phys. **11** (1969) 183–197; *Wieringa, J. D.*, Thesis, Groningen, 1970; *Haag, R., Kadison, R. V.* and *Kastler, D.*, Commun. math. Phys. **16** (1970) 81–104; *Takesaki, M.* and *Winnink, M.*, Commun. math. Phys. **30** (1973) 129–152; see also *Ruelle, D.*, Commun. math. Phys. **3** (1966) 133–150; *Lanford, O. E.* and *Ruelle, D.*, Journ. Math. Phys. **8** (1967) 1460–1463.

[14] We already mentioned [6]. See also *Robinson, D. W.*, Commun. math. Phys. **7** (1968) 337–348; *Ruelle, D.*, Statistical Mechanics, Benjamin, N.Y., 1969; *Manuceau, H.* and *Trottin, J. C.*, Ann. Inst. *Poincaré, Henri* **A10** (1969) 359–380; *Kishimoto, A.*, Commun. math. Phys. **47** (1976) 25–32; *Robinson, D. W.*, preprint, Bielefeld, 1977.

[15] For two early counterexamples, see: *Dubin, D. A.* and *Sewell, G.*, Journ. Math. Phys. **11** (1970) 2990–2998; *Emch, G. G.* and *Knops, J. F.*, Journ. Math. Phys. **11** (1970) 3008–3018. For more recent, and radical, c.-examples see: *Radin, Ch.*, Commun. math. Phys. **54** (1977) 69–79; *Battle, G., Thesis, Ph. D.*, Duke University, 1977; Journ. Math. Phys. (to appear); Commun. math. Phys. (to appear).

[16] See the last two entries in [14] above, and [15].

[17] See the second entry in [15] above.

[18] *Roepstorff, G.*, Commun. math. Phys. **46** (1976) 253–262; and preprint, Aachen, 1976; *Sewell, G.*, Commun. math. Phys. (to appear); *Araki, H., in* Japan – US Conference on C*-algebras and their Applications in Physics, Los Angeles, April 18–22, 1977; for a general review, see *Verbeure, A., in* Marseille International Colloquium on Algebras of Operators and their Applications to Mathematical Physics, June 20–24, 1977; the latter covers, amongst other things, several original contributions on the subject from Verbeure and coworkers.

[19] *Pusz, W.* and *Woronowicz, S. L., in* Rome International Conference on the Mathematical Problems in Theoretical Physics, June 6–15, 1977; and Marseille International Colloquium on Algebras of Operators and their Applications to Mathematical Physics, June 20–24, 1977. Their approach can actually be related to the approach via correlation-inequalities, as pointed out in the last entry of [18] above.

[20] η-clustering will bedefined in subsection 6 below.

[21] *Haag, R., Kastler, D,* and *Trych-Pohlmeyer, E.*, Commun. math. Phys. **38** (1974) 173–193; the cluster property, as given in this lecture, is an improvement of the original version, and is due to *Bratteli, O.* and *Kastler, D.*, Commun. math. Phys. **46** (1976) 37–42; this approach has been further extended to a theory of the chemical potential by *Araki, H., Haag, R., Kastler, D.* and *Takesaki, M.*, Commun. math. Phys. (to appear); see also *Kastler, D.*, Lecture Notes, UCLA 1977.

[22] *Gallavotti, G.* and *Verboven, E. J.*, Il Nuovo Cimento **28** (1975) 274–286; *Aizenmann, M., Goldstein, S., Gruber, C., Lebowitz, J. L.* and *Martin, Ph.*, preprint, Lausanne, 1976; *Pulvirenti, M.*, Journ. Math. Phys. (to appear); see also *Andersson, S. I.*, Fortschritte der Physik **24** (1976) 55–83.

[23] *Kossakowski, A., Frigerio, A., Gorini, V.* and *Verri, M.*, preprint, Milano, 1977.

[24] *Takesaki, M.*, Springer Lecture Notes in Mathematics No. 128, Berlin, 1970; and *in* Springer Lecture Notes in Physics No. 20, pp. 205–246, Berlin, 1973.

[25] *van Daele, A.*, Journ. Funct. Analysis **15** (1974) 378–393; see also *in* Proceedings of the Varenna Conference (1973) on C*-algebras and their Applications to Statistical Mechanics and Quantum Field Theory, *Kastler, D.*, ed.

[26] *Rieffel, M. A.* and *Daele van, A.*, Pac. J. Math. **69** (1977) 187–221.

[27] *Narnhofer, H.*, Acta Phys. Austriaca **31** (1970) 349–353.

[28] *Emch, G. G.* and *Radin, Ch.*, Journ. Math. Phys. **12** (1971) 2043–2046.

[29] *Robinson, D. W.*, Commun. math. Phys. **31** (1973) 171–189.

[30] *Araki, H.*, Commun. math. Phys. **28** (1972) 267–277.

[31] *Emch, G. G.*, Commun. math. Phys. **49** (1976) 191–215.

[32] *Davies, E. B.*, Commun. math. Phys. **19** (1970) 83–105; *Frigerio, A.*, preprint, Milano, 1977; *Krümmerer, B.*, *in* VIth International Colloquium on Group Theoretical Methods in Physics, Tübingen, July 18–22, 1977.

[33] *Emch, G.,G.*, *Knops, H. J. F.* and *Verboven, E. J.*, Commun. math. Phys. **8** (1968) 300–314.

[34] *Onsager, L.*, Phys. Rev. **65** (1944) 117–149; see also *Yang, C. N.* and *Lee, T. D.*, Phys. Rev. **87** (1952) 404–409 and 410–419; *Montroll, E. W.*, *Potts, R. B.* and *Ward, J. C.*, Journ. Math. Phys. **4** (1963) 308–322; *Schultz, T. D.*, *Mattis, D. C.* and *Lieb, E. H.*, Revs. Mod. Phys. **36** (1964) 856–871, recall also ref [3] above.

[35] *Lanford, O. E.* and *Ruelle, D.*, Commun. math. Phys. **13** (1969) 194–215; *Ruelle, D.*, *in* Cargése Lecture Notes, *Kastler, D.*, Ed., Gordon and Breach, New York, 1969; *Ruelle, D.*, Journ. Funct. Analysis **6** (1970) 116–151; see also *Robinson, D. W.*, Commun. math. Phys. **7** (1968) 337–348; *Araki, H.*, ref [6] above.

[36] *Sakai, S.*, C*-Algebras and W*-Algebras, Springer, Berlin, 1971; *Wils, W.*, C. R. Acad. Sc., Paris **267** (1968) 810–812, and Lecture Notes, Aarhus, 1969. See also ref. [24] above.

[37] See for instance the papers by *Ruelle, D.* in [35] above, by *Takesaki, M.* and *Winnink, M.* in [13] above, and by *Andersson, S. I.* in [22] above.

[38] *Fröhlich, J.*, *Simon, B.* and *Spencer, T.*, Phys. Rev. Letters **36** (1976) 804–806; and preprint Princeton, 1976; see also [7] above.

[39] *Bardeen, J. L.*, *Cooper, N.* and *Schrieffer, J. R.*, Phys. Rev. **108** (1957) 1175–1204; see also: *Schrieffer, J. R.*, Theory of Superconductivity, Benjamin, New York, 1964.

[40] *Haag, R.*, Nuovo Cimento **25** (1962) 287–298.

[41] *Araki, H.* and *Woods, E. J.*, Journ. Math. Phys. **4** (1963) 637–662; see also the first reference in [15] above.

[42] *Ezawa, H.*, Journ. Math. Phys. **5** (1964) 1078–1090; *Emch, G. G.* and *Guenin, M.*, Journ. Math. Phys. **7** (1966) 915–921; *Thirring, W.* and *Wehrl, A.*, Commun. math. Phys. **4** (1967) 303–314, *Thirring, W.*, Commun. math. Phys. **7** (1968) 181–189, and *in* The Many-Body Problem, *Garrido, L. M.* et al., Eds., Plenum Press, London, 1969; *Jelinek, F.*, Commun. math. Phys. **9** (1968) 169–175; see also pp. 28–32 in [4] above.

[43] See the second reference in [15] above.

[44] See for instance *Kac, M.*, *in* Brandeis Summer Institute, 1966 Lecture Notes, *Chretien, M.*, *Gross, E. P.* and *Deser, S.*, Eds., Gordon and Breach, New York 1968; and *in* Fundamental Problems in Statistical Mechanics II, *Cohen, E. G. D.*, Ed., North-Holland, Amsterdam, 1968.

[45] See the papers by Battle in [15] above.

[46] *Martin, Ph. A.*, Journ. Stat. Phys. **16** (1977) 149–168; *Giuffre, M. S.*, Ph. D. dissertation, University of Rochester, 1977, for a review, see *Emch, G. G.*, *in* Proceedings of the VIth International Colloquium on Group Theoretical Methods in Physics, Tübingen, July 18–22, 1977.

[47] *Mermin, D.* and *Wagner, H.*, Phys. Rev. Letters **17** (1966) 1133–1136.

[48] See in particular *Lieb, E. H.*, *in* Rome International Conference on the Mathematical Problems in Theoretical Physics, June 6–15, 1977.

[49] With the exception of the "critical slowing-down" obtained in the first reference of [46] above.

[50] *Emch, G. G.*, *Knops, H. J. F.* and *Verboven, E. J.*, Journ. Math. Phys. **11**(1970) 1655–1668.

[51] See *Takesaki, M.* and *Winnink, M.*, in [13] above.

[52] See for instance: *Uhlenbeck, G. E., in* Fundamental Problems in Statistical Mechanics II, *Cohen, E. G. D.*, Ed., North-Holland, Amsterdam, 1968.

[53] *Kastler, D.*, preprint, Marseille, 1967: this is a seminar talk in which the author reported on work jointly done with *Haag, R.* and *Michel, L.*

[54] *Kastler, D., Mebkhout, M., Loupias, G.* and *Michel, L.*, Commun. math. Phys. **27** (1972) 195–222.

[55] *Emch, G. G., Knops, H. J. F.* and *Verboven, E. J.*, Journ. Phys. Soc. Japan **26** (Suppl.) (1969) 301–303; see also *Knops, H. J. F.*, Dissertation, University of Nijmegen, 1969.

[56] *Gates, D. J.*, Journ. Math. Phys. **12** (1971) 766–778.

Chapter VI

Dynamical Groups for the Motion of Relativistic Composite Systems

A. O. Barut

1 Introduction

The relativistic quantum theory of composite systems is still largely an open problem. In quantum field theory, even for a two-body system, the method that has been studied most extensively, the Bethe-Salpeter equation, gives useful results only after drastic, mostly non-relativistic, approximations. The best way to understand the bound-state problems still seems to be via some effective potentials.

In the present work we propose to study relativistic composite systems globally, treat them as "elementary" relativistic objects with internal degrees of freedom. Whereas the field theory puts the emphasis on the constituents and on their local mutual interactions, and tries to deduce the existence of bound states, we shall put the emphasis on the global quantum numbers, observable and observed spectral properties of the system (e.g., mass levels and their widths, form factors, etc.), and then subsequently try to infer the nature and the interactions of the constituents.

There are good reasons for this *"inverse"* approach:

(i) For strongly coupled relativistic systems the notion of constituents becomes blurred; the identity of constituents is lost, due to virtual pairs and other effects.

(ii) We may not know what the constituents and the forces between them are. For example, hadrons surely have composite structures, but the nature of what they are made of is precisely the problem we wish to solve by the "inverse" method by probing the hadrons with external agents.

(iii) Composite systems may have internal quantized degrees of freedom, but no constituents, as in the case of extended elastic objects (strings, liquid drops, membranes, etc

In other words, we are interested in a relativistic quantum mechanical global description of the final nonperturbative solution of underlying internal dynamics [1].

Our approach has three main ingredients: (i) Relativity and Quantum Theory: Representations of the Poincaré Group Π; (ii) Internal Degrees of Freedom: Representations of Π Inducted from a Dynamical Group G; (iii) Infinite-Component Wave Equations: Spectral Properties of Interactions of Composite Systems.

The first two principles are formulated fairly precisely at the beginning. The third principle is introduced inductively after discussing various examples.

These three principles lead to a general framework in which we can discuss the global quantum numbers and intrinsic properties of composite systems, their wave functions and external couplings. The system is treated as though it is an elementary entity, characterized by internal degrees of freedom-hence by a mass spectrum, various dipole and higher moment form factors, structure functions, etc. The formalism is especially appropriate if we do not know the internal constitutions of composite objects (e.g., hadrons), or the internal structure is too complex to be able to permit a relativistic calculation based on the (assumed) interactions of the constituents. A number of specific examples are treated.

2 The general framework

The theory developed in this work is based on the following basic principles:

a) Relativity. The laws of particle physics are invariant with respect to the inhomogeneous Lorentz group G: G is the symmetry group not only for elementary systems but for a collection of many systems, or, in particular, for bound or unbound composite systems. Some more precise definitions of composite systems will be given later. For nonrelativistic considerations, G may be contracted to the galilean group G_{NR}. Note, however, that G may be a larger group, e.g., the *conformal group* of space-time, if proper physical interpretation of the additional transformations can be given [2].

b) Quantization. The states of quantum systems are given by the representation spaces of a collection of unitary (antiunitary), in general reducible, projective representations of G, i.e., by the maps

$$G \to \{U(G)\}.$$

As a very special case, we may define *elementary systems* by the irreducible representations $U^0(G)$ of G. For composite systems, we need reducible representations and therefore need additional physical principles to select the appropriate reducible representation, e.g., to determine the type of mass spectrum that a composite system can exhibit. It is here that the "dynamical groups" $\mathcal{G}$ will enter the scene. (Cf. Section 4.)

The principle b) can be elucidated as follows: Under a transformation of G we pass from one equivalent state of the system to another (e.g., two different choices of the coordinate systems). All such equivalent states also give all possible states of the system. A system is characterized by looking at it in all possible ways, or by the totality of observers for which it is the same. Hence, all the rays in the Hilbert space by the application of U(G) constitute all states of the system.

This principle might at first seem restrictive for it assumes that no *new* states of the system occur if we go, for example, to an accelerated frame, and an accelerated charge radiates. In such cases, the principle applies globally to the larger system including the acceleration mechanism. Thus, so far we are using global characterizations of the systems, not local, and globally an isolated system cannot accelerate.

c) **Physical identification and observables.** The abstract mathematical quantities specifying the representations U(G), such as Lie algebra elements, Casimir invariants, discrete symmetry operations must be given physical interpretation. Same or equivalent representations may represent different physical systems. A representation U(G) may also be *realized* in different forms or in different spaces. For example, it is well known that a free particle, and a particle in an external plane wave, may be described by the same representation of the Poincaré group. Thus again we need additional physical principles relating the group representations to physical systems.

3 Composite systems and reducible representations

Our physical system is a single moving relativistic composite system. Interactions of such systems will be considered later (Sections 12–14.)
In the rest frame of such a system, defined by the total momentum $\vec{P} = 0$, we have first to describe and classify all hte *rest-frame states* by internal quantum numbers of the system. Now the Casimir operators of the Poincaré group Π,

$$M^2 = P_\mu P^\mu$$

$$W^2 = (\tfrac{1}{2}\, \epsilon_{\mu\nu\lambda\sigma}\, J^{\nu\lambda} P^\sigma)^2 \tag{1}$$

$$\epsilon = \operatorname{sign}(P^0)\,,$$

will be just quantum numbers with a certain range together with other commuting quantum numbers among P_μ, $J_{\mu\nu}$, ..., and no longer invariants characterizing the system, because a composite system has in general many mass and spin states. We must find other quantities characterizing or naming a composite structure. If the reducible representation U(G) is decomposed into irreducible components as

$$U(G) = \oplus \sum_{m,j}\!\!\!\!\!\!\int\; U^{(m,j)}, \tag{2}$$

we must be able to relate the range of this decomposition to the structure of composite systems: Which and how many spin and mass values has the composite system and what transitions can occur between these states?

One might think that a larger group containing the Poincaré group could be applicable. However, this approach leads to infinite dimensional (pseudo) Lie groups. A more useful and physically interpretable method is via the induced representations.

4 The method of induced representations from dynamical groups

The unitary irreducible representations U of the Poincaré group Π can be viewed as induced representations from the representations L_K of certain subgroups K, the little groups, or the stability subgroup of a fixed momentum vector p_μ. Physically, this process has the following intuitive meaning: We are looking at the system at a fixed value of p_μ, for example, in the rest frame $p_\mu = (m, 0, 0, 0)$, for $m^2 > 0$. The group of transformations leaving this value of p_μ invariant is precisely K. Hence, L_K determines the states, the degrees of freedom of the system, at fixed momentum – for example, the spin degrees of freedom. The rest is just computation, namely to obtain states with all other values of momentum starting from the one by application of Lorentz transformations (boosts). The states of L_K are labelled by spin projection, for example, $|j, \sigma\rangle$. Hence, all the remaining states can be labelled by m, momentum and spin projection:

$$|m, j; \vec{p}, \sigma\rangle = U(L_p)\, |j; \sigma\rangle. \tag{3}$$

We shall assume that the reader is familiar with the classification of irreducible unitary representations of the Poincaré group [3]. Of course, the states can be labelled by other quantum numbers than the above set, depending on the applications. For example, we can diagonalize energy, or total angular momentum, etc., to label the states.

There are two other related mathematical forms of the induced representations: wave equations and fiber bundle structure. The theory of wave equations will be discussed in detail in Sections 5, 8 and 10, the fiber bundle structure will be discussed elsewhere.

For composite systems we shall generalize the method of induced representations to certain classes of reducible representations.

Definition 1: A quantum system is called a "reducible relativistic composite system" if its Hilbert space of states is a direct sum (or direct integral) of irreducible representations of the Poincaré group.

Globally isolated composite systems are of this kind. Open systems would not be of this type because their states do not have definite values of momentum, mass or spin.

The definition 1 is still very general: masses spins may occur as often as one pleases, discrete or continuous, even space-like and time-like representations may occur.

We first specialize as follows.

Definition 2: A composite quantum system admits a dynamical group G if its states at rest, $\vec{P} = 0$, are in a one-to-one correspondence with an irreducible (but not necessarily unitary) representation $U(G)$ of G.

In order to induce all representations of Π, the dynamical group G must contain all little groups K. Furthermore, for the purpose of writing down covariant wave equations, G must contain the homogeneous Lorentz group Λ:

Proposition 1: The dynamical group G contains the Lorentz group Λ.
The "elementary systems" are special cases of definitions 1 and 2, and even of proposition 1.

Corollary 1: The simplest relativistic dynamical group is the homogeneous Lorentz group, $G \equiv \Lambda$.

Corollary 2: Relativistic composite systems describable by a representation $D(\Lambda)$ of $G = \Lambda$ are given by the reduction of $D(\Lambda)$ with respect to the little group $K = SO(3)$, $E(2)$ or $SO(2, 1)$, respectively, for systems with time-like, light-like or space-like momentum, We are mainly concerned here with the time-like case.
Even if $D(\Lambda)$ is irreducible, there are, except for the trivial one-dimensional representation, always many mass and spin states. A single mass and a single spin case (elementary system) is obtained only by a covariant projection procedure via a wave equation.

5 Electron-positron complex

The Dirac "particle" is a prototype of many generalizations that we shall consider in this study. Here there is one mass m and one spin value $j = \frac{1}{2}$, but each occurring twice. Strictly speaking, it is, according to our terminology, a composite system. Because this simple system will be a prototype to other compositie systems, we give in detail some features of it-however, in the language of dynamical groups.
It is well known that the Dirac equation is equivalent to induced representations, induced by the reducible representation $D(\Lambda) = D^{0\frac{1}{2}} \oplus D^{\frac{1}{2}0}$ of the Lorentz group Λ. It can also be considered as induced by an *irreducible* representation $D(G)$ of the synamical group $G = SO(4, 2)$, the conformal group. This is established as follows:
We divide the fifteen Dirac matrices, for future purposes, into the following categories (in various representations) [4]

$$L_{ij} = \epsilon_{ijk} L_k = \frac{i}{2}\gamma_i\gamma_j = \frac{1}{2}\epsilon_{ijk}\frac{1}{2} I \otimes \sigma_k \left\{\frac{1}{2}\begin{pmatrix}\overline{\sigma} & 0 \\ 0 & \overline{\sigma}\end{pmatrix} = \overline{L}\right\}$$

$$L_{i4} = A_i = -\frac{i}{2}\gamma_i\gamma_5 = -\frac{1}{2}\rho_3 \otimes \sigma_i = -\frac{1}{2}\begin{pmatrix}\sigma_i & 0 \\ 0 & -\sigma_i\end{pmatrix}$$

$$L_{i5} = M_i = \frac{i}{2}\gamma_i\gamma_0 = -\frac{i}{2}\rho_1 \otimes \sigma_i = -\frac{i}{2}\begin{pmatrix}0 & \sigma_i \\ \sigma_i & 0\end{pmatrix}$$

$$L_{i6} = \Gamma_i = \frac{1}{2}\gamma_i = \frac{i}{2}\rho_2 \otimes \sigma_i = \frac{i}{2}\begin{pmatrix}0 & -i\sigma_i \\ i\sigma_i & 0\end{pmatrix}$$

$$L_{56} = \Gamma_0 = \frac{1}{2}\gamma_0 = \frac{1}{2}\rho_3 \otimes I = \frac{1}{2}\begin{pmatrix}I & 0 \\ 0 & -I\end{pmatrix}$$

$$L_{46} = \Gamma_4 = -\frac{1}{2}\gamma_5 = \frac{i}{2}\rho_1 \otimes I = \frac{i}{2}\begin{pmatrix} 0 & I \\ I & 0 \end{pmatrix}$$

$$L_{45} = T = -\frac{i}{2}\gamma_5\gamma_0 = \frac{i}{2}\rho_2 \otimes I = \frac{i}{2}\begin{pmatrix} 0 & -iI \\ iI & 0 \end{pmatrix} \tag{4}$$

with

$$\{\gamma_\mu, \gamma_\mu\} = 2g_{\mu\nu}, \gamma_5 = \gamma_0\gamma_1\gamma_2\gamma_3; \gamma_i^2 = \gamma_5^2 = -1, \gamma_0^2 = +1$$

$$\sigma_i\sigma_j = i\epsilon_{ijk}\sigma_k + \delta_{ij}$$

$$\rho_i\rho_j = i\epsilon_{ijk}\rho_k + \delta_{ij}, [\sigma_i, \rho_i] = 0\,,$$

where the last column indicates a matrix form in which ρ_3 is diagonal.
Hence, it can be verified that $L_{AB} = -L_{BA}$, $A, B = 1 \ldots 6$ satisfy the following commutation relations:

$$[L_{AB}, L_{CD}] = -i(g_{AC}L_{BD} + g_{BD}L_{AC} - g_{AD}L_{BC} - g_{BC}L_{AD})$$

or, more simply,

$$[L_{AB}, L_{AC}] = -ig_{AA}L_{BC}, \tag{5}$$

with

$$g_{AA} = (-,-,-,-,+,+)\,.$$

We define further

$$L^2 = L_1^2 + L_2^2 + L_3^2 \text{ and } Q^2 = L_{56}^2 - L_{46}^2 - L_{45}^2\,.$$

These are precisely the abstract commutation relations of the Lie algebra of the conformal group G, O(4, 2). Hence, Dirac matrices provide a particular, four-dimensional, nonunitary but irreducible representation of G. We shall use later infinite-dimensional representations of the same algebra (5).
The identification of the generators in (4) is as follows: L_{ij} are the generators of the O(3) subgroup identified with space-rotations; L_{ij} together with A_i generate the group O(4); M_i correspond to pure Lorentz transformations; hence, L_{ij} and M_i generate the Lorentz group Λ. This can be seen also from the signature of the metric g_{AA}. Having identified the Lorentz group Λ, one can see from the commutation relations (5) that $\Gamma_\mu = (\Gamma_0, \Gamma_i)$ is a four-vector operator relative to Λ, i.e.,

$$[L_{i5}, \Gamma_\mu] = i(\delta_{\mu 0}\Gamma_i + \delta_{\mu i}\Gamma_0)\,. \tag{6}$$

In any representation of O(4, 2), we may diagonalize L_{12}, L^2 and L_{56}, Q^2 (or L_{12}, L_{34} and L_{56}) and denote the corresponding eigenvalues by j_z, $j(j+1)$ and n, $\varphi(\varphi+1)$ (or j_z, $n_1 - n_2$ and n), respectively; j and j_z have the interpretation of spin quantum numbers (= total angular momentum in rest frame), and we shall call n the "principal quantum number". The reason for this last name will become evident when we come to infinite-dimensional representations.

For the four-dimensional representation (4) we have the four basic states:

$$|n, j, j_z\rangle$$

with

$$j_z = \pm \tfrac{1}{2},\ j = \tfrac{1}{2},\ \text{and}\ \varphi = n = \pm \tfrac{1}{2}\ . \qquad (7)$$

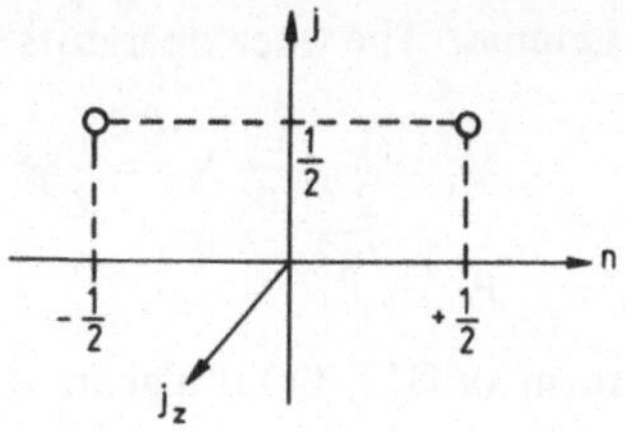

Fig. 1

The parity operator P commutes with $\vec{L}$ and anticommutes with $\vec{M}$; it can be chosen to be $P = \gamma_0$. Then the intrinsic parity of the states (7) is just the sign of n.
Figure 1 shows the diagram of quantum numbers which will be considerable generalized later.
We now take the states (7) as the "rest frame" states and perform the induction process given by Eq. (3). This can be done for the time-like case because we have representations of $K = SO(3)$. In fact, we have in (7) the direct sum of two unitary representations of K. The induction is given by

$$|njj_z; p\rangle = e^{i\vec{\xi}\cdot\vec{M}}\ |njj_z\rangle\ . \qquad (8)$$

Here $U(\vec{\xi}) = \exp(i\vec{\xi}\cdot\vec{M})$ is the representation of Lorentz transformations from rest frame $\dot{p} = (m, 0, 0, 0)$ to momentum

$$p^\mu = \left(m \cosh\frac{\xi}{2}, \hat{\xi}\, m \sin\frac{\xi}{2}\right),\ \xi = \sqrt{\vec{\xi}^2},\ \hat{\xi} = \frac{\vec{\xi}}{\xi}\ . \qquad (9)$$

Remark: The four-dimensional representations of O(4, 2) do not contain unitary representations of E(2) or SO(2, 1) except the trivial one-dimensional ones). Hence, there are no nontrivial unitary quantum systems based on (4) with light-like or space-like momenta.

Lemma: The states (8) satisfy the Dirac equation.

Proof: We abbreviate; $|njj_z; p\rangle \equiv |n, p\rangle$; $|njj_z\rangle \equiv |n\rangle$

$$\begin{aligned}\gamma^\mu p_\mu\ |n, p\rangle &= \gamma^\mu p_\mu e^{i\vec{\xi}\cdot\vec{M}}\ |n\rangle \\ &= e^{i\vec{\xi}\cdot\vec{M}} e^{-i\vec{\xi}\cdot\vec{M}} \gamma^\mu p_\mu e^{i\vec{\xi}\cdot\vec{M}}\ |n\rangle \\ &= e^{i\vec{\xi}\cdot\vec{M}} \gamma^0 p_0\ |n\rangle = nm e^{i\vec{\xi}\cdot\vec{M}}\ |n\rangle = nm\,|n, p\rangle\end{aligned}$$

or,

$$(\gamma^\mu p_\mu - nm)\ |n; p\rangle = 0\ . \qquad (10)$$

We shall now introduce by algebraic methods two other forms of states (8)–again for purposes of future analogy.

Lemma: The three operators

$$Y_1 = \frac{1}{2}\gamma^i \frac{p_i}{p},\ Y_2 = \frac{1}{2}\gamma^0 \frac{\gamma^i p_i}{p},\ Y_3 = \frac{1}{2}\gamma_0 \qquad p = \sqrt{\vec{p}^2} \tag{11}$$

form an O(2, 1)-Lie algebra: (Proof by direct computation.)

$$[Y_1, Y_2] = iY_3, [Y_2, Y_3] = iY_1, [Y_3, Y_1] = -iY_2 \,. \tag{12}$$

Hence, Eq. (10) can be written as a linear Lie algebra equation

$$(2p_0 Y_3 + 2pY_1 - nm)\,\psi(p) = 0\,. \tag{13}$$

Solution of the Lie-algebra Eq. (12): Y_3 and Y_1 do not commute. But we can solve (12) by tilting it in the 13-plane using (12): Let

$$\psi = e^{-i\theta Y_2}\psi',$$

then

$$(2p_0 e^{i\theta Y_2} Y_3 e^{-i\theta Y_2} + 2pe^{i\theta Y_2} Y_1 e^{-i\theta Y_2} - nm)\,\psi' = 0$$
$$[2p_0(\cosh\theta Y_3 + \sinh\theta Y_1) + 2p(\cosh\theta Y_1 + \sinh\theta Y_3) - nm]\,\psi' = 0$$
$$[(2p_0\cosh\theta + 2p\sinh\theta) Y_3 + (2p_0\sinh\theta + 2p\cosh\theta) Y_1 - nm]\,\psi' = 0\,.$$

(i) We first choose $\tanh\theta = p/p_0$ and obtain

$$[2\sqrt{p_0^2 - p^2}\,Y_3 - nm]\,\psi' = 0\,. \tag{14}$$

Clearly, eigenstates ψ' of $Y_3 = \frac{1}{2}\gamma_0$ solve this equation. Then

$$\psi = e^{-i\theta(p_i/p)\frac{1}{2}\gamma^0\gamma^i}\psi' = e^{-\tanh^{-1}(p/p_0)(p_i/p_0)\cdot M^i}\psi' = e^{+i\vec{\xi}\cdot\vec{M}}\psi'\,.$$

(ii) However, if we choose, in (14), $\tanh\theta = p_0/p$, we obtain by a similar calculation

$$\left(\sqrt{\vec{p}^2 - p_0^2}\,\frac{\vec{\gamma}\cdot\vec{p}}{p} - nm\right)\psi'' = 0\,. \tag{15}$$

Thus choosing ψ'' as eigenstates of $\frac{\vec{\gamma}\cdot\vec{p}}{p}$, we have

$$\psi = e^{-i\tanh^{-1}(p_0/p)(p^i/2)\frac{1}{2}\gamma^0\gamma^i}\psi'' \tag{15'}$$

Remark: These transformations are related to so-called Foldy-Wouthuysen and Cini-Touschek transformations. Indeed, multiplying (13) from the left by γ^0 gives the Hamiltonian form

$$(\tfrac{1}{2}p_0 + pY_2 - nmY_3)\,\psi = 0\,. \tag{16}$$

Now we tilt with Y_1:

$$\psi = e^{-i\theta Y_1}\phi\,.$$

If $\tanh\theta = p/m$, we get

$$(p_0 - \sqrt{m^2 + \vec{p}^2}\, Y_3)\,\phi = 0 \tag{16'}$$

The eigenstates ϕ of $\frac{1}{2}\gamma_0$ satisfying this equation are obtained from the Dirac function $\psi(p)$ by

$$\phi = e^{i\theta Y_1}\,\psi = e^{-i\vec{\eta}\cdot\vec{\Gamma}}\,\psi$$
$$\vec{\eta} = \frac{\vec{p}}{p}\tanh^{-1}\left(\frac{p}{m}\right). \tag{17}$$

The ϕ's are the Foldy-Wouthuysen states.
Similarly, if we choose $\tanh\theta = m/p$, we find

$$(\tfrac{1}{2}\,p_0 \pm \sqrt{\vec{p}^2 + (nm)^2}\,Y_2)\,\phi' = 0\,. \tag{18}$$

The solutions ϕ' are the eigenstates of $Y_2 = \frac{\vec{\alpha}\cdot\vec{p}}{p}$, helicity operator, and are therefore related to Dirac wave function ψ by

$$\phi' = e^{i\vec{\eta}'\cdot\vec{\Gamma}}\,\psi$$
$$\eta' = \frac{\vec{p}}{p}\tanh^{-1}(m/p)\,. \tag{18'}$$

Finally, using the successive O(2, 1)-transformations, we can relate directly rest frame states to Foldy-Wouthuysen states, etc.
Thus the dynamical algebra O(2, 1), Eqs. (11)–(12), give a unified treatment to all the basic transformations and/or boosts of the Dirac system.
The present treatment of the "electron-positron complex" via the dynamical group G = O(4, 2) is quite different from the standard arguments for the establishment of the Dirac equation which are:

(i) Dirac's argument for first-order wave equations.

(ii) The existence of a vector operator γ_μ necessitates the doubling of the representation space of the Lorentz group to $D^{0\frac{1}{2}} \oplus D^{\frac{1}{2}0}$, i.e., there is no four-vector current operator in either of the two-dimensional representations, but only in the direct sum (Gel' fand-Yaglom theorem [5]). The existence of a four-vector operator is related to the problem of minimal coupling, or to a geometric description of gauge transformations.

(iii) The existence of the parity operator also necessitates a doubling into $D^{0\frac{1}{2}} \oplus D^{\frac{1}{2}0}$, i.e. there is no parity operator in either of $D^{0\frac{1}{2}}$ or $D^{\frac{1}{2}0}$, but only in the direct sum. In other words, the two-dimensional Weyl equations contain only states of one-type of parity (the others are missing), and no-minimal covariant coupling is possible-hence, describe neutral spin-$\frac{1}{2}$ particles.

The External Interactions of the Electron-Positron Complex

The electromagnetic coupling of a composite system will be modeled after that of the Dirac particle. In the latter case, the gauge invariance of the second kind, or the minimal coupling $p^\mu \to p^\mu - eA^\mu$ gives the equation

$$[\gamma^\mu (p_\mu - eA_\mu) - m]\,\psi(x) = 0\,. \tag{19}$$

In a self-consistent theory (QED) we must also include part of the self-field of the electron into A_μ.
Thus the interaction operator is given by

$$j_\mu A^\mu = -\,e\gamma_\mu A^\mu\,. \tag{20}$$

For a Fourier component of the electromagnetic field

$$A_\mu = \left(\frac{2\pi}{qV}\right)^{\frac{1}{2}} \epsilon_\mu e^{-iqx}\,. \tag{21}$$

The first-order interaction matrix element (i.e., the so-called vertex $ee\gamma$) is

$$\begin{aligned} F_\mu &= \int \bar{\psi}_{p'n'}(x)\, j_\mu A^\mu \psi_{pn}\, d^4x \\ &= -\,e\left(\frac{2\pi}{qV}\right)\frac{1}{V}\int e^{i(p'-p-q)^\mu x_\mu}\epsilon_\mu \\ &\quad \times \langle \vec{p}' = 0, n' \,|\, e^{-i\vec{\xi}_{p'}\cdot\vec{M}}\gamma^\mu e^{-i\vec{\xi}_p\cdot\vec{M}} \,|\, n, p = 0\rangle\, d^4x\,, \end{aligned}$$

where we used (8). The ψ-function is related to (8) by some normalization factors:

$$\psi_{pn}(x) = \sqrt{\frac{m}{p_0}}\,\frac{1}{\sqrt{V}}\, e^{-ipx} e^{i\vec{\xi}\cdot\vec{M}}\,|\,n\rangle$$

such that $\int \psi^+ \psi\, d^3x = m/p_0$.
The photon vertex function on general grounds can be written, for a spin-$\frac{1}{2}$ particle, as

$$F_\mu = -\,e\bar{u}(p')\left[\gamma_\mu F_1(q^2) + \frac{i}{2m}\sigma_{\mu\nu}q^\nu F_2(q^2)\right]u(p)\,. \tag{22}$$

Comparing this with the above, we find that the Dirac particle to lowest order has the form factors

$$F_1(q^2) = 1,\; F_2(q^2) = 0\,. \tag{23}$$

In higher orders, we get more complicated nontrivial (even infinite) form factors. We shall compare these results with those of composite systems.

6 Composite systems. Inductive approach

We begin with a simple example which will show that our framework as applied to e^+e^--system also applies to realistic composite systems. This is for intuitive and pedagogical reasons. The general theory will be given subsequently.

Example: Nonrelativistic two-body system

We consider a two-body unrelativistic system described by the Hamiltonian

$$P_0 \psi = \left(\frac{p_1^2}{2m_1} + \frac{p_2^2}{2m_2} + \frac{\alpha}{|r_1 - r_2|} \right) \psi . \tag{24}$$

Or, passing to the center-of-mass and relative coordinates:

$$P_0 \psi = \left(\frac{P^2}{2M} + \frac{p^2}{2\mu} + \frac{\alpha}{r} \right) \psi \tag{25}$$

where $M = m_1 + m_2$, $\mu = m_1 m_2/(m_1 + m_2)$.
We rewrite Eq. (25) as

$$r \left(P_0 - \frac{P^2}{2M} \right) \psi = \left(\frac{1}{2\mu} r p^2 + \alpha \right) \psi , \tag{26}$$

valid everywhere except $r = 0$.
We introduce the following operators:

$$\begin{aligned} \Gamma_0 &= \tfrac{1}{2} [r_0 r p^2 + r/r_0] \\ \Gamma_4 &= \tfrac{1}{2} [r_0 r p^2 - r/r_0] \\ T &= \vec{r} \cdot \vec{p} - i \end{aligned} \tag{27}$$

where $r = \sqrt{\vec{r}^{\,2}}$, $p = \sqrt{\vec{p}^{\,2}}$ and r_0 is an arbitrary number. These operators satisfy the Lie algebra relations of $O(2, 1) \sim SU(1, 1) \sim SL(2, R) \sim Sp(2, R)$:

$$[\Gamma_0, \Gamma_4] = iT, \; [\Gamma_4, T] = -i\Gamma_0, \; [T, \Gamma_0] = i\Gamma_4 \tag{28}$$

just as the corresponding operators in Dirac case, in Eq. (4), or, just as the Y_i in Eqs. (11)–(12).
The invariant Casimir operator of the Lie algebra (27) is calculated to be

$$Q^2 = \Gamma_0^2 - \Gamma_4^2 - T^2 = (\vec{r} \times \vec{p})^2 = \vec{L}^2, \tag{29}$$

hence with eigenvalues $l(l+1)$, for all r_0. Note the important fact that the Casimir operator L^2 of the angular momentum Lie-algebra

$$\vec{L} = \vec{r} \times \vec{p} \tag{30}$$

coincides with that of the algebra (27) of "radial equation" (See below).

With the help of this Lie algebra (27), we can solve our dynamical problem (26) purely algebraically as follows:

Equation (26) using (27) becomes

$$\left[r_0(\Gamma_0 - \Gamma_4)\left(P_0 - \frac{\vec{P}^2}{2M}\right) - \frac{1}{2\mu}\frac{1}{r_0}(\Gamma_0 + \Gamma_4) - \alpha\right]\psi = 0 \tag{31}$$

or

$$(a\Gamma_0 + b\Gamma_4 + c)\psi = 0 \tag{32}$$

where

$$a = r_0\left(P_0 - \frac{\vec{P}^2}{2M}\right) - \frac{1}{2\mu r_0}$$

$$b = -r_0\left(P_0 - \frac{\vec{P}^2}{2M}\right) - \frac{1}{2\mu r_0}$$

$$c = -\alpha.$$

Equation (32) is a linear Lie algebra equation exactly of the same form that we sam in Eq. (13), and can be solved in the same way:
If we first tilt by putting

$$\psi = e^{i\theta T}\psi' \tag{33}$$

and choosing, for $(1 - b') > 0$,

$$\theta = \tanh^{-1}(-b/a),$$

we obtain for the new wave function ψ' the equation

$$[\sqrt{a^2 - b^2}\,\Gamma_0 + c]\psi' = 0\,. \tag{34}$$

In the representation (27), the Lie algebra (28) with (29) has a *unitary* infinite-dimensional representation (see Appendix A) in which Γ_0 has a discrete spectrum

$$\Gamma_0|n\rangle = n|n\rangle \tag{35}$$

with the range

$$n = l + 1 + s,\ s = 0, 1, 2, \dots. \tag{35'}$$

Thus, if ψ' is an eigenstate of Γ_0,

$$(\sqrt{a^2 - b^2}\,n - \alpha)\psi' = 0 \qquad \text{or} \qquad (a^2 - b^2)n^2 = \alpha^2\,.$$

Inserting the values of a and b from (32), we obtain finally (independent of r_0)

$$P_0 = -\frac{\mu\alpha^2}{2n^2} + \frac{\vec{P}^2}{2M}. \tag{36}$$

This is the total energy of the moving atom for discrete bound states. The scattering states are also contained in Eq. (31). In this case, we diagonalize, in (32), Γ_4 by choosing

$$\theta = \tanh^{-1}(a/-b) \tag{37}$$

which gives

$$(\sqrt{b^2 - a^2}\,\Gamma_4 + c)\,\psi' = 0\,.$$

The operator Γ_4 has a continuous spectrum in $[-\infty, +\infty]$:

$$\Gamma_4\,|\,\lambda\rangle = \lambda\,|\,\lambda\rangle \tag{38}$$

and, instead of (36), we obtain

$$P_0 = \frac{\mu\alpha^2}{2\lambda^2} + \frac{\vec{P}^2}{2M}, \tag{39}$$

the total energy of two scattering particles.

7 Dynamical algebras, their contraction and generalizations

The algebra (27), although it provides the energy spectrum of our system (24), is yet incomplete: It does not tell us, for example, the multiplicities of the levels. To complete the solution we must actually exhibit, the complete dynamical group G and the induced representations of the Poincaré group, or, in this example, of the Galilei group. We shall now complete (27) together with the angular momentum algebra (30) to a larger dynamical algebra G in terms of the relative variables $\vec{r}, \vec{p}$ which will give us the range of j-values in (35′) and all the quantum numbers in the *rest frame* of the whole system.
It will be convenient to present first the relativistic dynamical group; the unrelativistic case will be obtained by contraction.

Conformal Algebras

We start with two realizations of the conformal algebra-one in the coordinate space and the other in the momentum space [6]:

(1) Coordinate Space

$$\begin{aligned} L_{\mu\nu} &= X_\mu \Pi_\nu - X_\nu \Pi_\mu \\ P_\mu &\equiv L_{\mu 6} + L_{\mu 4} = \Pi_\mu \\ K_\mu &\equiv L_{\mu 6} - L_{\mu 4} = 2X_\mu(X_\nu \Pi^\nu + iH) - (X_\nu X^\nu)\Pi_\mu \\ D &\equiv L_{64} = X_\nu \Pi^\nu + iH \end{aligned} \tag{40}$$

where X_μ and Π_μ are conjugate variables, $[\Pi_\mu, X_\nu] = ig_{\mu\nu}$, and H is a number, the *homogeneity*, and is related to the second Casimir invariant of the algebra by

$$Q_2 = \frac{1}{2} L_{AB} L^{AB} = H^2 - 4H\,. \tag{41}$$

The generators L_{AB}, A, B = 1, 2, ..., 6, satisfy the same commutation relations (5) with the same metric g_{AA}. The indices μ, ν run over 5 = 0, 1, 2, 3 with $g_{\mu\mu} = (+,-,-,-)$.

(2) Momentum Space

We obtain from (40) another conformal algebra by the substitution $\Pi_\mu \to x_\mu$ and $X_\mu \to -\pi_\mu$, because then the relation

$$[\Pi_\mu, X_\nu] = [\pi_\mu, x_\nu] = ig_{\mu\nu} \tag{42}$$

is preserved. Performing this substitution in (40) and reordering the π's to the right by convention, we obtain

$$\begin{aligned} l_{\mu\nu} &= x_\mu \pi_\nu - x_\nu \pi_\mu \\ l_{\mu 6} + l_{\mu 4} &= x_\mu \\ l_{\mu 6} - l_{\mu 4} &= 2(x_\nu \pi^\nu + ih)\pi_\mu - x_\mu \pi_\nu \pi^\nu \\ l_{64} &= -(x_\nu \pi^\nu + ih)\,. \end{aligned} \tag{43}$$

The operators (40) or (43) may be thought of as differential operators acting on the space of functions $f(X_\mu)$, $X_\mu \in M^4$, or on $f(x_\mu)$, $x_\mu \in M^4$, respectively, or their conjugate momentum spaces. We shall now extract from these conformal algebras of differential operators acting on functions over $\mathbb{R}^3$.

Conformal algebras on $\mathbb{R}^3$

Consider the Lie algebra L given in (40). The quantity $P_\mu P^\mu$ is not an invariant of L. However, for H = 1, we have

$$[P_\mu P^\mu, K_\lambda] = 4x_\lambda P_\mu P^\mu;\ [P_\mu P^\mu, D] = 2P_\mu P^\mu\,. \tag{44}$$

Hence, if the space of functions f is such that $P_\mu P^\mu f = 0$, then $P_\mu P^\mu$ is an invariant. Similarly for (43): In the case h = 1, the equation

$$x_\mu x^\mu f = 0 \tag{45}$$

is an invariant. In particular, consider now functions $f(\vec{r})$ over $\vec{r} \in \mathbb{R}^3$. Then from (45) we represent $(\vec{x} = \vec{r})$

$$x_0^2 f = \vec{r}^{\,2} f, \text{ or } x_0 f = rf,\ r = \sqrt{\vec{r}^{\,2}} \tag{46}$$

and

$$\pi_0 f = -i \frac{\partial}{\partial t} f = 0\,.$$

With these two relations (46), the algebra (43) becomes

$$\begin{aligned}
l_{ij} &= r_i\pi_j - r_j\pi_i = \epsilon_{ijk}(\vec{r} \times \vec{\pi})^k \\
l_{i0} &= r\pi_i \\
l_{i6} + l_{i4} &= r_i \\
l_{06} + l_{04} &= r \\
l_{i6} - l_{i4} &= 2(\vec{r} \cdot \vec{\pi} - i)\pi_i - r_i\pi^2 \\
l_{06} - l_{04} &= + r\pi^2 \\
l_{64} &= \vec{r} \cdot \vec{\pi} - i\,.
\end{aligned} \tag{47}$$

One can directly verify that (47) still satisfies the commutation relations of O(4, 2). Similarly, starting from (40) and using

$$\Pi_0 f(\vec{\Pi}) = Pf(\vec{\Pi}),\ X_0 f(\vec{P}) = \frac{\partial}{\partial P^0} f(\vec{P}) = 0,\ P = \sqrt{\vec{P}^2}\,, \tag{48}$$

we obtain an algebra conjugate to (47), which can also be obtained directly from (47) by the substitution (42); i.e., $\vec{r} \to \vec{\Pi}$ and $\vec{\pi} \to -\vec{X}$, hence $r \to P$.

$$\begin{aligned}
L_{ij} &= X_i\Pi_j - X_j\Pi_i \\
L_{i0} &= X_iP \\
L_{i6} + L_{i4} &= \Pi_i \\
L_{06} + L_{04} &= P \\
L_{i6} - L_{i4} &= 2X_i(\vec{X} \cdot \vec{\Pi} + i) - \vec{X}^2\Pi_i \\
L_{06} - L_{04} &= -\vec{X}^2P \\
L_{64} &= -\vec{X} \cdot \vec{\Pi} + i\,.
\end{aligned} \tag{49}$$

The algebra (47) is the desired dynamical algebra *G* in terms of the relative coordinates and contains indeed the angular algebra (30) and "radial" algebra (27). The Casimir invariants of *G* are fixed numbers so that in an irreducible representation of *G* the range of the quantum numbers n and j are fully determined, as we shall see.
Note: Because of the signature of the metric g_{AB}, (41), the algebra (47) admits various automorphismus, among them one obtained by changing the indices 5–6. The algebra mostly quoted in the literature is obtained from (47) by this interchange 5–6. For example, $l_{i0} = r\pi_i \to l_{i6}$, etc. Most results only depend on the commutation relations and not on the explicit realization of the generators, hence remain unchanged.

The unrelativistic contraction of the conformal algebra

Because the unrelativistic energies are counted above the rest energy, we set in (40) [7]

$$\Pi_0 \to mc + \frac{1}{c}\tilde{\Pi}_0$$

and obtain

$$\begin{aligned}
L_{i0} &= X_i(mc + \frac{1}{c}\tilde{\Pi}_0) - X_0\Pi_i \\
L_{06} + L_{04} &= (mc + \frac{1}{c}\tilde{\Pi}_0) \\
L_{06} - L_{04} &= 2X_0(X_0(mc + \frac{1}{c}\tilde{\Pi}_0) - \vec{X}\cdot\vec{\Pi} + iH) - (X_0^2 - \vec{X}^2)(mc + \frac{1}{c}\tilde{\Pi}_0) \\
L_{64} &= X_0(mc + \frac{1}{c}\tilde{\Pi}_0) - \vec{X}\cdot\vec{\Pi} + iH\,.
\end{aligned} \tag{50}$$

The remaining elements are unchanged. Now we go to the three-dimensional representations, using (48), and perform contraction ($c \to \infty$) to obtain

$$\begin{aligned}
L_{ij} &= X_i\Pi_j - X_j\Pi_i \\
L_{i0} &= mX_i \\
L_{i6} + L_{i4} &= \Pi_i \\
L_{06} + L_{04} &= P \\
L_{06} - L_{04} &= \vec{X}^2 P \\
L_{i6} - L_{i4} &= 2X_i(\vec{X}\cdot\vec{\Pi} + iH) - \vec{X}^2\Pi_i \\
L_{64} &= -\vec{X}\cdot\vec{\Pi} + iH\,.
\end{aligned} \tag{51}$$

The conjugate algebra is

$$\begin{aligned}
l_{ij} &= r_i\pi_j - r_j\pi_i \\
l_{i0} &= -m\pi_i \\
l_{i6} + l_{i4} &= r_i \\
l_{06} + l_{04} &= r \\
l_{i6} - l_{i4} &= 2(\vec{r}\cdot\vec{\pi} - i\hbar)\pi_i - r_i\pi^2 \\
l_{06} - l_{04} &= r\,\pi^2 \\
l_{64} &= \vec{r}\cdot\vec{\pi} - i\,.
\end{aligned} \tag{52}$$

The quantum mechanical Galilei algebra is a subalgebra of (52) or (51) consisting of $\vec{L} = \vec{r} \times \vec{\pi}$, $\vec{K} = l_{0i} = -m\vec{\pi}$, $P_0 = r$ and $\vec{P} = -\vec{r}$ (or, $\vec{R} \times \vec{\Pi}$, $m\vec{X}$, P and $-\vec{P}$, respectively) with

$$[K_i, P_j] = m\delta_{ij}\,. \tag{53}$$

Equation (53) defines the so-called quantum mechanical Galilean algebra; in the pure geometric Galilean algebra the commutator (53) is zero.

The O(2, 1) radial algebra is still the same.

The remaining commutator which are different from (47) are:

$$[-m\pi_i, r] = mi\hat{r}_i$$
$$[-m\pi_i, r\pi^2] = mi\hat{r}_i\pi^2$$
$$[-m\pi_i, 2(\vec{r}\cdot\vec{\pi} - i)\,\pi_j - r_j\pi^2] = 2mi\left[\frac{1}{r}\pi_i\pi_j - \frac{1}{r^3}r_i(\vec{r}\cdot\vec{\pi})\,\pi_j - \frac{1}{2}\delta_{ij}\pi^2\right].$$

Thus the total contracted algebra does not close: it contains only the 12-dimensional Schrödinger algebra (Galilei + O(2, 1)); the remaining three generators do not form a Lie algebra with these.

A Generalization of the dynamical algebra

The algebra (27) can be generalized to a class of additional potentials as follows: ($r_0 = 1$)

$$\begin{aligned} \Gamma'_0 &= \tfrac{1}{2}(rp^2 + r) + rV(r) \\ \Gamma'_4 &= \tfrac{1}{2}(rp^2 - r) + rV(r) \\ T &= \vec{r}\cdot\vec{p} - i\,. \end{aligned} \tag{54}$$

The commutation relations (28) remain unchanged provided

$$[rV(r), T] = -\,irV(r)$$

or (55)

$$[V, T] = -\,i2V\,.$$

This is just a homogeneity condition of degree -2 on the potential V, because T is the dilatation operator. Examples are

$$V = a/r^2\,, \tag{56}$$

but also

$$V = a_1/x^2 + a_2/y^2 + a_3/z^2\,. \tag{57}$$

The Casimir operator Q^2, (29), is now:

$$\begin{aligned} Q'^2 &= (\Gamma_0 + rV)^2 - (\Gamma_4 + rV)^2 - T^2 \\ &= Q^2 + (\Gamma_0 rV - \Gamma_4 rV) + (rV\Gamma_0 - rV\Gamma_4) \\ &= Q^2 + r^2V + rVr = Q^2 + 2r^2V = \vec{L}^2 + 2r^2V\,. \end{aligned} \tag{58}$$

The wave equation (26), when written as

$$r\left(P_0 - \frac{P^2}{2M}\right)\psi = \left(\frac{1}{2\mu}rp^2 + \alpha + \frac{2rV}{2\mu}\right)\psi\,,$$

remains unchanged in the form (31)–(34). Hence the spectrum (36) or (39) is unchanged; only the connection between n and l (Eq. (35′)) is now different and depends on the eigenvalues of Q'^2. In the case $V = a/r^2$, the eigenvalues of Q'^2 are simple:

$$\begin{aligned} Q'^2 \psi &= [l(l+1) + 2a]\,\psi \\ &= \phi(\phi+1)\,\psi \end{aligned} \tag{59}$$

Then the range of n in the spectrum is

$$n = -\varphi + s, \quad s = 0, 1, 2, \dots .$$

In the unisotropic case,

$$V = \frac{a_1}{x^2} + \frac{a_2}{y^2} + \frac{a_3}{z^2}, \qquad \text{we have} \qquad Q^2 = L^2 + 2r^2\left(\frac{a_1}{x^2} + \frac{a_2}{y^2} + \frac{a_3}{z^2}\right),$$

and it is difficult to find the spectrum of the second term.

Note: The dynamical algebras that solve the Klein-Gordon and Dirac equations *with* a Coulomb potential are all of the type (54) with $V = a/r^2$, Eq. (56) [8]:

Klein-Gordon Kepler problem: $a = -\alpha^2$

Dirac Kepler problem: $a = -\alpha^2 - i\alpha\rho_1\vec{\sigma}\cdot\vec{r}$.

Dynamical algebra for a relative oscillatory motion

The dynamical O(2, 1) algebra for an oscillator-type *relative* motion is (in appropriate units)

$$\begin{aligned} \Gamma_0 &= \tfrac{1}{2}(p^2 + \tfrac{1}{4}r^2 + c/r^2) \\ \Gamma_4 &= \tfrac{1}{2}(p^2 - \tfrac{1}{4}r^2 + c/r^2) \\ T &= \tfrac{1}{2}(\vec{r}\cdot\vec{p} - \tfrac{3}{2}i), \end{aligned} \tag{60}$$

where c is arbitrary and $r = \sqrt{\vec{r}^{\,2}}$.

In this case Eq. (26) is even simpler

$$(P_0 - P^2/2M)\,\psi = \lambda\Gamma_0\psi, \tag{61}$$

the solution of which needs no tilting operation.

The invariant operator is

$$Q^2 = J^2 = (\vec{r}\times\vec{p})^2, \tag{62}$$

independent of c.

The energy spectrum is given by

$$P_0 = \frac{P^2}{2M} + \lambda n, \quad n = j+1, j+2, \dots . \tag{63}$$

The completion of (60) together with $\vec{L}$ into a bigger algebra containing the Lorentz group is an open problem. The bigger algebra can be taken to be SU(3, 1) because of the SU(3)-symmetry of the Hamiltonian), or again SO(4, 2) in which, however, certain states of definite parity properties are missing.

Generalization to many-body-problems

The previous procedure outlined for a two-body system can be generalized to certain many-body problems. Again we separate the center-of-mass motion and establish a dynamical algebra for the relative motion. In fact, there are exactly soluble N-body problems with dynamical group O(2, 1), namely of oscillator type [9] or of Coulomb-type [10]. In the latter case, we can even complete the O(2, 1) to a dynamical algebra giving us all the degeneracies and global quantum number of levels.

Dynamical algebras for arbitrary systems

A question often raised is this: Is there a dynamical group *G* for arbitrary potentials V in the two-body problem, or for arbitrary systems, besides the oscillator, Kepler, or similar symmetric cases? The answer is: The dynamical group *G* provides the quantum numbers of geometric origin, such as (njm), which can be used for any central problem. It is the connection between mass or energy and these quantum numbers which depends on the nature of the potential; and this connection is very simple for the above soluble problems, complicated in other cases. But for quite a variety of problems (e.g., Morse potential [11]), a dynamical algebra is known. We give next an example of a nonlinear system.

A class of nonlinear Hamiltonian systems

A class of nonlinear field theories [12] have special solutions corresponding to the Lagrangian

$$L = \tfrac{1}{2}\,\alpha(q)\,\dot{\vec{q}}^{\,2} + \tfrac{1}{2}\,\beta(q)\,(\vec{q}\cdot\dot{\vec{q}})^2$$

where $\alpha(q)$ and $\beta(q)$ are arbitrary functions. The corresponding Hamiltonian is

$$H = \frac{1}{2\alpha^2}\,p^2 - \frac{\beta}{2\alpha^2}\,\frac{(\vec{p}\cdot\vec{q})^2}{(\alpha^2+\beta q^2)}. \tag{64}$$

The quantization of such Hamiltonians poses problems such as the ordering of operators $\vec{p}$ and $\vec{q}$. The dynamical group quantization gives a definite answer to this problem and, at the same time, provides an exact solution. For example, for

$$H = \tfrac{1}{2}\,\vec{p}^{\,2}\,(1+\vec{q}^{\,2})^2\,,$$

the quantum mechanical Hamiltonian with the following ordering

$$\hat{H} = \frac{1}{2}(p + p\vec{q}^{\,2})(p + p\vec{q}^{\,2})$$

has a group-theoretic significance: Using now the algebra (49) [instead of the usual (47)], it can be written as

$$\hat{H} = \frac{1}{2}\Gamma_0^2 ,$$

with the spectrum $E = \frac{1}{2}n^2$, where the range of n is determined by the Casimir operator of the dynamical algebra G.

Boosting the composite system

Equation (31) in the rest-frame, $\vec{P} = 0$, is

$$\left[r_0(\Gamma_0 - \Gamma_4)P_0 - \frac{1}{2\mu r_0}(\Gamma_0 + \Gamma_4) - \alpha\right]\psi(0) = 0 . \tag{65}$$

Can we induce the moving wave function $\psi(p)$ from $\psi(0)$:

$$\psi(p) = e^{i\vec{\xi}\cdot\vec{K}}\psi(0)$$

analogous to (3) or (8)? Using the contracted algebra (51) with $\vec{K} = \vec{X}$ and boost parameter $\vec{\xi} = \vec{P}/M$, the rest-frame equation (65) is transformed into ($r_0 = 1$)

$$\left[(\Gamma_0 - \Gamma_4)(P_0 - \frac{1}{2M}\vec{P}^2) - \frac{1}{2\mu}(\Gamma_0 + \Gamma_4) + \vec{P}\cdot\vec{\Gamma} - \alpha\right]\psi(P) = 0 , \tag{66}$$

which can also be obtained from the corresponding relativistic equation. There are other forms of (66), where wave function is further appropriately tilted by $e^{i\theta\cdot T}$ or $e^{i\vec{\varphi}\cdot\vec{M}}$.

8 Passage to relativistic wave equations

From the 1-to-1 correspondence between the O(4, 2)-operators of Dirac-algebra (42), and the O(4, 2)-operators of the relative dynamics (47) *and* the correspondence between the rest-frame equations

$$\begin{aligned} &(\gamma^0 p_0 - m)\psi(0) = 0 \\ &\left[(\Gamma_0 - \Gamma_4)P_0 - \frac{1}{2\mu}(\Gamma_0 + \Gamma_4) - \alpha\right]\psi(0) = 0 , \end{aligned} \tag{67}$$

we now propose to write covariant wave equations for our composite systems.

One difference so far between the two above rest-frame equations is the lack of Γ_4-term in the Dirac equation. This is due to the parity properties of the two systems. There is surely a Lorentz scalar analog of Γ_4 in the four-dimensional representation, namely $\frac{1}{2}\gamma_5$, and such a term could be added to the Dirac equation, but it would violate parity [4]. In the infinite-dimensional case, Γ_4 is a scalar and invariant under parity.

It will be more convenient and economical to exhibit the covariant equation and to study its reduction to the nonrelativistic form (31). Various arguments have been described in the literature to arrive at this covariant form [13].

The relativistic equation is

$$(J_\mu P^\mu + \beta\Gamma_4 + \gamma)\,\psi(P) = 0$$

where (68)

$$J_\mu = \alpha_1\Gamma_\mu + \alpha_2 P_\mu + \alpha_3 P_\mu \Gamma_4 + \alpha_4 L_{\mu\nu} q^\nu .$$

Here α_i, β, γ are constants whose values we shall give below. $\Gamma_\mu, \Gamma_4, L_{\mu\nu}$ are elements of O(4, 2), Eq. (47), whose transformation properties under the Lorentz group Λ are determined if we identify the Lie algebra of Λ with L_{ij} and $L_{i0} = M_i$; P_μ is the total momentum of the two-body system, q_ν the difference of momenta of the two-body system.

The values of the coefficients have been found to be

$$\alpha_1 = 1, \alpha_2 = -\frac{\alpha}{2m_2}, \alpha_3 = \frac{1}{2m_2}, \beta = \frac{m_2^2 - m_1^2}{2m_2}, \gamma = \alpha\frac{m_1^2 + m_2^2}{2m_2} \tag{69}$$

where m_2 is the nucleon mass, m_1 the electron mass, and α is the five-structure constant. The passage to and from the rest frame is accomplished by the relativistic boost operators $\exp(i\vec{\xi}\cdot\vec{M})$ using the equations (8) and (9), exactly as in the case of Dirac equation. This process depends only on the commutation relations $[\vec{M}, \Gamma_\mu]$. The rest-frame wave equation is

$$\left[\left(\Gamma_0 - \frac{\alpha}{2m_2}M - \frac{1}{2m_2}M\Gamma_4\right)M + \frac{m_2^2 - m_1^2}{2m_2}\Gamma_4 + \alpha\frac{m_1^2 + m_2^2}{2m_2}\right]\psi(0) = 0\,. \tag{70}$$

Note that M here is the total relativistic mass of the atom including the rest mass of the nucleon. Hence we must subtract rest mass in order to compare with the nonrelativistic energy formula.

From (69) and (70), we obtain the relativistic Balmer formula

$$M^2 = m_1^2 + m_2^2 + 2m_1m_2[1 - \alpha^2/(n^2 + \alpha^2)]^{1/2}$$

which contains for the binding energy $B = M - (m_1 + m_2)$, the correct Dirac-spectrum with even the recoil correction to order $(\alpha)^4$:

$$\Delta B = -\frac{1}{8}\frac{m_1}{m_2}\left(\frac{\alpha}{n}\right)^4\frac{1}{m_r}, \quad m_r = \frac{m_1 m_2}{m_1 + m_2},$$

which is the same as obtained in quantum electrodynamics.

Summary

The structure of our framework can thus be summarized by the following diagram.

Space-time symmetry group G Little groups K ⊂ G with representations L(K)	⟶	(irreducible) Unitary/representations $U^{L(K)}(G)$ of G in some linear space. (elementary systems)
↓		↓
Dynamical groups $\mathcal{G} \supset G$ Representations $L(\mathcal{G})$	⟶	Induced (reducible) representations $U^{L(\mathcal{G})}(G)$ of G (composite systems)

The next step in this development is the formulation of covariant wave equations corresponding to composite systems, i.e., corresponding to $U^{L(\mathcal{G})}(G)$.

The theory of finite-dimensional wave equation is well-known. For example, for $\mathcal{G} = \Lambda = SO(3, 1)$ or $SL(2, C)$, and $L(\mathcal{G})$ = finite-dimensional representation of $SL(2, C)$, there is a unified formalism to derive covariant one-mass, one-spin equations (e.g., Dirac, Maxwell, Proca, Pauli-Fierz equations, etc., etc.), as well as many-mass and many-spin equations (e.g., Bhabba, Gel'fand-Yaglom equations, etc.) [14].

Important Note: There are in principle infinitely many wave equations, even for a given spin j and mass: Start from any $L(\mathcal{G})$, induce the representations of the Poincaré group and project out the particular spin j.

The choice of the proper wave equation really depends, besides on mass and spin, on the correct coupling of the wave equation to external fields, or to other particles. For example, different wave equations under the principle of minimal coupling give different electromagnetic properties (e.g., magnetic moment) to the same particle (m, j). Even in the case of "elementary" particles, the question as to which coupled equations occur in nature (e.g., coupled Maxwell-Dirac equations) can so far only be determined by the experimental consequences of the theory. Although there are some geometrical or group-theoretic, so-called "gauge-principles", which give coupled field equations between a set of particles and a set of vector mesons (generalizing electrodynamics), their choice too must be verified by experiment.

We shall now extend the methods of formulating wave equations to composite systems.

9 Principles on the choice of infinite component wave equations

Our framework so far had two basic principles:

(I) Relativity and Quantum Mechanics: Representations of Poincaré group G

(II) Compositeness: Dynamical group $\mathcal{G}$ and its representations $L(\mathcal{G})$ inducing reducible representations $U^{L(\mathcal{G})}(G)$ of G.

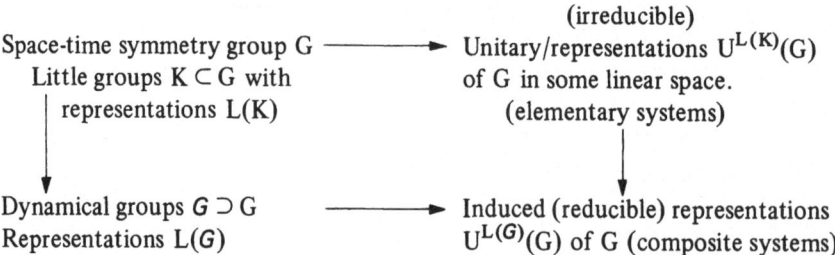

The second principle completely determines the spin content and multiplicities of the representations of G. These multiplicities are expressed by the (n–j)-diagrams of all quantum numbers. For discrete quantum numbers, each vertex in the (n–j)-diagram corresponds to an irreducible representation of the Poincaré group. It remains to assign $P_\mu P^\mu$ to each such irreducible representation. This will be accomplished by the third principle:

(III) Covariant wave equation assigns masses to the irreducible components of the reducible representation $U^{L(G)}(G)$.

In general, almost any mass values can be assigned to the irreducible component by writing a general wave equation formed out of the scalars from the generators of G and G:

$$f(P_\mu \Gamma^\mu, P_\mu P^\mu, W_\mu W^\mu, \Gamma_4, \Gamma_4 P_\mu P^\mu, \ldots)\,\psi = 0\,. \tag{71}$$

We further restrict the wave equation by the quantum mechanical requirement:

(III′) A probability (or matter) current

$$j_\mu^{\text{matter}} = \bar{\psi} J_\mu \psi \tag{72}$$

exists which is conserved by virtue of the wave equation which is derivable from a Lagrangian formalism. The mass spectrum can be derived from the requirement of current conservation alone [15].

The electromagnetic current j_μ^{em} may or may not be proportional to j_μ^{matter}. In the Dirac equation it is, but there are other examples where it is not.

Thus the formalism still allows a large class of systems. The postulated form of the wave equation determines the quantitative numbers as far as mass-spectrum and form factors are concerned, although some general features (such as the functional form of the form factors) are already determined by G and $L(G)$.

Because of the fact that the compositeness is already included in G, we may postulate wave equations which are linear in the generators of G. The previous examples show that Dirac, Majorana, H-atom eqs. are all of this type. It has been found that nucleons also can be described by equations of this type.

Higher order terms in the Dirac equation, e.g.,

$$\left(\gamma^\mu p_\mu + \sum_{n=0}^{\infty} c_n (p^2)^n\right)\psi = 0 \tag{73}$$

give a certain structure to the electron. However, instead of the infinitely many parameters in this equation, a single infinite-component wave equation $(\Gamma_\mu P^\mu + K)\,\psi = 0$ accomplishes the same objective in a much better way. Therefore, for a given G, we should keep the wave equation to a simple linear form $(J_\mu P^\mu + K)\,\psi = 0$. For more complicated systems, we have to choose different representations $L(G)$ of G, or even change the dynamical group G.

The final choice of the wave equation together with its coupling to interactions completes the theory, for we could then calculate all the properties of the composite system that experiment can give us.

10 The Majorana "particle"

There are simpler covariant infinite-component wave equations than (68). The simplest perhaps is the Majorana equation given in 1932, intended still to describe an electron and not a composite system. But we shall give a new composite-structure model to the Majorana "particle". The equation is

$$(\Gamma_\mu P^\mu - K)\,\psi(P) = 0\,. \tag{74}$$

Here Γ_μ is again a vector operator in the Lie algebra of the dynamical algebra G. In the original Majorana equation, G is a subgroup O(3, 2) of the dynamical group O(4, 2), Eq. (47), namely the subgroup generated by L_{ij}, $L_{i5}(\vec{M})$, $L_{i6}(\Gamma)$ and $L_{56}(\Gamma_0)$. We can use the same Eq. (74) also for O(4, 2)-case. The difference is in the spectrum and multiplicity of states, hence in the physical interpretation of the corresponding composite system.

For $G = O(3, 2)$, the spectrum of L^2, L_z and Γ_0 is given by

$$L^2\,|jm\rangle = j(j+1)\,|jm\rangle;\ L_z\,|jm\rangle = m\,|jm\rangle,\ \Gamma_0\,|jm\rangle = (j+\tfrac{1}{2})\,|jm\rangle\,. \tag{75}$$

The mass spectrum of (74) can be immediately derived by transforming to the rest frame:

$$(\Gamma_0 M - K)\,\psi(0) = 0\,. \tag{76}$$

Hence,

$$\left.\begin{array}{lll} M = K/(j+\tfrac{1}{2}), & \text{for} & O(3,2) \\ M = K/n, & \text{for} & O(4,2)\,. \end{array}\right. \tag{76'}$$

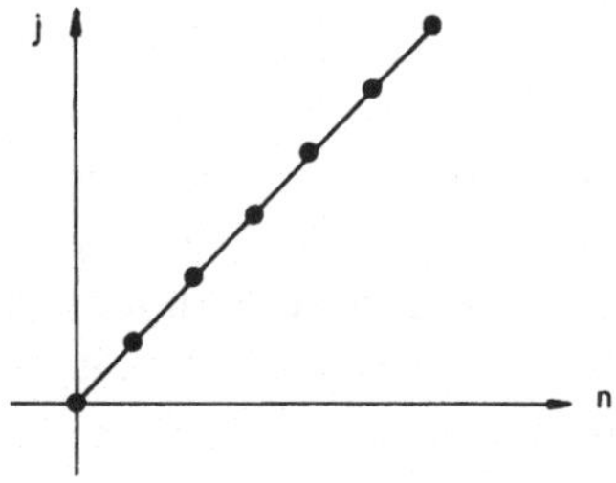

Fig. 2

The (n, j)-diagram in the first case, with $j_{min} = 0$ or $\frac{1}{2}$, is in the second case, the diagram is shown in Appendix B.

The mass spectrum depends only on G and its representation. Now the Lie algebra of G can be realized in terms of distinct relative coordinates. For example, either by

$$\Gamma_0 = -r\nabla^2 + r/4 \tag{77}$$

where $r = \sqrt{x_1^2 + x_2^2}$, $\nabla^2 = \partial/\partial_1^2 + \partial/\partial x_2^2$ are the two-dimensional radial coordinate and the two-dimensional Laplacian, or by

$$\Gamma_0 = \frac{1}{2}\left(q_1^2 + q_2^2 - \left(\frac{\partial^2}{\partial q_1^2} + \frac{\partial^2}{\partial q_2^2}\right)\right) \tag{78}$$

in some two-dimensional space. The remaining operators have similar representations in these spaces [16]. Consequently, the rest frame equation (76) can be written as either

$$\left[-\frac{\hbar^2}{2\mu}\nabla^2 - \frac{\alpha}{r} - E\right]\psi = 0,\ \alpha = \frac{K\hbar^2}{2M\mu},\ E = -\frac{\hbar^2}{8\mu}$$

or (79)

$$\left[-\frac{\hbar^2}{2\mu}\left(\frac{\partial^2}{\partial q_1^2} + \frac{\partial^2}{\partial q_2^2}\right) + \frac{1}{2}\mu\omega^2 q^2 - E\right]\psi = 0,\ \omega = \frac{2\hbar}{\mu},\ E = \frac{4\hbar^2 K}{\mu M}$$

i.e., we have either a relative Kepler-type motion or a relative oscillator-type motion. The internal relative dynamics can be even transformed from a nonrelativistic kinematics (79) to a relativistic kinematics by the dilatation of the wave function [17]:

$$\psi = e^{i\beta M_3}\phi,\ \tan\beta = \frac{r_0^2 - 1}{r_0^2 + 1},\ r_0 = \frac{\mu}{m_1}\left(1 + \frac{\alpha^2}{n^2}\right)^{\frac{1}{2}},$$

(80)

$$\left[-\nabla^2 - 2\alpha E_1/r - (E_1^2 - m_1^2)\right]\phi = 0,\ E_1 = m_1\left(1 + \frac{\alpha^2}{n^2}\right)^{-\frac{1}{2}}.$$

The introduction of relative coordinates for a given mass spectrum is therefore so far arbitrary. In order to "see" which of the motions the charge actually performs, we have coupled the system to the external electromagnetic field and calculated the form factors [16]. In fact, experimentally, we probe the unknown internal structure of a composite object by various probes (γ, e, ν, ...) and try to infer the motion of the constituents which react to the probes.

11 An inverse problem

The determination of the relative internal dynamics from the mass spectrum belongs to the class of "inverse problems", such as the inverse scattering problem which deals with the determination of the relative potential from the observed scattering data. Therefore, we propose the following inverse problem:

"Given an algebraic infinite-component wave equation (i.e., a mass spectrum), which internal realizations are compatible with it? "

We shall later see that we can sharpen the problem if we use the information on the composite systems that we obtain using electromagnetic and other probes. In fact, since we do not know the internal structure of a new composite system (say a hadron), we must infer from spectral information back to the internal dynamics. This is what the "spectroscopy" is all about: to study the structure of atoms, molecules, hadrons, ... from the analysis of spectral lines, or "to find out the shape of a bell by means of the sounds which it is capable of sending out" (Sir Arthur Schuster, 1882).

More generally, our third principle on the choice of wave equation is also an inverse problem. To choose the terms and the coefficients of the wave equation in such a way as to fit the observed spectra (first problem), then for the established wave equation to find the internal dynamics (second problem).

12 The Class of composite systems based on conformal group

For $G = SO(4, 2)$ and $L(G)$ = infinite-dimensional, we have already seen examples of quantum systems. The essential element is the group structure. The representation of the Lie algebra by differential operators in the relative coordinates, such as given in Eq. (47), is secondary, for the purpose of interpretation, or intuitive Schrödinger picture. Actually, all the final results, matrix elements, etc., hence comparison with experiment, do not depend on this coordinate realization but only on the type to representation and on the quantum numbers. In fact, the introduction of unobservable relative coordinates is ambiguous and can be arbitrary. For example, for the same mass spectrum and degeneracy of levels, different relative coordinates can be introduced [16].

A general class of composite systems may be defined by taking $G = O(4, 2)$, and more and more complicated representations $L(G)$ of G. In general, all three Casimir invariants of G will be different from zero.

13 Electromagnetic interactions of composite systems

So far we have described the motion of an isolated composite system as a whole. How does such a system interact with external fields, or with another similar composite system?

Elementary systems such as the electron-positron complex are known to couple to the external electromagnetic field by the principle of minimal coupling. This is a generalization of classical $j^\mu A_\mu$-coupling, and can be justified (if not uniquely derived) by various differential-geometrical arguments (such as the principle of gauge-invariance). For a composite-system, one could in principle couple the charged constituents minimally to the external field. This is, of course, done for nonrelativistic systems. The procedure is, first of all, complicated and, furthermore, in our case of relativistic systems, the separation of constituents is difficult and ambiguous. In fact, we have tried to get away from the picture of constituents and treat the composite system as a single indivisible relativistic object.

It is therefore natural to formulate the principle of minimal coupling also for our infinite-component wave equations:

Electromagnetic Coupling Principle for Infinite-Component Wave Equations: To first order in the fine-structure constant α, the coupling to quantized photon field A_μ is given by the interaction term $J^\mu A_\mu$, where J^μ is given by the wave equation (68). Thus the matrix elements are proportional to

$$F_\mu \equiv \langle n'p' | J_\mu | np \rangle \epsilon^\mu . \tag{81}$$

Remark: In this form this coupling principle differs from the minimal coupling in the form $P_\mu \to P_\mu - eA_\mu$ because of the quadratic terms $\alpha P_\mu P^\mu$ in the wave equation. The form $J_\mu A^\mu$ has been found to be adequate for most purposes, but the difference should be kept in mind. The terms proportional to P^μ in the current J_μ express the composite-nature of the system; they have been called "convective currents."

The coupling $J_\mu A^\mu$ can explicitly be justified for the two-body system starting fron non-relativistic problems and going over to covariant formulation, in the same way as we have arrived at the covariant wave equations [13].

Applications

The application of the coupling (81) to the interaction of composite systems are many [1].
For the minimal coupling of the Majorana-particle, the vertex function (81) (transition amplitudes between any two states) is

$$F_\mu = \langle j'm' | e^{-i\vec{\xi}'\cdot\vec{M}} \Gamma_\mu e^{i\vec{\xi}\cdot\vec{M}} | jm\rangle . \tag{82}$$

In particular, for the spin 0 ground state, writing

$$F_\mu = \frac{1}{2m} G(t)(p' + p)_\mu, \; t = (p' - p)^2 , \tag{83}$$

we obtain the form factor

$$G(t) = (1 - t^2/4m^2)^{-3/2} \tag{84}$$

which is already characteristic of a composite structure. For a spin-1/2 ground state, using the form of F_μ in (22), we find with

$$\begin{aligned} &G_E(t) = f_1 + \frac{t}{4m^2} f_2, \; G_M(t) = f_1 + 2mf_2, \\ &G_E(t) = \frac{1}{\mu} G_M(t) = (1 - t^2/4m^2)^{-3/2} \end{aligned} \tag{85}$$

and

$$\mu = G_M(0) = - 1/2 \; \text{(in units of } e\hbar/2mc)$$

the magnetic moment of the Majorana "particle."
The decreasing mass spectrum (76') and the negative magnetic moment (85) for the Majorana particle are not the expected behavior for a composite system. A generalized Majorana equation, more along the lines of our Eq. (68), changes completely these properties. The generalized equation

$$(\Gamma_\mu P^\mu - \alpha_2 P_\mu P^\mu - \kappa) \psi(P) = 0 \tag{86}$$

can again be solved in the rest frame:

$$M = \frac{1}{2\alpha_2} \left[(j + \tfrac{1}{2}) \pm [(j + \tfrac{1}{2})^2 - 4\kappa\alpha_2]^{\frac{1}{2}} \right]. \tag{87}$$

The minimal coupling to the electromagnetic field gives now also a positive magnetic moment

$$\mu = \tfrac{1}{2} . \quad [15] \tag{88}$$

Properties of O(4, 2) – composite systems

With a view to applications to relativistic H-atom and to hadrons, many properties of the O(4, 2)-systems have been studied in detail over the last decade, based on Eq. (68) and minimal coupling. The aim is to obtain a good overall picture of all properties of hadrons from a single equation and to determine the parameters of the wave equation. These properties are:

(i) Quantum numbers of excited states and mass spectrum

(ii) Decay rates of excited states

(iii) Elastic form factors $G_E(t)$, $G_M(t)$, ... and magnetic moments

(iv) All inelastic form factors and transition moments

(v) Structure functions in deep inelastic scattering

(vi) Electromagnetic polarizabilities.

All of these calculations are so far consistent with experiment for a system like proton (and, of course, H-atom). In particular, the observed *dipole form factor* of the proton is a characteristic property of O(4, 2)-systems and was predicted on this basis (instead of (85)).

14 Other interactions of composite systems

In the unrelativistic quantum theory the interactions of two composite systems (e.g., two H-atoms) will be treated as a many-body (e.g., four-body) problem. In some cases, certain approximations can reduce the process to an effective two-body problem; for example, dipole or van der Waals interactions between atoms. A many-body treatment is always very complicated, even problematical in the relativistic case.

The relativistic infinite-component wave equations for composite systems provide a new procedure to deal with the interactions of composite systems.

In fact, it is possible to develop diagrammatical methods similar to the Feynman diagrams for elementary particle scattering.

Let the infinite-component wave equation be

$$\Lambda\psi = 0 .$$

We introduce a general interaction by writing

$$\Lambda\psi = \Lambda_{int}\psi .$$

The general solution of this equation can be written as

$$\psi = \psi_0 + \Lambda^{-1}\Lambda_{int}\psi$$
$$= \psi_0 + \overline{\psi}_0 \cdot \psi_0\Lambda_{int}\psi \,,$$

where $\Lambda\psi_0 = 0$ and we have used a formal expansion of the Green's function Λ^{-1} in terms of the unperturbed function. The S-matrix is the coefficient of an outgoing solution ψ_0:

$$S = I + \int \overline{\psi}_0 \Lambda_{int}\psi$$

and can be expanded by iteration:

$$S = I + \int \overline{\psi}_0 \Lambda_{int}(\psi_0 + \Lambda^{-1}\Lambda_{int}\psi) \,, \text{ etc.} \ldots$$

Interaction of a composite particle with an elementary particle

In this case ψ_0 consists of the direct product of two states-one composite (double solid line), one elementary (dotted line), and the first two terms of the S-matrix are:

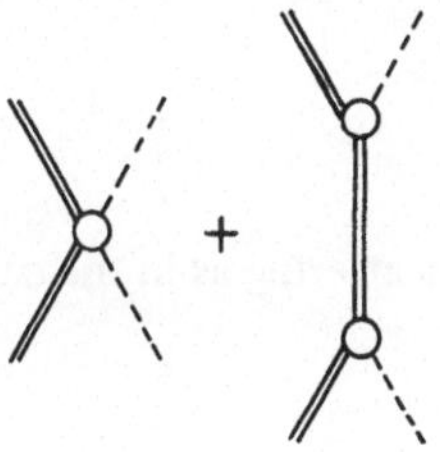

the corresponding amplitudes being

$$\left\{\langle n_3 p_3 | \Lambda_{int} | n_1 p_1 \rangle + \sum_n \langle n_3 p_3 | \Lambda_{int} | np\rangle\langle np | \Lambda_{int} | n_1 p_1\rangle\right\}$$
$$\times \langle p_4 | p_2 \rangle + \text{exchange terms } (3 \Leftrightarrow 4) \,.$$

Here $|np\rangle$ are infinite-multiplet states, including bound and scattering states, and $|p\rangle$ are just free-particle states. Examples of processes of this type are $\gamma H \to \gamma H$ (Compton effect on atoms) or $\gamma p \to \gamma p$ (proton Compton effect), or $\pi p \to \pi p$, if π is treated as an external scalar particle.

It is important to emphasize that a single diagram with infinite-multiplets contains infinitely many diagrams in terms of the constituents. In fact, we shall see that the diagrams above already contain all the salient features of composite particle scattering.

Interactions of two composite systems

In the direct product space of the states of composite systems, we can reduce the amplitudes to the products of transition matrix elements of the form

$F =$ 

$= \langle np \,|\, \Lambda_{int} \,|\, n'p' \rangle.$

These basic matrix elements can be explicitly calculated for a class of interactions, e.g., $\Lambda_{int} = gI$, $\Lambda_{int} = J_\mu A^\mu$. We call the above vertex the *"generalized form factor"* of the composite system. It is a function of the momentum transfer $t = (p' - p)^2$. The functional form is characteristic of the dynamical group G and the representation $L(G)$.

With the help of the generalized form factors F, the lowest order diagrams for the scattering of two systems are

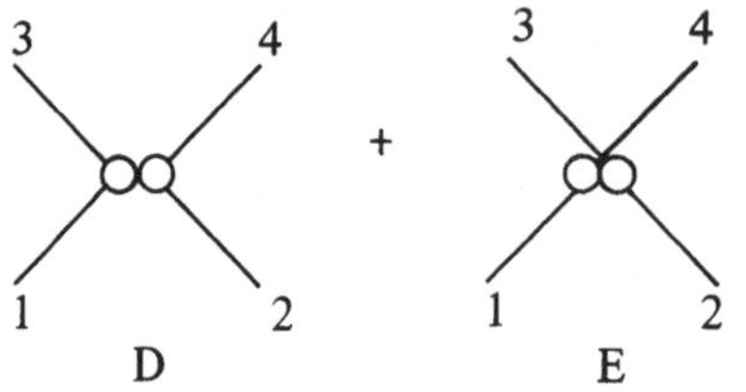

They correspond to direct and exchange scattering as in the case of the interactions of two H-atoms.

The next order terms are

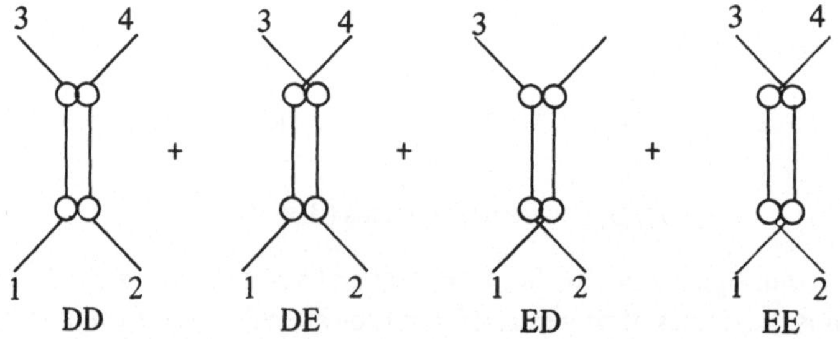

The diagram (EE) is equivalent to (DD) in the case of the factorized vertices as shown. More general nonfactorizable vertices are also possible.

15 A characteristic property of relativistic composite system: Space-like states

Relativistic infinite-component wave equations, such as (68) or (74), have, in addition to the time-like solutions that we have discussed, also space-like and light-like solutions in which the total momentum of the system satisfies $P_\mu P^\mu < 0$, or $P_\mu P^\mu = 0$, respectively. This fact in the past has often been considered to be a difficulty of the infinite-component wave equation. Therefore, we must elaborate and clarify this point. In the nonrelativistic limit of these equations, these space-like solutions move to minus infinity, and in a system like a nonrelativistic H-atom we do not have space-like or light-light solutions.

It turns out, however, that relativistically these additional states can be physically interpreted and play an important role. The reason lies in the "negative energy" solutions of the relativistic particles in the Dirac equation, Eq. (7), $n = -1$.

Now a state of a composite system, in which one of the relativistic constituents has negative energy, can have a space-like or light-like total momentum P_μ. In quantum-electrodynamics, when dealing with a Compton-effect on a bound electron, for example, the summation over all intermediary states must include these states. In this sense, we can give a physical interpretation to space-like solutions of infinite component wave equations [18]. The experimental detection of space-like states is a totally different problem than the usually assumed asymptotic time-like states (which we narrowly call particles) [19].

If negative energy-states are reinterpreted as positive energy antiparticles, in a second quantized theory, then a similar reinterpretation can also be given to space-like states of the composite particles.

Appendix A

List of unitary irreducible representations of the covering group $\overline{SO(2,1)} \sim \overline{SU(1,1)}$

We have denoted a basis of the Lie algebra by Γ_0, Γ_4 and T with the commutation relations (28). Let the eigenvalues of the invariant operator Q^2, Eq. (29), be denoted by

$$Q^2 = \varphi(\varphi + 1) \tag{A.1}$$

The values φ and $(-\varphi - 1)$ give equivalent representations. We have the following classes of unitary irreducible representations:

(i) Discrete series $D_+(\varphi) : Q^2 > 0, \varphi \leqslant -1$.
Eigenvalues of $\Gamma_0 : n = -\varphi, -\varphi + 1, -\varphi + 2, \ldots$. (A.2)

(ii) Discrete series $D_-(\varphi) : Q^2 > 0, \varphi \leqq -1$
$\Gamma_0 : n = \varphi, \varphi - 1, \varphi - 2, \ldots$. (A.3)

(iii) Supplementary series $D_s(\phi, \epsilon) : -\frac{1}{4} < Q^2 < 0; -\frac{1}{2} \varphi < 0$.
$\Gamma : n = \epsilon_0 + s, s = 0, \pm 1, \pm 2, \ldots; |\epsilon_0| < \frac{1}{2} - |\varphi + \frac{1}{2}|$. (A.4)

(iv) Principal series $D_p(\varphi, \lambda, \epsilon): Q^2 < -\frac{1}{4}; \varphi = -\frac{1}{2} + i\lambda, \lambda$ real
$$\Gamma_0 : n = \epsilon_0 + s, s = 0, \pm 1, \pm 2, \ldots -\frac{1}{2} \leqq \epsilon_0 \leqq \frac{1}{2}. \tag{A.5}$$

Here ϵ_0 is a new label occurring in the covering group of SO(2, 1). For the spinor group SO(2, 1) itself:

$$\begin{aligned} &D_s(\phi, \epsilon) \to D_s(\phi, 0) \\ &D_p(\varphi, \lambda, \epsilon_0) \to D_p(\varphi, \lambda, \epsilon_0 = 0 \text{ or } \epsilon_0 = \pm \tfrac{1}{2}). \end{aligned} \tag{A.6}$$

In dynamical problems, we evaluate for the given realization of Γ_0, Γ_4, T the eigenvalue of Q^2 and then select the representation from (A.2) to (A.5).

Appendix B

List of most degenerate unitary irreducible representations of SO (4, 2)

A basis for the Lie algebra of SO(4, 2) was denoted by L_{AB}, Eq. (41). Some unitary realizations were given in Eqs. (40), (43), (47), (49) as well as a nonunitary realization, Eq. (4). We consider first a class of representations for which the second order Casimir operator

$$Q^2_{O(4,2)} = \tfrac{1}{2} L_{AB}L^{AB}, \ A, B = 1 \ldots 6 \tag{B.1}$$

has the following eigenvalues:

$$Q^2 = -3(1 - \mu^2), \ \mu = 0, \pm \tfrac{1}{2}, \pm 1, \ldots. \tag{B.2}$$

These representations are characterized completely by a single "representation relation" [20].

$$\{L_{AB}, L^A_C\} = -2(1 - \mu^2) g_{BC} \tag{B.3}$$

with the metric given in (41). The number μ has the physical interpretation as the minimum of spin (Casimir operator of SO(3) subgroup);

$$|\mu| = j_{min}\,. \tag{B.4}$$

The O(4, 1)-subgroup has the invariant operators

$$\begin{aligned} Q^2_{O(4,1)} &= \tfrac{1}{2} L_{ab}L^{ab}, \ a, b = 1, 2, \ldots 5 \\ &= -2(1 - \mu^2) \end{aligned} \tag{B.5}$$

and

$$W = -\tfrac{1}{64} \epsilon^{abcde} \epsilon_a^{fghi} L_{bc} L_{de} L_{fg} L_{hi} = \mu^2 (1 - \mu^2)$$

which are fixed for fixed μ, showing that these representations of SO(4, 2) remain irreducible when reduced to the subgroup O(4, 1).

However, the invariant operators of the O(4)-subgroup have the spectra

$$\begin{aligned} Q^2_{O(4)} &= \tfrac{1}{2} L_{\alpha\beta} L^{\alpha\beta}, \ \alpha, \beta = 1, 2, 3, 4 \\ &= -(1 - \mu^2) + \Gamma_0^2 \\ P_{O(4)} &= \tfrac{1}{8} \epsilon^{\alpha\beta\gamma\delta} L_{\alpha\beta} L_{\gamma\delta} \\ &= \pm \mu \Gamma_0 . \end{aligned} \qquad (B.6)$$

From these relations one can deduce the representations which have the multiplicity pattern shown below. Here n is the eigenvalue of Γ_0.

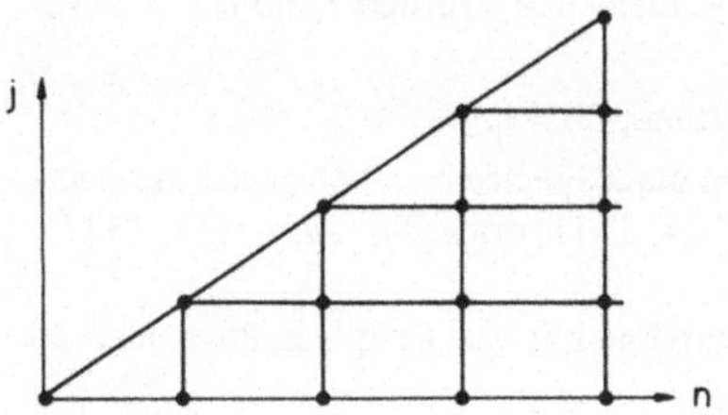

Fig. 3

References

A) Books

For more information on group theoretic background as well as for more details, examples, and references on dynamical groups, see

Wybourne, B. G., Classical Groups for Physicists (John Wiley, New York, 1974), p. 414.

Barut, A. O., Dynamical Groups and Generalized Symmetries in Quantum Theory (University of Canterbury Press, Christchurch, N. Z., 1972), p. 98.

Barut, A. O. and *Raczka, R.*, Theory of Group Representations and Applications (Polish Scientific Publisher, Warsaw, 1977), p. 717.

B) Quoted References

[1] For more detailed discussion on these points and the internal structure of hadrons, cf. *Barut, A. O.*, Review of Hadron Symmetries: Multiplets, Supermultiplets, Infinite Multiplets, in Proceedings of the XVth International High Energy Physics Conference, Naukova Dumka Puhl., Kiew, 1972.

[2] *Barut, A. O.* and *Haugen, R.*, The theory of Conformally Invariant Mass, Ann. of Phys. **71**, 519–541 (1970); Nuovo Cim. **18A** 495–510; 511–531 (1973).

[3] *Wigner, E. P.*, Unitary Representations of the Inhomogeneous Lorentz Group, Ann. Math. **40**, 149–204 (1939).

[4] *Barut, A. O.*, Reformulation of the Dirac Theory of the Electron, Phys. Rev. Lett. **20**, 893 (1968).

[5] *Gel'fand, I. M.* and *Yaglom, A. M.*, General Lorentz Invariant Equations and Infinite-Dimensional Representations of the Lorentz Group, Z. Eksp. Theor. Fiz. **8**, 703–733 (1948).

[6] *Barut, A. O.* and *Bornzin, G.*, Unification of External Conformal Symmetry Group and the Internal Dynamical Group, J. Math. Phys. **15**, 1000–1006 (1974).

[7] *Barut, A. O.*, Conformal Group → Schrödinger Group → Dynamical Group, Helv. Phys. Acta. **46,** 496–503 (1973).

[8] *Barut, A. O.* and *Bornzin, G.*, SO(4, 2) Formulation of the Symmetry Breaking in Relativistic Kepler Problem with or without Magnetic Charges, J. Math. Phys. **12,** 841–846 (1971).

[9] *Gambardello, P. J.*, Exact results in quantum many-body systems of interacting particles in many dimensions with SU(1, 1) on the dynamical group, J. Math. Phys. **16,** 1172–1187 (1975).

[10] *Barut, A. O.*, to be published.

[11] *Cordero, P.* and *Hojman, S.*, "Algebraic solution of a short-range potential problem," Lett. N. C. **4,** 1123 (1970).

[12] *Barut, A. O.*, *Girardello, L.* and *Wyss, W.*, Nonlinear (n + 1)-Symmetric Field Theories, Helv. Phys. Acta. **49,** 807–813 (1976).

[13] *Barut, A. O.* and *Rasmussen, W.*, "H-atom as a relativistic Elementary Particle I and II," J. Phys. **B6,** 1695–1712; 1713–1740.

[14] "Theory of Group Representations and Applications," c.e.a., Ch. 21.

[15] *Barut, A. O.*, *Corrigan, D.* and *Kleinert, H.*, Derivation of Mass Spectrum and Magnetic Moments from Current Conservation in Relativistic O(3, 2) and O(4, 2)-Theories, Phys. Rev. **167,** 1527–1534 (1968).

[16] *Barut, A. O.* and *Duru, H.*, Relativistic Composite Systems and Minimal Coupling, Phys. Rev. **10,** 3448–3454 (1974).

[17] *Barut, A. O.*, "Dilatation of Nonrelativistic Wavefunctions into Relativistic Form," J. Phys. **B8,** L205–208 (1975).

[18] *Barut, A. O.* and *Nagel, J.*, Interpretation of Space-Like Solutions of Infinite-Component Wave Equations and Grodsky-Streater "No-Go", Theorem, J. Phys. **A10,** 1233–1242 (1977).

[19] *Barut, A. O.*, Space-Like States in Relativistic Quantum Theory, in Proc. Erice Conference 1976, edited by E. Recami (North Holland, 1978).

[20] *Barut, A. O.* and *Böhm, A.*, Reduction of a Class of O(4, 2) Representations with Respect to SO(4, 1) and SO(3, 2), J. Math. Phys. **11,** 2938–2945 (1970).

Chapter VII

New Representation Spaces of the Poincaré Group and Functional Quantum Theory

H. Stumpf

Introduction

Group theory and the structure of linear metrical spaces used in quantum theory are in close connection. This is established by the theorem that a linear selfadjoint operator being forminvariant with respect to a symmetry group has eigenstates which must be base states of the corresponding representations of this group. Since the quantum observables have to be represented by selfadjoint operators and since the infinitesimal generators of a symmetry group are selfadjoint, it follows that they have to be themselves quantum observables. From the set of symmetry observables a subset of complete compatible observables can be chosen which fixes the structure of the corresponding representation space. Thus, all results and calculation methods of quantum theory depend strongly upon the representation spaces of symmetry groups under consideration. For example, the representation spaces of the Poincaré group which were first investigated by Wigner and Bargmann [1, 2] and which are used in ordinary quantum field theory lead in connection with the principle of microcausality and locality to divergent results. Out of the various attempts to remove these defects we discuss those which are closely connected with the introduction of new representation spaces of the Poincaré group.

Studying the Lee-model, Heisenberg introduced so-called dipole ghosts in order to regularize the field theory of this model [3]. Guided by these results, Heisenberg proposed to regularize an unrenormalizable selfcoupled spinor field by dipole ghosts in order to obtain a model for a field theory of matter which is free of divergencies [4]. But the introduction of such dipole ghosts into a field theory is equivalent to the use of corresponding new representation spaces of the Poincaré group which are nonunitary and that destroys almost completely all calculation methods of ordinary quantum field theory. This effect of a new representation space in quantum field theory is not restricted to the dipole ghost situation but occurs also in other cases.

The fact that the type of representation space influences the mathematical as well as the physical content of a theory has been discovered on the physical level by studying various types of field equations. As any representation space has its genuine field equations, cf. [5], new representations are found by the investigation of various types of field equations. So Bopp [6] and Podolski [7] discovered the regularizing effect of a higher order field equation on the classical electron mass. On the quantum level the Pauli-Villars regularization [8] really corresponds to the use of higher order field equations and the dipole ghost regularization of Heisenberg is a special case of a Pauli-Villars regularization.

For scalar fields Froissart [9] derived higher order free field equations leading to dipole ghosts and higher order ghosts. Dürr [10] studied corresponding nonlinear spinor field equations. Since nonunitary representations are embedded in indefinite linear spaces the use of such representations is equivalent to the introduction of indefinite metric. We do not give here a complete account of these results but refer to literature [11], [12]. In particular we do not discuss here gauge theories which also contain indefinite metric, cf. [13], but which in general are not regularized by it. We only emphasize that regularizing Poincaré representations, higher order field equations and indefinite metric are in very close connection.

In general, these cases cannot be treated by means of the calculational methods which are adapted to the Bargmann-Wigner representations. Hence, methods to deal with more general representations in quantum field theory are required. Such methods were developed by Stumpf and coworkers and we call this new version of quantum theory a functional quantum theory. It can be applied to nonunitary Poincaré representations as well as, of course, to ordinary quantum theory with unitary representations, provided it gives finite results. Hence, functional quantum theory is an attempt to work with field equations of higher order, i.e. to go beyond the scope of conventional quantum field theory. To have a definite model we shall discuss functional quantum theory for the example of the nonlinear spinor field which has special advantages as will be demonstrated in section 5.

1 Quantum theoretical fundamentals

In order to understand the construction of functional quantum theory, it is necessary to discuss the general fundamentals of conventional quantum theory. We consider the class of physical systems which allow a quantum theoretical description. Because the quantum theory is a statistical theory, each system has to be associated with an ensemble of systems of the same kind. Here and in the following we use the term "system" as an abbreviation for the associated ensemble. The physical information about any system of this class can be gained from the set of observables R attached to the system. For simplicity we assume that we deal with an enumerable set of observables $R := \{R_k, k = 1 \ldots\}$. In contrast to the classical theory the observables are not numbervalued functions, but operatorvalued functions. Their operator properties are determined by (anti-)commutator relations. For $R, R' \in R$ one has in general $[R, R']_\pm \neq 0$, where in so-called canonical theories the (anti-)commutators are determined in a unique way by the quantization rules.

For the realization of the observables, introduced in an algebraic way, one uses linear spaces. The elements of these linear spaces are called state vectors. From the statistical interpretation follows that one has to deal with Hilbert spaces. The Hilbert space H is spanned by the set of base states $\{|\varphi_l\rangle \in H, l = 1, 2, \ldots\}$ and has (besides others) the following properties:

With $|\varphi_l\rangle$ and $|\varphi_m\rangle \in H$, we have

$$\begin{aligned} &\alpha)\ |\varphi_l\rangle + |\varphi_m\rangle = |\varphi\rangle \in H \\ &\beta)\ a|\varphi_l\rangle = |\varphi_l\rangle' \in H, a \in \mathbb{C}. \end{aligned} \tag{1.1}$$

That means, the elements of *H* can be added and they can be multiplied by complex numbers. Furthermore, a positive definite bilinear form can be defined which serves as a scalarproduct and has the property

$$\gamma)\ \langle\varphi|\psi\rangle = \langle\psi|\varphi\rangle^{x} \in \mathbb{C}, |\psi\rangle, |\varphi\rangle \in H\,.$$

The states of such a Hilbert space do not only have a mathematical meaning. The Hilbert space *H* we need for the representation of the set of observables associated with a system, characterizes the physical states of the system by its vectors. The characterization of a physical state for an ensemble means to specify the frequency of the observed values of the observables. Depending on the preparation of the system, one distinguishes between pure and mixed states. A state is called pure if there is a complete set of simultaneously measurable observables with values which are without fluctuations in this state. If such a set of observables cannot be found for a given state, the state is called mixed.
We restrict our discussion to the pure states, because only these states are of importance for our considerations. Let $|\varphi\rangle$ be a pure state. Then in quantum statistics the mean value of an observable R is defined by

$$\bar{R}_{\varphi} = \langle\varphi|R|\varphi\rangle\,. \tag{1.2}$$

One can choose in *H* such base systems that certain operators of physical interest become diagonal. Take $|\varphi_n\rangle$ as an eigenvector of the operator R. Then a number r_n exists, that

$$R|\varphi_n\rangle = r_n|\varphi_n\rangle \tag{1.3}$$

is valid. One gets for the mean value:

$$\bar{R}_{\varphi_n} = r_n\langle\varphi_n|\varphi_n\rangle = r_n \tag{1.4}$$

with $|\varphi_n\rangle$ normalized to unity.
For the mean square deviation one gets

$$\langle\varphi_n|(R - r_n)^2|\varphi_n\rangle = 0\,. \tag{1.5}$$

The eigenvalues r_n of R are with absolute certainty the result of a measurement of the physical quantity corresponding to R, if the system is in the state $|\varphi_n\rangle$. For physical reasons the results of measurements and therefore the eigenvalues have to be real. From this follows that all operators which are interpreted as observables must be selfadjoint.
Analogous to the classical theory, the sharp measurements in quantum theory are used for the definition of the quantum theoretical event space. It must be specified, how many sharp measurements are necessary and sufficient for the unique characterization of the event space of a certain system. In general, one needs several observables $R_1 \ldots R_n$. If their eigenvalues are simultaneously measurable, then a base $|\varphi_1\rangle \ldots$ exists which simultaneously diagonalizes the operators $R_1 \ldots R_n$. This is only possible if all R_i mutually commute. If all other observables which commute with the set $\{R_i\}$ are functions of the R_i, the set $\{R_i\}$ of observables is called complete. For a complete set $\{R_i\}$, only the eigenvalues r_n^i can be measured simultaneously. The set $\{r_n^i\}$ of possible measurement results, therefore, determines the space of events. Because the eigenstates $\{|\varphi_i\rangle\}$ are without fluctuation with respect to the complete set of observables $\{R_i\}$, the base system of eigen-

states is a system of pure states. The Hilbert space *H* can therefore be spanned by the set of all pure states of a system. But in general the observables $R_1 \dots R_n$ of a system are not sufficient to determine the state of a system uniquely, because the pure states are degenerate with respect to $R_1 \dots R_n$, i.e. for one point $r_m^1 \dots r_m^n$ in the space of events several pure states exist. This degeneration is by no means trivial or uninteresting, because by means of these states the special dynamic of the system becomes effective as is shown in the following.

The first question we put is, how to construct a complete set of observables $R_1 \dots R_n$ for a given system. The foundation for the construction of maximal commuting sets of observables is constituted by the transformation theory. Such a transformation theory is necessary to calculate the change of physical measuring results, if the observer changes his frame of reference. For each theory, there is a class of equivalent, special frames of reference. The set of transformations which connect such reference frames constitutes a group, the so-called symmetry group of the system. Formally, a symmetry group can be characterized by the fact that the dynamical equations of a system are forminvariant under the action of the group elements. We confine ourselves to the space-time symmetries, because they are, together with some other conditions, fundamental for the construction of quantum theories. For nonrelativistic systems, the corresponding symmetry group is the Galilei group, for a relativistic quantum theory the Poincaré group. Because Heisenberg's nonlinear spinor equation is a relativistic equation, only the Poincaré group will be discussed in the following.

The change from the frame S to the frame S′ in Minkowski space is described by the Poincaré transformation (Λ, d)

$$x'_\mu = \Lambda^\nu_\mu x_\nu + d_\mu \tag{1.6}$$

with respect to the four coordinates $\{x_\mu, 1 \leqslant \mu \leqslant 4\}$. The observation in the frame S yields a certain physical state characterized by a state vector $|\varphi\rangle$ in accordance with the rules of quantum theory. For the same physical state an observation in the frame S′ gives another state vector $|\varphi\rangle'$. Because of the postulated equivalence of the frames S and S′, the physical relevant quantities have to remain unchanged. This fact is expressed by the forminvariance of the equations of motion. The dynamics of a system in quantum theory is characterized sufficiently by the set of eigenvalue equations

$$(R_\alpha - r_n^\alpha)\,|\varphi_n\rangle = 0,\ 1 \leqslant \alpha \leqslant n \tag{1.7}$$

in the frame S. In S′ the equations read

$$(R'_\alpha - r_n^{\alpha'})\,|\varphi_n\rangle' = 0,\ 1 \leqslant \alpha \leqslant n\,. \tag{1.8}$$

For simplicity, we only discuss scalar observables. In this case the requirement of forminvariance leads to the conditions $R_\alpha = R'_\alpha\ r_n^\alpha = r_n^{\alpha'}$. Therefore, $|\varphi_n\rangle'$ must be a solution of (1.7). But then it is possible to expand $|\varphi_n\rangle'$ in a series in the subspace of eigensolutions of (1.7) and the expansion coefficients depend on the special transformation (1.6) from S to S′. Therefore we have

$$|\varphi_n\rangle' = U(\Lambda, d)\,|\varphi_n\rangle\,. \tag{1.9}$$

Because in both frames the same statistical rules have to be valid, the operator $U(\Lambda, d)$ has to be a unitary operator in H. The space of states therefore is a representation space for the unitary representations of the Poincaré group. Substitution of (1.9) in (1.8) yields $R'_\alpha = U^{-1} R_\alpha U$, $1 \leqslant \alpha \leqslant n$ and because of the forminvariance $[R_\alpha, U] = 0$, i.e. the scalar observables commute with the group operations. Taking only into account nontrivial observables, i.e. neglecting all that does not depend on the space degrees of freedom, the only operators commuting with all $U(\Lambda, d)$ are the Casimir operators of the representation, which in the case of the Poincaré group are denoted by $\mathbb{P}^2$ and $\mathbb{\Gamma}^2$. Therefore, the Casimir operators are identical with the scalar physical observables. They are of mathematical and physical significance. They classify the irreducible representations of the Poincaré group, on the other hand, they are numbers in an irreducible representation and represent mass and spin of a system, e.g. of an elementary particle, but just as well of an arbitrarily complex relativistic system.

Additionally, one can show that together with mass $\mathbb{P}^2$ and spin $\mathbb{\Gamma}^2$ the total momentum $\mathbb{P}$ and the spin projection $\mathbb{S}_3$ can be measured simultaneously, i.e. the corresponding operators are diagonal in an irreducible representation, So we have the eigenvalue equations

$$\begin{aligned} &\mathbb{P}\,|\varphi\rangle = \mathbf{p}\,|\varphi\rangle;\ \ \mathbb{P}^2\,|\varphi\rangle = m^2\,|\varphi\rangle \\ &\mathbb{S}_3\,|\varphi\rangle = s_3\,|\varphi\rangle;\ \mathbb{\Gamma}^2\,|\varphi\rangle = s(s+1)\,|\varphi\rangle. \end{aligned} \tag{1.10}$$

With respect to the Poincaré group, the complete set of commuting observables is given by $\{\mathbb{P}, \mathbb{S}_3, \mathbb{P}^2, \mathbb{\Gamma}^2\}$.

We now come to the construction of a complete state space. As already mentioned above, due to the degeneracy, a complete set of simultaneously measurable observables is not sufficient for the construction of a complete state space. The eigenvalue equations (1.10) are only necessary, but not sufficient conditions for the characterization of this state space. To characterize the state space completely, the states $|\varphi\rangle$ have to satisfy additional boundary conditions. These boundary conditions lead to the distinction between scattering- and boundstates. For the boundstates, we have a one-to-one correspondence between the set of measurable quantum numbers and the states. With respect to the scattering states, the boundary conditions give an additional information, not contained in the set of the quantum numbers, and thus giving the degeneracy a physical meaning. This degeneracy describes the freedom we have in the experimental preparation of initial states and the measurement of final states. The initial and final states are characterized by state vectors $|\varphi^\pm(n)\rangle$ where n denotes the initial respectively final configuration. They satisfy the equations (1.10), but as an additional relevant observable the S-matrix is introduced:

$$S_{nm} := \langle\varphi^-(n)\,|\,\varphi^+(m)\rangle \tag{1.11}$$

which characterizes the result of scattering processes.

Summarizing it can be said that a complete characterization of the state space is only possible my means of the set of commuting observables and the S-matrix. These types of observables are called global observables, because they are connected with global properties of the system. It is interesting to note that this mathematical completeness, resulting from eigenvalues and boundary conditions, is also physically sufficient. As Heisenberg has pointed out, the global observables are sufficient to describe microscopic experimental observations [14]. Therefore, both for physical and mathematical reasons the discussion can be restricted to the global observables of a system.

2 Poincaré group representations in field theory

The general considerations of section 1 shall now be applied to quantum field theory. To obtain the connection with the treatment of the nonlinear spinor field in the following sections we consider a general quantized spinor field operator $\psi_\alpha(x)$. To work successfully with this operator, it has to be explained in which linear spaces this operator is defined. A general answer can be given by the Lehmann-Källen theorem [15]. As the symmetry group is the Poincaré group, $\psi_\alpha(x)$ is defined in a corresponding representation space of this group. Due to the general transformation rules of section 1 we have for a Poincaré transformation

$$\psi'_\alpha(x') = L_{\alpha\beta}(\Lambda)\psi_\beta(\Lambda^{-1}(x+d)) = U^{-1}(\Lambda, d)\,\psi_\alpha(x)\,U(\Lambda, d)\,. \tag{2.1}$$

Considering now one-particle states $|\,p, m, \sigma\rangle$ of momentum p, mass m and spin projection σ, which are assumed to be eigenstates of the set of observables (1.10), we obtain by means of (2.1) and (1.9)

$$\langle 0\,|\,\psi_\alpha(x)\,|\,p, m, \sigma\rangle = u_\sigma(p)\,e^{-ip\,x} := \varphi(x\,|\,p, m, \sigma), \tag{2.2}$$

where $|\,0\rangle$ is the invariant groundstate of the theory.
We confine ourselves to the case $p_0 > 0$, therefore u_σ is a positive frequency spinor. Now the Lehmann-Källen theorem tells us that

$$\begin{aligned}\langle 0\,|\,\psi_\alpha(x)\,\bar\psi_\beta(x)\,|\,0\rangle &= \sum_{p,m,\sigma} \varphi(x\,|\,p, m, \sigma)\,\bar\varphi(x'\,|\,p, m, \sigma)\\ &= \int dm^2\,[\rho_1(m^2)\gamma^\mu\partial_\mu + \rho_2(m^2)]\,\Delta^+(x-x', m^2)\end{aligned} \tag{2.3}$$

holds, with $\rho_1(m^2)$, $\rho_2(m^2) \geqslant 0$ and with

$$\begin{aligned}\Delta^+(x-x', m) &= -\frac{i}{(2\pi)^3}\int d^4k\,\Theta(k_0)\,\delta(k^2-m^2)\,e^{-ikx}\\ &= \frac{1}{(2\pi)^4}\int d^3k \int_{C^+} dk_0\,\frac{e^{ikx}}{k^2-m^2}\end{aligned} \tag{2.4}$$

as the positive frequency solution of the homogeneous K-G-equation.
The spectral representation (2.4) can be derived on the sole assumption that the underlying representation space is positive definite. Similar spectral decompositions can be performed for expectation values of n field operators for $n > 2$. To introduce the new Poincaré representations it is, however, sufficient to discuss the two-point function (2.3)

respectively its generalizations. The matrix elements (2.2) which occur in (2.3) satisfy the equations

$$(-i\gamma^\nu \partial_\nu + m)\langle 0 | \psi(x) | p, m, \sigma\rangle = 0 , \qquad (2.5)$$

From the group theoretical point of view these equations are the characteristic equations for the Poincaré representations which occur. They have nothing to do with the system dynamics, but are a purely kinematical group theoretical consequence. The effect of the system dynamics, however, is to mix such representations and to select certain representations in the course of the time evolution of the system.

Performing dynamical calculations, especially S-matrix calculations, one recognizes that for nontrivial models in the course of calculation just (2.3) or strongly related expressions occur and that just these expressions lead to divergencies. To avoid these divergencies, Heisenberg [4] proposed to use instead of (2.3) an expression which contains additional good ghosts and dipole ghosts. Such expressions lead to finite values in various calculations but they simultaneously produce an indefinite state space. We demonstrate this in more detail.

We assume $\Psi_\alpha(x)$ to be the Hermitean spinor field operator and the set of states $\{|\rho\rangle\}$ to be the one-particle fermion sector of the nonlinear field including ghost and dipole ghost states. Then we have the relations

$$\langle \rho' | \rho \rangle = g_{\rho'\rho} \qquad (2.6)$$

and

$$\langle 0 | \Psi_\alpha(x') \Psi_\beta(x) | 0 \rangle =: \overset{+}{F}_{\alpha\beta}(x', x) \qquad (2.7)$$

where $g_{\rho'\rho}$ is the metrical tensor of the one-particle fermion sector and $|0\rangle$ is the groundstate of the field. If we further define

$$\langle 0 | \Psi_\alpha(x) | \rho \rangle =: \varphi_\alpha(x|\rho) \qquad (2.8)$$

and assume the expansion

$$|\rho\rangle = \int \sigma^\alpha(x|\rho) \Psi_\alpha(x) |0\rangle \, d^4x + \ldots \qquad (2.9)$$

then we obtain

$$\varphi_\beta(x|\rho) = \int \overset{+}{F}_{\beta\alpha}(x', x) \sigma^\alpha(x|\rho) \, d^4x + \ldots . \qquad (2.10)$$

The dots in (1.4) and (1.5) indicate that in general higher order contributions have to be used in order to describe the state correctly, cf. [16]. Here we only consider the "free field" version of the propagator and omit these dots. Then the following theorem holds:

Theorem: If the sets $\{\Psi_\alpha(x)|0\rangle\}$ and $\{|\rho\rangle\}$ are equivalent, i.e. if the linear space L_1 of the one-particle fermion sector is given by $L_1 = lh\{\Psi_\alpha(x)|0\rangle\} = lh\{|\rho\rangle\}$, then for the two-point function $\overset{+}{F}_{\alpha\beta}(x, x')$ the spectral decomposition

$$\overset{+}{F}_{\alpha\beta}(x, x') = \sum_{\rho\rho'} \varphi_\alpha(x|\rho)^{x} g^{\rho\rho'} \varphi_\beta(x'|\rho') \qquad \text{holds.} \qquad (2.11)$$

Proof: Due to the equivalence not only (1.4) is valid but also the inversion

$$\Psi_\alpha(x)|0\rangle = \sum_\rho b_\alpha(x|\rho)|\rho\rangle . \tag{2.12}$$

From this follows due to the Hermitecity of the field

$$\langle\rho'|\Psi_\alpha(x)|0\rangle = \varphi_\alpha(x|\rho)^x = \sum_\rho b_\alpha(x|\rho)\, g_{\rho'\rho} \tag{2.13}$$

and

$$\overset{+}{F}_{\beta\alpha}(x', x) = \sum_\rho b_\alpha(x|\rho)\,\langle 0|\Psi_\beta(x')|\rho\rangle = \sum_\rho \varphi_\beta(x'|\rho)\, b_\alpha(x|\rho) . \tag{2.14}$$

By definition we have

$$\sum_{\rho'} g^{\lambda\rho'} g_{\rho'\rho} = \delta^\lambda_\rho \tag{2.15}$$

and applying this to (1.8) we obtain

$$b_\alpha(x|\lambda) = \sum_{\rho'} g^{\lambda\rho'} \varphi_\alpha(x|\rho')^x . \tag{2.16}$$

Substitution of (2.16) into (2.14) gives (2.11), q.e.d.
The quantum number ρ of the one-particle fermion states contains the momentum and the spin quantum numbers and additionally the specification of the particles by mass etc. Only the latter is of interest, as the momentum and spin parts are always diagonal in the model considered. Hence we have $g_{\rho\rho'} = 1\!\!1 \otimes \widetilde{g}_{rr'}$, with $|\rho\rangle = |p_\rho, s_\rho, r\rangle$. In the Lee-model version of Heisenberg $g_{rr'}$ is given by $\left(\begin{smallmatrix}0&1\\1&0\end{smallmatrix}\right)$ describing a good ghost and a dipole ghost. Assuming that this couple should accompany a physical fermion the metrical fundamental tensor has to have the structure

$$\widetilde{g}_{rr'} = \begin{pmatrix} 1 & 0 & 0 \\ 0 & 0 & 1 \\ 0 & 1 & 0 \end{pmatrix} \tag{2.17}$$

Unfortunately, the spinor field propagator proposed by Heisenberg has not this simple structure but rather a more complicated one, since additional bad ghosts and parity mixtures appear [4]. In order to get a clear insight into the mechanism of unitarization and to obtain a parity invariant theory, we use a modified spinor field propagator which only contains that minimum of indefiniteness being acceptable for a successful solution of the problem. We consider the general propagator

$$\begin{aligned} F(x-x') = {} & \frac{1}{(2\pi)^4} \int \rho_1(m'^2)\, \frac{(-\not p + \mu)^2 (-\not p + m')}{(p_2^2 - \mu^2)^2 (p^2 - m'^2)}\, e^{ip(x-x')}\, dm'^2\, d^4p \\ & + \frac{1}{(2\pi)^4} \int \rho_2(m'^2)\, \frac{(-\not p - \mu)^2 (-\not p - m')}{(p^2 - \mu^2)^2 (p^2 - m'^2)}\, e^{ip(x-x')}\, dm'^2\, d^4p \end{aligned} \tag{2.18}$$

If we only admit one parity by putting $\rho_2 = 0$ and if we assume $\rho_1(m'^2) = \delta(m^2 - m'^2)$ then the propagator (2.18) corresponds to the propagator of a free field $\chi(x)$ with the equation

$$(-i\gamma^\nu\partial_\nu + \mu)(-i\gamma^\mu\partial_\mu + m)(-i\gamma^\mu\partial_\mu + \mu)\chi(x) = 0 \tag{2.19}$$

For this field we have

$$F(x-x') = \langle 0|T\chi(x)\bar{\chi}(x')|0\rangle = \frac{1}{(2\pi)^4}\int \frac{(-\not{p}+\mu)(-\not{p}+m)(-\not{p}+\mu)}{(p^2-\mu^2+i\epsilon)^2(p^2-m^2+i\epsilon)} e^{ip(x-x')}d^4p \tag{2.20}$$

with

$$(-i\gamma^\nu\partial_\nu + \mu)(-i\gamma^\mu\partial_\mu + m)(-i\gamma^\nu\partial_\nu + \mu)F(x-x') = -\delta^4(x-x') \;. \tag{2.21}$$

The length dimension of $\chi(x)$ is $-\frac{1}{2}$, i.e. the local interaction is renormalizable and additionally the field is noncanonical as in contrast to ordinary canonical quantum theory the anticommutator is given by

$$[\chi(x)\bar{\chi}(x')]_{+/x_0 = x_0'} = 0 \;. \tag{2.22}$$

This noncanonical property of the anticommutator (2.22) can be directly derived from the set of first order equations

$$\begin{aligned} &(-i\gamma^\nu\partial_\nu + \mu)\psi_1(x) = 0 \\ &(-i\gamma^\mu\partial_\mu + m)\psi_2(x) = \psi_1(x) \\ &(-i\gamma^\nu\partial_\nu + \mu)\psi_3(x) = \psi_2(x) \end{aligned} \tag{2.23}$$

which is equivalent to equation (2.19) if we put $\chi(x) \equiv \psi_3(x)$. If the common Lagrange formalism of canonical quantization is applied to the fields $\psi_i(x)$ and their canonical conjugates $\Pi_i(x)$, then for $\chi(x)$ itself the noncanonical relation (2.22) results. We do not give here a detailed account of this field with a higher order field equation, but rather consider the properties of the propagator. From (2.20) follows

$$F^+(x-x') \equiv \langle 0|\chi(x)\bar{\chi}(x')|0\rangle = \frac{1}{(2\pi)^4}\int_{C^+} \frac{(-\not{p}+\mu)^2(-\not{p}+m)}{(p^2-\mu^2)^2(p^2-m^2)} e^{ip(x-x')}d^4p \tag{2.24}$$

We decompose the fraction in (2.24) into partial fractions and obtain

$$\frac{(-\not{p}+\mu)^2(-\not{p}+m)}{(p^2-\mu^2)^2(p^2-m^2)} = -\frac{1}{(\mu-m)^2}\frac{(-\not{p}+\mu)}{(p^2-\mu^2)} - \frac{1}{(\mu-m)}\frac{(-\not{p}+\mu)^2}{(p^2-\mu^2)^2} + \frac{1}{(\mu-m)^2}\frac{(-\not{p}+m)}{(p^2-m^2)} \tag{2.25}$$

Substitution of (2.25) into (2.24) and integration over p_0 gives by observing

$(p-\mu) = 2\mu \sum_{s=1}^{2} v_\alpha^s \bar{v}_\beta^s$ the expression

$$
\begin{aligned}
F^+(x-x') = &- \frac{1}{(\mu-m)^2}\sum_s \int \frac{d^3p}{\omega(m)}\, m\, \langle 0|\chi(x)|p,s,n\rangle \langle p,s,n|\bar{\chi}(x')|0\rangle \frac{i}{(2\pi)^3} \\
&+ \frac{1}{(\mu-m)}\sum_s \int \frac{d^3p}{\omega(\mu)}\, \mu\, \langle 0|\chi(x)|p,s,g\rangle \langle p,s,d|\bar{\chi}(x')|0\rangle \frac{i}{(2\pi)^3} \\
&+ \frac{1}{(\mu-m)}\sum_s \int \frac{d^3p}{\omega(\mu)}\, \mu\, \langle 0|\chi(x)|p,s,d\rangle \langle p,s,g|\bar{\chi}(x')|0\rangle \frac{i}{(2\pi)^3}
\end{aligned}
\tag{2.26}
$$

with $\omega(m) := (p^2 + m^2)^{1/2}$ and

$$
\begin{aligned}
\langle 0|\chi(x)|p,s,n\rangle &:= v^s(p)\, e^{ip(m)x} \\
\langle 0|\chi(x)|p,s,g\rangle &:= v^s(p)\, e^{ip(\mu)x}
\end{aligned}
\tag{2.27}
$$

$$
\langle 0|\chi(x)|p,s,d\rangle := \left[\frac{1}{\mu-m} + \frac{1}{2\omega(\mu)^2} + \frac{\gamma^0}{2\omega(\mu)} - \frac{ix_0}{\omega(\mu)}\right] v^s(p)\, e^{ip(\mu)x} .
$$

In this case the metrical tensor $g_{\rho\rho'} \equiv g(psr, p's'r')$ can be divided into a diagonal and a nondiagonal part with

$$
g(p,s,r,p',s',r') = \delta_{pp'}\delta_{ss'}\tilde{g}_{rr'} , \tag{2.28}
$$

where $\tilde{g}_{rr'}$ is given by (2.17) with $r \equiv n, g, d$; i.e. a physical fermion ($\equiv$ nucleon), a good ghost and a dipole ghost. The scalar products can be written

$$
\tilde{g}_{rr'} = \begin{pmatrix} 1 & 0 & 0 \\ 0 & 0 & 1 \\ 0 & 1 & 0 \end{pmatrix} = \begin{pmatrix} \langle n|n\rangle & \langle n|g\rangle & \langle n|d\rangle \\ \langle g|n\rangle & \langle g|g\rangle & \langle g|d\rangle \\ \langle d|n\rangle & \langle d|g\rangle & \langle d|d\rangle \end{pmatrix} . \tag{2.29}
$$

This means that $\langle g|g\rangle = 0 = \langle d|d\rangle$ and that the physical part of $\tilde{g}_{rr}$ can be defined by

$$
(\tilde{g}_{rr'})_{Ph.} := \begin{pmatrix} \langle n|n\rangle & \langle n|g\rangle \\ \langle g|n\rangle & \langle g|g\rangle \end{pmatrix} = \begin{pmatrix} 1 & 0 \\ 0 & 0 \end{pmatrix} \tag{2.30}
$$

as the good ghost states do not contribute to the norm.

As can be easily seen, the one-particle matrix elements (2.27) are not solutions of the first order equation (2.6), but are solutions of the third order equation

$$
(-i\gamma^\mu\partial_\mu + m)(-i\gamma^\nu\partial_\nu + \mu)^2 \langle 0|\chi(x)|\rho\rangle = 0 . \tag{2.31}
$$

Thus, instead of the solutions of (2.5) the representation space is now constructed by solutions of (2.31). Due to the special integration procedure applied for the evaluation of (2.24) the solutions (2.27) are not correctly covariant functions. But the state representations can be calculated directly by means of (2.24) and give completely covariant base functions. Additionally, also the integration procedure can be modified to give relativistic correct results; this will not be discussed here in detail.

Summarizing the results of this discussion, we notice that the dipole regularized propagator possesses a Lehmann spectral representation which leads to indefinite metric in the quantum state space. But this has a serious consequence from the group theoretical point of view. Because of the required invariance of the theory under Poincaré transformations, the corresponding representations are nonunitary and the theory cannot be interpreted. To restore the physical content of the theory, one has to perform a procedure called "unitarization". This is well-known in quantum field theories, where two types of indefinite metric appear. In quantum electrodynamics the indefinite metric is used to preserve explicit relativistic invariance and leads to the Gupta-Bleuler formalism. From the group theoretical point of view in this case the indefinite metric appears as a consequence of the utilization of a nonunitary representation of the Poincaré group, caused by the requirement that A_μ has to transform like a 4-vector. In this case, the unitarization procedure is easy, because the corresponding representation space is explicitly known. Then the space of physical states can be constructed by forming the factor space modulo the ghost states.

In the second case, indefinite metric is used to avoid infinities and influences deeply the dynamical properties of the theory, and no general rule exists to eliminate unphysical features. A review of various attempts in this field is given by Nagy [11] and Nakanishi [12]. Out of the various concepts to regularize divergent quantum field theories by indefinite metric, there is a special concept introduced by Heisenberg, namely the dipole ghost regularization of the Lee-model. But the extension of the dipole ghost regularized Lee-model and of subsequent other comparable models e.g. of Froissart, Nagy etc. to nonlinear spinor theory is not straight-forward. While in all these models Fockspace methods are applied, nonlinear spinor theory has to be evaluated beyond the Fock representation by using coordinate free general relativistic calculation methods, e.g. the new Tamm-Dancoff method or the Green function method in order to avoid physically inadmissible restrictions. Combining these methods with dipole regularization, it can be shown that serious drawbacks result:

i) Because of noncanonical quantization the derivation of the prepared Green functions hierarchy breaks down. [17]
ii) The derivation of the S-matrix from the Green functions is not possible as in LSZ formalism.
iii) For unitarization explicit state representations are required.

For the removal of these serious drawbacks the present axiomatic quantum field theory gives no hint. Neither Heisenberg nor his coworkers made any proposals to overcome these difficulties. Therefore, if one wishes to work with the new representations given above, one has to observe that the conventional approaches fail and that they require a completely new approach to quantum theory. This new approach is given by the functional quantum theory of the nonlinear spinor field developed by Stumpf and coworkers [16, 18–26]. Summarizing the results of this approach it can be said that functional quantum theory is a method to calculate the global observables, i.e. quantum numbers and the S-matrix for any quantum system, provided the theory leads to no divergencies. Although especially developed for the needs of the dipole regularized spinor theory, it can be also applied to other models. In brevity, it consists in the direct construction of the state space of the corresponding system beyond the Fockspace representation and it explicitly solves the problems i)–iii). In the following, a short review of the present state of functional quantum theory is given which explains explicitly how to work with the new Poincaré representations.

3 Functional map

As has already been mentioned, functional quantum theory is universally applicable. We discuss its application to the nonlinear spinor field, regularized by ghosts and dipole ghosts, i.e. by noncanonical quantization or equivalently by the use of indefinite metric. This theory is the attempt for an unified microscopic description of matter. The starting point is the spinor field equation

$$D_\alpha^\beta(x)\,\Psi_\beta(x) + V_\alpha^{\beta\gamma\delta} : \Psi_\beta(x)\,\Psi_\gamma(x)\,\Psi_\delta(x) : = 0\,. \tag{3.1}$$

$\Psi_\alpha(x)$ denotes a hermitean spinor field operator, $D_\alpha^\beta(x)$ a hermitean Dirac operator with $m = 0$, and $V_\alpha^{\beta\gamma\delta}$ a general Vertex operator [27]. We discuss the definition and calculation of the observables in this theory by means of functional quantum theory.

To evaluate equation (3.1) in the sense of general quantum theory of fields, it is not allowed to use a formulation in the framework of coupling theories, as e.g. to use the interaction representation.

Coupling theories use the masses of elementary particles and their mutual coupling constants as the starting point for the calculation of scattering cross sections. In contrast to this, nonlinear spinor theory only postulates the nonlinear equation (3.1), from which all informations about elementary particles have to be extracted, especially masses and cross sections. The calculations have to be performed under the strict condition: the calculation method has to be relativistically invariant.

Generally, there exist two complementary calculation methods: the algebraic method and the method of states. The advantage of the algebraic method is its explicit invariance under the symmetry groups, the advantage of the method of states is the possibility to incorporate the boundary conditions. But the advantage of one method is just the disadvantage of the complementary method. Thus, there does not exist a general explicit relativistic invariant method for the calculation of states, on the other hand, the algebraic construction of the field operators under commutation and boundary conditions for the equation (3.1) is an unsolved problem. One attempt to resolve these problems is the introduction of invariant field functionals with the corresponding functional equations by Schwinger [28]. In a sense this attempt is a synthesis between the algebraic method and the method of states and allows relativistically invariant calculations. For coupling theories, the Schwinger functionals (Green functions) contain all information needed. A coupling theory can be regarded as being resolved, when the Green functions are known. As discussed above, this does not apply to the nonlinear spinor theory. This failure enforces the development of functional quantum theory.

Before discussing the formal aspects, it is useful to realize the physical content of the theory. There exists a nonrelativistic analogue to this theory which is thoroughly understood. This analogon is the field model of nucleons and nuclear forces. The field equation of this model reads

$$\left(i\frac{\partial}{\partial t} + \Delta\right)\psi_\rho(\mathbf{r},t) + \lambda\int \psi^{+}_{\rho'}(\mathbf{r}',t)\, V(\mathbf{r}-\mathbf{r}')\, \psi_{\rho'}(\mathbf{r}',t)\, \psi_\rho(\mathbf{r},t)\, d^3r' = 0 . \tag{3.2}$$

The corresponding states can be calculated in a Fock space representation. The nucleon bound states are given by nonlinear functionals of the field operators applied to the groundstate. They are called "clusters", and all information about bound and scattering states must be given as cluster representations. In nonlinear spinor theory, no particle is more elementary than the others, all particles have to be regarded as composite. Therefore, any particle has to be described by a nonlinear functional of the field operators, determined by its symmetries and the field equations, and all observables have to be defined with respect to relativistic cluster representations. This has to be achieved by functional quantum theory.

As discussed above, it is sufficient to consider global observables only, because they allow to compare theory and experiment. To define them, we introduce the time ordered matrix elements

$$\tau_n(\underset{\alpha_1}{x_1} \ldots \underset{\alpha_n}{x_n}\,|a) := \langle 0|\,T\,\Psi_{\alpha_1}(x_1)\ldots\Psi_{\alpha_n}(x_n)\,|\,a\rangle,\ 1 \leqslant n < \infty \tag{3.3}$$

of spinor field operators between the ground state $\langle 0\,|$ and the state $|a\rangle$. The τ_n-functions transform under Poincaré transformations like the direct product of n Dirac spinors and were already used by Heisenberg to perform so-called Tamm-Dancoff calculations. Such calculations can be considered as special cases of functional calculations. We give the following

Definition: Functional quantum theory of the nonlinear spinor field is a map of the state space H on a suitable functional state space $\mathbf{H}$.
With the set of base functionals

$$\{\,|D_n(\overset{\alpha_1}{x_1} \ldots \overset{\alpha_n}{x_n})\rangle,\ 1 \leqslant n < \infty\} \tag{3.4}$$

this map is generated by the definition of physical state functionals

$$|\mathbf{T}(j,a)\rangle := \sum_{n=1}^{\infty}\int \tau_n(\underset{\alpha_1}{x_1} \ldots \underset{\alpha_n}{x_n}\,|\,a)\,|\,D_n(\overset{\alpha_1}{x_1} \ldots \overset{\alpha_n}{x_n})\rangle\, d^4x_1\, d^4x_n . \tag{3.5}$$

Then the map is given by

$$|\,a\rangle \rightarrow |\,\mathbf{T}(j,a)\rangle . \tag{3.6}$$

In order to formulate a quantum theory in $\mathbf{H}$, we have to introduce a scalarproduct with the requirement

$$\langle a\,|\,b\rangle = (\mathbf{T}(j,a)\,|\,\mathbf{T}(j,b)) , \tag{3.7}$$

at which we assume that this map is an isomorphism.

We first discuss the state functionals (3.5). The base functionals are given by

$$|D_n(\overset{\alpha_1}{x_1} \dots \overset{\alpha_n}{x_n})\rangle := \frac{1}{n!} j^{\alpha_1}(x_1) \dots j^{\alpha_n}(x_n)|\varphi_0\rangle \tag{3.8}$$

and

$$\langle D^n(\underset{\alpha_1}{x_1} \dots \underset{\alpha_n}{x_n})| := \frac{1}{n!}\langle\varphi_0|\partial_{\alpha_1}(x_1) \dots \partial_{\alpha_n}(x_n) \tag{3.9}$$

with the functional source operators $j(x)$, $\partial(x)$ which satisfy the anticommutation relations

$$\begin{aligned} &[j(x), j(x')]_+ = [\partial(x), \partial(x')]_+ = 0 \\ &[j(x), \partial(x')]_+ = \delta(x - x'). \end{aligned} \tag{3.10}$$

At this and in the following we always omit for brevity the spinorial indices.
$|\varphi_0\rangle$ is the functional groundstate with $\partial(x)|\varphi_0\rangle = 0$. $\partial(x)$ has to be identified with $j^+(x)$. To the representation of the Poincaré group in H a representation corresponds in $\mathbf{H}$ by means of the map (3.6). Because we take the states $|a\rangle$ as a base system for the representations which diagonalize the complete set of observables $\mathbb{P}$, $\mathbb{S}_3$, $\mathbb{P}^2$, $\mathbb{\Gamma}^2$, the same must hold for the functional states $|\mathbf{T}(j, a)\rangle$. If the sources transform like

$$V^{-1}(\Lambda, d)\, j(x)\, V(\Lambda, d) = L\, j(\Lambda^{-1}(x' - d)) \tag{3.11}$$

it can be shown that the transformation law for $|\mathbf{T}(j, a)\rangle$ reads

$$|\mathbf{T}(j, a)\rangle' = V(\Lambda, d)|\mathbf{T}(j, a)\rangle \tag{3.12}$$

for a Poincaré transformation (Λ, d) and its representation $V(\Lambda, d) \subset \mathbf{H}$.

The realization of sources transforming like (3.11) is possible by means of a Jordan-Wigner construction [21]. The state functionals (3.5) are the suitable objects for the definition and the calculation of the global observables.
First, we have to write down all dynamical and symmetry conditions in the space $\mathbf{H}$. The fundamental equation (3.1) for the spinor field operator $\Psi(x)$ reads in the functional space $\mathbf{H}$ [18]

$$[D(x)\,\partial(x) + V\,[\partial^3(x) + 3F(0)\,\partial(x)] - i\rho_0\, j(x)]\,|\mathbf{T}(j, a)\rangle = 0 \tag{3.13}$$

Every physical state functional has to satisfy this equation.
The eigenvalue conditions in H are given by (1.10). They read in $\mathbf{H}$ [22]

$$\begin{aligned} &\mathbf{P}_\mu|\mathbf{T}(j, a)\rangle = p_\mu|\mathbf{T}(j, a)\rangle \qquad 1 \leqslant \mu \leqslant 4 \\ &\mathbf{P}^2|\mathbf{T}(j, a)\rangle = m^2|\mathbf{T}(j, a)\rangle \\ &\mathbf{G}_\mu\mathbf{G}^\mu|\mathbf{T}(j, a)\rangle = s(s+1)|\mathbf{T}(j, a)\rangle \\ &\mathbf{S}_3|\mathbf{T}(j, a)\rangle = s_3|\mathbf{T}(j, a)\rangle \end{aligned} \tag{3.14}$$

with

$$\begin{aligned} \mathbf{P}_\mu &: = \int j^\alpha(x)\, P_\mu(x)\, \partial_\alpha(x)\, d^4x \\ \mathbf{M}_{kl} &: = \int j^\alpha(x)\, M_{kl}(x)\, \partial_\alpha(x)\, d^4x \\ \mathbf{G}_\mu &: = \frac{1}{2m}\, \epsilon^{\mu\nu\rho\sigma}\, \mathbf{P}_\mu\, \mathbf{M}_{\rho\sigma}\,, \end{aligned} \tag{3.15}$$

where P_ν, M_{kl} are the generators of the Poincaré group in ordinary function space [22]. To complete the framework of functional quantum theory we define the S-matrix by

$$S_{mn} = \langle a^-(m) \,|\, a^+(n) \rangle = (\mathbf{T}^{(-)}(j, m) \,|\, \mathbf{T}^{(+)}(j, n)) \tag{3.16}$$

with

$$|\,\mathbf{T}^{(\pm)}(j, m)\rangle : = |\,\mathbf{T}(j, a^{(\pm)}(m))\rangle\,. \tag{3.17}$$

Comparing these results with the general treatment of quantum theory in section 1, we observe that we have formally obtained a quantum theory of the spinor field in functional space, namely a state equation (3.13) and the global observables (3.14) and (3.16). But in order to derive physically meaningful results from this formulation of the spinor field problem, additional discussions of this approach are necessary.

It is convenient to start with a discussion of the effect of the introduction of the new Poincaré representations with respect to the dynamical equation (3.13). In order to obtain finite results for the nonrenormalizable spinor field equation (3.1), Heisenberg proposed the vanishing of the anticommutator on the light cone, leading thus to a noncanonical quantization. An immediate consequence of this noncanonical quantization is the vanishing of the term $i\rho_0 j(x)$ in the functional equation (3.13). The consequences can be seen by projecting equation (3.13) on to the τ_n-representation from which it has been derived. Using canonical quantization, i.e. $\rho_0 \neq 0$, this projection gives an infinite dimensional system of coupled partial-differential equations for the τ_n-functions, following from the fundamental equation (3.1). But if one admits noncanonical quantization with $\rho_0 \equiv 0$ this system degenerates to a recursion formula for the τ_n-functions. Therefore, the introduction of noncanonical quantization causes an indefiniteness of the functional equation (3.13). To remove this, the lost information about the quantization has to be restored, but in a specifically noncanonical way. This can be done by a transformation of equation (3.13) which is valid both for the canonical and the noncanonical case. Denoting the time ordered two-point function of the spinor field by $F(x-y)$ and defining the generating functional for a set of transformed functions $\{\varphi_n, 1 \leqslant n < \infty\}$ by

$$|\,\mathbf{F}(j, a)\rangle : = \sum_{n=1}^{\infty} \frac{1}{n!} \int \varphi_n(x_1 \ldots x_n \,|\, a)\, |\, D_n(x_1 \ldots x_n)\rangle\, d^4x_1 \ldots d^4x_n \tag{3.18}$$

we define the following transformation

$$|\,\mathbf{T}(j, a)\rangle = \exp[\tfrac{1}{2} \int j(x)\, F(x-y)\, j(y)\, d^4x\, d^4y]\, |\,\mathbf{F}(j, a)\rangle \tag{3.19}$$

Applying now the transformation (3.19) to the equation (3.13), one gets the functional equation for the $|\mathbf{F}\rangle$-functional

$$[D(x)\, d(x) + V : d^3(x) :]\,|\,\mathbf{F}(j, a)\rangle = : D\,|\,\mathbf{F}(j, a)\rangle = 0 \tag{3.20}$$

with $d(x) : = \partial(x) + \int F(x, x')\, j(x')\, d^4x'$ where $: d^3(x) :$ means normalordering with respect to $j(x)$ and $\partial(x)$. The subsidiary conditions (3.14) remain unchanged:

$$\begin{aligned} &\mathbf{P}_\mu\,|\,\mathbf{F}(j, a)\rangle = p_\mu\,|\,\mathbf{F}(j, a)\rangle \\ &\mathbf{P}^2\,|\,\mathbf{F}(j, a)\rangle = m^2\,|\,\mathbf{F}(j,a)\rangle \\ &\mathbf{G}_\mu\mathbf{G}^\mu\,|\,\mathbf{F}(j, a)\rangle = s(s + 1)\,|\,\mathbf{F}(j, a)\rangle \\ &\mathbf{S}_3\,|\,\mathbf{F}(j, a)\rangle = s_3\,|\,\mathbf{F}(j, a)\rangle \end{aligned} \tag{3.21}$$

Having obtained this equation, a projection to the system of coupled φ_n-functions reveals that this system remains meaningful even in the case of noncanonical quantization. Thus, the functionals (3.18) and their equation (3.20) are a completely equivalent formulation of the equation (3.13) in the case of canonical quantization while they have the advantage of remaining well-defined in the noncanonical case, too. This stems from the fact that in contrast to (3.13), equation (3.20) contains the information about quantization in the time-ordered two-point function, whereas (3.13) refers to the anticommutator. While the latter degenerates for noncanonical quantization, the time ordered functions allow a smooth transition between both methods, thus conserving the form of the corresponding equation.

Let us make a final comment on the occurence of the term with $3F(0)$ in equation (3.13) which can be shown to be infinite in the canonical case. This is a consequence of having regularized the interaction term $\Psi(x)^3$ in the fundamental equation (3.1) by formally applying Wick's rule to $\Psi(x)^3$, i.e. by replacing it by $: \Psi(x)^3 :$ with

$$: \Psi(x)^3 : : = \Psi(x)^3 - 3F(0)\,\Psi(x) \tag{3.22}$$

Applying now the Wick-transformation to (3.13), the $3F(0)$ term is cancelled. In this way, we achieve equation (3.20) to be free from infinities.

From now on, we only regard the φ_n-functions as physically significant. The isomorphism which defines functional quantum theory, is now given by the map from H onto $\mathbf{H}$ by

$$|\,a\rangle \to |\,\mathbf{F}(j, a)\rangle \tag{3.23}$$

together with a scalarproduct in $\mathbf{H}$ with the property

$$\langle a\,|\,b\rangle = (\mathbf{F}(j, a)\,|\,\mathbf{F}(j, b)) \tag{3.24}$$

We also define the S-matrix by

$$S_{mn} = \langle a^-(m)\,|\,b^+(n)\rangle = (\mathbf{F}^{(-)}(j, m)\,|\,\mathbf{F}^{(+)}(j, n)) \tag{3.25}$$

with

$$|\,\mathbf{F}^{(\pm)}(j, m)\rangle : = |\,\mathbf{F}(j, a^{(\pm)}(m))\rangle$$

(3.18), (3.19) and (3.23) are the defining relations for the functional quantum theory of the nonlinear spinor field with noncanonical quantization. Having reached this point, we can classify the remaining problems to be solved as follows:

i) a solution procedure for equation (3.20) has to be developed, allowing the treatment of relativistic cluster processes;
ii) the scalarproduct operation ( ) has to be defined in order to allow the calculation of scalarproducts,
iii) the unitarization has to be performed to reduce the unphysical large state space **H** to its physical relevant part.

We first discuss the scalarproduct. To form an idea of it, we examine how the scalarproduct in H can be expressed by the expansion coefficients φ_n of the state functional (3.16) The $\varphi_n(x_1 \dots x_n)$ can be considered to be projections of $|a\rangle \in H$ on the vectors

$$ {}_n\langle x_1 \dots x_n | := \langle 0 | [N\, \Psi(x_1) \dots \Psi(x_n)] \,. \tag{3.26} $$

The symbol 'N' reminds us of normalordering, but one should keep in mind that it is defined by the Wick transformation applied to the time ordered functions and means normalordering only in the case of free fields.

The set (3.26) can be considered to be the natural base set for the description of states in nonlinear spinor theory. But the elements of the set (3.26) cannot be expected to be orthonormal. We define

$$ \begin{aligned} &{}_n\langle x_1 \dots x_n | x_1' \dots x_m' \rangle_m =: g_{nm}(x_1 \dots x_n | x_1' \dots x_m') \\ &= \langle 0 | N\Psi(x_1) \dots \Psi(x_n) | N\Psi(x_1') \dots \Psi(x_m') | 0 \rangle \,, \end{aligned} \tag{3.27} $$

where g_{nm} is unequal to the unity operator. Nevertheless, this set has to be used for the explicit state representation, giving

$$ |a\rangle = \sum_{n=1}^{\infty} \frac{1}{n!} \int \sigma_n(x_1 \dots x_n | a) | x_1 \dots x_n \rangle_n \, d^4x_1 \dots d^4x_n \,. \tag{3.28} $$

Then the scalarproduct in H can be written

$$ \langle a | b \rangle = \sum_{n=1}^{\infty} \frac{1}{(n!)^2} \int \varphi_n(x_1 \dots x_n | a)\, \sigma_n(x_1 \dots x_n | b)\, d^4x_1 \dots d^4x_n \,. \tag{3.29} $$

On the other hand, one gets from (3.28) by multiplication with ${}_m\langle x_1' \dots x_m' |$

$$ \begin{aligned} &\varphi_m(x_1' \dots x_m' | a) = \\ &\sum_{n=1}^{\infty} \frac{1}{n!} \int \sigma_n(x_1 \dots x_n | a)\, g_{mn}(x_1' \dots x_m' | x_1 \dots x_n)\, d^4x_1 \dots d^4x_n \,. \end{aligned} \tag{3.30} $$

Inverting g_{mn}, we get

$$ \sum_{m=1}^{\infty} \int g_{em}^{-1}(x_1 \dots x_l | x_1' \dots x_m' | \varphi_m(x_1' \dots x_m' | a)\, d^4x_1' \dots d^4x_m' = \sigma_l(x_1 \dots x_l | a) \tag{3.31} $$

Inserting (3.31) into (3.29) we obtain

$$\langle a \mid b \rangle = \sum_{n,m=1}^{\infty} \frac{1}{n!} \int \varphi_n^x (x_1 \ldots x_n \mid a) g_{nm}^{-1} (x_1 \ldots x_n \mid x_1' \ldots x_m') \varphi_m (x_1' \ldots x_m' \mid b) \, dx \, dx' . \tag{3.32}$$

Observing now the orthonormality relations for the base functionals (3.4)

$$\langle D_n(x_1 \ldots x_n) \mid D_m(x_1' \ldots x_m') \rangle = {}^{P}\sum_{\lambda_1 \ldots \lambda_n} \delta_{nm} (-1)^P \delta(x_1 - x_{\lambda_1}') \ldots \delta(x_n - x_{\lambda_n}'), \tag{3.33}$$

which follow from the definitions (3.8), (3.9) and (3.10) we can write the scalarproduct (3.32) in functional space

$$\langle a \mid b \rangle = \langle F(j, a) \mid G^{-1}(j) \mid F(j, b) \rangle \tag{3.34}$$

with

$$G^{-1}(j) := \sum_{n,m} \int g_{nm}^{-1} (x_1 \ldots x_n \mid x_1' \ldots x_m') \mid D_n(x_1 \ldots x_n) \rangle \langle D_m(x_1' \ldots x_m') \mid dx \, dx' . \tag{3.35}$$

According to our definition of the map we define the formal scalarproduct by

$$(F(j, a) \mid F(j, b)) . = \langle F(j, a) \mid G^{-1}(j) \mid F(j, b) \rangle . \tag{3.36}$$

This shows that we have to use a weighted scalarproduct, due to the nonorthogonality of the base vectors (3.26). It is, therefore, convenient to define dual base sets $|x_1 \ldots x_n\rangle^n$ by the relation

$${}_m\langle x_1 \ldots x_m \mid x_1' \ldots x_n' \rangle^n = \delta_{nm} \frac{1}{(n!)^2} {}^{P}\sum_{\lambda_1 \ldots \lambda_n} (-1)^P \delta(x_1 - x_{\lambda_1}') \ldots \delta(x_n - x_{\lambda_n}'). \tag{3.37}$$

Then we deduce from (3.28)

$$\frac{1}{n!} \sigma_n (x_1 \ldots x_n \mid a) = {}^n\langle x_1 \ldots x_n \mid a \rangle \tag{3.38}$$

and we have

$$\mid b \rangle = \sum_{n=1}^{\infty} \int \varphi_n (x_1 \ldots x_n \mid b) \mid x_1 \ldots x_n \rangle^n \, d^4x_1 \ldots d^4x_n . \tag{3.39}$$

Therefore, the set $\{\sigma_n\}$ resp. $\{\varphi_n\}$ can be considered as the co- resp. contravariant components of a state $|a\rangle$, as already recognized by Nishijima [29] for the special case of a state expansion on a spacelike hyperplane.

Defining now the dual functionals

$$\mid S(j, a) \rangle := \sum_n c_n \int \sigma_n (x_1 \ldots x_n \mid a) \mid D_n(x_1 \ldots x_n) \rangle \, d^4x_1 \ldots d^4x_n \tag{3.40}$$

the relations (3.29) and (3.30) read in functional space as follows:

$$\langle a | b \rangle = \langle \mathbf{F}(j, a) | \mathbf{S}(j, b) \rangle$$
$$| \mathbf{F}(j, a) \rangle = \mathbf{G}(j) | \mathbf{S}(j, a) \rangle \tag{3.41}$$

with

$$\mathbf{G}(j) = \sum_{bb'} | \mathbf{F}(j, b) \rangle \langle \mathbf{F}(j, b) | g^{bb'}, \tag{3.42}$$

where $| b \rangle$ is some complete base system.

Now it can clearly be seen how to proceed: First, one has to solve the functional equation (3.20) to get the explicit state representation. In order to calculate $| \mathbf{S}(j, a) \rangle$, one has to derive and solve an equation for the kernel $\mathbf{G}(j)$. For the derivation of such an equation see [16]. Then the dual functionals are determined and the scalarproducts and the S-matrix can be calculated. This program is pursued in the following sections

4 Functional calculation methods

In this section we consider calculational methods leading to specified states and their scalarproducts. We first discuss state calculations. Generally, there are two classes of states, namely bound states and scattering states. These states differ in the choice of the boundary conditions. The procedure to incorporate the boundary conditions in the calculational scheme is more complicated for scattering states than for bound states. In developing such calculational schemes it becomes obvious that the bound state calculations are a subprogram of the scattering calculations. Thus, we concentrate on the scattering calculations. As has already been emphasized, these calculations have to be done for relativistic clusters. Comparing them which nonrelativistic clusters, it can be seen that the relativistic clusters are more complicated. Roughly speaking, any relativistic cluster can be considered to be composed of two parts, namely an "irreducible" kernel cluster with a finite number of degrees of freedom which corresponds to its nonrelativistic analogon and a polarization cloud of an infinite number of degrees of freedom which reflects the influence of the relativistic field. The method of handling these infinite clusters is a reduction procedure which first eliminates the polarization clouds.

After this has been completed, an analogous treatment to the nonrelativistic case is possible. We start the discussion of this method by defining the projection operators

$$P_k := k! \int | D_k(x_1' \ldots x_k') \rangle \langle D_k(x_1' \ldots x_k') | \, d^4x_1' \ldots d^4x_k' \tag{4.1}$$

with

$$P_k | \mathbf{F}(j, a) \rangle := | \mathbf{F}_k(j, a) \rangle . \tag{4.2}$$

For a given eigenstate $| \mathbf{F}(j, a) \rangle$, there exists a smallest $k = \lambda$ for which $| \mathbf{F}_k(j, a) \rangle \neq 0$, $\lambda \leqslant k < \infty$ while $| \mathbf{F}_k(j, a) \rangle = 0$, $1 \leqslant k \leqslant \lambda - 1$ is valid. λ is fixed uniquely by group theoretical reasons, depending on the quantum numbers of the eigenstate $|a\rangle$, like angular momentum, baryonnumber etc. Then the state functional (3.17) can be written

$$|F(j,a)\rangle = \sum_{l=0}^{\infty} |F_{\lambda+l}(j,a)\rangle . \tag{4.3}$$

Substitution of (4.3) into (3.20) and application of $P_{\lambda+l}$ gives the set of equations

$$\begin{aligned} &D(x)\,\partial(x)|F_{\lambda+l}\rangle + D(x)\int F(x,x')j(x')\,d^4x'|F_{\lambda+l-2}\rangle \\ &+ V\,\partial^3(x)|F_{\lambda+l+2}\rangle + 3V\int F(x,x')j(x')\,d^4x'\partial^2(x)|F_{\lambda+l}\rangle \\ &+ 3V\left[\int F(x,x')j(x')\,d^4x'\right]^2 \partial(x)|F_{\lambda+l-2}\rangle + V\left[\int F(x,x')j(x')\,d^4x'\right]^3 |F_{\lambda+l-4}\rangle = 0 \end{aligned} \tag{4.4}$$

$l = 2\nu,\ 0 \leqslant \nu < \infty$; ν integer.

Defining the operators

$$\begin{aligned} &A_1 := (D_x\partial_x + 3\,V\,F_x j\,\partial_x^2); \qquad A_2 := V\,\partial_x^3 \\ &A_3 := (D_x F_x j + 3\,V\,(F_x j)^2\,\partial x); \quad A_4 := V\,(F_x j)^3 \end{aligned} \tag{4.5}$$

we obtain

$$\sum_{i=1}^{4} A_i = D , \tag{4.6}$$

where D is defined by (3.18). Then the system (4.4) can be written with $|F_{\lambda+l}\rangle =: |\lambda + l\rangle$

$$l = 0: A_1\,|\lambda\rangle + A_2\,|\lambda + 2\rangle = 0 \tag{4.7a}$$

$$l = 2, 4, \ldots : A_1\,|\lambda + l\rangle + A_2\,|\lambda + l - 2\rangle + A_3\,|\lambda + l - 2\rangle + A_4\,|\lambda + l - 4\rangle = 0, \tag{4.7b}$$

where all $|\kappa\rangle$ with $\kappa < \lambda$ vanish. Defining the auxiliary functional

$$|\chi\rangle := \sum_{l=2,4,\ldots} |\lambda + l\rangle \tag{4.8}$$

and summing over equations (4.7b) for $l = 2, 4, \ldots$ by suitable redefinitions of the summations from (4.7) one obtains the set of equations

$$A_1\,|\lambda\rangle + A_2\,|\lambda + 2\rangle = 0 \tag{4.9a}$$

$$D|\chi\rangle - A_2\,|\lambda + 2\rangle + (A_3 + A_4)|\lambda\rangle = 0 . \tag{4.9b}$$

Adding (4.9a) to (4.9b) gives

$$A_1\,|\lambda\rangle + A_2\,|\lambda + 2\rangle = 0 \tag{4.10a}$$

$$D|\chi\rangle + (A_1 + A_3 + A_4)|\lambda\rangle = 0 . \tag{4.10b}$$

Equivalently one can write for (4.10)

$$P_\lambda(A_1\,|\lambda\rangle + A_2\,|\chi\rangle) = 0 \tag{4.11a}$$

$$D\,|\chi\rangle + (A_1 + A_3 + A_4)\,|\lambda\rangle = 0\,. \tag{4.11b}$$

Comparing equations (4.11) with our intentions for the solution procedure, it is obvious that the state functional $|\chi\rangle$ contains the polarization cloud which has to be eliminated. To achieve this we write (4.11b) in the form

$$(D_x d_x + V : d_x^3 :)\,|\chi\rangle = |f_x\rangle : = (A_1 + A_3 + A_4)\,|\lambda\rangle \tag{4.12}$$

and define

$$d_x\,|\chi\rangle = : |K_x\rangle \tag{4.13}$$

Then (4.12) goes over into

$$(D_x + V\; d_x d_x)\,|K_x\rangle = |f_x\rangle \tag{4.14}$$

or

$$(1 + G\; V\; d_x d_x)\,|K_x\rangle = G\,|f_x\rangle \tag{4.15}$$

with $G : = D^{-1}$. The iterative solution of this equation is

$$|K_x\rangle = \sum_{\nu=0}^{\infty} (G\; V\; d_x d_x)^\nu\, G\,|f_x\rangle\,. \tag{4.16}$$

In order to substitute this functional into (4.11a) we observe

$$P_\lambda A_2\,|\chi\rangle = P_\lambda\; V\; \partial_x^3\,|\chi\rangle = P_\lambda\; V\; d_x^3\,|\chi\rangle = P_\lambda\; V\; d_x^2\,|K_x\rangle\,. \tag{4.17}$$

Therefore, we obtain

$$P_\lambda(A_1\,|\lambda\rangle + A_2\,|\chi\rangle) = P_\lambda(A_1\,|\lambda\rangle + V\; d_x^2\,|K_x\rangle) = 0\,. \tag{4.18}$$

and from this follows with (4.16)

$$P_\lambda\left[A_1\,|\lambda\rangle + \sum_{\nu=0}^{\infty} V\; d_x^2\,(G\; V\; d_x d_x)^\nu\, G\,(A_1 + A_3 + A_4)\,|\lambda\rangle\right] = : P_\lambda \widetilde{D}\,|\lambda\rangle = 0 \tag{4.19}$$

which is the needed equation for the irreducible part of the relativistic clusters. The essential feature of this technique is that the kernels of the relevant equations are derived by an iteration procedure, while the solutions have to be obtained by noniterative techniques.

When scattering states are discussed in order to incorporate the appropriate boundary conditions, the equation (4.19) has to be prepared to give the corresponding channel equations. In this case, the "irreducible" cluster $|\lambda\rangle$ is assumed to be a scattering configuration which is built up by subclusters defining the different ingoing resp. outgoing composite particles. The notation "irreducible", therefore, refers only to the global group theoretical requirements of the configuration and does not exclude that the "irreducible" cluster can be interpreted in this way. To prepare channel equations for such a configuration means that a separation of the selfenergy and the interaction energy is required. This task is by no means trivial. As can be observed from (4.19); the prize to be paid for the elimination of the polarization cloud is a potential with an infinite number of interaction terms. In this case a separation of selfenergy and interaction energy cannot be achieved by simple inspection. Therefore, one has to develop a method solving this task analytically.

As has been shown in [24], in functional quantum theory like in ordinary quantum mechanics, the introduction of channels produces an antisymmetry breaking. Therefore, the construction of the full antisymmetric scattering states has to be performed by using virtual channel scattering states with a subsequent superposition restoring full antisymmetry. In this intermediate step of channel definition and calculation we are, therefore, allowed to break antisymmetry. It is just this fact which leads to an explicit calculation of channel equations. To do this, we consider equations (4.16) and the corresponding functional state $|F_\lambda\rangle$ with

$$|F_\lambda\rangle = \int \varphi_\lambda(x_1 \ldots x_\lambda | a) | D_\lambda(x_1 \ldots x_\lambda)\rangle \, d^4x_1 \ldots d^4x_\lambda \,. \tag{4.20}$$

To define a channel state, we introduce a partition of $x_1 \ldots x_\lambda$, namely

$$x_1 \ldots x_\lambda \to (x_1 \ldots x_{\lambda_1} | x_{\lambda_1+1} \ldots x_{\lambda_1+\lambda_2} | \ldots | x_{\lambda_1+\ldots+\lambda_{n-1}+1} \ldots x_{\lambda_n}) \tag{4.21}$$

which corresponds to a division of $\varphi_\lambda(x_1 \ldots x_\lambda)$ into n cluster states with λ_i, $1 \leqslant i \leqslant n$ coordinates. Within any cluster, the coordinates are assumed to be antisymmetric in the corresponding state functions, while between different cluster coordinates no antisymmetry is present. As has been shown in [24], the full antisymmetry can be restored by a superposition of n clusters with λ_i coordinates, where the partitions are taken of suitable permutations of $x_1 \ldots x_\lambda$. Formally we express this by an operator $Q_\kappa(n)$. If an unsymmetric state function is assumed

$$\varphi_\lambda^n := \varphi_\lambda(x_1 \ldots x_{\lambda_1} | \ldots | x_{\lambda_1+\ldots+\lambda_{n-1}+1} \ldots x_{\lambda_n}) \tag{4.22}$$

the totally antisymmetric state function is given by superposition

$$\varphi_\lambda(x_1 \ldots x_\lambda) = \sum_\kappa Q_\kappa(n)\, \varphi_\lambda(x_1 \ldots x_{\lambda_1} | \ldots | x_{\lambda_1+\ldots+\lambda_{n-1}+1} \ldots x_{\lambda_n}) \,. \tag{4.23}$$

After the channel equation for one representative n-cluster channel has been derived, the equations of the permuted channels follow by application of $Q_\kappa(n)$. Thus, we can restrict ourselves to the discussion of one representative channel for a n-cluster system with the partition $\lambda_1 \ldots \lambda_n$. For simplicity, we choose as a representative partition for this case (4.21). As for (4.21) the state functions (4.22) are antisymmetric only for the internal groupings of the n-cluster, it is reasonable to use corresponding auxiliary functional spaces which reflect this property. This can be achieved by introducting n different source operators j_α, $1 \leqslant \alpha \leqslant n$ and ∂_α, $1 \leqslant \alpha \leqslant n$ acting in different functional spaces $\mathbf{H}_\alpha$ with

$$[j(x|\alpha), j(x'|\alpha)]_+ = [\partial(x|\alpha), \partial(x'|\alpha)]_+ = 0 \tag{4.24}$$

$$[j(x|\alpha), \partial(x'|\alpha)]_+ = 0 \tag{4.25}$$

while the operators from different spaces $\mathbf{H}_\alpha$ commute. Forming the direct product $\mathbf{H} = \mathbf{H}_1 \otimes \ldots \otimes \mathbf{H}_n$ of these space, the base functionals are given by

$$|D_n(x_1^1 \ldots x_{\lambda_1}^1 | \ldots | x_1^n \ldots x_{\alpha_n}^n)\rangle = |D_{\alpha_1}(x_1^1 \ldots x_{\alpha_1}^1)\rangle \otimes \ldots \otimes |D_{\alpha_n}(x_1^n \ldots x_{\alpha_n}^n)\rangle . \tag{4.26}$$

Now the channel equations can be derived in the following way: We substitute in (4.19) for ∂_x, d_x and j_x the expressions

$$\partial_x \to \sum_{\alpha=1}^{n} \partial_{\alpha n};\ d_x \to \sum_{\alpha=1}^{n} d_{\alpha n};\ j_x \to \sum_{\alpha=1}^{n} j_{\alpha n} . \tag{4.27}$$

Then the functional operator of (4.19) is given by

$$\begin{aligned} &\{D_x(\Sigma\,\partial_{\alpha x}) + 3\,V\,(\Sigma\partial_{\alpha x})^2\,F_x(\Sigma\,j_{\alpha x}) + \\ &\sum_{\nu=0}^{\infty} V\,(\Sigma\,d_{\alpha x})^2\,[G\,V\,(\Sigma\,d_{\alpha x})^2\,]^\nu\,G\,[A_1(\Sigma) + A_3(\Sigma) + A_4(\Sigma)]\} \\ &= \widetilde{D}\,[\Sigma\,j_\alpha, \Sigma\,\partial_\alpha] = \sum_\alpha \widetilde{D}\,[j_\alpha, \partial_\alpha] + \text{mixed terms} . \end{aligned} \tag{4.28}$$

As for a single eigenstate, the operator of (4.19) is defined by

$$K_{\lambda_i}(x_1^i \ldots x_{\lambda_i}^i | y_1^i \ldots y_{\lambda_i}^i) := \langle D_{\lambda_i}(x_1^i \ldots x_{\lambda_i}^i)|\widetilde{D}\,[j_i\,\partial_i]\,|D_{\lambda_i}(y_1^i \ldots y_{\lambda_i}^i)\rangle . \tag{4.29}$$

we obtain the channel equation

$$\begin{aligned} &\langle D_\lambda(x_1^1 \ldots x_{\lambda_1}^1 | \ldots | x_1^n \ldots x_{\lambda_n}^n)|\widetilde{D}\,[\Sigma\,j_\alpha, \Sigma\,\partial_\alpha]\,|D_\lambda(y_1^1 \ldots y_{\lambda_1}^1 | \ldots | y_1^n \ldots y_{\lambda_n}^n)\rangle \\ &\quad\times \varphi_\lambda(y_1^1 \ldots y_{\lambda_1}^1 | \ldots | y_1^n \ldots y_{\lambda_n}^n) \\ &\equiv \left[\sum_{i=1}^{n} K_{\lambda_i}(x_1^i \ldots x_{\lambda_i}^i | y_1^i \ldots y_{\lambda_i}^i) + V\,(x_1^1 \ldots x_{\lambda_1}^1 | \ldots | x_1^n \ldots x_{\lambda_n}^n | y_1^1 \ldots y_{\lambda_1}^1 | \ldots | y_1^n \ldots y_{\lambda_n}^n)\right] \\ &\quad\times \varphi_\lambda(y_1^1 \ldots y_{\lambda_1}^1 | \ldots | y_1^n \ldots y_{\lambda_n}^n) = 0 \end{aligned} \tag{4.30}$$

where V is the matrix element of the mixed terms in (4.28). From (4.30) the permuted equations follow after [24]

$$\left[\sum_i Q_\kappa(n)\,K_{\lambda_i}\,Q_\kappa(n)^{-1} + Q_\kappa(n)\,V\,Q_\kappa(n)^{-1}\right] Q_\kappa(n)\,\varphi_\lambda(y_1^1 \ldots y_{\lambda_1}^1 | \ldots | y_1^n \ldots y_{\lambda_n}^n) = 0 \tag{4.31}$$

and the first solution is given by (4.23). It can be shown that in ordinary nonrelativistic quantum mechanics, the derivation of channel equations by generating channel functionals just reproduces the well-known nonrelativistic channel equations, if one identifies the operators of the Fock representation with the sources $j_\alpha(x)$ used here [30].

We do not discuss the detailed treatment of these channel equations, as we intended to give only a general outlook on the treatment of the functional equation. Finally, it should be remarked that the bound states occur as solutions of the selfenergy terms and that just these selfenergy terms are also given by equations (4.19) if one specifies the proper quantum numbers. The method can be extended in order to treat inelastic processes with multiparticle production which neither we will discuss here.

As has been pointed out, for the calculation of scalar products the kernel **G**(j) is required. Therefore, if calculation methods are discussed, also this problem has to be solved. Doing this, one has to observe that the kernel of (3.30) is meaningless as long as the normal ordering procedure is defined only by the transformation (3.19). To remove this difficulty, additional assumptions are required. We define

$$\overset{\circ}{\varphi}_{n,m}(x_1 \dots x_n \mid x_1 \dots x_m') := \langle 0 \mid N\,\Psi(x_1) \dots \Psi(x_n)\,N\,\Psi(x_1') \dots \Psi(x_m') \mid 0 \rangle \tag{4.32}$$

and simplify the following calculations by using the diagonal approximation of (3.35), namely

$$\mathbf{G}_d(j) = \sum_n \int \overset{\circ}{\varphi}_{n,n}(x_1 \dots x_n \mid x_1' \dots x_n') \mid D_n(x_1 \dots x_n)\rangle \langle D_n(x_1' \dots x_n') \mid dx\,dx' \tag{4.33}$$

Then we postulate that normal ordering can be applied to (4.32) which leads to

$$\overset{\circ}{\varphi}_{n,n}(x_1 \dots x_n \mid x_1' \dots x'_n) = \sum_{\alpha=0}^{n-1} \sum F^+(x_{i_1} x_{j_1}) \dots F^+(x_{i_\alpha} x_{j_\alpha})\, \overset{\circ}{\varphi}_{2n-2\alpha}(i_1 \dots i_\alpha, j_1 \dots j_\alpha)\,, \tag{4.34}$$

where $F^+(x, x')$ is the two-point function of the nonlinear spinor theory in the representation (2.20) and $\overset{\circ}{\varphi}_{2n-2\alpha}(i_1 \dots i_\alpha, j_1 \dots j_\alpha)$ means the completely normalordered expression of $(2n-2\alpha)$ spinor field operators with the coordinates $\{x_k, x_l', 1 \leqslant k \leqslant n, 1 \leqslant l \leqslant n, k \neq i_1 \dots i_\alpha, l \neq j_1 \dots j_\alpha\}$. Considering the time ordered Schwinger functional state $\mid \mathbf{T}(j, 0)\rangle$ for the vacuum by (3.19) the normalordered functional $\mid \mathbf{F}(j, 0)\rangle$ of the vacuum follows exactly from it. Independently of any special representation of the spinor field operators for $\mid \mathbf{T}(j, 0)\rangle$ a functional equation can be derived leading by (3.19) to a functional equation for $\mid \mathbf{F}(j, 0)\rangle$ which can be written formally

$$\mathbf{D}_0(j, d) \mid \mathbf{F}(j, 0)\rangle = \mid \mathbf{F}(j, 0)\rangle \tag{4.35}$$

with

$$\mid \mathbf{F}(j, 0)\rangle = \sum_{n=0}^{\infty} \int \overset{\circ}{\varphi}_{2n}(x_1 \dots x_{2n}) \mid D_{2n}(x_1 ,.. x_{2n})\rangle\, d^4x_1 \dots d^4x_n\,. \tag{4.36}$$

Equation (4.35) can be integrated according to [24]. Introducing projection operators by

$$P_2 \mid \mathbf{F}(j, 0)\rangle = \mid \mathbf{F}_2(j, 0)\rangle; \; \Pi_2 \mid \mathbf{F}(j, 0)\rangle = \mid \mathbf{F}_r(j, 0)\rangle \tag{4.37}$$

with $\mid \mathbf{F}(j, 0)\rangle = \mid \mathbf{F}_2(j, 0)\rangle + \mid \mathbf{F}_r(j, 0)\rangle$ where $\mid \mathbf{F}_2(j, 0)\rangle$ represents the two-point function functional, one obtains from (4.35)

$$[(1 - P_2 \mathbf{D}_0 P_2) - P_2(1 - \mathbf{D}_0)\,\Pi_2(1 - \Pi_2 \mathbf{D}_0 \Pi_2)^{-1}\,\Pi_2(1 - \mathbf{D}_0) P_2] \mid \mathbf{F}_2(j, 0)\rangle = 0 \tag{4.38}$$

and

$$|\mathbf{F}_r(j,0)\rangle = -(1-\Pi_2\mathbf{D}_0\Pi_2)^{-1}\,\Pi_2(1-\mathbf{D}_0)\,P_2\,|\mathbf{F}_2(j,0)\rangle \tag{4.39}$$

Representing the two-point functional by

$$|\mathbf{F}_2(j,0)\rangle := \int \overset{\circ}{\varphi}_2(x_1x_2)|D_2(x_1x_2)\rangle\,dx_1\,dx_2 \tag{4.40}$$

we assume (4.38) to be solved and identify $F(x,x') \equiv \overset{\circ}{\varphi}_2(x,x')$. Then we have only to consider relation (4.39). Performing a Neumann expansion of the kernel of (4.39), we obtain for the normalordered functions of (4.37)

$$\overset{\circ}{\varphi}_{2n}(x_1\ldots x_{2n}) = \sum_{l=0}^{\infty}(-1)^{l+1}\,\langle D_{2n}(x_1\ldots x_{2n})|(\Pi_2\mathbf{D}_0\Pi_2)^l(1-\mathbf{D}_0)\,P_2\,|\mathbf{F}_2(j,0)\rangle\,. \tag{4.41}$$

Defining the kernel

$$K_{2n}(x_1\ldots x_n|x_1'\ldots x_n') :=$$

$$\sum_{l=0}^{\infty}(-1)^{l+1}\sum_{\alpha=0}^{n-2}\sum F^+(x_{i_1}x_{j_1}')\ldots F^+(x_{i_\alpha}x_{j_\alpha}')\,\langle D_{2n-2\alpha}|(\Pi_2\mathbf{D}_0\Pi_2)^l(1-\mathbf{D}_0)P_2\,|\mathbf{F}_2\rangle \tag{4.42}$$

by substitution of (4.41) into (4.34), the relation

$$\overset{\circ}{\varphi}_{n,n}(x_1\ldots x_n|x_1'\ldots x_n') = \sum F^+(x_{i_1}x_{j_1}')\ldots F^+(x_{i_n}x_{j_n}') + K_{2n}(x_1\ldots x_n|x_1'\ldots x_n') \tag{4.43}$$

results. Therefore, we have obtained a definite representation of the kernel. If we only consider the lowest approximation of the kernel (4.43), we obtain

$$G_d(j) = \sum_n\int\sum_{\substack{i_1\ldots i_n\\ j_1\ldots j_n}} F^+(x_{i_1}x_{j_1}')\ldots F^+(x_{i_n},x_{j_n}')|D_n(x_1\ldots x_n)\rangle\,\rangle D_n(x_1'\ldots x_n')|\,dx\,dx' \tag{4.44}$$

and from this follows by projection of (3.41) into configuration space

$$\varphi_m(x_1'\ldots x_m'|a) = \int F^+(x_1'x_1)\ldots F^+x_m'x_m)\,\sigma_m(x_1\ldots x_m|a)\,dx_1\ldots dx_m', \tag{4.45}$$

which is the lowest approximation of formula (3.30). Due to the spectral representation of F^+ given by (1.6) it is possible to construct an inverse operator $(F^+)^{-1}$ and thus to solve equations (4.45). Hence

$$\sigma_m(x_1\ldots x_m|a) = \int F^+(x_1x_1')^{-1}\ldots F^+(x_mx_m')^{-1}\,\varphi_m(x_1'\ldots x_m'|a)\,dx_1'\ldots dx_m' \tag{4.46}$$

is an explicit calculable expression, and, therefore, an explicit state representation according to (3.28) can be obtained.

5 Noncanonical quantization and unitarization

In the preceding sections we derived the basic formulae of the functional quantum theory of the nonlinear spinor field under the supposition of a noncanonical anticommutator for the field operator $\Psi(x)$, i.e.

$$[\Psi(x), \Psi(x')]_{+/x_0 = x'_0} = 0 . \tag{5.1}$$

In section 2 it was demonstrated that such a noncanonical anticommutator can be directly derived by means of the canonical Lagrangean quantization formalism, if this formalism is applied to suitably coupled first order equations, or equivalently to a higher order field equation. As the fields of section 2 are free fields, the proof of the existence of a noncanonical quantization holds only for such fields. Hence the crucial question is: Is noncanonical quantization (5.1) for a general interacting spinor field $\Psi(x)$ compatible with the field equation (3.1) or not? No general proof of this compatibility has been given so far. As the anticommutator (5.1) concerns the operator property of $\Psi(x)$ and as the state space of cluster processes is very complicated for higher sectors, it can be expected that a general proof will be extremely complicated. Therefore, the question may be put: should this model be rejected as long as this proof is not available? As will be shown in the following the ghost – dipole ghost constellation leads to a very attractive unitarization procedure which cannot be obtained for other types of a Pauli-Villars regularization. Hence it seems to be worthwile to investigate the consequences of such a theory even if the general proof of selfconsistency is not available. Within the scope of such an investigation it is indeed not possible to detect sufficient conditions for selfconsistency, however, necessary conditions can be derived. This can be seen from the spectral decomposition of F^+. If we assume that (2.11) holds also for the interacting fields, then the one-particle solutions of equation (3.20) must coincide with those appearing in (2.11). This necessary condition for selfconsistency has not yet been studied. Therefore, in the first step we postulate that the F^+-function of the nonlinear equation (3.1) is identical with (2.24). Then, according to selfconsistency the one-particle sector is characterized by (2.29). We investigate the consequences of this assumption.

As can be seen by diagonalization of (1.12), the introduction of ghost states (and dipole ghost states) leads to indefinite metric in state space. This prevents a direct probabilistic interpretation of the state space like in conventional theories with positive definite metric. To remove this unphysical feature of the theory special procedures have to be performed. As the global observables given by a complete set of commuting symmetry observables and the S-matrix are sufficient for the description of microphysical experiments, only these observables are considered. While the symmetry observables can simply be projected into the positive part of the state space, the S-matrix in general becomes nonunitary and requires a unitarization procedure in order to obtain physical meaning. As in the nonlinear spinor theory the various particle sectors have to be derived by construction, i.e. direct calculation, and are not given a priori, there does not exist a unitarization procedure which solves the problem for all sectors simultaneously. One rather has to discuss the state vectors of any sector separately. To demonstrate the method we give a discussion of elastic nucleon-nucleon scattering. The in- and outgoing states are defined in this case by the direct product of the one-fermion sector discussed in section 1, namely

$$\{|p_1 s_1 n\rangle, |p_1 s_1 g\rangle, |p_1 s_1 d\rangle\} \otimes \{|p_2 s_2 n\rangle, |p_2 s_2 g\rangle, |p_2 s_2 d\rangle\}. \quad (5.2)$$

But only the set $\{|np_1 s_1\rangle \otimes |np_2 s_2\rangle\}$ has a physical meaning. However, due to (1.24) also the enlarged space $\mathbf{H_{Ph}} := \{|np_1 s_1\rangle, |gp_1 s_1\rangle\} \otimes \{|np_2 s_2\rangle, |gp_2 s_2\rangle\}$ can be considered as a physical one.

In general, the scattering states have to be calculated by means of channel equations which have been derived in the preceding section 4. These states contain unphysical admixtures produced by the special form of the dynamics in consideration. That means if one starts in the physical sector by defining the initial states to be real nucleons, the dynamical evolution causes the occurrence of ghost particles which appear in the scattering states, and this is equivalent to the loss of unitary of the S-matrix. We discuss this situation for elastic nucleon-nucleon scattering and denote the advanced and retarded scattering states of two ingoing respectively outgoing nucleons by $|nk_1 nk_2\rangle^{(\pm)}$ where we omit for brevity the spin variables and other quantum numbers. Then the S-matrix is given by

$$S(nk_1 nk_2 | nk_1' nk_2') := {}^{(-)}\langle nk_1 nk_2 | nk_1' nk_2' \rangle^{(+)} \quad (5.3)$$

where the scalarproduct has to be evaluated according to section 4. This evaluation shall not be discussed here; we only study the unitarization procedure. From the properties of the scalarproduct (5.2) can be written

$$S(nk_1 nk_2 | nk_1' nk_2') = \delta(k_1 - k_1')\,\delta(k_2 - k_2') + \delta(k_1 + k_2 - k_1' - k_2')\,T(s,t) \quad (5.4)$$

with the usual definition of the variables s and t corresponding to (5.3). In general, neither S nor T satisfy the usual unitary conditions.

Heisenberg performed a dipole ghost regularization for the Leemodel and achieved a unitary S-matrix by admixing suitable nonphysical ghost states to the scattering states [3, 4]. This method requires an explicit state construction. Karowski proposed an extension of the Heisenberg method to the dipole ghost regularized nonlinear spinor field [31]. He tried to restore unitarity by a suitable admixture of nonphysical S-matrix elements to the nonunitary S-matrix and in this way to avoid the direct construction of states. This method, however, underlies a kinematical restriction. An admixture is only possible if the quantum numbers of the admixed elements coincide with those of the original S-matrix element (5.2) resp. (5.3). If this condition cannot be satisfied, an interpretation of the process in consideration is not possible. It was shown by Stumpf and Scheerer [32] that such S-matrix admixtures violate the kinematical conditions and hence this method cannot be applied without leading to contradictions. Hence we return to the state unitarization. In this method the state space itself is separated into a positive and an indefinite part which do not interfere as a result of construction. Afterwards with the positive part the physical observables are formed.

We assume to have the proper physical scattering states in the Heisenberg representation $|nk_1 nk_2\rangle_{ph}^{(\pm)}$ which result from a unitarization procedure. Then, if we start in the physical sector, the scattering process must not lead into the unphysical sector. As the good ghosts are normalized to zero, the physical sector is given by $\mathbf{H}_{Ph} := \{n, g\} \otimes \{n, g\}$, while the unphysical sector follows to be $\mathbf{G}_U := \{d\} \otimes \{d\}$, i.e. contains the dipole ghost states. Then an advanced scattering state starting from $\mathbf{H}_{Ph}$ must not contain outgoing bad ghosts. Due to the orthonormality relations (1.23) no bad ghosts leave the scattering process if

$$^{f}\langle nk' gq \mid nk_1 nk_2 \rangle^{(+)}_{Ph} = 0 \tag{5.5}$$

and

$$^{f}\langle g q_1' g q_2' \mid nk_1 nk_2 \rangle^{(+)}_{Ph} = 0 \tag{5.6}$$

are satisfied where f means free two-particle state.
In the same way one concludes

$$^{(-)}_{Ph}\langle nk_1' nk_2' \mid nkgq \rangle^{f} = 0 \tag{5.7}$$

and

$$^{(-)}_{Ph}\langle nk_1' nk_2' \mid gq_1 gq_2 \rangle^{f} = 0 . \tag{5.8}$$

Additionally, we have the normalization condition

$$^{(-)}_{Ph}\langle nn \mid nn \rangle^{(-)}_{Ph} = {}^{(+)}_{Ph}\langle nn \mid nn \rangle^{(+)}_{Ph} = 1 . \tag{5.9}$$

Imposing these conditions, the scalarproducts of free states with scattering states have to be evaluated properly, what can be done according to the methods of functional quantum theory. The conditions themselves can be satisfied, if to the ordinary scattering states $|n, n\rangle^{(\pm)}$ suitable admixtures are added which are elements of $\mathbf{H}_{Ph}$. We only give the results: The unitarized part of the Hilbert space $\{|nn\rangle^{(\pm)}_{ph}\}$ for two nucleon scattering is given by

$$|nn\rangle^{(+)}_{Ph} = \rho_1 \left\{ |nn\rangle^{(+)} + \frac{d_1(U)}{d_2(U)} |ng\rangle^{(+)} + \frac{d_1'(U)}{d_2(U)} |gg\rangle^{(+)} \right\} \tag{5.10}$$

with the definitions

$$U^{xy}_{zw} := {}^{f}\langle xy \mid zw \rangle^{(+)}$$

$$d_1(U) := \begin{vmatrix} U^{nn}_{ng} & U^{nn}_{gg} \\ U^{gg}_{ng} & U^{gg}_{gg} \end{vmatrix} ; \quad d_1'(U) := \begin{vmatrix} U^{nn}_{ng} & U^{nn}_{gg} \\ U^{ng}_{ng} & U^{ng}_{gg} \end{vmatrix} \tag{5.11}$$

$$d_2(U) := \begin{vmatrix} U^{gg}_{gg} & U^{gg}_{ng} \\ U^{gg}_{gg} & U^{ng}_{ng} \end{vmatrix} .$$

The direct unitarization of the state space has the great advantage that no kinematical subsidiary conditions have to be imposed. Hence, this method is not subject to restrictions which prevent its practical application. Either the unitarity conditions (5.8), (5.7), (5.6) are satisfied by themselves, namely the process in consideration does not admit the occurrence of bad ghosts, then nothing at all has to be done, or the conditions are nontrivial and the procedure can be performed. The physical S-matrix then follows in the usual way by the formation of scalarproducts between the sets of states $\{|\,n\,n\rangle_{ph}^{(+)}\}$ and $\{{}_{ph}^{(-)}\langle n\,n\,|\}$. By this procedure the physical interpretation of the asymtotic states can be maintained as only states of norm zero are admixed, i.e. the ingoing and outgoing states can be uniquely identified. This is the advantage of the dipole ghost regularization compared with other regularization procedures.

References

[1] *Wigner, E. P.:* Ann. Math. **40,** 149 (1939).

[2] *Bargmann, E. P., Wigner, E. P.:* Proc. Nat. Acad. Sci. (USA) **34,** 211 (1948).

[3] *Heisenberg, W.:* Nucl. Phys. **4,** 532 (1957).

[4] *Heisenberg, W.:* Introduction to the Unified Field Theory of Elementary Particles, Wiley, London 1967.

[5] *Fonda, L., Ghirardi, G. I.:* Symmetry Principles in Quantum Physics, M. Dekker Inc., New York 1970.

[6] *Bopp, F.:* Ann. d. phys. **38,** 345 (1940).

[7] *Podolski, B.:* Phys. Rev. **62,** 68 (1941).

[8] *Pauli, W., Villars, F.:* Rev. Mod. Phys. **21,** 434 (1949).

[9] *Froissart, M.:* Suppl. Nuovo Cim. **14,** 197 (1959).

[10] *Dürr, H. P.:* Nuovo Cim. **27A,** 305 (1975), **22A,** 386 (1974).

[11] *Nagy, K. L.:* State Vector Spaces with Indefinite Metric in Quantum Field Theory, P. Noordhoff, Groningen 1966.

[12] *Nakanishi, N.:* Suppl. Progr. Theor. Phys. **51,** 1 (1972).

[13] *t'Hooft, G., Feldmann, H.:* CERN, reprint 73/9.

[14] *Heisenberg, W.:* Z. Phys. **120,** 513, 673 (1943).

[15] *Lehmann, H.:* Nuovo Cim. **11,** 342 (1954).

[16] *Stumpf, H.:* Z. Naturforsch. **30a,** 708 (1975).

[17] *Mitter, H.:* Z. Naturforsch. **20a,** 1505 (1965).

[18] *Stumpf, H.:* Acta Phys. Austr. Suppl. **9,** 195 (1972).

[19] *Stumpf, H.:* Z. Naturforsch. **31a,** 528 (1976).

[20] *Stumpf, H.:* Naturforsch. **25a,** 575 (1970).

[21] *Stumpf, H., Scheerer, K.:* Z. Naturforsch. **30a,** 1361 (1975).

[22] *Stumpf, H., Scheerer, K., Märtl, H. G.:* Z. Naturforsch. **25a,** 1561 (1970).

[23] *Stumpf, H.:* Z. Naturforsch. **27a,** 1058 (1972).

[24] *Stumpf, H.:* Z. Naturforsch. **29a,** 549 (1974).

[25] *Stumpf, H., Engeser, W., Illig, K.:* Z. Naturforsch. **26a,** 1723 (1971).

[26] *Stumpf, H.:* Z. Naturforsch. **26a,** 623 (1971).

[27] *Dürr, H. P., Wagner, F.:* Nuovo Cim. **46,** 21 (1966).

[28] *Schwinger, I.:* Proc. Nat. Acad. Sci. USA **37,** 452 (1951).

[29] *Nishijima, K.:* Progr. Theor. Phys. **10,** 549 (1953).

[30] *Mautner, J. P.:* Diplomarbeit, Inst. Theor. Physik, Universität Tübingen 1978.

[31] *Karowski, M.:* Nuovo Cim. **23A,** 126 (1974).

[32] *Stumpf, H., Scheerer, K.:* Z. Naturforsch. **34a,** 284 (1979).

Chapter VIII
The Algebraic Theory of Automata

M. Dal Cin

Introduction

Much scientific work today is directed towards understanding complexity – the complexity of numerical algorithms, of the English syntax, of living organisms or ecological systems, to cite only a few examples. The aim of this chapter is to introduce the reader to the theory of discrete information processing systems (automata) and to develop an algebraic framework within which we can talk about their complexity.

Section 1 provides the proper setting for the application of (semi-)group theoretical methods in discrete system theory. Section 2 deals with the triangulation of state transition monoids. In Section 3 the decomposition of group automata is discussed. This discussion will lead to a theorem of K. B. Krohn and J. L. Rhodes which says that decomposing loopfree any finite automaton into simpler automata corresponds roughly to taking the wreath product of certain associated groups and semigroups.

1 Systems, automata and switching networks

A system is, according to a dictionary definition, a collection of objects united by some form of interaction and interdependence. Physicists have always been concerned with complex systems. Their systems, however, belong to only a few of the many different classes of systems that are the subject of system theory [1], [2]. We shall focus our attention on time discrete input-output systems.

1.1 Input-output systems

Consider a black box which reacts on sequences of inputs producing outputs, Fig. 1. Lasers, control systems, and computers are typical examples. In general, the output of a system depends not only on its present input but also on past inputs. The term "state of the system at time t" denotes the information on past inputs which the system is capable of

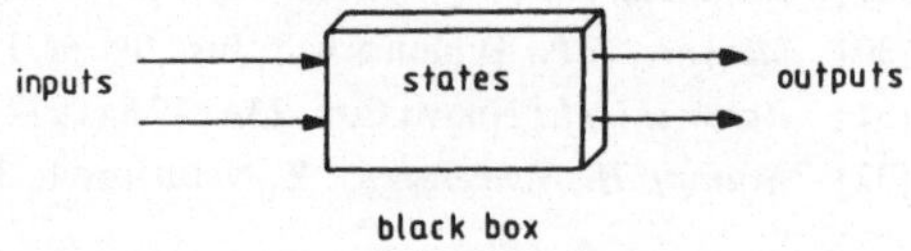

Fig. 1

storing at time t. This information and the input determines the system's output (at time t) and its next state [3].

The transition from physics to the theory of input-output systems comes when we observe that the evolution of the state of a physical system depends on certain parameters (inputs) representing forces, which can be controled from outside the system, and that only certain properties of the system's state are measurable quantities (i.e. outputs) at a given time [1], [4]. It is then natural to ask:

- Is it possible to apply control inputs in such a way as to drive the system from any given state to any desired final state? (The reachability problem of system theory)
- Is it possible to determine the state of a system by applying certain control inputs to the system and observing its outputs? (The observability problem of system theory)
- Given the input-output behavior of a physical system, find the proper set of states and equations which correctly describe this behavior. (The realization problem of system theory)
- Determine a collection of standard modules for a given class of systems such that any system of this class can be decomposed into these standard modules. (The decomposition problem of system theory)

Clearly, decompositions make only sense, if the complexity of a module's interaction with other modules is less than the complexity of the module's internal structure. In other words, the modules should have simple interfaces. The number and types of standard modules which appear in the decomposition determine then the complexity of the system.

Two important classes of input-output systems are linear systems and automata. Linear systems are encountered, e.g., in control theory and chemistry. Automata are encountered in computer science and digital electronics. Linear automata are subjects of coding and communication theory.

This chapter deals with the decomposition problem of finite automata. Chapter 9 by M. Hazewinkel addresses the first three problems for linear systems.

1.2 Automata

Automata are input-output systems which operate on discrete time. Their sets of inputs and outputs are finite. Furthermore, their storage capacity is determined by a countable number of different internal states. Automata are idealized models of a large number of physical devices and the theory of automata constitutes one of the most important parts of system theory and computer science. Every (abstract) finite automaton can be realized by a physical device. There is a canonical correspondence between finite automata and (transformation) monoids. This correspondence will give us a natural setting for the application of group and semigroup theoretical methods to automata theory.

To begin with, we give an abstract definition of automata.

1.2.1 Definition: An automaton is specified by a quintuple $A = |X, Y, Q, \delta, \omega|$ where

- X, Y are finite non empty sets (alphabets) of inputs and outputs, respectively.
- Q is the set of internal states.
- $\delta : Q \times X \to Q$ is the next-state function, and
- $\omega : Q \times X \to Y$ is the output function of A.

If at any given time $t \in \mathbb{N}$ automaton A is in some state q and receives input $x \in X$, its output is $\omega(q, x)$ and at time $t + 1$ it will be in state $\delta(q, x)$. Automaton A is said to be finite, if Q is finite.

The wellknown Turing-Machine is an example of an infinite automaton [3]. In the following, only finite automata will be considered. Automaton A is a linear finite automaton if X, Y, Q are finite dimensional vector spaces over a finite field and if δ and ω are linear maps.

There are several ways to describe finite automata,

(1) state- and output tables for δ and ω (Tab. 1)

(2) state graphs (Fig. 2)

(3) sequential logical networks (Fig. 11)

(4) special system description languages such as Regular Expressions or Chomsky Grammars [5], [6].

1.2.2 Examples: (1) Flip-Flop:

Automaton $D = |X, X, X, pr_2, pr_1|$ is called a delay automaton ($pr_i : X \times X \to X : (x_1, x_2) \to x_i$ are projections). If $|X| = 2$, D is called a delay flip-flop. Tab. 1 is the state-table of D and Fig. 2 shows its state graph.

The behavior of a D-flip-flop can be read off its state graph. For example, given input sequence uuvuvvu, the automaton (starting in state v) reaches the states vuuvuvvu and produces the same sequence as output sequence. This two-state automaton is very simple but nevertheless very important as we will see later on. Clearly, D is reachable and observable. Its state at a given time is identical to its output.

Tab. 1: State- and output table of a D-Flip-Flop

input	u	v	u	v
u	u	v	u	u
state				
v	u	v	v	v
	next state		output	

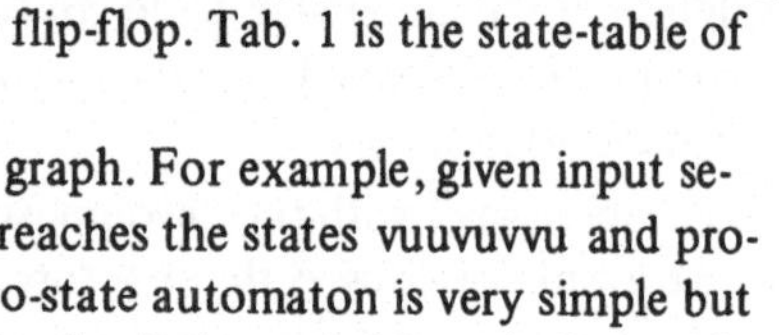

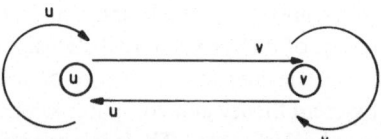

Fig. 2 State graph of a D-Flip-Flop

(2) Counter: Let $Z_m = \{0, 1, 2, ..., m-1\}$; automaton $A(Z_m) = |Z_2, Z_m, Z_m, \delta_m, \omega_m|$ with δ_m the addition modulo m and $\omega_m(n, x) = \delta_{m-1,n}$ is called a counter modulo m. $A(Z_m)$ is reachable, is it also observable?

(3) Coder: Let X and Y be finite sets and $F : X \to Y$ a partial map. Choose two symbols q and $\square$ not in $X \cup Y$. The coder for F is

$F = |X, Y \cup \square, q, i_q, \omega|$ with $i_q : q \to q$ and

$\omega(q, x) = F(x)$ if $F(x) \neq \emptyset$ and $\omega(q, x) = \square$ else.

This one-state automaton F "computes" the partial function F.

Fig. 3 shows a so called *cascade* of three automata A_1, A_2, A_3. The interfaces between these automata are per definitionem coders. Coders are simple enough interfaces, since one-state automata are not capable of storing any specific information. The flow of information (signals) within a cascade is unidirectional.

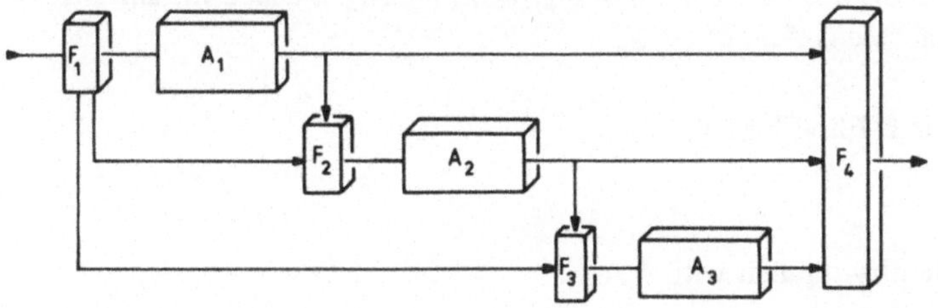

Fig. 3
A cascade

We extend the next-state function δ on X^* as follows: The input monoid X^* of automaton A is defined in Chap. 1 (Sec. 3.6). Let λ be the empty input, $w \in XX^*$ (s. Chap. 1 Sec. 3.1) and $x \in X$. Then we define $\delta(q, \lambda) = q$ and $\delta(q, xw) = \delta(\delta(q, x), w)$. The extension of the output function on X^+ is given by: $\omega(q, xw) = \omega(\delta(q, x), w)$. That is, automata process inputs sequentially. (Therefore, they are sometimes also called sequential machines).

1.2.3 Example. Consider the D-Flip-Flop and $w = x_1 x_2 \dots x_n \in XX^*$.

$$\begin{aligned}\delta(x_0, w) &= \delta(\delta(x_0, x_1), x_2 \dots x_n) = \delta(pr_2(x_0, x_1), x_2 \dots x_n) \\ &= \delta(x_1, x_2 \dots x_n) = \dots = \delta(x_{n-1}, x_n) = x_n\end{aligned}$$

and

$$\begin{aligned}\omega(x_0, w) &= \omega(\delta(x_0, x_1), x_2 \dots x_n) = \omega(x_1, x_2 \dots x_n) \\ &= \dots \omega(x_{n-1}, x_n) = pr_1(x_{n-1}, x_n) = x_{n-1} .\end{aligned}$$

Thus, the output of D is the next to last input. That is, D *stores* (or delays) its input exactly one time step.

The sequential performance of automata induced by sequences of inputs (input words) is best described by transformation monoids.

1.2.4 Definition: Let M be a monoid of (partial) maps acting on a finite set W. The tuple M(W) = (M, W) is a transformation monoid (tm) if the identity transformation on W, i_W, is the neutral element of M.

Clearly, for any given automaton A every input x of A induces a map $\delta_x : Q \to Q : q \mapsto \delta(q, x)$ These maps generate a monoid M^A and $M_A := M^A(Q) = (M^A, Q)$ is a transformation monoid; M_A is the tm, M^A the monoid of automaton A.

1.2.5 Examples.

(a) D-Flip-Flop: The tm of a D-Flip-Flop is shown in Tab. 2.

(b) The monoid of a counter modulo m is the cyclic group C(m).

(c) A finite group G is the monoid of a nonsingular (i.e. δ_0 is invertible), finite, linear automaton iff G is the semidirect product of an abelian group all of whose elements (except e) have the same prime order by a cyclic group [13].

(d) Given a tm M(W) = (M, W), define automaton $A_{M(W)}$ (or A_M) as $|M, W, W, \delta_M, pr_1|$ where δ_M is the action of M on W. The tm of A_M is M(W). A_M is called the automaton of M(W); a special case is A_{M_A}.

Tab. 2 The monoid of the D-Flip-Flop. T

M_D	λ	u	v
λ	λ	u	v
u	u	u	v
v	v	u	v

We identify input u with δ_u

Our aim will be to parametrize the transformation monoids of automata in such a way that their action on the state set becomes triangular (cf. Chap. 1, Sec. 3.23). We will then be able to decompose these automata into cascades.

1.3 Sequential switching networks

In this section we briefly show that any finite automaton can be realized by a physical object.

1.3.1 Definition: Automaton A realizes automaton B if B has the same next-state and output tables as a subautomaton $\hat{A}$ of A, where $\hat{A} = |\hat{X}, \hat{Y}, \hat{Q}, \hat{\delta}, \hat{\omega}|$ is a subautomaton of $A = |X, Y, Q, \delta, \omega|$ if $\hat{X} = X, \hat{Y} = Y, \hat{Q} \subseteq Q$ and $\hat{\delta} = \delta|_{\hat{Q} \times X}$ and $\hat{\omega} = \omega|_{\hat{Q} \times X}$.

Let $B_2 = \{(0, L), +, \cdot, {}^-\}$ be a Boolean algebra (the switching algebra). Its operations are given in Tab. 3. Given automaton A, the first step is to code the alphabets X, Y, Q of A into 0-L-tuples. When this is done, the state tables can be expressed by a system of Boolean equations. This system of equations can then be transposed into a logical network. Its building elements are gates (Fig. 4) and delay elements (Fig. 5), e.g., the D-Flip-Flop.

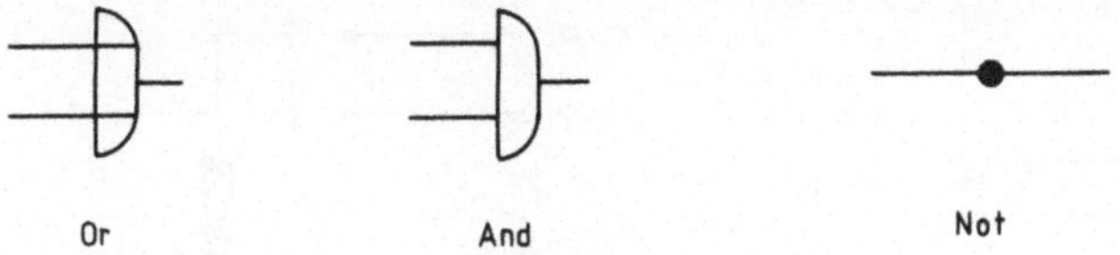

Fig. 4 Gates

Tab. 3

·	0	L	+	0	L	–	
0	0	0	0	0	L	0	L
L	0	L	L	L	L	L	0

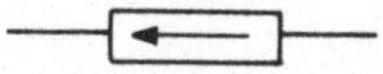

Fig. 5 Delay element

It is wellknown that logic gates can be physically realized by electronic modules. Hence, any finite automaton can be realized by electronic devices (switching networks). Examples are given below (Figs. 8–11). A difficult task which arises when automata are realized by physical structures is the synchronization of feedback loops (cf. Fig. 6). An interconnection of automata without feedback lines such as a cascade is called a loopfree interconnection. Fig. 11 shows an interconnection of two networks realizing $A(Z_3)$ that is not loopfree.

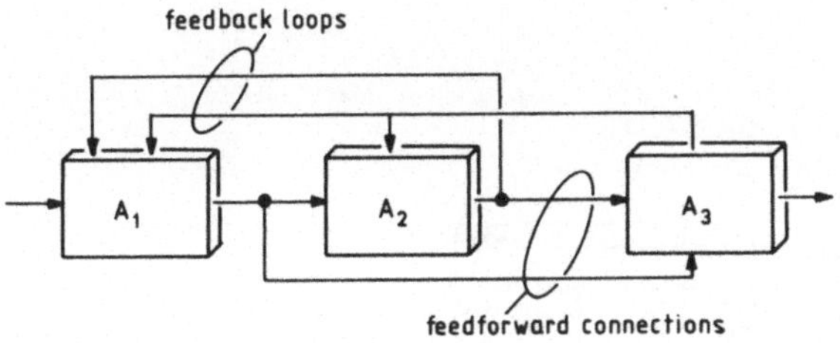

Fig. 6
Feedback and feedforward

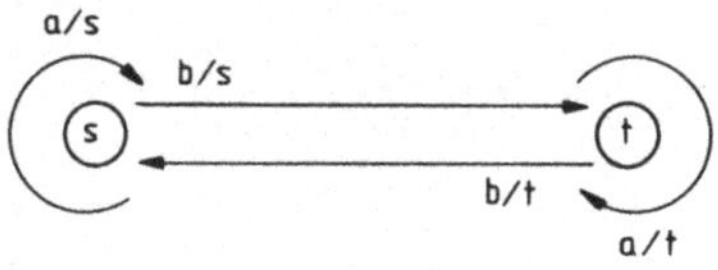

Fig. 7 T, T-Flip-Flop, $M^T = C(2)$

1.3.2 Examples: (a) The T-Flip-Flop

Its state graph is given in Fig. 7.

Coding. $s \mapsto 0, t \mapsto L; a \mapsto 0; b \mapsto L$.

Boolean equations:

Next state: $q(t+1) = \overline{x(t)} \cdot q(t) + x(t) \cdot \overline{q(t)}$

Output: $y(t) = q(t)$. This flip-flop is a linear automaton.

Voltage 0 (+ U) corresponds to the logical value 0(L).

(b) A counter

1.3.3 Exercise: Determine the next-state and output table of the sequential network and of its modules (A, B) given in Fig. 11.

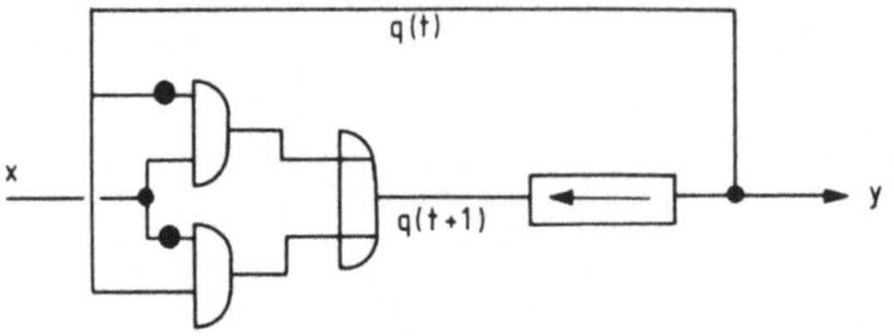

Fig. 8 T-Flip-Flop (logical net)

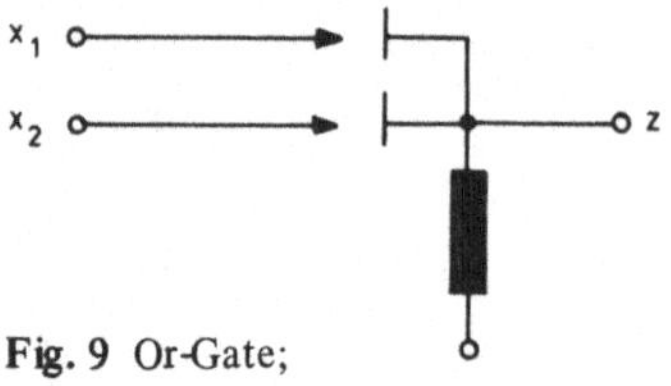

Fig. 9 Or-Gate;
$U_z = \max(0, U_{x_1}, U_{x_2})$

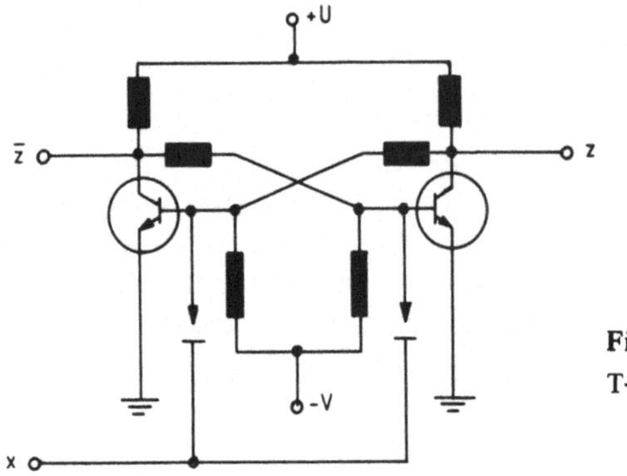

Fig. 10
T-Flip-Flop (realized by an electronical network)

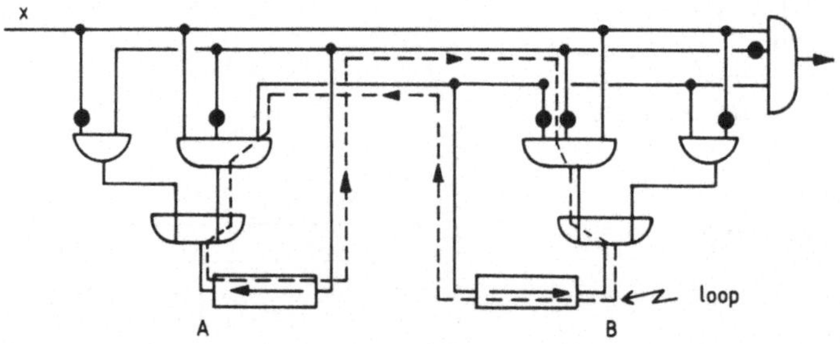

Fig. 11
Counter (modulo 3) (realization)

2 Transformation monoids and cascades

2.1 Simulation of automata and covering of transformation monoids

Automaton A *simulates* automaton B if A processes inputs in the same way as automaton B does provided the inputs and outputs of A are appropriately translated by coders, c.f. Fig. 12.

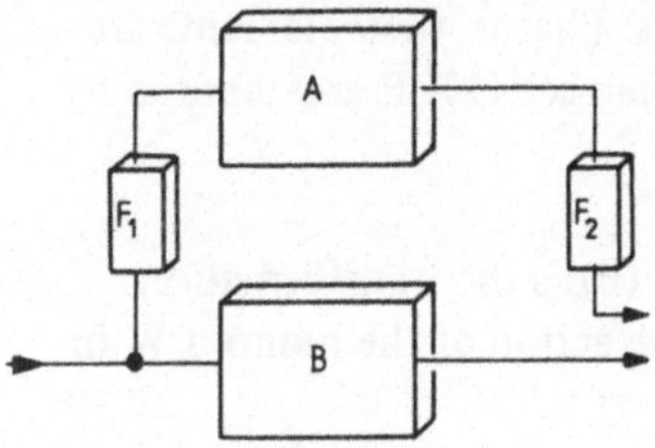

Fig. 12 Simulation of B by A
Coders: $F_1 : X_B \to X_A$, $F_2 : Y_A \to Y_B$

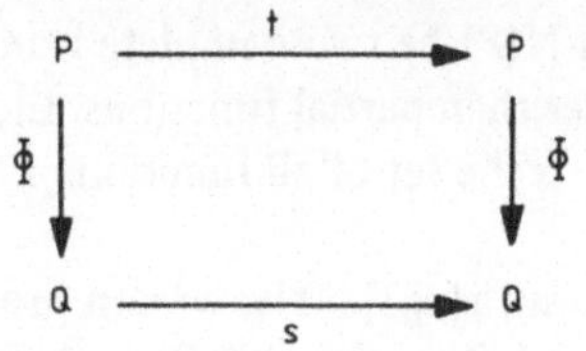

Fig. 13 Covering of transformation monoids

A simulates B if, roughly, the computational power of A is not less than that of B. The computational power of coders is neglected since coders only "translate" from one alphabet into another alphabet and do not process information.

The precise definition of the concept of simulation requires a quite formal set up. Moreover, there are many slightly different definitions. Instead of giving a formal definition we will present examples that clarify what is meant.

Note, that realization (as defined in Sec. 1.3) is a special case of simulation. It can also be shown that automaton A_{M_A} simulates automaton A [5].

In Sec. 1.2.2 we defined coders and cascades of automata. In a more abstract setting coders and cascades correspond to coverings and wreath products of transformation monoids.

2.1.1 Definition [5]: Let $M(Q) = (M, Q)$ and $N(P) = (N, P)$ be two transformation monoids. A surjective partial function $\Phi : P \to Q$ is called a covering if for each $s \in M$ there is a $t \in N$ such that for $p \in P$:

$$s(p\Phi) = (t(p))\,\Phi \tag{1}$$

whenever $s(p\Phi) \neq \emptyset$.

That is, Φ is a covering if the diagram of Fig. 13 commutes. (Note, that coverings, like morphisms, are applied from the right side in order to save brackets. For monoid actions we use the notations $t : p \mapsto t(p)$ or $t : p \mapsto tp$.)

We say that t covers s and write $s \prec t$. If Φ is a covering (as in Fig. 13) we write $M(Q) \underset{\Phi}{\prec} N(P)$ or $M(Q) \prec N(P)$. Covering is a transitive relation between tm's. (Exercise)

Observe, that the covering of $M(Q)$ by $N(P)$ can also be considered as a simulation of the automaton A_M of $M(Q)$ by the automaton A_N of $N(P)$, where Φ defines the appropriate coders for Fig. 12.

2.2 Cascades and wreath products

We now introduce a (loopfree) combination of transformation monoids which gives rise to a new transformation monoid.

Let M(Q) and N(P) be two complete transformation monoids. That is, their elements are functions rather than partial functions. Elements of the cartesian set $Q \times P$ are denoted by $\langle q, p \rangle$ and M^P is the set of all functions $f : P \to M : p \mapsto pf$.

2.2.1 Definition [5], [7]: The wreath product of M(Q) by N(P) is the transformation monoid $M(Q) \cdot N(P) = (W, Q \times P)$ with $W = (M^P \times N, *)$. The action of the monoid W on $Q \times P$ is given by

$$(f, t) \langle q, p \rangle = \langle (pf)(q), t(p) \rangle \qquad (2)$$

$f \in M^P$, $t \in N$. This action determines also the multiplication law of W.

For $f, g \in M^P$ we define $p(g + f) := (pg) \cdot (pf)$. However, note, that the operation + is not necessarily commutative. The multiplication law in W is then obtained by observing that $(g, n)((f, t) \langle q, p \rangle) = (g, n) \langle (pf)(q), t(p) \rangle = \langle (t(p)g)[(pf)(q)], n(t(p)) \rangle =$ $(tg + f, nt) \langle q, p \rangle$, where $p(tg) := t(p)g$ and $tg + f \in M^P$. Hence, (cf. Chap. 1, Sec. 3.3)

$$(g, n) * (f, t) = (tg + f, nt) \qquad (3)$$

Thus, the multiplication in W and the action of W are triangular. Clearly, if $(f, t) \langle q, p \rangle = (g, n) \langle \bar{q}, p \rangle$ for all $\langle q, p \rangle \in Q \times P$ then $(f, t) = (g, n)$.

2.2.2 Exercise. Show, that the multiplication law in W is associative. The neutral element of W is $(p \mapsto i_Q, i_P)$. It is the identity transformation on $Q \times P$.

2.2.3 Cascade decompositions: Let Z be the wreath product of M(Q) by N(P) as defined above and consider the automaton

$$C := A_Z = | W, Q \times P, Q \times P, \delta_C, pr_{Q \times P} | \qquad (4)$$

with δ_C the action of W on $Q \times P$. Automaton C is a cascade of the automata A_N and A_M, This cascade is shown in Fig. 14 where $F : P \times M^P \to M : (p, f) \mapsto pf : q \mapsto pf(q)$.
Observe, that $\delta_C : (Q \times P) \times W \to Q \times P$: (state : $\langle q, p \rangle$, input : (f, t)) $\mapsto$ next state : $\langle (pf)(q), t(p) \rangle$.
Now, a wreath product covering the tm M_A of automaton A leads to a simulation of A_{M_A} – and hence, of A (s. 2.1.1) – by a cascade. Therefore, it is of outmost importance for our purpose to find relations like

$$M_A \blacktriangleleft Z_1 \cdot (Z_2 \cdot (\ldots \cdot (Z_{n-1} \cdot Z_n) \ldots)) \qquad (5)$$

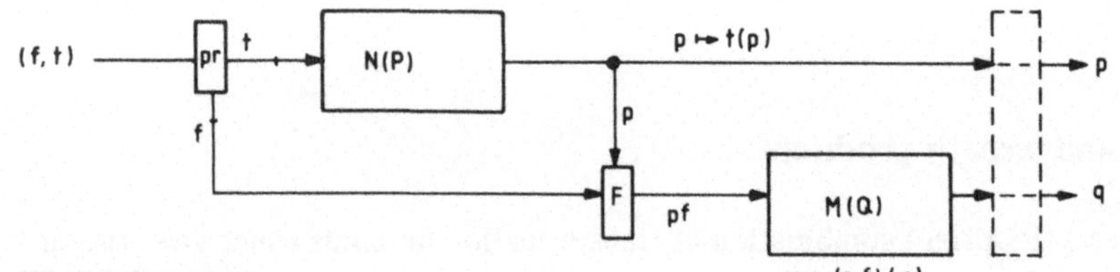

Fig. 14 Cascade C ·

where the tm's Z_i are in some way smaller or simpler than M_A and no longer decomposable in any meaningful way. If, in addition, each module of the cascade arising from (5) has fewer states than automaton A itself, then this decomposition of A into a cascade (or more precisely the simulation of A by the cascade) is called effective, [8], [9], [10], [14]. Thus, the decomposition (5) of the monoid of automaton A uncovers the internal loop-free structure of A and shows the flow of information within A. Therefore, the computational power of A – or, in modern terms, its complexity – is best characterized by this decomposition. Examples are given in Sec. 3.2. Moreover, an effective decomposition may help the system designer to break up a complex discrete information processing system (e.g., computer parts) into many smaller and more easily understood modules. These modules are also fairly decoupled such that no feedback of information between them arises. A system designed in this way can be more easily adapted to changes in its objectives.

3 Structures of finite automata

3.1 Simulation of group automata

A permutation automaton is any finite automaton $A = |X, Y, Q, \delta, \omega|$ with $M^A \leqslant \mathrm{Sym}\, Q$. Let G be a finite group and G(G) its right regular representation. The automaton of G(G) is a permutation automaton, viz. $\hat{A}_G := A_{G(G)}$; it will be called a group automaton. We first show that G(G) is covered by a wreath product and then conclude, that $\hat{A}_G$ can effectively be decomposed into a cascade consisting of a group automaton and an automaton of an associated transformation group.

3.1.1 Let $H \leqslant G$ and $H_G = \bigcap_{g \in G} g^{-1} Hg$ (see Fig. 15). H_G is the largest subgroup of H which is invariant in G and $G_{|H} := (G/H_G, H/G)$ is a transformation group; H/G is the set of right cosets of H and G/H_G the factor group of G modulo H_G. The action of the factor group G/H_G on the cosets Hg of H is given by $Hg \mapsto (Hg)(H_G\bar{g}) = HH_G g\bar{g} = Hg\bar{g}$. This action is faithfull. If $Hg\bar{g} = Hg\bar{\bar{g}}$ for all $Hg \in H/G$ then $g\bar{g} \in Hg\bar{\bar{g}}$ for all $g \in G$. Hence, $\bar{g} \in H_G\bar{\bar{g}}$. Clearly, H/G is a homogeneous space of G/H_G.

3.1.2 Theorem: Let $H \leqslant G$, then $G(G) \blacktriangleleft H(H) \cdot G_{|H}$. Special cases are:
(1) $G_{|H} = (G/H, H/G)$ if $H \trianglelefteq G$, (2) $G_{|H} = (G, H/G)$ if G is simple.

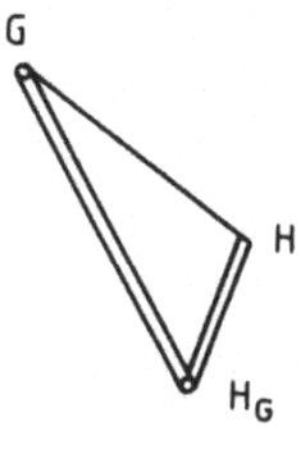

Fig. 15

Proof: Let c_i be coset leaders of H. Clearly, the map $\Phi : H \times H/G \to G : (h, c_i) \mapsto hc_i$ is bijective. Let $\Phi^{-1}(g) = (h_g, c_g)$. ($\Phi^{-1}$ is sometimes called the Lagrange map and (h_g, c_g) are Lagrange coordinates of g with respect to H.) Given $g \in G$ and c_i, the map $f_g : H/G \to H : c_i \mapsto h_{c_i g} = c_i g\, c_{c_i g}^{-1}$ is well defined. We show that $g \blacktriangleleft (f_g, d_g)$ where d_g is the second Lagrange coordinate of g with respect to H_G. $[(f_g, d_g)\langle h, c_i\rangle]\,\Phi = [\langle h(c_i f_g), d_g(c_i)\rangle]\,\Phi = [\langle hh_{c_i g}, c_{c_i g}\rangle]\,\Phi = hc_i g = [\langle h, c_i\rangle]\,\Phi g = g(\langle h, c_i\rangle\,\Phi)$. ■

It is now easy to see that the cascade C of Fig. 16 and automaton $\hat{A}_G$ are given by the same tables (if we identify (h_g, c_g) with g). Hence, C realizes $\hat{A}_G$. However, $\hat{A}_G$ simulates any automaton A with monoid G. Thus, the cascade C simulates these automata. The decomposition is effective if H is a nontrivial subgroup of G.

Observe, that $K : g \mapsto (f_g, d_g)$, $F : H/G \times H^{H/G} \to H$ and $\delta(\bar{g}\Phi^{-1}, g) = (f_g, d_g)(\langle h_{\bar{g}}, c_{\bar{g}}\rangle) = \langle h_{\bar{g}}(c_{\bar{g}} f_g), d_g(c_{\bar{g}})\rangle = \langle h_{\bar{g}}\, c_{\bar{g}}\, g\, c_{c_{\bar{g}} g}^{-1}, c_{c_{\bar{g}} g}\rangle = \langle \bar{g}g\, c_{\bar{g}g}^{-1}, c_{\bar{g}g}\rangle = \bar{g}g\Phi^{-1}$.

Thus, if C is in state $\bar{g}\Phi^{-1}$ and receives input g, its output is $\bar{g}g$ and its next state is $\bar{g}g\Phi^{-1}$.

3.1.3 Exercise: Consider the diagram of Fig. 17 and show that $G(G) \blacktriangleleft K(K) \cdot G_{|H} \cdot H_{|K \cap H}$.

Now, let G be a finite group and $N = \{G = G_0, G_1, \ldots, G_m = e\}$ a subnormal series of G. It follows that G(G) is covered by an iterated wreath product of $H_i(H_i)$, where $H_i = G_{i-1}/G_i$, and the Jordan-Hölder theorem tells us that the factors H_i can be chosen so that they are simple groups. Furthermore, $\hat{H}_i(H_i) \blacktriangleleft G(G)$. This wreath product corresponds to an iterated cascade decomposition of $\hat{A}_G$.

3.1.4 Corollary: Any finite permutation automaton can be simulated by an iterative cascade of automata whose monoids are simple groups. Hence, automata of simple groups can be considered as standard modules for finite automata. Particularly, an automaton can entirely be hooked up from counters modulo p (p prime) if its monoid is a solvable group (cf. Chap. 1, Sec. 3.6).

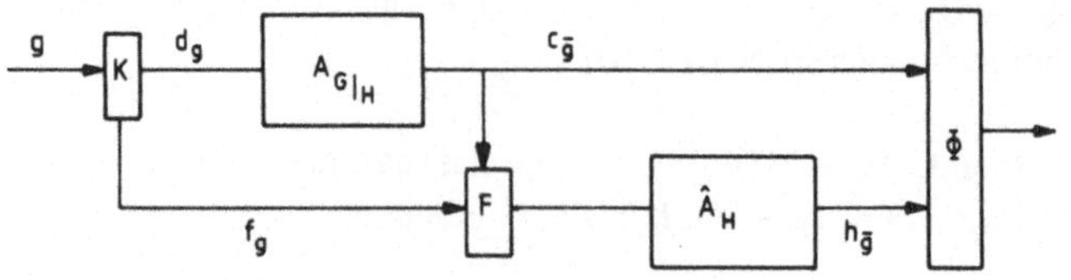

Fig. 16 Cascade $C = |G, G, H \times H/G, \delta, \omega|$

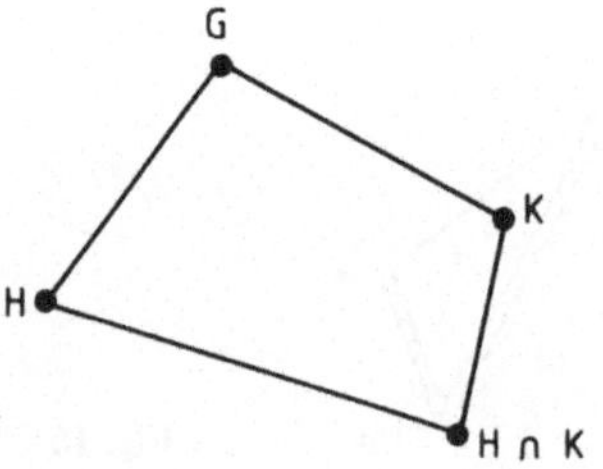

Fig. 17

3.2 Examples

(1) *The counter modulo* m. Its monoid is the cyclic permutation group of order m acting on $\{0, 1, \ldots, m-1\}$. For m = 6, $C(6) = C(2) \times C(3) \blacktriangleleft C(2) \cdot C(3)$ where C(2), C(3) are simple groups. Now, $A_{C(2)}$ has the same state table as the T-Flip-Flop and $A_{C(3)}$ is a counter modulo 3. Thus, $A(Z_6)$ can be realized as a cascade of a counter modulo 3 (Fig. 11) and a T-Flip-Flop. This cascade is a *parallel* connection (Fig. 18).
(2) Consider the automaton given by Fig. 19. Its monoid is $C(4) \blacktriangleleft C(2) \cdot C(2)$. Thus, this automaton can be realized by a cascade of two T-Flip-Flops. This cascade is a *series* connection (Fig. 20).

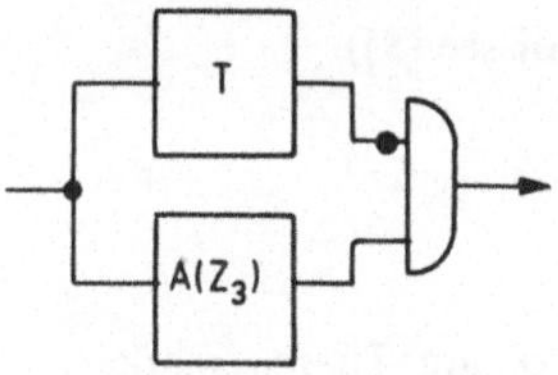

Fig. 18 Counter modulo 6

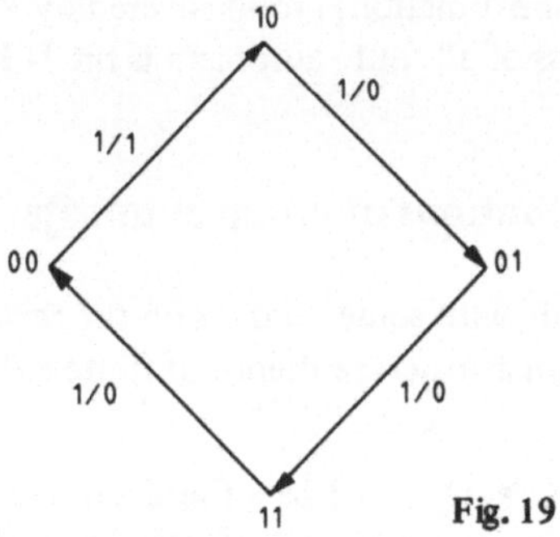

Fig. 19 Example 2

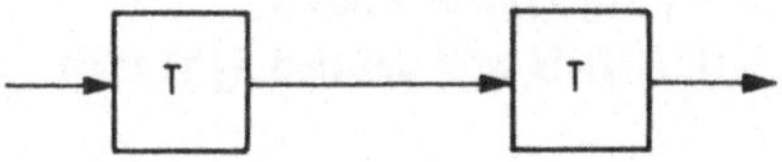

Fig. 20 Example 2

3.3 The Krohn-Rhodes-Eilenberg decomposition of finite automata

Finally, we consider the general case. The following theorem provides another step in the proof of the general decomposition theorem. Remember, that U = U(M), viz. the set of invertible elements of the monoid M, is a maximal subgroup of M (Chap. 1, Sec. 3.6). Let 1 be the unit of M and $V = (M - U) \cup 1$.

3.3.1 Theorem: Let M(Q) be a transformation monoid. Then

$$M(Q) \blacktriangleleft V(Q) \cdot U(U) \qquad (6)$$

where V(Q) is the induced transformation monoid on Q.

Proof [5]: (1) M − U is an ideal of M. Hence, V is a monoid.
(2) $\Phi : Q \times U \to Q : \langle q, g \rangle \mapsto g(q)$ is a covering. Clearly, Φ is surjective since $Q1 = Q$. Let $f_s : U \to V : g \mapsto gsg^{-1}, s \in V$. Since V is an ideal, $gsg^{-1} \in V$. We show, that $m \blacktriangleleft (f_m, 1)$ if $m \in V$ and $m \blacktriangleleft (f_1, m)$ if $m \in U$. Now,

(i) $m(\langle q, g\rangle \Phi) = m(g(q)), g \in U, m \in M, q \in Q$. On the other hand,
(ii) $m \in V : ((f_m, 1)\langle q, g\rangle)\Phi = \langle g^{-1}(gm(q)), g\rangle\Phi = m(g(q))$.
(iii) $m \in U : ((f_1, m)\langle q, g\rangle)\Phi = \langle q, gm\rangle\Phi = m(g(q))$. ■

3.3.2 Exercise: Show, that $C_{m,r} \blacktriangleleft C_{1,r} \times C(m)$.
The building elements of A_U are automata of simple groups (Sec. 3.1). It remains to find the building elements of A_V. (Note, that V has no invertible elements except 1).

3.3.3 The Decomposition Theorem: Each finite transformation monoid Z admits a decomposition into an iterated wreath product where each factor is either M_D or a simple (transformation) group covered by Z. Thus, the only additional standard module for the class of all finite automata is the D-Flip-Flop. (For a proof see [5]).

3.4 Realizations of group automata

We conclude with some remarks on the *realizations* of group automata. These remarks are based on a structure theory of finite automata developed by Hartmanis and Stearns [11].

Let $A = |X, Y, Q, \delta, \omega|$ be a finite automaton. An *A-partition* is a partition Π of Q such that: If $q \equiv \bar{q} \pmod{\Pi}$ and $s \in M^A$, then $s(q) \equiv s(\bar{q}) \pmod{\Pi}$. An A-partition gives rise to an automaton A/Π (the *factor semi-automaton* of A modulo Π) as follows
$A/\Pi := |X, \Pi, \Pi, \delta_\Pi, pr_1|$ where $\delta_\Pi([q], x) = [\delta(q, x)]$; A/Π is well defined. ([q] is the equivalence class of q modulo Π.)

3.4.1 Theorem: [11]

(1) Given automaton A, the A-partitions form a lattice denoted by L(A);
(2) Automaton A can be realized as series connections of automata with fewer states iff L(A) is nontrivial;
(3) Automaton A can be realized as a parallel connection of two automata A_1 and A_2, each with fewer states than A, iff there are nontrivial A-partitions Π_1 and Π_2 in L(A) such that $\Pi_1 \cup \Pi_2 = Q$. Then A_i is the factor semi-automaton A/Π_i.

Thus, the lattice L(A) gives us information on all possible series-parallel realizations of A.

3.4.2 Now, assume that the monoid of A is a transformation group G(Q) which acts transitively on Q. The state space Q of A can be identified with the coset space G/G_q. (G_q is the isotropy subgroup at an arbitrary state q of A, see Chap. 1). The lattice L(A) is (lattice-)isomorphic to the lattice $U_q(G) = \{U \mid G_q \leqslant U \leqslant G\}$. Hence, the subgroup lattice U(G) of G gives us the same information on the series-parallel structures of A as the lattice L(A) does.

For example, the lattice of Fig. 3.4 (Chap. 1, Sec. 3.2) tells us that no four-state automaton with transformation group S(4) has an effective series-parallel realization since $G_q \cong S^i(3)$ is a maximal subgroup of S(4). Hence, $|U_q(S(4))| = 2$, and L(A) is trivial.

References

[1] *Zadeh, L. A., Polak, E.;* System Theory, McGraw-Hill Inc., New York, 1968.

[2] *Pichler, F.;* Mathematische Systemtheorie, Walter de Gryter, Berlin, 1975.

[3] *Arbib, M. A.;* Theories of Abstract Automata, Prentice Hall, Englewood Cliffs, 1969.

[4] *Arbib, M. A., Manes, E. G.;* A Category-theoretic approach to systems in a fuzzy world, in Systems: Approaches, Theories, Applications, W. E. Harnett (ed.), D. Reidel Pub. Comp., Dordrecht, 1, 26, 1975.

[5] *Eilenberg, S.;* Automata, Languages, and Machines, Vol. A, B, Academic Press, New York, 1974/75.

[6] *Salomaa, A.;* Formal Languages, Academic Press, New York, 1973.

[7] *Wells, C.;* Some applications of the wreath product construction, Am. Math. Month. **83**, 317–338, 1976.

[8] *Dal Cin, M., Dilger, E.;* On effective structures of automata, Group Theoretical Methods in Physics, P. Kramer and A. Rieckers (eds.), Springer Lecture Notes in Physics, 79, 467–469, Springer Verlag Berlin-Heidelberg-New York, 1978.

[9] *Dilger, E.;* On permutation reset automata, Information and Control, Vol. 30, 86–95, 1976.

[10] *Nozaki, A.;* Practical decomposition of automata, Information and Control **36**, 275–291, 1978.

[11] *Hartmanis, J., Stearns, R. E.;* Algebraic Strucutre Theory of Sequential Machines, Prentice-Hall, Englewood Cliffs, 1968.

[12] *Jürgensen, H.;* Some Applications of the Theory of Semigroups to Automata; in Group Theoretical Methods in Physics, P. Kramer and A. Rieckers eds., Springer Lecture Notes in Physics, 79, 307–322, Springer Verlag, Berlin-Heidelberg-New York, 1978.

[13] *Zalcstein, Y.;* On the semigroup of linear sequential machines, Int. Journ. of Computer and Information Sciences Vol. **2**, 25–28, 1973.

[14] *Hotzel, E.;* Zur schleifenfreien Zerlegung von Mealy-Automaten, Mitteilungen der Ges. f. Mathem. u. Datenverarb. Nr. 29, Bonn, 1974.

The author wishes to acknowledge his debt to M. A. Arbib, M. Conrad, E. Dilger, W. Güttinger, M. Hazewinkel and H. Jürgensen.

Chapter IX

On the (Internal) Symmetry Groups of Linear Dynamical Systems

M. Hazewinkel

1 Introduction and statement of the main definitions and results

A time invariant linear dynamical system is a set of equations

$$(1.1)\quad \begin{array}{lcl} \dot{x} = Fx + Gu & & x(t+1) = Fx(t) + Gu(t) \\ y = Hx & (\Sigma) & y(t) = Hx(t) \\ \text{(continous time)} & & \text{(discrete time)}\,, \end{array}$$

where $x \in X = \mathbb{R}^n$, $u \in U = \mathbb{R}^m$, $y \in Y = \mathbb{R}^p$ and where F, G, H are matrices with coefficients in $\mathbb{R}$ of the dimensions $n \times n$, $n \times m$, $p \times n$ respectively. We speak then of a system of dimension n, $\dim(\Sigma) = n$, with m inputs and p outputs. Of cource the discrete time case also makes sense over any field k, (instead of $\mathbb{R}$). The spaces X, U, Y are respectively called state space, input space and output space. The usual picture is a "black box".

(1.2)

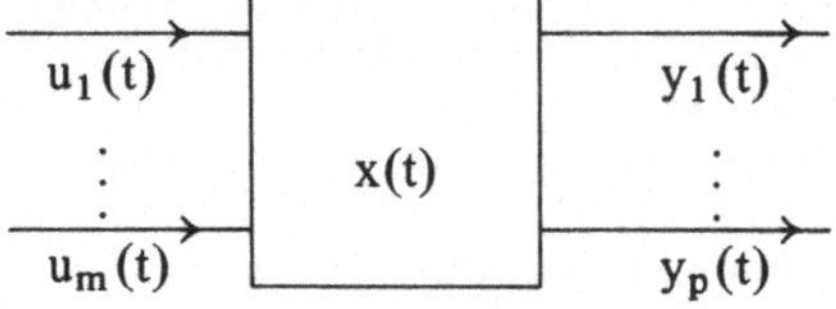

That is, the system Σ is viewed as a machine which transforms an m-tuple of input or control functions $u_1(t), \ldots, u_m(t)$ into a p-tuple of output or observation functions $y_1(t), \ldots, y_p(t)$. Many physical systems can be viewed as such a "black box". For instance the box may be a chemical reaction vat. The $u_1(t), \ldots, u_m(t)$ may be concentrations of various chemicals which are inserted and the $y_1(t), \ldots, y_p(t)$ represent certain series of measurements serving as indicators that everything goes as we wish (or not). Especially the output aspect (represented by the matrix H) captures something very often encountered in physics, electronics, chemistry, and also astronomy: only certain functions of the state variables $x_1(t), \ldots, x_n(t)$ are directly observable! Thus in astronomy one has to make do with certain projections (against the sky sphere) of the space variables describing, e.g., the solar system, in atomic physics one may have to rely only on scattering data, and, as a last example, in economics one uses socalled economic indices, which, hopefully, reflect more or less accurately the goings on of the "real" (largely unknown) underlying economic processes.

The formulas expressing y(t) in terms of the u(t) are

$$y(t) = He^{Ft}x(0) + \int_0^t He^{F(t-\tau)}Gu(\tau)d\tau,$$

(1.3)

$$y(t) = HF^t x(0) + \sum_{i=0}^{t-1} HF^{t-i-1}Gu(i),$$

where x(0) is the state of the system at time 0 (and where we start putting in input at time t = 0). Thus the input-output behaviour of our box depends of course on the initial state x(0). One is particularly interested in the input-output behaviour of Σ when x(0) = 0. We shall write $f(\Sigma)$ for the associated input-output operator. Thus

(1.4) $f(\Sigma) : u(t) \mapsto \int_0^t He^{F(t-\tau)}Gu(\tau)d\tau,\ f(\Sigma) : u(t) \mapsto \sum_{i=0}^{t-1} HF^{t-i-1}Gu(i)$

It is now an important fact that the input-output behaviour description of the machine (1.2) is degenerate, much as, say, energy levels in atomic physics may be degenerate. More precisely the matrices F, G, H (and the initial state x(0)) depend on the choice of a basis in state space and from the input-output behaviour of the machine there is (without changing the machine) no way of deciding on a "canonical" basis for the state space $X = \mathbb{R}^n$. More mathematically we have the following. Let $GL_n(\mathbb{R})$ be the group of all invertible real $n \times n$ matrices and let $L_{m,n,p}(\mathbb{R})$ be the space of all triples of matrices (F, G, H) of dimensions $n \times n$, $n \times m$, $p \times n$ respectively. The group $GL_n(\mathbb{R})$ acts on $L_{m,n,p}(\mathbb{R})$ and $\mathbb{R}^n$ = space of initial states, as

(1.5) $(F, G, H)^S = (SFS^{-1}, SG, HS^{-1}),\ x(0)^S = Sx(0)$

and as is easily checked the associated input-output behaviour of the corresponding machine as given by (1.3) and (1.4) is invariant under this action of $GL_n(\mathbb{R})$; i.e., in particular $f(\Sigma^S) = f(\Sigma)$. This action corresponds to base change in state space. Indeed if $x' = Sx$ and $\dot{x} = Fx + Gu$, $y = Hx$ then $S^{-1}\dot{x}' = FS^{-1}x' + Gu$, $y = HS^{-1}x'$ so that $\dot{x}' = SFS^{-1}x' + SGu$, $y = HS^{-1}x'$ and $x'(0) = Sx(0)$.

This chapter is concerned with those aspects of the theory of linear dynamical systems which are more or less directly related to the presence of the internal symmetry group $GL_n(\mathbb{R})$ of the internal description of linear dynamical systems by triples of matrices (cf. (1.1)) as compared to the degenerate external description by means of the operator $f(\Sigma)$ (or (1.3)). This is not really a research paper (though it does in fact contain a few new results) but rather a graduate level expository account of some of the material of [3–8] and immediately related matters.

In the remaining part of this introduction we give a slightly informal description of most of the main results of sections 2–8 below.

We shall concentrate on the continuous time case.

1.6 Feedback and how to resolve the external description degeneracy. In the case of atomic physics a degenerate energy level may be split by means of, e.g., a suitable magnetic field. One can ask whether there exists something analogous in our case of degenerate external (= observable) descriptions of linear dynamical systems. There does in fact exist some such thing. It is called state space feedback. Consider the system (1.1). Introduction of state space feedback L changes it to the system $\Sigma(L)$

$$(1.7)\quad \begin{aligned} \dot{x} &= (F + GL)x + Gu \\ y &= Hx \end{aligned}$$

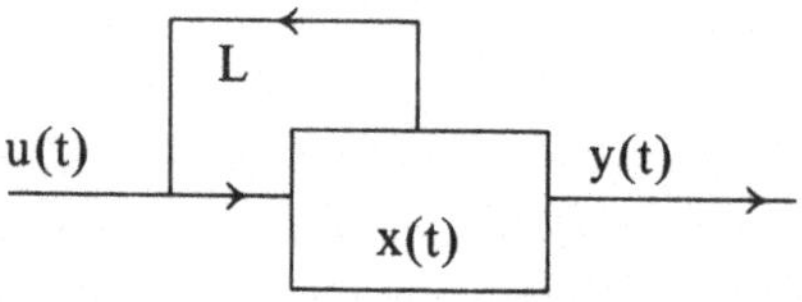

In thinking about these things the author has found it helpful to visualize a linear dynamical system with (variable) feedback as a set of n-integrators, 1, ..., n, interconnected by means of the matrix F, a set of m input points connected to the integrators by means of the matrix G, a set of p output points connected to the integrators by means of the màtrix H and a set of connections from the integrators to the input points (feedback) which may be varied in strength by the experimentator (as in atomic physics the splitting magnetic field may be varied). Cf. also the picture below.

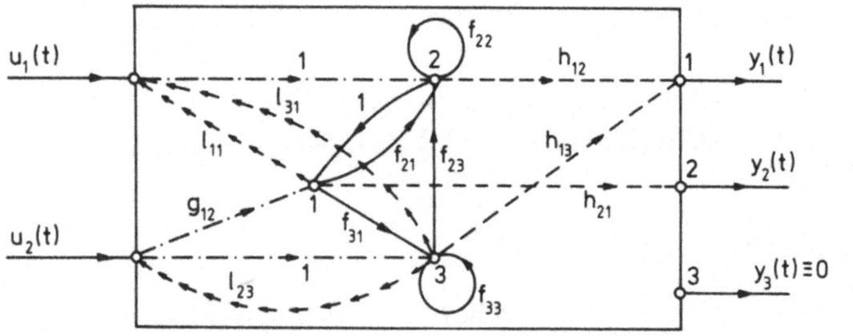

Fig. 1

——→—— interconnections between the integrators as given by the matrix F

$$F = \begin{pmatrix} 0 & 1 & 0 \\ f_{21} & f_{22} & f_{23} \\ f_{31} & 0 & f_{33} \end{pmatrix}$$

–·–·→·– connections from the input points to integrators as given by the matrix G

$$G = \begin{pmatrix} 0 & g_{12} \\ 1 & 0 \\ 0 & 1 \end{pmatrix}$$

- - →- - connections from the integrators to the output points as given by the matrix H

$$H = \begin{pmatrix} 0 & h_{12} & h_{13} \\ h_{21} & 0 & 0 \\ 0 & 0 & 0 \end{pmatrix}$$

→ → → connections from the integrators to the input points (can be varied in strength by the experimentator) as given by the matrix L

$$L = \begin{pmatrix} l_{11} & 0 & l_{13} \\ 0 & 0 & l_{23} \end{pmatrix}$$

Now let $\Sigma = (F, G, H)$ and $\Sigma' = (F', G', H')$ be two linear dynamical systems, and suppose that Σ and Σ' are completely reachable and completely observable. (This is an entirely natural restriction in this context, cf. 1.12 below; for a precise definition of the notions, cf. 2.6 below). Suppose that $\Sigma \neq \Sigma'$ but $f(\Sigma) = f(\Sigma')$. Let $\Sigma(L)$, $\Sigma'(L)$ be the systems obtained by introducing the feedback L, i.e. $\Sigma(L) = (F + GL, G, H)$, $\Sigma'(L) = (F' + G'L, G', H')$. Then there is a suitable feedback matrix L, which can be taken arbitrarily small (so that $\Sigma(L)$ and $\Sigma'(L)$ are still completely reachable and observable) such that $f(\Sigma(L)) \neq f(\Sigma'(L))$. I.e. feedback splits the $GL_n(\mathbb{R})$ – degenerate external description of linear dynamical systems.

1.8 Realization theory. Let Σ be a linear dynamical system (1.1). Then, if we leave Σ unchanged, from our observations we can deduce the operator $f(\Sigma)$ or, equivalently, we can find the sequence of matrices $A(\Sigma) = (A_0, A_1, A_2, \ldots)$, $A_i = HF^iG$. To obtain these use δ-functions and derivates of δ-functions as inputs. Another way to see this is to apply Laplace transforms to (1.1). This gives

(1.9) $s\hat{x}(s) = F\hat{x}(s) + G\hat{u}(s), \hat{y}(s) = H\hat{x}(s)$,

so that the relation between the Laplace transforms $\hat{y}(s)$, $\hat{u}(s)$ of the outputs $y(t)$ and inputs $u(t)$ is given by multiplication with the socalled transfer matrix $T(s)$

(1.10) $\hat{y}(s) = T(s)\hat{u}(s)$, $T(s) = H(s - F)^{-1}G$.

The power series development of $T(s)$ in powers of s^{-1} (around $s = \infty$) is now

(1.11) $T(s) = A_0 s^{-1} + A_1 s^{-2} + A_2 s^{-3} + \ldots$.

The question now naturally arises: when does a sequence of $p \times m$ matrices $A = (A_0, A_1, \ldots)$ come from a linear dynamical system (1.1), or, as we shall say, when is *A realizable.*

1.12 Theorem (cf. [10]):

(i) If A is realizable by an n-dimensional system Σ then it is also realizable by an $n' \leqslant n$ dimensional system Σ' which is moreover completely reachable and completely observable.

(ii) The sequence A is realizable by an n dimensional system Σ if and only if rank $(H_s(A)) \leqslant n$ for all $s \in \mathbb{N} \cup \{0\}$.

Here $H_s(A)$ is the block Hankel matrix

$$H_s(A) = \begin{pmatrix} A_0 & A_1 & \cdots & A_s \\ A_1 & & & \vdots \\ \vdots & & & \\ A_s & \cdots & & A_{2s} \end{pmatrix}.$$

1.13 Invariants and the structure of $M^{cr,co}_{m,n,p}(\mathbb{R}) = L^{co,cr}_{m,n,p}(\mathbb{R})/GL_n(\mathbb{R})$.

Let $L_{m,n,p}(\mathbb{R})$ be the space of all triples of matrices (F, G, H) of dimensions $n \times n$, $n \times m$, $p \times n$ respectively. The group $GL_n(\mathbb{R})$ acts on $L_{m,n,p}(\mathbb{R})$ as in (1.5). The input-output matrices $A_i = HF^iG$ are clearly invariants for this action and the question arises whether these are the only invariants. Here an invariant is defined as a function ρ: $L_{m,n,p}(\mathbb{R}) \to \mathbb{R}$ (or possibly a function defined on an invariant open dense subset of $L_{m,n,p}(\mathbb{R})$) such that $\rho((F, G, H)^S) = \rho(F, G, H)$ for all triples (F, G, H) (in the open dense subset).

1.14 Theorem: Every continuous invariant of $GL_n(\mathbb{R})$ acting on $L_{m,n,p}(\mathbb{R})$ is a function of the entries of $A_0, \ldots, A_{2n-1}$.

Let $L^{co,cr}_{m,n,p}(\mathbb{R})$ be the subspace of all triples $(F, G, H) \in L_{m,n,p}(\mathbb{R})$ which are both completely observable and completely reachable. This is an open and dense subspace of $L_{m,n,p}(\mathbb{R})$. On this subspace $GL_n(\mathbb{R})$ acts faithfully and a more precise version of theorem 1.14 describes the quotient space $M^{co,cr}_{m,n,p}(\mathbb{R}) = L^{co,cr}_{m,n,p}(\mathbb{R})/GL_n(\mathbb{R})$ explicitly and gives an algorithm for recovering (F, G, H) up-to-$GL_n(\mathbb{R})$-equivalence from $A_0, \ldots, A_{2n-1}$ (cf. 4.25 below). It turns out that $M^{co,cr}_{m,n,p}(\mathbb{R})$ is a smooth differentiable manifold and that the projection $L^{co,cr}_{m,n,p}(\mathbb{R}) \to M^{co,cr}_{m,n,p}(\mathbb{R})$ is a principal $GL_n(\mathbb{R})$-bundle (cf. 6.4 below).

1.15 Canonical forms. For many purposes (prediction, construction of feedbacks, identification and, not least, for proving theorems) an internal description of a black box by means of a triple of matrices (F, G, H) is preferable over knowledge of the input-output operator $f(\Sigma)$. As was remarked in section 1.14 above there do exist algorithms for calculating some $\Sigma = (F, G, H)$ which realizes $f(\Sigma)$ or $A(\Sigma)$ from the matrices $A_0, \ldots, A_{2n-1}$. One such algorithm is described in 4.25 below. All these algorithms have the drawback that they are discontinuous in general. This is a nontrivial difficulty, because after all one calculates the (F, G, H) because one wants to use them as a basis for further calculations, design, predictions etc., and the $A_0, \ldots, A_{2n-1}$ are after all subject to (small) measurement errors. Thus the question arises whether there exist continuous methods of recovering (F, G, H) up-to-$GL_n(\mathbb{R})$-equivalence from $A_0, \ldots, A_{2n-1}$. Or, in other words, because $M^{co,cr}_{m,n,p}(\mathbb{R})$ is an explicitly describable subspace of the space of all sequences of $2np \times m$ matrices and $M^{co,cr}_{m,n,p}(\mathbb{R}) = L^{co,cr}_{m,n,p}(\mathbb{R})/GL_n(\mathbb{R})$, the question arises whether there exist continuous canonical forms on $L^{co,cr}_{m,n,p}(\mathbb{R})$, where a continuous canonical form is defined as follows.

1.16 Definition: A continuous canonical form on a $GL_n(\mathbb{R})$-invariant subspace $L' \subset L_{m,n,p}(\mathbb{R})$ is a continuous map $c : L' \to L'$ such that

(i) $c((F, G, H)^S) = c((F, G, H))$ for all $(F, G, H) \in L'$,

(ii) if $c((F, G, H)) = c((F', G', H'))$ then there is a $S \in GL_n(\mathbb{R})$ such that $(F', G', H') = (F, G, H)^S$, and

(iii) for all $(F, G, H) \in L'$ there is an $S \in GL_n(\mathbb{R})$ such that $c(F, G, H) = (F, G, H)^S$.

For some additional remarks on the desirability of *continuous* canonical forms cf. [2] and also [15]. Also our proof of the "feedback suspends degeneracy" theorem mentioned in 1.6 above is based on the use of a suitable canonical form. It turns out that there exist open dense subspaces $U_\alpha \subset L_{m,n,p}(\mathbb{R})$, which together cover $L^{co,cr}_{m,n,p}(\mathbb{R})$, on which continuous canonical forms exist. Cf. 3.10 below. On the other hand.

1.17 Theorem: There exists a continuous canonical form on all of $L^{co,cr}_{m,n,p}(\mathbb{R})$ if and only if $m = 1$ or $p = 1$.

1.18 On the geometry of $M^{co,cr}_{m,n,p}(\mathbb{R})$. Holes. Now suppose we have a black box (1.2) which is to be modelled by a linear dynamical system of dimension n. Then the input-output data give us a point of $M^{co,cr}_{m,n,p}(\mathbb{R})$ and as more and more data come in we find (ideally) a sequence of points in $M^{co,cr}_{m,n,p}(\mathbb{R})$ representing better and better linear dynamical system approximations to the given black box. The same thing happens when one is dealing with a slowly varying black box or linear dynamical system. If this sequence approaches a limit we have "identified" the black box. Unfortunately the space $M^{co,cr}_{m,n,p}(\mathbb{R})$ is never compact so that a sequence of points may fail to converge to anything whatever. There are holes in $M^{co,cr}_{m,n,p}(\mathbb{R})$. Consider for example the following family of 2-dimensional, one input, one output systems

$$(1.19) \quad g_z = \binom{1}{1},\ F_z = \begin{pmatrix} z & -z \\ 0 & -z \end{pmatrix},\ H_z = (z^2, 0),\ z = 1, 2, 3, \ldots .$$

Let $u(t), 0 \leqslant t \leqslant t_0$ be a smooth input function, then $y(t) = \lim_{z \to \infty} f(\Sigma_z)u(t)$ exists and is equal to $y(t) = \frac{d}{dt} u(t)$. This operator can not be of the form $f(\Sigma)$ for any system Σ of the form (1.1) (because the $f(\Sigma)$ are always bounded operators and $\frac{d}{dt}$ is an unbounded operator). A characteristic feature of this example is that the individual matrices F_z, G_z, H_z do not have limits as $z \to \infty$. (A not unexpected phenomenon, because after all we are taking quotients by the noncompact group $GL_n(\mathbb{R})$). This sort of situation is actually important in practice, e.g. in the study of very high gain state feedback systems $\dot{x} = Fx + Gu$, $u = cLx$, where c is a large scalar gain factor. Cf. [12].

Another type of hole in $M^{co,cr}_{m,n,p}(\mathbb{R})$ corresponds to lower dimensional systems, and in a way these two holes and combinations of them are all the holes there are in the sense of the following definitions and theorems for the case $p = m = 1$. There are similar theorems in the more input/more output cases.

1.20 Definition: We shall say that a family of systems $\Sigma_z = (F_z, G_z, H_z)$ converges in input-output behaviour to an operator B if for every m-vector of smooth input functions $u(t)$ with support in $(0, \infty)$ we have $\lim_{z \to \infty} f(\Sigma_z)\, u(t) = Bu(t)$ uniformly in t on bounded t intervals.

1.21 Definition: A differential operator of order r is an operator of the form

$u(t) \mapsto y(t) = Dy(t) = a_0 u(t) + a_1 \frac{d}{dt} u(t) + \ldots + a_r \frac{d^r}{dt^r} u(t)$, where the $a_0, \ldots, a_r$ are $p \times m$ matrices with coefficients in $\mathbb{R}$, and $a_r \neq 0$. We write $\operatorname{ord}(D)$ for the order of D. By definition $\operatorname{ord}(0) = -1$.

1.22 Theorem: Let $(\Sigma_z)_z$ be a family of systems in $L_{1,n,1}(\mathbb{R})$ which converges in input-output behaviour. Let B be the limit input-output operator. Then there exist a system Σ' and a differential operator D such that

$$Bu(t) = f(\Sigma')u(t) + Du(t)$$

and $\operatorname{ord}(D) + \dim(\Sigma') \leqslant n-1$.

1.23 Theorem: Let D be a linear differential operator and $\Sigma' \in L_{1,n,1}(\mathbb{R})$ and suppose that $\operatorname{ord}(D) + \dim(\Sigma') \leqslant n-1$. Then there exists a family of systems $(\Sigma_z)_z$, $\Sigma_z \in L^{co,cr}_{1,n,1}(\mathbb{R})$ such that for every smooth input vector $u(t)$

$$\lim_{z \to \infty} f(\Sigma_z)u(t) = f(\Sigma')u(t) + Du(t)$$

uniformely on bounded t-intervals.

1.24 Concluding introductory remarks. Many of the results described above have their analogues in the discrete case and/or the time varying case, cf. [3–8, 9–11, 14]. But not all. For instance the obvious analogues of theorems 1.23 and 1.22 fail utterly in the discrete time case. In this case $\lim_{z \to \infty} f(\Sigma_z)u(t)$ exists for all inputs $u(t)$ if and only if the individual matrices $A_i(z) = H_z F_z^i G_z$ converge for $z \to \infty$. This means that in the case of input-output convergence the limit operator is necessarily of the form $f(\Sigma')$ for some, possibly lower dimensional, system Σ'. The same answer obtains in the continuous time case if besides input-output convergence one also requires that the F_z, G_z, H_z (or more generally the $A_i(z)$) remain bounded.

A number of sections have been marked with a *: these contain additional material and can without endangering one's understanding be omitted the first time through.

2 Complete reachability and complete observability

Let F, G, H) $\in L_{m,n,p}(\mathbb{R})$ be a real linear dynamical system of state space dimension n, with m inputs and p outputs. We define

(2.1) $R_s(F, G) = (G\ FG \dots F^sG)$, $s = 0, 1, 2, \dots$, $R(F, G) = R_n(F, G)$

the $n \times (s + 1)m$ matrices consisting of the blocks $G, FG, \dots, F^sG$, and dually

$$(2.2)\quad Q_s(F, H) = \begin{pmatrix} H \\ HF \\ \vdots \\ HF^s \end{pmatrix}, s = 0, 1, 2, \dots, Q(F, H) = Q_n(F, H).$$

We also define

$$(2.3)\quad \mathcal{H}_s(F, G, H) = \mathcal{H}_s(\Sigma) = \begin{pmatrix} A_0 & A_1 & \dots & A_s \\ A_1 & & \cdot & \vdots \\ \vdots & & \cdot & \vdots \\ & \cdot & & \\ A_s & & \dots & A_{2s} \end{pmatrix} = Q_s(F, H)R_s(F, G), s = 0, 1, 2, \dots,$$

where $A_i = HF^iG$, $i = 0, 1, 2, \dots$.

It is useful to notice that

(2.4) $R_k((F, G)^S) = SR_k(F, G)$, $Q_k((F, H)^S) = Q_k(F, H)S^{-1}$,

where of course $(F, G)^S = (SFS^{-1}, SG)$, $(F, H)^S = (SFS^{-1}, HS^{-1})$. It follows that

(2.5) $\mathcal{H}_k(\Sigma^S) = \mathcal{H}_k((F, G, H)^S) = \mathcal{H}_k((F, G, H)) = \mathcal{H}_k(\Sigma)$

for all $S \in GL_n(\mathbb{R})$, which is of course also immediately clear from (2.3).

2.6 Definitions of complete reachability of complete observability. The system $(F, G, H) \in L_{m,n,p}(\mathbb{R})$ is said to be completely reachable iff rank $(R(F, G)) = n$. The system (F, G, H) is said to be completely observable iff rank $(Q(F, H)) = n$. These are generic conditions; in fact the subspace $L^{co,cr}_{m,n,p}(\mathbb{R})$ of $L_{m,n,p}(\mathbb{R})$ consisting of all systems which are both completely reachable and completely observable is open and dense. We note that (F, G, H) is co (= completely observable) and cr (= completely reachable) iff the matrix $H_n(F,G,H) = Q(F,H)\,R(F,G)$ is of rank n.

***2.7 Termilogical justification.** Let $(F,G,H) \in L_{m,n,p}(\mathbb{R})$. Then (F, G, H) is completely reachable iff for every $x_1 \in \mathbb{R}^n$ there is an input function u(t) such that the unique solution of

$$\dot{x} = Fx + Gu(t),\ x(0) = 0$$

passes through x_1; i.e. every state is reachable from zero. For a proof cf., e.g., [17, theorem 3.5.3 on page 66] or [10, section 2.3]. Instead of completely reachable one also often finds the terminology (completely state) controllable in the literature.

Dually the system (F, G, H) is completely observable iff the initial state x(0) at time zero is deducible from y(t), $0 \leq t \leq t_1, t_1 > 0$ (using zero inputs). Equivalently (F, G, H) is completely observable if the initial state x(0) is deducible from the input-output behaviour of the system on an interval $[0, t_1]$, $t_1 > 0$. Cf., e.g., [14, Ch. V, section 3] or [17, theorem 3.5.26 on page 75].

The following theorem says that as far as input-output behaviour goes every system can be replaced by a system which is co and cr. Thus it is natural to concentrate our investigations on this class of systems.

2.8 Theorem ([10]): Let $\Sigma = (F,G,H) \in L_{m,n,p}(\mathbb{R})$ with input-output operator $f(\Sigma)$. Let $n' = \mathrm{rank}(H_n(\Sigma))$. Then there exists an

$$\Sigma' = (F',G',H') \in L^{co,cr}_{m,n;p}(\mathbb{R}) \text{ such that } f(\Sigma) = f(\Sigma').$$

Proof: Let $X = \mathbb{R}^n$ be the state space of Σ. Let X^{reach} be the linear subspace of X spanned by the columns of R(F, G). Then, clearly, $G(\mathbb{R}^m) \subset X^{reach}$ and $F(X^{reach}) \subset X^{reach}$ (Because $F^n = a_0 I + a_1 F + \ldots + a_{n-1} F^{n-1}$ for certain $a_i \in \mathbb{R}$ by the Cayley-Hamilton theorem). Taking a basis for X^{reach} and completing this to a basis for X we see that for suitable $S \in GL_n(\mathbb{R})$, Σ^S is of the form

$$\Sigma^S = \left(\begin{pmatrix} G'' \\ 0 \end{pmatrix}, \left(\begin{array}{c|c} F'' & F_{12} \\ \hline 0 & F_{22} \end{array}\right), \left(H'' \mid H''_2 \right) \right)$$

where the partition blocks are respectively of the sizes:
$n'' \times m$, $n - n'' \times m$, $n'' \times n''$, $n'' \times n - n''$, $n - n'' \times n''$, $(n-n'') \times (n-n'')$, $p \times n''$, $p \times (n - n'')$ for $G'', 0, F'', F_{12}, 0, F_{22}, H'', H''_2$ respectively if $n'' = \dim X^{reach}$.
Now clearly

$$He^{F\tau}G = (HS^{-1})\,e^{SFS^{-1}\tau}SG = H''e^{F''\tau}G''$$

and rank $R(F'', G'')$ = rank $(R(SFS^{-1}, SG))$ = rank $(SR(F, G))$ = rank $R(F, G) = n''$. It follows, cf. (1.4), that Σ and $\Sigma'' = (F'', G'', H'')$ have the same input-output operator. Thus to prove the theorem it now suffices to prove the theorem under the extra hypothesis that (F, G, H) is cr. Let X_0 be the subspace of all $x \in X$ such that $HF^i x = 0$ for all $i = 0, 1, \ldots, n$; i.e., $X_0 = \mathrm{Ker}(Q(F, H))$. Then $HF^i x = 0$ for all $i = 1, 2, \ldots$, using the Cayley-Hamilton theorem. Hence $FX_0 \subset X_0$ and $HX_0 = 0$. Taking a basis for X_0 and completing it to a basis for X we see that for a suitable $S \in GL_n(\mathbb{R})$, Σ^S is of the form

$$\Sigma^S = \left(\begin{pmatrix} G_1' \\ \hline G' \end{pmatrix}, \left(\begin{array}{c|c} F_{11}' & F_{12}' \\ \hline 0 & F' \end{array} \right), (0, H') \right),$$

where G', F', H' are respectively of the sizes $n' \times m$, $n' \times n'$, $p \times n'$, n' = rank $(Q(F, H))$, which is also equal to rank $H_n(F, G, H)$ if (F, G, H) is cr.

Clearly

$$He^{F\tau}G = (HS^{-1})e^{SFS^{-1}\tau}SG = H'e^{F'\tau}G'$$

$$\mathrm{rank}(Q(F, H)) = \mathrm{rank}(Q(SFS^{-1}, SHS^{-1}) = \mathrm{rank}(Q(F', H')),$$

so that $\Sigma' = (F', G', H')$ is completely observable and $f_{\Sigma'} = f_{\Sigma}$. Also $R(SFS^{-1}, SG)$ is of the form

$$R(SFS^{-1}, SG) = \left(\frac{R'}{R(F', G')} \right).$$

But rank $R(F, G) = n$ so that the n rows of $R(SFS^{-1}, SG) = SR(F, G)$ are independent. It follows that the n' rows of $R(F', G')$ are also independent, proving that Σ' is also completely reachable.

***2.9 Pole Assignment**. A set Λ of complex numbers with multiplicities is called symmetric if with $\beta \in \Lambda$ also $\bar{\beta} \in \Lambda$ with the same multiplicity. Here $\bar{\beta}$ is the complex conjugate of β. If A is a real $n \times n$ matrix then $\sigma(A)$, the spectrum of A, is a symmetric set.

2.10 Theorem: The pair of matrices (F, G), $F \in \mathbb{R}^{n \times n}$, $G \in \mathbb{R}^{n \times m}$ is completely reachable iff every symmetric set with multiplicities of size n occurs as the spectrum of $F + GL$ for a suitable (state feedback) matrix L.

I.e. the system (F, G, H) is cr iff we can by means of suitable state feedback arbitrarily reassign the poles of the system. For a proof cf., e.g., [18, section 2.2].

3 Nice Selections and the Local Structure of $L^{cr}_{m,n,p}(\mathbb{R})/GL_n(\mathbb{R})$

3.1 Nice Selections. Let $(F, G, H) \in L_{m,n,p}(\mathbb{R})$. We use $I(n, m)$ to denote the ordered set of indices of the columns of the matrix $R(F, G)$.
I.e. $I(n, m) = \{(i, j) \mid i = 0, \ldots, n; j = 1, \ldots, m\}$ with the ordening $(0, 1) < (0, 2) < \ldots < (0, m) < (1, 1) < \ldots < (1, m) < \ldots < (n, 1) < \ldots < (n, m)$. A *nice selection* $\alpha \subset I(n, m)$ is a subset of $I(n, m)$ of size $n = \dim \Sigma$ such that $(i, j) \in \alpha \Rightarrow (i\text{-}1, j) \in \alpha$ if $i \geqslant 1$. Pictorially we represent $I(n, m)$ as an $(n + 1) \times m$ rectangular array of which the first row represents the indices of the columns of G, the second row the indices of the columns of FG, ... etc .... We indicate the elements of a subset α with crosses. The subset of the picture on the left is then a nice selection ($m = 4, n = 5$) and the subset α' of the picture on the right below is not a nice selection

.	x	.	x		.	.	x	.
.	x	.	x		.	x	.	x
.	x	.	.		.	.	x	x
.	.	.	.		.	.	.	.
.	.	.	.		.	.	.	.
.	.	.	.		.	.	.	.

If β is a subset of $I(n, m)$ we denote with $R(F, G)_\beta$ the matrix obtained from $R(F, G)$ by removing all columns whose index is not in β.
We use $L_{m,n}(\mathbb{R})$ to denote the space of all pairs of real matrices (F, G) of dimensions $n \times n$, $n \times m$ respectively.

3.2 Lemma: Let $(F, G) \in L_{m,n}(\mathbb{R})$ be a completely reachable pair of matrices. Then there is a nice selection α such that $R(F, G)_\alpha$ is invertible.

Remark: Complete reachabilitiy means that $\operatorname{rank} R(F, G) = n$, so that there is in any case some subset β of size n of $I(n, m)$ such that $R(F, G)_\beta$ is invertible. The lemma says that in that case there is also a *nice* selection for which this holds.

Proof of the lemma: Define a nice subselection of $I(n, m)$ as any subset β (of size $\leqslant n$) such that $(i, j) \in \beta, i \geqslant 1 \Rightarrow (i - 1, j) \in \beta$. Let α be a maximally large nice subselection of $I(n, m)$ such that the columns in $R(F, G)_\alpha$ are linearly independent. We shall show that $\operatorname{rank}(R(F, G)_\alpha) = \operatorname{rank}(R(F, G))$, which will prove the lemma because by assumption $\operatorname{rank} R(F, G) = n$.
Let $\alpha = \{(0, j_1), \ldots, (i_1, j_1); \ldots, (0, j_s), \ldots, (i_s, j_s)\}$.. Then by the maximality of α we know the columns of $R(F, G)$ with indices $(0, j), j \in \{1, \ldots, m\} \setminus \{j_1, \ldots, j_s\}$ and the columns of $R(F, G)$ with indices $(i_t + 1, j_t)$, $t = 1, \ldots, s$ are linearly dependent on the columns of $R(F, G)_\alpha$. With induction assume that all columns with indices $(i_t + k, j_t)$, $k \leqslant r$,

$t = 1, \ldots, s$ and $(k-1, j), k \leqslant r, j \in \{1, \ldots, m\} \setminus \{j_1, \ldots, j_s\}$ are linearly dependent on the columns of $R(F, G)_\alpha$. So we have relations

$$F^{r-1} g_j = \sum_{(i,j) \in \alpha} a(i,j) F^i g_j, \; j \in \{1, \ldots, m\} \setminus \{j_1, \ldots, j_s\}$$

$$F^{i_t + r} g_{j_t} = \sum_{(i,j) \in \alpha} b(i,j) F^i g_j, \; t = 1, \ldots, s,$$

where g_j denotes the j-th column of G. Multiplying on the left with F we find

$$F^r g_j = \sum_{(i,j) \in \alpha} a(i,j) F^{i+1} g_j$$

$$F^{i_t + r + 1} g_{j_t} = \sum_{(i,j) \in \alpha} b(i,j) F^{i+1} g_j.$$

We have already seen that the $F^{i+1} g_j$, $(i,j) \in \alpha$ are linear combinations of the columns of $R(F, G)_\alpha$. It follows that also the $F^r g_j$ and $F_t^{i+r+1} g_{j_t}$ are linear combinations of the columns of $R(F, G)_\alpha$. This finishes the induction and hence the proof of the lemma.

3.3 Successor indices. Let $\alpha \subset I(n, m)$ be a nice selection. The successor indices of α are those elements $(i, j) \in I(n, m) \setminus \alpha$ for which $i = 0$ or for which $(i', j) \in \alpha$ for all $i' < i$ if $i \geqslant 1$. For every $j_o \in \{1, \ldots, m\}$ there is precisely one successor index of α of the form (i, j_o); this successor index is denoted $s(\alpha, j_o)$. In the picture below the successor indices of α are indiced by *'s (and the elements of α with x's).

Columns of G	*	x	*	x	x_1	e_1	x_3	e_2
Columns of FG	.	x	.	x	.	e_3	.	e_4
.	.	x	.	*	.	e_5	.	x_4
.	.	*	.	.	.	x_2	.	
.	.	.	.	.	.	.	.	.
Columns of F^5G	.	.	.	.	.	.	.	.

3.4 Lemma: Let $\alpha \subset I(n, m)$ be a nice selection and $x_1, \ldots, x_m$ an m-tuple of n-vectors. Then there is precisely one pair $(F, G) \in L_{m,n}(\mathbb{R})$ such that

$R(F, G)_\alpha = I_{n \times n}$, the $n \times n$ unit matrix

$R(F, G)_{s(\alpha, j)} = x_j$ for all $j = 1, \ldots, m$.

Proof: Let f_i be the i-th column of the matrix $F, i = 1, 2, \ldots, n$. Then in the example given above the values of the $g_j, j = 1, \ldots, m$ and $f_i, i = 1, \ldots, n$ can simply be read of from the diagram. One has in this case

$$g_1 = x_1, g_2 = e_1, g_3 = x_3, g_4 = e_2$$
$$f_1 = e_3, f_2 = e_4, f_3 = e_5, f_4 = x_4, f_5 = x_2.$$

It is easy to see that this works in general and to write down the general proof though it tends to be notationally cumbersome.

3.5 Local structure of $L^{cr}_{m,n,p}(\mathbb{R})/GL_n(\mathbb{R})$. Let $\alpha \subset I(n, m)$ be a nice selection. We define

$$(3.6)\quad \begin{aligned} U_\alpha &= \{(F, G, H) \in L_{m,n,p}(\mathbb{R}) \mid \det R(F, G)_\alpha \neq 0\} \\ V_\alpha &= \{(F, G, H) \in L_{m,n,p}(\mathbb{R}) \mid R(F, G)_\alpha = I_{n \times n}\}\ . \end{aligned}$$

3.7 Lemma:

(i) $U_\alpha \simeq V_\alpha \times GL_n(\mathbb{R})$
(ii) $V_\alpha \simeq \mathbb{R}^{mn + np}$

Proof: (i) Let $(F, G, H) \in U_\alpha$. We assign to (F, G, H) the pair $((F, G, H)^S, S^{-1})$ where $S = R(F, G)_\alpha^{-1}$. Then $(F, G, H)^S \in V_\alpha$ because $R(SFS^{-1}, SG) = SR(F, G)$ and hence $R(SFS^{-1}, SG)_\alpha = SR(F, G)_\alpha$. Inversely given $((F, G, H), S) \in V_\alpha \times GL_n(\mathbb{R})$ we assign to it the element $(F, G, H)^S$. This proves (i). Assertion (ii) follows immediately from lemma 3.4. Indeed, let $z \in \mathbb{R}^{mn + np}$ and view z as an m + p tuple of n-vectors $z = (x_1, \ldots, x_m; y_1, \ldots, y_p)$. Then there are unique F, G, H such that $R(F, G)_\alpha = I_{n \times n}$, $R(F, G)_{s(\alpha, j)} = x_j$, $h_l = y_l$ where h_l is the l-th row of H.

3.8 Local structure of $L^{co,cr}_{m,n,p}(\mathbb{R})/GL_n(\mathbb{R})$. Let again α be a nice selection. Then we define in addition.

$$(3.9)\quad U^{co}_\alpha = U_\alpha \cap L^{co,cr}_{m,n,p}(\mathbb{R}),\ V^{co}_\alpha = V_\alpha \cap L^{co,cr}_{m,n,p}(\mathbb{R})$$

Then one has clearly that V^{co}_α is an open dense (algebraic) subset of V_α and that $U^{co}_\alpha \simeq V^{co}_\alpha \times GL_n(\mathbb{R})$.

3.10 The local nice selection canonical forms c_α. Lemma 3.7 defines us a (local) continuous form on U_α for each nice selection α. It is

$$(3.11)\quad c_\alpha((F, G, H)) = (F, G, H)^{S_\alpha} \in V_\alpha,\ S_\alpha = R(F, G)_\alpha^{-1},\ (F, G, H) \in U_\alpha$$

The U_α are open dense subsets of $L^{cr}_{m,n,p}(\mathbb{R})$, and by lemma 3.2 the union of all the U_α, α a nice selection, covers all of $L^{cr}_{m,n,p}(\mathbb{R})$. This is thus a set of local canonical forms which can be useful in identification problems (it leads to statistically and numerically well posed problems, cf. [15, section II].

3.12 The dual results. Dually we consider the set $I(n, p)$ of all row indices of $Q(F, H)$, which we also picture as an $(n+1) \times p$ array of dots. Now the first row represents the rows of H, the second row the rows of HF, ... A nice selection is defined as before and one has the obvious analogues of all the results given above. In particular if $(F, G, H) \in L^{co}_{m,n,p}(\mathbb{R})$ there is a nice selection $\beta \subset I(n, p)$ such that $Q(F, H)_\beta$ is invertible. Here $Q(F, H)_\beta$ is the matrix obtained from $Q(F, H)$ by removing all rows whose index is not in β.
One also has of course local canonical forms $\bar{c}_\beta$ (defined on $\bar{U}_\beta$) for every nice selection $\beta \subset I(n, p)$:

(3.13) $\bar{c}_\beta((F, G, H)) = (F, G, H)^{S_\beta}$, $S_\beta = Q(F, H)_\beta$, $(F, G, H) \in \bar{U}_\beta$

(3.14) $\bar{U}_\beta = \{(F, G, H) \in L_{m,n,p}(\mathbb{R}) \mid Q(F, H)_\beta \text{ is invertible}\}$.

4 Realization theory

Let $A = (A_0, A_1, A_2, \ldots)$ be a sequence of $p \times m$ matrices. We shall say that the sequence A is realizable by an n-dimensional linear system if there exist a system $(F, G, H) \in L_{m,n,p}$(iR
$L_{m,n,p}(\mathbb{R})$ such that $A_i = HF^iG$, $i = 0, 1, 2, \ldots$. It follows immediately from (the proof of) theorem 2.8 above that if A is realizable by means of (F, G, H), then there is also a possible lower dimensional system $\Sigma' = (F', G', H') \in L^{co,cr}_{m,n',p}(\mathbb{R})$, $n' \leqslant n$. which also realizes A and which is moreover completely reachable and completely observable.
For each sequence of $p \times m$ matrices A we define the block Hankel matrices

$$(4.1)\quad H_s(A) = \begin{pmatrix} A_0 & A_1 & \cdots & A_s \\ A_1 & & \cdot & \\ \cdot & \cdot & & \cdot \\ \vdots & & & \vdots \\ & \cdot & & \\ A_s & & \cdots & A_{2s} \end{pmatrix}, \quad s = 0, 1, 2, \ldots .$$

4.2 Theorem: The sequence of real $p \times m$ matrices $A = (A_0, A_1, \ldots)$ is realizable by means of a completely reachable and completely observable n-dimensional system if and only if rank $H_s(A) = n$ for all large enough s. Moreover if both $\Sigma, \Sigma' \in L^{co,cr}_{m,n,p}(\mathbb{R})$ realize A then $\Sigma' = \Sigma^S$ for some $S \in GL_n(\mathbb{R})$.
This theorem will be proved below. First, however, we mention a consequence.

4.3 Corollary: If the sequence of $p \times m$ matrices A is such that rank $H_s(A) = n$ for all sufficiently large s, then rank $H_s(A) = n$ for all $s \geq n-1$.

Proof. If $\Sigma = (F, G, H)$ realizes A and Σ is co and cr and of dimension n, then rank $R_{n-1}(F, G) = \text{rank } Q_{n-1}(F, H) = n$, so that rank $H_{n-1}(A) = \text{rank } (R_{n-1}(F,G) Q_{n-1}(F, H)) = n$.

A first step in the proof of theorem 4.2 is now the following lemma which says that if rank $H_s(A) = n$ for all $s \geq r-1$, then the A_i for $i \geq 2r$ are uniquely determined by the 2r matrices $A_0, \ldots, A_{2r-1}$.

4.4 Lemma: Let $A = (A_0, A_1, \ldots)$ be a series of $p \times m$ matrices such that rank $H_s(A) = n$ for all $s \geq r-1$. There are $m \times m$ matrices $S_0, \ldots, S_{r-1}$ and $p \times p$ matrices $T_0, \ldots, T_{r-1}$ such that for all $i = 0, 1, 2, \ldots$.

$$(4.5)\quad A_{i+r} = A_i S_0 + A_{i+1} S_1 + \ldots + A_{i+r-1} S_{r-1} = \\ = T_0 A_i + T_1 A_{i+1} + \ldots + T_{r-1} A_{i+r-1}.$$

Proof: Because rank $H_{r-1}(A) = n$ and rank $H_r(A) = n$ we have

$$n = \text{rank } H_{r-1}(A) = \text{rank} \left(\begin{array}{cccc|c} A_0 & A_1 & \cdots & A_{r-1} & A_r \\ A_1 & & & \vdots & \vdots \\ \vdots & & & & \\ A_{r-1} & & \cdots & A_{2r-2} & A_{2r-1} \end{array} \right)$$

so that there are $m \times m$ matrices $S_0, \ldots, S_{r-1}$ such that

$$A_{i+r} = A_i S_0 + \ldots + A_{i+r-1} S_{r-1},\ i = 0, \ldots, r-1 .$$

Similarly, it follows from

$$n = \text{rank } H_{r-1}(A) = \text{rank} \left(\begin{array}{ccc} A_0 & \cdots & A_{r-1} \\ \vdots & & \vdots \\ A_{r-1} & \cdots & A_{2r-2} \\ \hline A_r & \cdots & A_{2r-1} \end{array} \right)$$

that there are matrices $T_0, \ldots, T_{r-1}$ such that

$$(4.6)\quad A_{r+i} = T_0 A_i + \ldots + T_{r-1} A_{i+r-1},\ i = 0, \ldots, r-1 .$$

Suppose with induction we have already proved (4.5) for $i \leq k-1$, $k \geq r$.

Consider the following submatrix of $H_k(A)$

$$(4.7)\quad \left(\begin{array}{cccc|ccc} A_0 & A_1 & \cdots & A_{r-1} & A_r & \cdots & A_k \\ A_1 & & & \vdots & \vdots & & \vdots \\ \vdots & & & & & & \\ A_{r-1} & & \cdots & A_{2r-2} & A_{2r-1} & \cdots & A_{k+r-1} \\ \hline A_r & & \cdots & A_{2r-1} & A_{2r} & \cdots & A_{k+r} \end{array}\right).$$

Using the relations (4.5) for $i \leq k-1$ we see that the rank of 4.7 is equal to the rank of

$$(4.8)\quad \left(\begin{array}{cccc|ccc} A_0 & A_1 & \cdots & A_{r-1} & 0 & \cdots & 0 \quad 0 \\ A_1 & & & \vdots & \vdots & & \vdots \quad \vdots \\ \vdots & & & & & & \\ A_{r-1} & & \cdots & A_{2r-2} & 0 & \cdots & 0 \quad 0 \\ \hline A_r & & \cdots & A_{2r-1} & 0 & \cdots & 0 \quad X \end{array}\right),$$

where $X = A_{k+r} - A_k S_0 - \ldots - A_{k+r-1} S_{r-1}$. Using (4.6) we see by means of row operations on (4.8) that the rank of (4.7) is also equal to the rank of

$$\left(\begin{array}{ccc|cccc} A_0 & \cdots & A_{r-1} & 0 & \cdots & 0 & 0 \\ \vdots & & \vdots & \vdots & & \vdots & \vdots \\ A_{r-1} & & A_{2r-2} & 0 & \cdots & 0 & 0 \\ \hline 0 & \cdots & 0 & 0 & \cdots & 0 & X \end{array}\right).$$

Now the rank of (4.7) is $n = \text{rank } H_{r-1}(A)$. Hence $X = 0$ which proves the induction step. This proves the first half of (4.5); the second half is proved similarly.
More generally one has the following result (which we shall not need in the sequel).

***4.9 Lemma:** Let $A_0, \ldots, A_s$ be a finite series of matrices and suppose there are $i, j \in \mathbb{N} \cup \{0\}$ such that $i + j = s - 1$ and

$$\text{rank}\begin{pmatrix} A_0 & \cdots & A_i \\ \vdots & & \vdots \\ A_j & \cdots & A_{i+j} \end{pmatrix} = \text{rank}\left(\begin{array}{ccc|c} A_0 & \cdots & A_i & A_{i+1} \\ \vdots & & \vdots & \vdots \\ A_j & \cdots & A_{i+j} & A_{i+j+1} \end{array}\right) = \text{rank}\left(\begin{array}{ccc} A_0 & \cdots & A_i \\ \vdots & & \vdots \\ A_j & \cdots & A_{i+j} \\ \hline A_{j+1} & \cdots & A_{i+j+1} \end{array}\right) = n$$

for some $n \in \mathbb{N} \cup \{0\}$, then there are unique $A_{s+1}, A_{s+2}, \ldots$ such that

$$\operatorname{rank} H_t(A) = n$$

for all $t \geqslant \max(i, j)$.

Proof: By hypothesis we know that there exist matrices $S_0, \ldots, S_i$

$$(4.10) \quad A_{i+r+1} = A_r S_0 + \ldots + A_{r+i} S_i,\ r = 0, \ldots, j\,.$$

Now define A_t for $t > s$ by the formula

$$(4.11) \quad A_t = A_{t-i-1} S_0 + \ldots + A_{t-1} S_i\,.$$

Also by hypothesis we know that there exist $T_0, \ldots, T_j$ such that

$$(4.12) \quad A_{j+r+1} = T_0 A_r + \ldots + T_j A_{j+r},\ r = 0, \ldots, i\,.$$

To prove that $\operatorname{rank} H_t(A) = n$ for all $t \geqslant \max(i, j)$ it now clearly suffices to show that (4.12) holds in fact for all $r \geqslant 0$. Suppose this has been proved for $r \leqslant q-1$, $q \geqslant i+1$. Consider the matrix

$$(4.13) \quad \left(\begin{array}{ccc|ccc} A_0 & \cdots & A_i & A_{i+1} & \cdots & A_q \\ \vdots & & \vdots & \vdots & & \vdots \\ A_j & \cdots & A_{i+j} & A_{i+j+1} & \cdots & A_{j+q} \\ \hline A_{j+1} & \cdots & A_{i+j+1} & A_{i+j+2} & \cdots & A_{j+q+1} \end{array}\right).$$

By means of column operations, the hypothesis of the lemma, and (4.10)–(4.11) we see that the rank of the matrix (4.13) is n. Using row operations and (4.12) for $r \leqslant q-1$ (induction hypothesis) we see that the rank of (4.13) is also equal to the rank of

$$(4.14) \quad \left(\begin{array}{ccc|ccc} A_0 & \cdots & A_i & A_{i+1} & \cdots & A_q \\ \vdots & & \vdots & \vdots & & \vdots \\ A_j & \cdots & A_{i+j} & A_{i+j+1} & \cdots & A_{j+q} \\ \hline 0 & \cdots & 0 & 0 & \cdots & 0 \ X \end{array}\right)$$

where X is the matrix $A_{j+q+1} - T_0 A_q - \ldots - T_j A_{j+q}$. Now use column operations and (4.10), (4.11) to see that the rank of (4.14) is also equal to the rank of

$$(4.15) \quad \left(\begin{array}{ccc|cccc} A_0 & \cdots & A_i & 0 & \cdots & 0 & 0 \\ \vdots & & \vdots & \vdots & & \vdots & \vdots \\ A_j & \cdots & A_{i+j} & 0 & \cdots & 0 & 0 \\ \hline 0 & \cdots & 0 & 0 & \cdots & 0 & X \end{array}\right).$$

It follows that $X = 0$.

4.16 Proof of theorem 4.2 (first step: existence of a co and cr realization; [10]): Let $r \in \mathbb{N}$ be such that $r \geq n$ and rank $H_s(A) = n$ for all $s \geq r-1$. We write

$$H = H_{r-1}(A) = \begin{pmatrix} A_0 & \dots & A_{r-1} \\ \vdots & & \vdots \\ A_{r-1} & \dots & A_{2r-2} \end{pmatrix}, H^{(k)} = \begin{pmatrix} A_k & \dots & A_{r+k-1} \\ \vdots & & \vdots \\ A_{r+k-1} & \dots & A_{2r+k-1} \end{pmatrix}$$

and for all $s, t \in \mathbb{N}$ we define

$$E_{s\times t} = (I_{s\times s} | 0_{s\times(t-s)}) \text{ if } s < t$$

$$E_{s\times s} = I_{s\times s} \qquad \text{if } s = t$$

$$E_{s\times t} = \left(\frac{I_{t\times t}}{0_{(s-t)\times t}}\right) \qquad \text{if } s > t,$$

where $I_{a\times a}$ is the $a \times a$ identity matrix and $0_{a\times b}$ is the $a \times b$ zero matrix. Because H is of rank n, there exist an invertible $pr \times pr$ matrix P and an invertible $mr \times mr$ matrix M such that

$$(4.17) \quad PHM = \left(\begin{array}{c|c} I_{n\times n} & 0_{n\times(mr-n)} \\ \hline 0_{(pr-n)\times n} & 0_{(pr-n)\times(mr-n)} \end{array}\right) = E_{pr\times n}E_{n\times mr}.$$

Now define

$$(4.18) \quad F = E_{n\times pr}PH^{(1)}M\,E_{mr\times n},\ G = E_{n\times pr}PHE_{mr\times m},$$

$$H = E_{p\times pr}HME_{mr\times n}$$

We claim that then (F, G, H) realizes A, i.e. that

$$(4.19) \quad A_i = HF^iG,\ i = 0, 1, 2, \dots .$$

To prove this we define

$$D = \begin{pmatrix} 0 & \dots & 0 & S_0 \\ I & & \vdots & \vdots \\ 0 & \ddots & \vdots & \vdots \\ \vdots & \ddots & 0 & \vdots \\ 0 & \dots 0 & I & S_{i-1} \end{pmatrix} \qquad C = \begin{pmatrix} 0' & I' & 0' & \dots & 0' \\ \vdots & \ddots & & \ddots & \vdots \\ \vdots & & & & 0' \\ 0' & \dots & 0' & & I' \\ T_0 & \dots & & & T_{r-1} \end{pmatrix}$$

where $0, I, 0', I'$ are respectively the $m \times m$ zero matrix, the $m \times m$ identity matrix, the $p \times p$ zero matrix and the $p \times p$ identity matrix and where the $S_0, \dots, S_{r-1}$ and $T_0, \dots, T_{r-1}$ are such that (4.5) holds for all i. Then

$$(4.20) \quad H^{(k)} = C^kH = HD^k,\ k = 1, 2, \dots .$$

Let $H^* = ME_{mr \times n}E_{n \times pr}P$. Then H^* is a pseudoinverse of H in that

(4.21) $HH^*H = H$

(Indeed using (4.17) we have $HH^*H = P^{-1}E_{pr \times n}E_{n \times mr}M^{-1}ME_{mr \times n}E_{n \times pr}P$ $P^{-1}E_{pr \times n}E_{n \times mr}M^{-1} = H$ because $M^{-1}M = I$, $PP^{-1} = I$, $E_{n \times mr}E_{mr \times n} = I_{n \times n}$, $E_{n \times pr}E_{pr \times n} = I_{n \times n}$.).
We now first prove that

(4.22) $E_{n \times pr}PC^kHME_{mr \times n} = F^k$, $k = 1, 2, \dots$.

In view of (4.20) this is the definition of F(cf. (4.18)) in the case k = 1. So assume (4.22) has been proved for $k \leqslant t$. We then have

$$\begin{aligned}
E_{n \times pr}PC^{t+1}HME_{mr \times n} &= E_{n \times pr}PC^tHDME_{mr \times n} \text{ (by (4.20)}\\
&= E_{n \times pr}PC^tHH^*HDME_{mr \times n} \text{ (by (4.21))}\\
&= E_{n \times pr}PC^tHME_{mr \times n}E_{n \times pr}PHDME_{mr \times n}\\
&\qquad \text{(by the definition of } H^*)\\
&= F^tE_{n \times pr}PCHME_{mr \times n} \text{ (by the induction hypothesis and (4.20))}\\
&= F^tF \text{ (by the definition of F, cf. (4.18) and (4.20))}.
\end{aligned}$$

We now have for all $k \geqslant 0$

$$\begin{aligned}
A_k &= E_{p \times pr}H^{(k)}E_{mr \times m} \text{ (definition of } H^{(k)})\\
&= E_{p \times pr}C^kHE_{mr \times m} \text{ (by (4.20))}\\
&= E_{p \times pr}C^kHH^*HE_{mr \times m} \text{ (by (4.21))}\\
&= E_{p \times pr}C^kHME_{mr \times n}E_{n \times pr}PHE_{mr \times m} \text{ (by the definition of } H^*)\\
&= E_{p \times pr}HD^kME_{mr \times n}G \text{ (by the definition of G and (4.20))}\\
&= E_{p \times pr}HH^*HD^kME_{mr \times n}G \text{ (by (4.21))}\\
&= E_{p \times pr}HME_{mr \times n}E_{n \times pr}PHD^kME_{mr \times n}G \text{ (by the definition of } H^*)\\
&= HE_{n \times pr}PC^kHME_{mr \times n}G \text{ (by the definition of H and (4.20))}\\
&= HF^kG \text{ (by (4.22))}.
\end{aligned}$$

This proves the existence of an n-dimensional system $\Sigma = (F, G, H)$ which realizes A. Now for all s = 0, 1, 2, ...

$$H_s(A) = Q_s(F, H)R_s(F, G),$$

where

$$Q_s(F, H) = \begin{pmatrix} H \\ HF \\ \vdots \\ HF^s \end{pmatrix}, \quad R_s(F, G) = (G \; FG \; \dots \; F^sG).$$

Both $Q_s(F, H)$ and $R_s(F, G)$ have necessarily rank $\leq n$. It follows via the Cayley-Hamilton theorem that (F, G, H) is completely reachable and completely controllable, because rank $\mathcal{H}_s(A) = n$ for $s \geq r-1$.

4.23 Proof of the uniqueness statement of theorem 4.2: Let $\Sigma = (F, G, H)$ and $\bar{\Sigma} = (\bar{F}, \bar{G}, \bar{H})$ be two co and cr realizations of A. Then $\dim(\Sigma) = \operatorname{rank} \mathcal{H}_{n-1}(A) = \dim(\bar{\Sigma})$. By hypothesis we have

(4.24) $A_i = HF^iG = \bar{H}\bar{F}^i\bar{G}$, $i = 0, 1, 2, \ldots$.

According to lemma 3.2 and 3.11 there exists a nice selection α (of size n) of $I(n-1, m)$, the set of column indices of $R_{n-1}(F, G)$ and $\mathcal{H}_{n-1}(F, G, H)$, and there exists a nice selection β (of size n) of $I(n-1, p)$, the set of row indices of $Q_{n-1}(F, H)$ and $\mathcal{H}_{n-1}(F, G, H)$, such that

$$\operatorname{rank}(R_{n-1}(F, G)_\alpha) = \operatorname{rank}(Q_{n-1}(F, H)_\beta) = n\,.$$

(Note that a nice selection in $I(n, m)$ (or $I(n, p)$) is always contained in $I(n-1, m)$ (or $I(n-1, p)$).) Let $\mathcal{H}_{n-1}(F, G, H)_{\alpha,\beta}$ be the matrix obtained from $\mathcal{H}_{n-1}(F, G, H)$ by removing all rows whose index is not in β and all columns whose index is not in α. Then

$$\mathcal{H}_{n-1}(F, G, H)_{\alpha,\beta} = Q_{n-1}(F, H)_\beta\, R_{n-1}(F, G)_\alpha$$

so that $\mathcal{H}_{n-1}(F, G, H)_{\alpha,\beta}$ is an invertible $n \times n$ matrix. Also

$$\mathcal{H}_{n-1}(F, G, H)_{\alpha,\beta} = \mathcal{H}_{n-1}(\bar{F}, \bar{G}, \bar{H})_{\alpha,\beta} = Q_{n-1}(\bar{F}, \bar{H})_\beta R_{n-1}(\bar{F}, \bar{G})_\alpha$$

so that $Q_{n-1}(\bar{F}, \bar{H})_\beta$ and $R_{n-1}(\bar{F}, \bar{G})_\alpha$ are also invertible. Now let

$$\Sigma_1 = (F_1, G_1, H_1) = (F, G, H)^T,\ T = Q_{n-1}(F, H)_\beta$$
$$\bar{\Sigma}_1 = (\bar{F}_1, \bar{G}_1, \bar{H}_1) = (\bar{F}, \bar{G}, \bar{H})^{\bar{T}};\ \bar{T} = Q_{n-1}(\bar{F}, \bar{H})_\beta\,.$$

Then of course Σ_1 and $\bar{\Sigma}_1$ also realize A. Moreover, using (2.4) we see

$$Q_{n-1}(F_1, H_1)_\beta = I_n = Q_{n-1}(\bar{F}_1, \bar{H}_1)_\beta\,.$$

It follows that

$$R(F_1, G_1) = \mathcal{H}_n(\Sigma_1)_\beta = \mathcal{H}_n(\Sigma)_\beta = \mathcal{H}_n(\bar{\Sigma})_\beta = \mathcal{H}_n(\bar{\Sigma}_1)_\beta = R(\bar{F}_1, \bar{G}_1)$$

and, in turn, this means that $F_1 = \bar{F}_1$ and $G_1 = \bar{G}_1$ by lemma (3.7) (i) combined with lemma (3.4). Further the matrix consisting of the first p rows of $\mathcal{H}_n(\Sigma_1) = \mathcal{H}_n(\bar{\Sigma}_1)$ is equal to

$$H_1 R(F_1, G_1) = \bar{H}_1 R(\bar{F}_1, \bar{G}_1)$$

so that also $H_1 = \bar{H}_1$ because $R(F_1, G_1) = R(\bar{F}_1, \bar{G}_1)$ is of rank n. This proves that indeed $\bar{\Sigma} = \Sigma^S$ with $S = \bar{T}^{-1}T$.

4.25 A realization algorithm. Now that we know that A is realizable by a co and cr system of dimension n iff rank $H_s(A) = n$ for all large enough s it is possible to give a rather easier algorithm for calculating a realization than the one used in 4.16 above (which is the algorithm of B.L. Ho). It goes as follows. Because A is realizable by a $\Sigma \in L^{co,cr}_{m,n,p}(\mathbb{R})$ there exist a nice selection $\alpha \subset I(n, m)$, the set of column indices of $R(F, G)$ and $H_n(\Sigma)$, and a nice selection $\beta \subset I(n, p)$, the set of row indices of $Q(F, H)$ and $H_n(\Sigma)$, such that

$$(4.26) \quad H_n(A)_{\alpha,\beta} = S$$

is an invertible $n \times n$ matrix. Consider

$$S^{-1} H_n(A)_\beta .$$

This $n \times (n + 1)\, m$ matrix is necessarily of the form $R(F, G)$ for some $(F, G) \in L^{cr}_{m,n}(\mathbb{R})$ and moreover by (4.26)

$$(S^{-1} H_n(A)_\beta)_\alpha = I_n$$

so that F, G can simply be written down from $S^{-1} H_n(A)_\beta$ as in the proof of lemma 3.4. The matrix H is now obtained as the matrix consisting of the first p rows of $H_n(A)_\alpha$. After choosing α, this algorithm describes the unique triple (F, G, H) which realizes A such that moreover $R(F, G)_\alpha = I_n$.

***4.27 Relation with rational functions.** Suppose that $H_k(A)$ is of rank n for all sufficiently large k. Then by theorem 4.2 the sequence A is realizable. Using Laplace transforms (cf. 1.8 above) we see that this means that the $p \times m$ matrix of power series $\sum_{i=0}^{\infty} A_i s^{-i-1}$ is in fact a matrix of rational functions.

$$(4.28) \quad \sum_{i=0}^{\infty} A_i s^{-i-1} = (s^n - a_{n-1} s^{n-1} - \ldots - a_1 s - a_0)^{-1} B(s) = d(s)^{-1} B(s) ,$$

where B(s) is a $p \times m$ matrix of polynomials in s of degree $\leqslant n - 1$.
Inversely if

$$(4.29) \quad \sum_{i=0}^{\infty} A_i s^{-i-1} = d'(s)^{-1} B'(s)$$

for a matrix of polynomials $B'(s)$ and a polynomial $d'(s) = s^r - a'_{r-1} s^{r-1} - \ldots - a'_1 s - a'_0$ with $r = \text{degree}\,(d'(s)) > \text{degree}\, B'(s)$, then

$$A_{i+r} = a'_0 A_i + a'_1 A_{i+1} + \ldots + a'_{r-1} A_{i+r-1}$$

for all $i = 0,1,2,\ldots$. And this, in turn implies that

$$\text{rank}\, H_k(A) = \text{rank}\, H_{r-1}(A)$$

for all $k \geq r-1$, so that A is realizable. It follows that A is realizable iff $\Sigma A_i s^{-i-1}$ represents a rational function which goes to zero as $s \to \infty$.

5 Feedback splits the external description degeneracy

In this section we shall prove the result described in section 1.6. To do this we first discuss still another local canonical form.

5.1 The Kronecker nice selection of a system. Let $(F, G, H) \in L^{cr}_{m,n,p}(\mathbb{R})$. We proceed as follows to obtain a "first" nice selection κ such that $(F, G, H) \in U_\kappa$.
Consider the set of column indices $I(m, n)$ in the order $(0, 1) < (0, 2) < \ldots < (0, m) < (1, 1) < \ldots < (1, m) < \ldots < (n, 1) < \ldots < (n, m)$. For each (i, j) we set $(i, j) \in \kappa \Leftrightarrow F^i g_j$ is linear independent of the $F^{i'} g_{j'}$ with $(i', j') < (i, j)$. We shall call the subset κ of $I(n, m)$ thus obtained, the Kronecker selection of (F, G, H) and denote it with $\kappa(F, G, H)$. It is obvious that κ has n elements if $(F, G, H) \in L^{cr}_{m,n,p}(\mathbb{R})$.

5.2 Lemma: The Kronecker selection κ defined above is a nice selection.

Proof: Let $(i, j) \in \kappa$ and suppose $i \geq 1$. Suppose that $(i', j) \notin \kappa$, $i' < i$. This means that there is a relation

$$F^{i'} g_j = \sum_{(k, l) < (i', j)} b(k, l) F^k g_l \, .$$

Multiplying with $F^{i-i'}$ on the left one obtains

$$F^i g_j = \sum_{(k, l) < (i', j)} b(k, l) F^{i-i'+k} g_l$$

showing that $F^i g_j$ is linearly dependent on the $F^s g_{j'}$, with $(s, j') < (i, j)$. A contradiction, q.e.d.

5.3 Lemma. Let $(F, G, H) \in L^{cr}_{m,n,p}(\mathbb{R})$ and $S \in GL_n(\mathbb{R})$, then

$$\kappa(F, G, H) = \kappa((F, G, H)^S) \, .$$

5.4 Lemma. Let $(F, G, H) \in L^{cr}_{m,n,p}(\mathbb{R})$ and let L be an $m \times n$ matrix. Then

$$\kappa(F, G, H) = \kappa(F + GL, G, H) \, .$$

The proof of lemma 5.3 is immediate, because the dependency relations between the $(SFS^{-1})^i(Sg_j) = S(F^ig_j)$, $(i, j) \in I(n, m)$, are precisely the same as those between the F^ig_j, $(i, j) \in I(n, m)$. As to lemma 5.4 we define

$$
(5.5)\quad
\begin{aligned}
X_0(\Sigma) &= \text{subspace of } X = \mathbb{R}^n \text{ generated by } g_1, \ldots, g_m \\
X_1(\Sigma) &= \text{subspace of } X = \mathbb{R}^n \text{ generated by } g_1, \ldots, g_m, Fg_1, \ldots, Fg_m \\
&\vdots \\
X_n(\Sigma) &= \text{subspace of } X \in \mathbb{R}^n \text{ generated by } g_1, \ldots, g_m, \\
&\qquad Fg_1, \ldots, Fg_m, \ldots, F^ng_1, \ldots, F^ng_m .
\end{aligned}
$$

Let $\Sigma(L) = (F + GL, G, H)$ and let $\hat{F} = F + GL$. Then one easily obtains by induction that

(5.6) $X_i(\Sigma(L)) = X_i(\Sigma)$, $i = 0, \ldots, n$

and that

(5.7) $\hat{F}^ig_j \equiv F^ig_j \bmod X^{i-1}(\Sigma)$, $i = 0, 1, \ldots, n$

(where, by definition, $X^{-1}(\Sigma) = \{0\}$). Lemma 5.4 is an immediate consequence of (5.7). (Note that a basis for $X^i(\Sigma)$ is formed by the vectors F^kg_l with $(k, l) \in \kappa(\Sigma)$ and $k \leqslant i$; the classes of the F^kg_l with $(k, l) \in \kappa(\Sigma)$, $k = i$ are a basis for the quotient space $X^i(\Sigma)/X^{i-1}(\Sigma)$, $i = 0, \ldots, n$).
If $\Sigma = (F, G, H) \in L^{cr,co}_{m,n,p}(\mathbb{R})$ then $\kappa(F, G, H)$ can be calculated from $H_n(F, G, H)$. Indeed in that case $Q(F, H)$ is of rank n. Therefore, because $H_n(F, G, H) = Q(F, H)R(F, G)$, the dependency relations between the columns of $H_n(F, G, H)$ and between the columns of $R(F, G)$ are exactly the same.

5.8 Remark: If $(F, G, H) \in L^{cr}_{m,n,p}(\mathbb{R})$ then also $(F + GL, G, H) \in L^{cr}_{m,n,p}(\mathbb{R})$ as is easily checked. But if $(F, G, H) \in L^{co}_{m,n,p}(\mathbb{R})$, then $(F + GL, G, H)$ need not also be completely observable. Though of course this will be the case for sufficiently small L (because $L^{co}_{m,n,p}(\mathbb{R})$ is an open subset of $L_{m,n,p}(\mathbb{R})$).

***5.9 The Kronecker control invariants.** The invariant $\kappa(F, G, H)$ depends only on F and G, so that we can also write $\kappa(F, G)$. For each $j = 1, \ldots, m$, let k_j be the number of elements (i, l) in $\kappa(F, G)$ such that $l = j$. Let $\kappa_1(F, G) \geqslant \ldots \geqslant \kappa_{m'}(F, G)$, $m' = \text{rank}(G)$, be the sequence of those k_j which are $\neq 0$ ordered with respect to size. It follows from lemma's 5.3 and 5.4 that the $\kappa_i(F, G)$ are invariant for the transformations

(5.10) $(F, G) \mapsto (F, G)^S = (SFS^{-1}, SG)$ (base change in state space)

(5.11) $(F, G) \mapsto (F + GL, G)$ (feedback).

One easily checks that the $\kappa_i(F, G)$ are also invariant under

(5.12) $(F, G) \mapsto (F, GT)$, $T \in GL_m(\mathbb{R})$ (base change in input space).

This can, e.g., be seen as follows. Let $\lambda_i(\Sigma) = \dim X^i(\Sigma) - \dim X^{i-1}(\Sigma)$ for $i = 0, 1, \ldots, n$. Consider an rectangular array of $(n+1) \times m$ boxes with the rows labelled $0, \ldots, n$. Now put a cross in the first $\lambda_i(\Sigma)$ boxes of row i for $i = 0, \ldots, n$. Then $\kappa_j(\Sigma), j = 1, \ldots, m'$ is the number of crosses in column j of the array. Obviously the $\lambda_i(\Sigma)$ do not change under a transformation of type (5.12), proving that also the $\kappa_j(F, G)$ are invariant under 5.12. The group generated by all these transformations is called the *feedback group,* Thus the $\kappa_i(F, G)$ are invariants of the feedback group acting on $L^{cr}_{m;n}(\mathbb{R})$. It now turns out that these are in fact the only invariants. I.e. if $(F, G), (\bar{F}, \bar{G}) \in L^{cr}_{m,n}(\mathbb{R})$ and $\kappa_i(F, G) = \kappa_i(\bar{F}, \bar{G})$, $i = 1, \ldots, m'$, then $(\bar{F}, G)$ can be obtained from (F, G) by means of a series of transformations from (5.10)–(5.12). Cf. [11] for a proof, or cf. 5.30 below.

The $\kappa_i(F, G)$ are also identifiable with Kronecker's minimal column indices of the singular matrix pencil $(zI_n - F \mid G)$, cf. [11].

Still another way to view the $\kappa_i(F, G)$ is a follows.

Consider the transfer matrix $T(s) = H(sI_n - F)^{-1}G$ of the cr and co linear dynamical system $\Sigma = (F, G, H)$ considered as a $p \times m$ matrix valued function of the complex variable s. One can now prove (cf. [14]):

Theorem: There exist matrices $N(s)$ and $D(s)$ of polynomial functions of s such that (i) $T(s) = N(s)D(s)^{-1}$, (ii) there exist matrices of polynomials such that $X(s)N(s) + Y(s)D(s) = I_m$, (iii) $N(s)$ and $D(s)$ are unique up to multiplication on the right by a unit from the ring of polynomial $m \times m$ matrices. Moreover degree $(\det D(s)) = n = \dim(\Sigma)$. Now for each $s \in \mathbb{C}$, one defines

$$\phi_\Sigma(s) = \{(N(s)u, D(s)u) \mid u \in \mathbb{C}^m\} \subset \mathbb{C}^{p+m}.$$

If $s \in \mathbb{C}$ is such that $D(s)^{-1}$ exists, then also $\phi_\Sigma(s) = \{(T(s)u, u) \mid u \in \mathbb{C}^m\} \subset \mathbb{C}^{p+m}$. In any case $\phi_\Sigma(s)$ is a p-dimensional subspace of $\mathbb{C}^{p+m}$. In addition one defines $\phi_\Sigma(\infty) = \{(0, u) \mid u \in \mathbb{C}^m\} \subset \mathbb{C}^{p+m}$, which is entirely natural because $\lim_{s \to \infty} T(s) = 0$. This gives a continuous map of the Riemann sphere $\mathbb{C} \cup \{\infty\} = S^2$ to the Grassmann manifold $G_{m,p+m}(\mathbb{C})$ of m-planes in $p + m$ space. Let $\xi_m \to G_{m,p+m}(\mathbb{C})$ be the canonical complex vector bundle whose fibre over $z \in G_{m,p+m}(\mathbb{C})$ is the m-plane represented by z. Pulling back ξ_m along ϕ_Σ gives us a holomorphic complex vector bundle $\xi(\Sigma)$ over S^2.

Now holomorphic vectorbundles over the sphere S^2 have been classified by Grothendieck. The classification result is: every holomorphic vectorbundle over S^2 is isomorphic to a direct sum of line bundles and line bundles are classified by their degrees.

It now turns out that the numbers classifying $\xi(\Sigma)$, the bundle over S^2 defined by the system Σ, are precisely the $-\kappa_i(\Sigma)$, $i = 1, \ldots, m$, where $\kappa_i(\Sigma) = 0$ for $i > m' = \text{rank}(G)$. One also recovers $n = \dim(\Sigma)$, if $\Sigma \in L^{co,cr}_{m,n,p}(\mathbb{R})$, as the intersection number of $\phi_\Sigma(S^2)$ with a hyperplane in $G_{m,m+p}(\mathbb{C})$.

These observations are due to Clyde Martin and Bob Hermann, cf. [13].

As we have seen the $\kappa_i(\Sigma)$ are invariants for the transformations (5.10), (5.11), (5.12). Being defined in terms of F and G alone they are also obviously invariant under base change in output space: $(F, G, H) \mapsto (F, G, SH)$, $S \in GL_p(\mathbb{R})$. The $\kappa_i(\Sigma)$ are, however, definitely not a full set of invariants for the group $\mathcal{G}$ acting on $L_{m,n,p}(\mathbb{R})$, where $\mathcal{G}$ is the group generated by base changes in state space, input space and output space and the feedback transformations.

5.13 The canonical input base change matrix $T(\Sigma)$. Let $\Sigma = (F, G, H) \in L^{cr}_{m,n,p}(\mathbb{R})$ and let $\kappa = \kappa(\Sigma)$ be the Kronecker nice selection of Σ. Let $(i, j) = s(\kappa, j)$ be a successor index of κ. By the definition of κ we have a unique expression of the form

(5.14) $$F^i g_j = \sum_{\substack{(i,j')\in\kappa \\ j'<j}} a_j(j') F^i g_{j'} + \sum_{\substack{(k,l)\in\kappa \\ k<i}} a(k,l) F^k g_l$$

(where the $a(k, l)$ in the second sum also depend on i and j of course). Now define recursively

(5.15) $$\hat{g}_j = g_j - \sum_{j'<j} a_j(j') g_{j'}, \quad \hat{G} = (\hat{g}_1, \ldots, \hat{g}_m)$$

and

(5.16) $$T(\Sigma) = (b_{jk}),$$

where $b_{jk} = 1$ if $j = k$, $b_{jk} = -a_k(j)$, if $j < k$, and $b_{jk} = 0$ if $j > k$.
Then $\hat{G} = GT(\Sigma)$, and $T(\Sigma)$ is an upper triangular matrix of determinant 1.

5.17 Lemma: Let $\Sigma \in (F, G, H) \in L^{cr}_{m,n,p}(\mathbb{R})$, then

$$T(\Sigma) = T(\Sigma^S), \; T(\Sigma(L)) = T(\Sigma)$$

for all $S \in GL_n(\mathbb{R})$ and all feedback matrices $L \in \mathbb{R}^{m \times n}$.

Proof. Obvious. (Use (5.7)).

5.18 Example: Let m = 5, n = 9, and let $(F, G, H) \in L^{cr}_{5,9,p}(\mathbb{R})$ have Kronecker selection $\kappa(F, G, H)$ equal to

$$\kappa = \begin{matrix} \times & \times & \cdot & \times & \times \\ \times & \times & \cdot & \times & \cdot \\ \cdot & \times & \cdot & \cdot & \cdot \\ \cdot & \times & \cdot & \cdot & \cdot \\ \cdot & \cdot & \cdot & \cdot & \cdot \end{matrix}$$

where we have omitted the last five rows of dots.

Then $T(\Sigma)$ is an upper triangular matrix of the form

$$T(\Sigma) = \begin{pmatrix} 1 & 0 & * & 0 & * \\ 0 & 1 & * & * & * \\ 0 & 0 & 1 & 0 & 0 \\ 0 & 0 & 0 & 1 & * \\ 0 & 0 & 0 & 0 & 1 \end{pmatrix}$$

Note that $T(\Sigma)^{-1}$ is of precisely the same form.
This is a general phenomon. Indeed by (5.14) and (5.15) (cf. also example (5.18)) $\hat{g}_j$ is of the form

$$\text{(5.19)} \quad \hat{g}_j = g_j + \sum_{\substack{k_i > k_j \\ i<j}} b_{ij} g_i, \quad T(\Sigma) = (b_{ij}) .$$

So that $b_{ij} = 0$ unless $i = j$ (and then $b_{ij} = 1$) or $i<j$ and $k_i > k_j$
Let $t_1, \ldots, t_m$ be the columns of $T(\Sigma)$ and $e_1, \ldots, e_m$ the standard basis for $\mathbb{R}^m$. Then

$$\text{(5.20)} \quad t_j = e_j + \sum_{\substack{k_i > k_j \\ i<j}} b_{ij} e_i .$$

Using induction with respect to an ordening of the $\{1, \ldots, m\}$ satisfying $i<j \Rightarrow k_i \geqslant k_j$ it readily follows that

$$e_j = t_j + \sum_{\substack{i<j \\ k_i > k_j}} b'_{ij} t_i .$$

which proves that $T(\Sigma)^{-1}$ also has zero entries at all spots (i, j) with $i > j$ or ($i<j$ and $k_i \leqslant k_j$).

5.21 The block companion canonical form. Let κ be a nice selection. We are going to construct a canonical form on the subspace W_κ of all $\Sigma \in L^{cr,co}_{m,n,p}(\mathbb{R})$ with $\kappa(\Sigma) = \kappa$. We shall do this only in full detail for the case that κ is the nice selection of example 5.18. This special case is, however, general enough to see that this construction works in general. Let $(F, G, H) \in W_\kappa$ and let $\hat{G} = GT(\Sigma)$. Now consider the system $(F, \hat{G}, H)$ which is also in W_κ as is easily checked. This system has the property that for each successor index $s(\kappa, j) = (i, j)$ of κ with $i \neq 0$ we have

$$\text{(5.22)} \quad F^i \hat{g}_j = \sum_{\substack{(k,l) \in \kappa \\ k<i}} a'(k, l) F^k \hat{g}_l$$

(i.e. $T(F,\hat{G},H) = I_m$). Indeed, using (5.14)

$$F^i\hat{g}_j = F^i g_j - \sum_{j'<j} a_j(j')F^i g_{j'} = \sum_{\substack{(k,l)\in\kappa \\ k<i}} a(k,l)F^k g_l = \sum_{\substack{(k,l)\in\kappa \\ k<i}} a'(k,l)F^k\hat{g}_l$$

because, clearly, $X_i(F,G,H) = X_i(F,\hat{G},H)$ for all $i = 0, 1, 2, \ldots, n$, cf. (5.5), and cf. also the remarks just below (5.7).

Now define a new basis for $\mathbb{R}^n$ as follows. Let $\kappa = \{(0,j_1), \ldots, (i_1,j_1); \ldots; (0,j_r), \ldots, (i_r,j_r)$. Then $k_t = i_t + 1$, $t = 1, \ldots, r$, and $k_1 + \ldots + k_r = n$. For the successor indices $s(\kappa,j) = (k_t, j_t)$, $t = 1, \ldots, r$, write

$$F^{k_t}\hat{g}_{j_t} = -\sum_{\substack{(k,l)\in\kappa \\ k<k_t}} b_t(k,l)F^k\hat{g}_l\,. \tag{5.23}$$

Setting $b_t(k,l) = 0$ for all $(k,l) \notin \kappa$ we now define a new basis for $\mathbb{R}^n$ by

$$\begin{aligned}
e_1 &= F^{k_1-1}\hat{g}_{j_1} + \sum_{j=1}^{m} b_1(k_1-1,j)\,F^{k_1-2}\,\hat{g}_j + \ldots + \sum_{i=1}^{t} b_1(1,j)\,\hat{g}_j \\
e_2 &= F^{k_1-2}\hat{g}_{j_1} + \sum_{j=1}^{m} b_1(k_1-1,j)\,F^{k_1-3}\,\hat{g}_j + \ldots + \sum_{i=1}^{t} b_1(2,j)\,\hat{g}_j \\
&\;\;\vdots \\
e_{k_1} &= \hat{g}_{j_1} \\
e_{k_1+1} &= F^{k_2-1}\hat{g}_{j_2} + \sum_{j=1}^{m} b_2(k_2-1,j)\,F^{k_2-2}\,\hat{g}_j + \ldots + \sum_{i=1}^{t} b_2(1,j)\,\hat{g}_j \\
&\;\;\vdots \\
e_{k_1+k_2} &= \hat{g}_{j_2} \\
&\;\;\vdots \\
e_{k_1+\ldots+k_r} &= \hat{g}_{j_r}\,.
\end{aligned} \tag{5.24}$$

Let $X_0 \subset \mathbb{R}^n$ be the space spanned by the vectors $\hat{g}_{j_1}, \ldots, \hat{g}_{j_r}$ i.e. $X_0 = X_0(F,\hat{G},H) = X_0(\Sigma)$. Then we see from (5.23) that for the vectors defined by (5.24) above we have

$$Fe_1 \in X_0,\; F(e_i) \equiv e_{i-1} \bmod X_0 \quad \text{for } i = k_1, k_1-1, \ldots, 2$$

$$Fe_{k_1+1} \in X_0,\; F(e_i) \equiv e_{i-1} \bmod X_0 \quad \text{for } i = k_1+k_2, \ldots, k_1+2$$

$$\vdots \qquad\qquad \vdots$$

$$Fe_{k_1+\ldots+k_{r-1}} \in X_0,\; F(e_i) \equiv e_{i-1} \bmod X_0 \quad \text{for } i = k_1+\ldots+k_r, \ldots, k_1+\ldots+k_{r-1}+2.$$

It follows that which respect to the basis $e_1, \ldots, e_n$, F and $\hat{G}$ are of the form

$$
(5.25)\qquad F = \left(\begin{array}{ccc|ccc|c|ccc}
0\ 1\ 0 & \cdots & 0 & 0 & \cdots & 0 & & 0 & \cdots & 0 \\
 & \ddots & 0 & \vdots & & \vdots & \cdots & \vdots & & \vdots \\
0 & \cdots & 0\ 1 & 0 & \cdots & 0 & & 0 & \cdots & 0 \\
* & \cdots & * & * & \cdots & * & & * & \cdots & * \\
\hline
0 & \cdots & 0 & 0\ 1\ 0 & \cdots & 0 & & 0 & \cdots & 0 \\
\vdots & & \vdots & & \ddots & 0 & \cdots & \vdots & & \vdots \\
0 & \cdots & 0 & 0 & \cdots & 0\ 1 & & 0 & \cdots & 0 \\
* & \cdots & * & * & \cdots & * & & * & \cdots & * \\
\hline
 & \vdots & & & \vdots & & \ddots & & \vdots & \\
\hline
0 & \cdots & 0 & 0 & \cdots & 0 & & 0\ 1\ 0 & \cdots & 0 \\
\vdots & & \vdots & \vdots & & \vdots & \cdots & \vdots & \ddots & 0 \\
0 & \cdots & 0 & 0 & \cdots & 0 & & 0 & \cdots & 0\ 1 \\
* & \cdots & * & * & \cdots & * & & * & \cdots & *
\end{array}\right)
\begin{array}{l}
\left.\begin{array}{c}\\ \\ \\ \\ \end{array}\right\} k_1 \\
\left.\begin{array}{c}\\ \\ \\ \\ \end{array}\right\} k_2 \\
\\
\left.\begin{array}{c}\\ \\ \\ \\ \end{array}\right\} k_3
\end{array}
$$

$\hat{G} = (\hat{g}_1, \hat{g}_2, \ldots, \hat{g}_m)$, with

(5.26) $\hat{g}_{j_1} = e_{k_1}, \hat{g}_{j_2} = e_{k_1+k_2}, \ldots, \hat{g}_{j_r} = e_{k_1+\ldots+k_r} = e_n$,

$\hat{g}_j = 0$ for $j \in \{1, \ldots, m\} \setminus \{j_1, \ldots, j_r\}$.

In particular in the case that κ is the nice selection of example 5.18 we see that with respect to the basis $e_1, \ldots, e_n$ defined by 5.24 the matrices F and G take the form (cf. 5.18, the inverse of $T(\Sigma)$ is of the same form as $T(\Sigma)$).

$$
F' = \left(\begin{array}{cc|cccc|cc|c}
0 & 1 & 0 & 0 & 0 & 0 & 0 & 0 & 0 \\
a_1 & a_2 & a_3 & a_4 & a_5 & a_6 & a_7 & a_8 & a_9 \\
\hline
0 & 0 & 0 & 1 & 0 & 0 & 0 & 0 & 0 \\
0 & 0 & 0 & 0 & 1 & 0 & 0 & 0 & 0 \\
0 & 0 & 0 & 0 & 0 & 1 & 0 & 0 & 0 \\
b_1 & b_2 & b_3 & b_4 & b_5 & b_6 & b_7 & b_8 & b_9 \\
\hline
0 & 0 & 0 & 0 & 0 & 0 & 0 & 1 & 0 \\
c_1 & c_2 & c_3 & c_4 & c_5 & c_6 & c_7 & c_8 & c_9 \\
\hline
d_1 & d_2 & d_3 & d_4 & d_5 & d_6 & d_7 & d_8 & d_9
\end{array}\right)
$$

(5.27)

$$G' = \begin{pmatrix} 0 & 0 & 0 & 0 & 0 \\ 1 & 0 & * & 0 & * \\ 0 & 0 & 0 & 0 & 0 \\ 0 & 0 & 0 & 0 & 0 \\ 0 & 0 & 0 & 0 & 0 \\ 0 & 1 & * & * & * \\ 0 & 0 & 0 & 0 & 0 \\ 0 & 0 & 0 & 1 & * \\ 0 & 0 & 0 & 0 & 1 \end{pmatrix}$$

This does not yet define a canonical form on W_κ. True, for every $\Sigma \in W_\kappa$ there exists an $S \in GL_n(\mathbb{R})$ such that $(F, G)^S$ takes the form (5.27). But for two pairs $(F, G) \neq (\bar{F}, \bar{G})$, both of the form (5.27), there may very well exists an $S \neq I_n$ such that $(F, G)^S = (\bar{F}, \bar{G})$.

In fact, it is now not difficult to check that if S is an $n \times n$ matrix of the form

$$S = \left(\begin{array}{cc|cccc|cc|c} 1 & 0 & s_{13} & s_{14} & 0 & 0 & 0 & 0 & 0 \\ 0 & 1 & 0 & s_{13} & s_{14} & 0 & 0 & 0 & 0 \\ \hline 0 & 0 & 1 & 0 & 0 & 0 & 0 & 0 & 0 \\ 0 & 0 & 0 & 1 & 0 & 0 & 0 & 0 & 0 \\ 0 & 0 & 0 & 0 & 1 & 0 & 0 & 0 & 0 \\ 0 & 0 & 0 & 0 & 0 & 1 & 0 & 0 & 0 \\ \hline 0 & 0 & s_{73} & s_{74} & 0 & 0 & 1 & 0 & 0 \\ 0 & 0 & 0 & s_{73} & s_{74} & 0 & 0 & 1 & 0 \\ \hline s_{91} & 0 & s_{93} & s_{94} & s_{95} & 0 & s_{97} & 0 & 1 \end{array}\right)$$

then $SG = G$ and SFS^{-1} is of the same general form as F, if F and G are of the form (5.27). Choosing $s_{13}, s_{14}, s_{73}, s_{74}, s_{91}, s_{93}, s_{94}, s_{95}$ and s_{97} judiciously we see that for every $\Sigma = (F, G, H) \in W_\kappa$, there exists an $S \in GL_n(\mathbb{R})$ such that SFS^{-1} and SG take the forms

$$SFS^{-1} = \left(\begin{array}{cc|cccc|cc|c} 0 & 1 & 0 & 0 & 0 & 0 & 0 & 0 & 0 \\ a_1 & a_2 & a_3 & a_4 & 0 & 0 & a_7 & a_8 & a_9 \\ \hline 0 & 0 & 0 & 1 & 0 & 0 & 0 & 0 & 0 \\ 0 & 0 & 0 & 0 & 1 & 0 & 0 & 0 & 0 \\ 0 & 0 & 0 & 0 & 0 & 1 & 0 & 0 & 0 \\ b_1 & b_2 & b_3 & b_4 & b_5 & b_6 & b_7 & b_8 & b_9 \\ \hline 0 & 0 & 0 & 0 & 0 & 0 & 0 & 1 & 0 \\ c_1 & c_2 & c_3 & c_4 & 0 & 0 & c_7 & c_8 & c_9 \\ \hline d_1 & 0 & d_3 & 0 & 0 & 0 & d_7 & 0 & d_9 \end{array}\right) \tag{5.28}$$

$$SG = \begin{pmatrix} 0 & 0 & 0 & 0 & 0 \\ 1 & 0 & c_{13} & 0 & c_{15} \\ 0 & 0 & 0 & 0 & 0 \\ 0 & 0 & 0 & 0 & 0 \\ 0 & 0 & 0 & 0 & 0 \\ 0 & 1 & c_{23} & c_{24} & c_{25} \\ 0 & 0 & 0 & 0 & 0 \\ 0 & 0 & 0 & 1 & c_{45} \\ 0 & 0 & 0 & 0 & 1 \end{pmatrix}$$

where

$$T(\Sigma)^{-1} = \begin{pmatrix} 1 & 0 & c_{13} & 0 & c_{15} \\ 0 & 1 & c_{23} & c_{24} & c_{25} \\ 0 & 0 & 1 & 0 & 0 \\ 0 & 0 & 0 & 1 & c_{45} \\ 0 & 0 & 0 & 0 & 1 \end{pmatrix} .$$

The general pattern should be clear: the off-diagonal blocks have zero's in the last row iff there are more columns than rows, in fact in that case the last row ends with (number of columns) – (number of rows) zero's; the structure of the diagonal blocks is clear.
Now suppose that (F', G', H') and (F'', G'', H'') are two systems such that $(F', G')^S = (F'', G'')$ for some S and such that (F', G') and (F'', G'') are both of the forms (5.28). One checks easily that then necessarily $S = I_n$. We have shown

5.29 Proposition: Let κ be the nice selection of example 5.18. Then for every $\Sigma = (F, G, H) \in W_\kappa$ there is precisely one $S \in GL_n(\mathbb{R})$ such that SFS^{-1} and SG have the forms (5.28).

This means in particular (in view of the results of section 4 above) that if $\Sigma \in W_\kappa \cap L^{co,cr}_{n,m,p}(\mathbb{R})$, then the real numbers $a_1, \ldots, a_4, a_7, \ldots, a_9, b_1, \ldots, b_9, c_1, \ldots, c_4, c_7, \ldots, c_9, d_1, d_3, d_7, d_9$ can be calculated from $f(\Sigma)$ (or $A_0, \ldots, A_{2n-1}$). Of course these results hold quite generally for all nice selections κ. We note that in general W_κ is not an open subspace of $L^{cr}_{n,m,p}(\mathbb{R})$. In fact $W_\kappa/GL_n(\mathbb{R})$ is a linear subspace of $U_\kappa/GL_n(\mathbb{R}) = \mathbb{R}^{mn+np} \simeq V_\kappa$. In case κ is the nice selection of example 5.18 the codimension of $W_\kappa/GL_n(\mathbb{R})$ in $U_k/GL_n(\mathbb{R})$ is 12. (This number can immediately be read off from κ: g_3 linearly dependent on g_1, g_2 causes $9 - 2 = 7$ linear restrictions; Fg_5 linearly dependent on $g_1, g_2, g_4, g_5, Fg_1, Fg_2, Fg_4$ causes $9 - 7 = 2$ extra linear restrictions; F^2g_1 linearly dependent on $g_1, g_2, g_4, g_5, Fg_1, Fg_2, Fg_4$ causes $9 - 7 = 2$ more linear restrictions; and finally F^2g_4 dependent on $g_1, g_2, g_4, g_5, Fg_1, Fg_2, Fg_4, F^2g_2$ causes $9 - 8 = 1$ more linear restriction; $7 + 2 + 2 + 1 = 12$).

***5.30.** Using the results above, it is now easy to prove that the $\kappa_1(F,G), \ldots, \kappa_{m'}(F,G)$ are the only invariants of the feedback group acting on $L^{cr}_{m,n}(\mathbb{R})$. Indeed, we have already shown that the $\kappa_i(F,G)$, $i = 1, \ldots, m'$ are invariants.
Inversely, using first of all a transformation of type (5.12) we can see to it that (F, GT) has $k_1 \geqslant k_2 \geqslant \ldots \geqslant k_m$, and then $\kappa_1(F,G) = k_1, \ldots, \kappa_{m'}(F,G) = k_{m'}$, $k_i = 0$ for $i > m'$. Then, using transformations of type (5.10) and (5.12), we can change (F, GT) into a pair (F', G') with F' and G' of the type (5.25), (5.26). A final transformation of type (5.11) then changes F' into a matrix of type (5.25) with all stars equal to zero. The final pair (F'', G'') thus obtained depends only on the numbers $\kappa_1(F,G), \ldots, \kappa_{m'}(F,G)$.

5.31 Feedback breaks all symmetry: We are now in a position to prove the result mentioned in 1.6 that feedback splits the degenerate external description of systems. We shall certainly have proved this if we have proved.

5.32 Theorem: Let $\Sigma \in L^{co,cr}_{m,n,p}(\mathbb{R})$. Then Σ is completely determined by the input-output maps $f(\Sigma(L))$ for small L. More precisely let $\Sigma = (F, G, H)$ and $A_i(L) = H(F + GL)^i G$ for $i = 0, 1, \ldots, 2n-1$. Then the entries of $A_i(L)$ are differentiable functions of L, and F, G and H can be calculated from $A_0, \ldots, A_{2n-1}$ and the numbers

$$\frac{\partial A_i(L)}{\partial l_{jk}}\bigg|_{L=0}, \quad i = 0, \ldots, 2n-1, j = 1, \ldots, m, k = 1, \ldots, n .$$

Proof: Let $\kappa = \kappa(\Sigma)$. Recall that κ can be calculated from $A_0, \ldots, A_{2n-1}$ (because Σ is co and cr). Now assume that κ is the nice selection of example 5.18. (This is sufficiently general, I hope, to make it clear that the theorem holds in general). Let $\Sigma' = (F', G', H')$ be the block companion canonical form of (F, G, H) (Σ' is obtained as follows: first calculate any realization $\Sigma'' = (F'', G'', H'')$ of $A_0, \ldots, A_{2n-1}$, e.g. by means of the algorithm of 4.25 above and then put Σ'' in block companion canonical form as in 5.21 above). Then

$$\Sigma' = \Sigma^{S^{-1}}$$

for a certain $S \in GL_n(\mathbb{R})$, and it remains to calculate S. With this aim in mind we examine $\Sigma(L) = (F + GL, G, H)$ and its block companion canonical form. Consider

$$\begin{aligned} \Sigma(L)^{S^{-1}} &= (S^{-1}FS + S^{-1}GLS, S^{-1}G, HS) \\ &= (F' + G'LS, G', H') . \end{aligned}$$

Now assume that L is of the form

$$(5.33)\ L = \begin{pmatrix} 0 & . & . & . & 0 \\ l_{21} & . & . & . & l_{29} \\ 0 & . & . & . & 0 \\ 0 & . & . & . & 0 \\ 0 & . & . & . & 0 \end{pmatrix} .$$

Then if F′ is of the form (5.28) we see that if $S = (s_{ij})$

$$F' + G'LS = \left(\begin{array}{cc|cccc|cc|c} 0 & 1 & 0 & 0 & 0 & 0 & 0 & 0 & 0 \\ a_1 & a_2 & a_3 & a_4 & 0 & 0 & a_7 & a_8 & a_9 \\ \hline 0 & 0 & 0 & 1 & 0 & 0 & 0 & 0 & 0 \\ 0 & 0 & 0 & 0 & 1 & 0 & 0 & 0 & 0 \\ 0 & 0 & 0 & 0 & 0 & 1 & 0 & 0 & 0 \\ b'_1 & b'_2 & b'_3 & b'_4 & b'_5 & b'_6 & b'_7 & b'_8 & b'_9 \\ \hline 0 & 0 & 0 & 0 & 0 & 0 & 0 & 1 & 0 \\ c_1 & c_2 & c_3 & c_4 & 0 & 0 & c_7 & c_8 & c_9 \\ \hline d_1 & 0 & d_3 & 0 & 0 & 0 & d_7 & 0 & d_9 \end{array}\right)$$

with $b'_i = b_i(L) = b_i + \sum_{j=1}^{9} l_{2j}s_{ji}$, $i = 1, \ldots, 9$. Thus the block companion canonical form of $\Sigma(L)$ is always $\Sigma(L)^{S^{-1}}$ if L is of the form (5.33). Note that the number of the row which has nonzero entries is determined by $\kappa(\Sigma)$; it is the smallest i for which k_i is maximal; note also that if j is such that k_j is maximal then the j-th vector of G′ is always the $(k_1 + \ldots + k_j)$-th standard basis vector (cf. just below (5.19).

So to find S we proceed as follows. Calculate the block companion canonical forms of $\Sigma(L)$ from $A_0(L), \ldots, A_{2n-1}(L)$ for small L. (This can be done because for small enough L, $\Sigma(L)$ is still co). This gives us in particular the functions $b_i(L)$. Then

$$s_{ji} = \frac{\partial b_i(L)}{\partial l_{2j}}\Big|_{L=0} .$$

This determines S and gives us Σ as $\Sigma = (\Sigma')^S$. q.e.d.

6 Description of $L^{co,cr}_{m,n,p}(\mathbb{R})/GL_n(\mathbb{R})$. Invariants

6.1 Local structure of $L^{co,cr}_{m,n,p}(\mathbb{R})$. Let $\alpha \subset I(n, m)$ be a nice selection. We recall that $U_\alpha = \{(F, G, H) \in L_{m,n,p}(\mathbb{R}) \mid \det R(F, G)_\alpha \neq 0\}$, that $V_\alpha = \{(F, G, H) \in L_{m,n,p}(\mathbb{R}) \mid R(F, G)_\alpha = I_n\}$ and that $U_\alpha/GL_n(\mathbb{R}) \simeq V_\alpha \cong \mathbb{R}^{nm+np}$, cf. section 3.

For each $x \in \mathbb{R}^{nm+np}$ let $(F_\alpha(x), G_\alpha(x), H_\alpha(x)) \in V_\alpha$ be the unique system corresponding to x according to the isomorphism of 3.7 above.

6.2 The quotient manifold $M^{cr}_{m,n,p}(\mathbb{R}) = L^{cr}_{m,n,p}(\mathbb{R})/GL_n(\mathbb{R})$. Now that we know what $U_\alpha/GL_n(\mathbb{R})$ looks like it is not difficult to describe $L^{cr}_{m,n,p}(\mathbb{R})/GL_n(\mathbb{R})$. Recall that the union of the U_α for α nice covers $L^{cr}_{m,n,p}(\mathbb{R})$). We only need to figure out how the $V_\alpha \simeq \mathbb{R}^{mn+np}$ should be glued together. This is not particularly difficult because if $(F, G, H)^S = (F', G', H')$ for some S and $(F, G, H) \in U_\alpha$ then $S = R(F', G')_\alpha R(F, G)_\alpha^{-1}$. It follows that the quotient space $M^{cr}_{m,n,p}(\mathbb{R}) = L^{cr}_{m,n,p}(\mathbb{R})/GL_n(\mathbb{R})$ can be constructed as follows.

For each nice selection α let $\overline{V}_\alpha = \mathbb{R}^{mn+np}$ and for each second nice selection β let

$$\overline{V}_{\alpha\beta} = \{x \in \overline{V}_\alpha \mid \det R(F_\alpha(x), G_\alpha(x))_\beta \neq 0\} .$$

We define

$$\phi_{\alpha\beta} : \overline{V}_{\alpha\beta} \to \overline{V}_{\beta\alpha}$$

by the formula

(6.3) $\phi_{\alpha\beta}(x) = y \Leftrightarrow R(F_\alpha(x), G_\alpha(x))_\beta^{-1} R(F_\alpha(x), G_\alpha(x)) = R(F_\beta(y), G_\beta(y))$.

Let $M^{cr}_{m,n,p}(\mathbb{R})$ be the topological space obtained by glueing together the $\overline{V}_\alpha$ by means of the isomorphisms $\phi_{\alpha\beta}$.

Then $M^{cr}_{m,n,p}(\mathbb{R}) = L^{cr}_{m,n,p}(\mathbb{R})/GL_n(\mathbb{R})$. If we denote also with $\overline{V}_\alpha$ the isomorphic image of $\overline{V}_\alpha$ in $M^{cr}_{m,n,p}(\mathbb{R})$ then the quotient map $\pi : L^{cr}_{m,n,p}(\mathbb{R}) \to M^{cr}_{m,n,p}(\mathbb{R})$ can be described as follows. For each $\Sigma = (F, G, H) \in L^{cr}_{m,n,p}(\mathbb{R})$, choose a nice selection α such that $\Sigma \in U_\alpha$. Then $\pi(\Sigma) = x \in \overline{V}_\alpha \subset M^{cr}_{m,n,p}(\mathbb{R})$ where x is such that $\Sigma^S = (F_\alpha(x), G_\alpha(x), H_\alpha(x))$ with $S = R(F, G)_\alpha^{-1}$.

6.4 Theorem: $M^{cr}_{m,n,p}(\mathbb{R})$ is a differentiable manifold and $\pi : L^{cr}_{m,n,p}(\mathbb{R}) \to M^{cr}_{m,n,p}(\mathbb{R})$ is a principal $GL_n(\mathbb{R})$ fibre bundle.

For a proof, cf. [5].

6.5 The quotient manifold $M^{co,cr}_{m,n,p}(\mathbb{R}) = L^{co,cr}_{m,n,p}(\mathbb{R})/GL_n(\mathbb{R})$. Let $M^{co,cr}_{m,n,p}(\mathbb{R}) = \pi(L^{co,cr}_{m,n,p}(\mathbb{R}))$. Then $M^{co,cr}_{n,n,p}(\mathbb{R})$ is an open submanifold of $M^{cr}_{m,n,p}(\mathbb{R})$. It can be described as follows. For each nice selection α let $\overline{V}^{co}_\alpha = \{x \in \overline{V}_\alpha \mid (F_\alpha(x), G_\alpha(x), H_\alpha(x))$ is completely observable$\}$, and for each second nice selection β let $\overline{V}^{co}_{\alpha\beta} = \overline{V}^{co}_\alpha \cap \overline{V}_{\alpha\beta}$. Then $\phi_{\alpha\beta}(\overline{V}^{co}_{\alpha\beta}) = \overline{V}^{co}_{\beta\alpha}$ and $M^{co,cr}_{m,n,p}(\mathbb{R})$ is the differentiable manifold obtained by glueing together the $\overline{V}^{co}_\alpha$ by means of the isomorphisms $\phi_{\alpha\beta} : \overline{V}^{co}_{\alpha\beta} \to \overline{V}^{co}_{\beta\alpha}$.

6.6 $M^{co,cr}_{m,n,p}(\mathbb{R})$ as a submanifold of $\mathbb{R}^{2nmp}$. Let $(F, G, H) \in L^{co,cr}_{n,m,p}(\mathbb{R})$. We associate to (F, G, H) the sequence of 2n $p \times m$ matrices $(A_0, \ldots, A_{2n-1}) \in \mathbb{R}^{2nmp}$, where $A_i = HF^iG$, $i = 0, \ldots, 2n-1$. The results of section 4 above (realization theory) prove that this map is injective and prove that its image consists of those elements $(A_0, \ldots, A_{2n-1}) \in \mathbb{R}^{2nmp}$ such that rank $H_{n-1}(A)$ = rank $H_n(A)$ = n. We thus obtain $M^{co,cr}_{m,n,p}(\mathbb{R})$ as a (nonsingular algebraic) smooth submanifold of $\mathbb{R}^{2nmp}$.

6.7 Invariants. By definition a smooth invariant for $GL_n(\mathbb{R})$ acting on $L_{m,n,p}(\mathbb{R})$ is a smooth function $f : U \to \mathbb{R}$, defined on an open dense subset $U \subset L_{m,n,p}(\mathbb{R})$ such that $f(\Sigma) = f(\Sigma^S)$ for all $\Sigma \in U$ and $S \in GL_n(\mathbb{R})$ such that $\Sigma^S \in U$.

Now $L^{co,cr}_{m,n,p}(\mathbb{R})$ is open and dense in $L_{m,n,p}(\mathbb{R})$. It now follows from 6.6 that every invariant can be written as a smooth function of the entries of the invariant matrix valued functions $A_0, \ldots, A_{2n-1}$ on $L_{m,n,p}(\mathbb{R})$.

7 On the (non) existence of canonical forms

7.1 Canonical forms: Let L' be a $GL_n(\mathbb{R})$-invariant subspace of $L_{m,n,p}(\mathbb{R})$. A canonical form for $GL_n(\mathbb{R})$ acting on L' is a mapping $c: L' \to L'$ such that the following three properties hold

(7.2) $c(\Sigma^S) = c(\Sigma)$ for all $\Sigma \in L'$, $S \in GL_n(\mathbb{R})$

(7.3) for all $\Sigma \in L'$ there is an $S \in GL_n(\mathbb{R})$ such that $c(\Sigma) = \Sigma^S$.

(7.4) $c(\Sigma) = c(\Sigma') \Rightarrow \exists S \in GL_n(\mathbb{R})$ such that $\Sigma' = \Sigma^S$

(Note that (7.4) is implied by (7.3)).
Thus a canonical form selects precisely one element out of each orbit of $GL_n(\mathbb{R})$ acting on L'. We speak of a continuous canonical form if c is continuous.
Of course, there exist canonical forms on, say $L^{co,cr}_{m,n,p}(\mathbb{R})$, e.g. the following one, $\bar{c}_\kappa : L^{co,cr}_{m,n,p}(\mathbb{R}) \to L^{co,cr}_{m,n,p}(\mathbb{R})$ which is defined as follows: let $\Sigma \in L^{co,cr}_{m,n,p}(\mathbb{R})$, calculate $\kappa(\Sigma)$ and let $\bar{c}_\kappa(\Sigma)$ be the block companion canonical form of Σ as described in section 5.21 above.
This canonical form is not continuous, however (, though still quite useful, as we saw in section 5.31). As we argued in 1.15 above, for some purposes it would be desirable to have a continuous canonical form (cf. also [2]). In this connection let us also remark that the Jordan canonical form for square matrices under similarity transformations ($M \to SMS^{-1}$) is also not continuous, and this causes a number of unpleasant numerical difficulties, cf. [16].

***7.5 Continuous canonical forms and sections.** Let L' be a $GL_n(\mathbb{R})$-invariant subspace of $L^{cr}_{m,n,p}(\mathbb{R})$. Let $M' = \pi(L') \subset M^{cr}_{m,n,p}(\mathbb{R})$ be the image of L' under the projection π (cf. 6.2 above). Now let $c: L' \to L'$ be a continuous canonical form on L'. Then $c(\Sigma^S) = c(\Sigma)$ for all $\Sigma \in L'$ so that c factorizes through M' to define a continuous map $s: M' \to L'$ such that $c = s \circ \pi$. Because of (7.3) we have $\pi \circ c = \pi$ so that $\pi = \pi \circ s \circ \pi$. Because π is surjective it follows that $\pi \circ s = \mathrm{id}$, so that s is a continuous section of the (principal $GL_n(\mathbb{R})$) fibre bundle $\pi: L' \to M'$. Inversely let $s: M' \to L'$ be a continuous section of π. Then $s \circ \pi: L' \to L'$ is a continuous canonical form on L'.

7.6 (Non) existence of global canonical forms. In this section we shall prove theorem 1.17 which says that there exists a continuous canonical form on all of $L^{cr,co}_{m,n,p}(\mathbb{R})$ if and only if $m = 1$ or $p = 1$.

First suppose that $m = 1$. Then there is only one nice selection in $I(n, m)$, viz. $((0, 1), (1, 1), \ldots, (n-1, 1))$. We have already seen that there exists a continuous canonical form $c_\alpha : U_\alpha \to U_\alpha$ for all nice selections α. (cf. 3.10). This proves the theorem for $m = 1$. The case $p = 1$ is treated similarly (cf. 3.11). It remains to prove that there is no continuous canonical form on $L^{co,cr}_{m,n,p}(\mathbb{R})$ if $m \geqslant 2$ and $p \geqslant 2$. To do this we construct two families of linear dynamical systems as follows for all $a \in \mathbb{R}$, $b \in \mathbb{R}$ (We assume $n \geqslant 2$; if $n = 1$ the examples must be modified somewhat).

$$G_1(a) = \left(\begin{array}{cc|ccc} a & 1 & 0 & \ldots & 0 \\ 1 & 1 & 0 & \ldots & 0 \\ \hline 2 & 1 & & & \\ \vdots & \vdots & & B & \\ 2 & 1 & & & \end{array}\right) \qquad G_2(b) = \left(\begin{array}{cc|ccc} 1 & b & 0 & \ldots & 0 \\ 1 & 1 & 0 & \ldots & 0 \\ \hline 2 & 1 & & & \\ \vdots & \vdots & & B & \\ 2 & 1 & & & \end{array}\right),$$

where B is some (constant) $(n-2) \times (m-2)$ matrix with coefficients in $\mathbb{R}$

$$F_1(a) = \begin{pmatrix} 1 & 0 & \ldots & 0 \\ 0 & 2 & \ddots & \vdots \\ \vdots & \ddots & \ddots & 0 \\ 0 & \ldots & 0 & n \end{pmatrix} = F_2(b)$$

$$H_1(a) = \left(\begin{array}{cc|ccc} y_1(a) & 1 & 2 & \ldots & 2 \\ y_2(a) & 1 & 1 & \ldots & 1 \\ \hline 0 & 0 & & & \\ \vdots & \vdots & & C & \\ 0 & 0 & & & \end{array}\right) \qquad H_2(b) = \left(\begin{array}{cc|ccc} x_1(b) & 1 & 2 & \ldots & 2 \\ x_2(b) & 1 & 1 & \ldots & 1 \\ \hline 0 & 0 & & & \\ \vdots & \vdots & & C & \\ 0 & 0 & & & \end{array}\right)$$

where C is some (constant real $(p-2) \times (n-2)$ matrix. Here the continuous functions $y_1(a), y_2(a), x_1(b), x_2(b)$ are e.g. $y_1(a) = a$ for $|a| \leqslant 1$, $y_1(a) = a^{-1}$ for $|a| \geqslant 1$, $y_2(a) = \exp(-a^2)$, $x_1(b) = 1$ for $|b| \leqslant 1$, $x_1(b) = b^{-2}$ for $|b| \geqslant 1$, $x_2(b) = b^{-1}\exp(-b^{-2})$ for $b \neq 0$, $x_2(0) = 0$. The precise form of these functions is not important. What is important is that they are continuous, that $x_1(b) = b^{-1}y_1(b^{-1})$, $x_2(b) = b^{-1}y_2(b^{-1})$ for all $b \neq 0$ and that $y_2(a) \neq 0$ for all a and $x_1(b) \neq 0$ for all b.

For all $b \neq 0$ let $T(b)$ be the matrix

(7.7)
$$T(b) = \begin{pmatrix} b & 0 & \ldots & 0 \\ 0 & 1 & & \vdots \\ \vdots & & & 0 \\ 0 & \ldots & 0 & 1 \end{pmatrix}.$$

Let $\Sigma_1(a) = (F_1(a), G_1(a), H_1(a))$, $\Sigma_2(b) = (F_2(b), G_2(b), H_2(b))$. Then one easily checks that

(7.8) $ab = 1 \Rightarrow \Sigma_1(a)^{T(b)} = \Sigma_2(b)$.

Note also that $\Sigma_1(a), \Sigma_2(b) \in L^{co,cr}_{m,n,p}(\mathbb{R})$ for all $a, b \in \mathbb{R}$; in fact

(7.9) $\Sigma_1(a) \in U_\alpha, \alpha = ((0,2),(1,2),\dots,(n-1,2))$ for all $a \in \mathbb{R}$

(7.10) $\Sigma_2(b) \in U_\beta, \beta = ((0,1),(1,1),\dots,(n-1,1))$ for all $b \in \mathbb{R}$

which proves the complete reachability. The complete observability is seen similarly.

Now suppose that c is a continuous canonical form on $L^{co,cr}_{m,n,p}(\mathbb{R})$. Let $c(\Sigma_1(a)) = (\bar{F}_1(a), \bar{G}_1(a), \bar{H}_1(a))$, $c(\Sigma_2(b)) = (\bar{F}_2(b), \bar{G}_2(b), \bar{H}_2(b))$. Let $S(a)$ be such that $c(\Sigma_1(a)) = \Sigma_1(a)^{S(a)}$ and let $\bar{S}(b)$ be such that $c(\Sigma_2(b)) = \Sigma_2(b)^{\bar{S}(b)}$.
It follows from (7.9) and (7.10) that

$$\text{(7.11)}\quad \begin{aligned} S(a) &= R(\bar{F}_1(a), \bar{G}_1(a))_\alpha\, R(F_1(a), G_1(a))_\alpha^{-1} \\ \bar{S}(b) &= R(\bar{F}_2(b), \bar{G}_2(b))_\beta\, R(F_2(b), G_2(b))_\beta^{-1}\,. \end{aligned}$$

Consequently $S(a)$ and $\bar{S}(b)$ are (unique and are) continuous functions of a and b.
Now take $a = b = 1$. Then $ab = 1$ and $T(b) = I_n$ so that (cf (7.7), (7.8) and (7.11)) $S(1) = \bar{S}(1)$. It follows from this and the continuity of $S(a)$ and $\bar{S}(b)$ that we must have

(7.12) $\operatorname{sign}(\det S(a)) = \operatorname{sign}(\det \bar{S}(b))$ for all $a, b \in \mathbb{R}$.

Now take $a = b = -1$. Then $ab = 1$ and we have, using (7.8),

$$\begin{aligned} \Sigma_1(-1)^{(\bar{S}(-1)\,T(-1))} &= (\Sigma_1(-1)^{T(-1)})^{\bar{S}(-1)} \\ &= \Sigma_2(-1)^{\bar{S}(-1)} = c(\Sigma_2(-1)) \\ &= c(\Sigma_1(-1)) = \Sigma_1(-1)^{S(-1)}\,. \end{aligned}$$

It follows that $S(-1) = \bar{S}(-1)T(-1)$, and hence by (7.7), that

$$\det(S(-1)) = -\det(\bar{S}(-1))$$

which contradicts (7.12). This proves that there does not exists a continuous canonical form on $L^{co,cr}_{m,n,p}(\mathbb{R})$ if $m \geq 2$ and $p \geq 2$.

***7.13 Acknowledgement and remarks.** By choosing the matrices B and C in $G_1(a)$, $G_2(b)$, $H_1(a)$, $H_2(b)$ judiciously we can also ensure that $\operatorname{rank}(G_1(a) = m = \operatorname{rank} G_2(b)$ if $m < n$ and $\operatorname{rank} H_1(a) = p = \operatorname{rank} H_2(b)$ if $p < n$.
As we have seen in 7.5 above there exists a continuous canonical form on $L^{co,cr}_{m,n,p}(\mathbb{R})$ if and only if the prinicpal $GL_n(\mathbb{R})$ fibre bundle $\pi: L^{co,cr}_{m,n,p}(\mathbb{R}) \to \mathcal{M}^{co,cr}_{m,n,p}(\mathbb{R})$ admits a section. This, in turn is the case if and only if this bundle is trivial. The example on which the proof in 7.6 above is based is precisely the same example we used in [5] to prove that

the fibre bundle π is in fact nontrivial if $p \geq 2$ and $m \geq 2$, and from this point of view the example appears somewhat less "ad hoc" than in the present setting. The idea of using the example to prove nonexistence as done above is due to R. E. Kalman.

8 On the geometry of $M^{co,cr}_{m,n,p}$ (IR). Holes and (partial) compactifications

As we have seen in the introduction (cf. 1.19) the differentiable manifold $M^{co,cr}_{m,n,p}$(IR) is full of holes, a situation which is undesirable in certain situations. In this section we prove theorems 1.22 and 1.23 but, for the sake of simplicity, only in the case m = 1 and p = 1.[1])

8.1 An addendum to realization theory. Let $T(s) = d(s)^{-1}b(s)$ be a rational function, with degree $d(s) = n >$ degree $b(s)$. Then we know by 4.27 that there is a one input, one output system Σ with transfer function $T_\Sigma(s)$. We claim that we can see to it that $\dim(\Sigma) \leq n$. Indeed if

$$T_\Sigma(s) = a_0 s^{-1} + a_1 s^{-2} + a_2 s^{-3} + \ldots$$

then, if $d(s) = s^n - d_{n-1}s^{n-1} - d_1 s - d_0$, we have

$$a_{i+n} = d_0 a_i + d_1 a_{i+1} + \ldots + d_{n-1} a_{i+n-1}$$

for all $i \geq 0$. It follows that if $A = (a_0, a_1, a_2, \ldots)$, then rank $H_r(A) =$ rank $H_{n-1}(A)$ for all $r \geq n-1$. But $H_{n-1}(A)$ is an $n \times n$ matrix and hence rank $H_r(A) \leq n$ for all s, which by section 4 means that there is a realization of A (or T(s)) of dimension $\leq n$.
It follows that a cr and co system Σ of dimension n has a transfer function $T_\Sigma(s) = d(s)^{-1}b(s)$ with degree $(d(s)) = n$ and no common factors in d(s) and b(s), and inversely if $T(s) = d(s)^{-1}b(s)$, degree $b(s) < n =$ degree $(d(s))$, and b(s) and d(s) have no common factors, then all n-dimensional realizations of T(s) are co and cr.
Indeed if d(s) and b(s) have a common factor, then $T_\Sigma(s) = d'(s)^{-1}b'(s)$ with degree $(d'(s)) \leq n-1$ and it follows as above that rank $H_r(A) \leq n-1$ so that Σ is not cr and co. Inversely if Σ is not cr and co there is a Σ' of dimension $\leq n-1$ which also realizes A so that $T(s) = T_{\Sigma'}(s) = h'(sI - F')^{-1}g' = \det(sI - F')^{-1}B(s) = d'(s)^{-1}B(s)$ with degree $(d'(s)) \leq n-1$.

***8.2.** There is a more input, more output version of 8.1. But it is not perhaps the most obvious possibility. E.g. the lowest dimensional realization of $s^{-1}\binom{1\,2}{1\,1}$ has dimension 2. The right generalization is: Let $T(s) = D(s)^{-1}N(s)$, where D(s) and N(s) are as in the theorem mentioned in section 5.9. Then there is a co and cr realization of T(s) of dimension degree (det (D(s)).

---

1) Added in proof. For the analogous results in the multivariable case and a more careful, easier and more detailed treatment cf. M. Hazewinkel, "Families of systems: degeneration phenomena", Report 7918. Econometrie Inst., Erasmus Univ. Rotterdam.

8.3 Theorem: Let $D = a_0 + a_1 \frac{d}{dt} + \ldots + a_{n-1} \frac{d^{n-1}}{dt^{n-1}}$, $a_i \in \mathbb{R}$ be a differential operator of order $\leqslant n-1$. Then there exists a family of systems $(\Sigma_z)_z \subset L^{co,cr}_{1,n,1}(\mathbb{R})$ such that the $f(\Sigma_z)$ converge to D in the sense of definition 1.20.

To prove this theorem we need to do some exercises concerning differentiation, determinants and partial integration. They are

(8.4) Let $k \in \mathbb{Z}$, $k \geqslant -1$ and let $B_{n,k}$ be the $n \times n$ matrix with (i, j)-th entry equal to the binomial coefficient $\binom{i+j+k}{i+k+1}$. Then $\det(B_{n,k}) = 1$.

(8.5) Let $u^{(i)}(t) = \dfrac{d^i u(t)}{dt^i}$. Then $\displaystyle\int_0^t z^n e^{-z(t-\tau)} u(\tau) d\tau =$

$$= z^{n-1} u(t) + \ldots + (-1)^{n-1} u^{(n-1)}(t) + 0(z^{-1})$$

if $\operatorname{supp}(u) \subset (0, \infty)$, where 0 is the Landau symbol.

(8.6) Let $\phi(\tau) = (t-\tau)^m u(\tau)$, $\phi^{(i)}(\tau) = \dfrac{d^i \phi(\tau)}{d\tau^i}$. Then $\phi^{(i)}(t) = 0$ for $i < m$ and

$$\phi^{(i)}(t) = (-1)^m i(i-1) \ldots (i-m+1) u^{(i-m)}(t) \text{ if } i \geqslant m.$$

And finally, combining (8.5) and (8.6),

$$(8.7) \quad \int_0^t e^{-z(t-\tau)} z^n (t-\tau)^m u(\tau) d\tau = (-1)^m m! \sum_{i=m+1}^{n} (-1)^{i+1} z^{n-i} \binom{i-1}{m} u^{(i-1-m)}(t) + 0(z^{-1}).$$

8.8 Proof of theorem 8.3: We consider the following family of n dimensional systems (with one output and one input),

$$g_z = \begin{pmatrix} 0 \\ \vdots \\ 0 \\ z^m \end{pmatrix}, \quad F_z = \begin{pmatrix} -z & z & 0 & \ldots & 0 \\ 0 & -z & \ddots & & \vdots \\ \vdots & \ddots & \ddots & \ddots & 0 \\ \vdots & & \ddots & & z \\ 0 & \ldots & 0 & & -z \end{pmatrix}, h_z = (0, \ldots, 0, x_m, \ldots, x_1)$$

where the $x_1, \ldots, x_m$, $m \leqslant n$, are some still to be determined real numbers. One calculates

$$e^{sF_z} = \begin{pmatrix} 1 & sz & \frac{s^2 z^2}{2!} & \ldots & \frac{(sz)^{n-1}}{(n-1)!} \\ 0 & 1 & \ddots & & \vdots \\ \vdots & & \ddots & \ddots & \frac{s^2 z^2}{2!} \\ \vdots & & & 1 & sz \\ 0 & \ldots & & 0 & 1 \end{pmatrix}$$

Hence

$$h_z e^{(t-\tau)F_z} g_z = \sum_{i=1}^{m} x_i z^{m+i}(i!)^{-1}(t-\tau)^i e^{-z(t-\tau)}$$

and, using (8.7),

$$\int_0^t h_z e^{(t-\tau)F_z} g_z u(\tau)\,d\tau = \sum_{i=1}^{m} (i!)^{-1} x_i \sum_{j=i+1}^{m+i} (-1)^i (i!)(-1)^{j+1} \binom{j-1}{i} z^{m+i-j}$$

$$u^{(j-i-1)}(t) + O(z^{-1})$$

$$= \sum_{l=0}^{m-1} (-1)^{m-l+1} z^l \left(\sum_{i=1}^{m} x_i \binom{m+i-l-1}{i} u^{(m-l-1)}(t) + O(z^{-1}) \right.$$

Now, by (8.4) we know that $\det\left(\left(\binom{m+i-l-1}{i}\right)_{i,l}\right) = 1$, so that we can choose $x_1, \ldots, x_m$ in such a way that

$$\int_0^t h_z e^{(t-\tau)F_z} g_z u(\tau)\,d\tau = a_{m-1} u^{(m-1)}(t) + O(z^{-1})$$

where a_{m-1} is any pregiven real number.

It follows that $\lim_{z\to\infty} f(\Sigma_z) = a_{m-1} \frac{d^{m-1}}{dt^{m-1}}$

Let $\Sigma_z(i) = (F_z(i), g_z(i), h_z(i)), i = 0, \ldots, n-1$ be systems constructed as above with limiting input/output operator equal to $a_i \frac{d^i}{dt^i}$. Now consider the n^2-dimensional systems $\hat{\Sigma}_z$ defined by

$$\hat{F}_z = \begin{pmatrix} F_z(0) & 0 & \cdots & 0 \\ 0 & \ddots & & \vdots \\ \vdots & & \ddots & 0 \\ 0 & \cdots & 0 & F_z(n-1) \end{pmatrix}, \hat{g}_z = \begin{pmatrix} g_z(0) \\ \vdots \\ g_z(n-1) \end{pmatrix}, \hat{h}_z = (h_z(0), \ldots, h_z(n-1)).$$

Then clearly $\lim_{z\to\infty} f(\hat{\Sigma}_z) = D$. Let $T_z^{(i)}(s)$ be the transfer function of $\Sigma_z(i)$. Then for certain polynomials $B_z^{(i)}(s)$ we have

(8.9) $T_z^{(i)}(s) = d_z(s)^{-1} B_z^{(i)}(s)$, with $d_z(s)$ *independent of i*

The transfer function of $\hat{\Sigma}_z$ is clearly equal to

(8.10) $T_z(s) = \sum_{i=0}^{n-1} T_z^{(i)}(s) = d_z(s)^{-1} B_z(s),\ B_z(s) = \sum_{i=0}^{n-1} B_z^{(i)}(s)$

By 8.1 it follows from (8.10) that $T_z(s)$ can also be realized by an n-dimensional system, Σ'_z. Then also $\lim_{z\to\infty} f(\Sigma'_z) = D$. Finally we can change Σ'_z slightly to Σ_z for all z to find a family $(\Sigma_z)_z \subset L^{co,cr}_{1,n,1}$ (IR) such that $\lim_{z\to\infty} f(\Sigma_z) = D$. This proves the theorem.

8.11 Corollary: Let Σ' be a system of dimension i and let D be a differential operator of order $n-i-1$ (where order $(0) = -1$). Then there exists a family $(\Sigma_z)_z \subset L^{co,cr}_{1,n,1}$ (IR) such that $\lim_{z\to\infty} f(\Sigma_z) = D + f(\Sigma')$

Proof: Let $\Sigma''_z = (F''_z, g''_z, h''_z)$ be a family in $L_{1,n-i,1}$ (IR) such that $\lim_{z\to\infty} f(\Sigma''_z) = D$. Let $\Sigma' = (F', g', h')$. Let $\hat{\Sigma}_z$ be the n-dimensional system defined by the triple of matrices

$$\hat{F}_z = \begin{pmatrix} F''_z & 0 \\ 0 & F' \end{pmatrix}, \quad g_z = \begin{pmatrix} g''_z \\ g' \end{pmatrix}, \quad h_z = (h''_z, h') .$$

Then $\lim_{z\to\infty} f(\hat{\Sigma}_z) = D + f(\Sigma')$. Now perturb $\hat{\Sigma}_z$ slightly for each z to Σ_z, to find a completely reachable and completely observable family $(\Sigma_z)_z$ such that $\lim_{z\to\infty} f(\Sigma_z) = D + f(\Sigma')$.

8.12 Theorem: Let $(\Sigma_z)_z \subset L_{1,n,1}$ (IR) be a family of systems which converges in input-output behaviour in the sense of definition 1.20. Then there exist a system Σ' and a differential operator D such that $\dim(\Sigma') + \mathrm{ord}(D) \leqslant n-1$ and $\lim_{z\to\infty} f(\Sigma'_z) = f(\Sigma') + D$

Proof: Consider the relation

$$y_z(t) = f(\Sigma_z)u(t)$$

for smooth input functions u(t). Let $\hat{u}(s)$ and $\hat{y}_z(s)$ be the Laplace transforms of u(t) and $y_z(t)$. Then we have

$$\hat{y}_z(s) = T_z(s)\hat{u}(s) ,$$

where $T_z(s)$ is the transferfunction of Σ_z. Because the $f(\Sigma_z)$ converge as $z \to \infty$ (in the sense of definition 1.20), and because the Laplace transform is continuous, it follows that there is a rational function $T(s) = d(s)^{-1} b(s)$ with degree $d(s) \leqslant n$, degree $b(s) \leqslant n-1$ such that

$$\lim_{z\to\infty} T_z(s) = T(s)$$

pointwise in s for all but finitely many s. Write

$$T(s) = e_0 + e_1 s + \ldots + e_{n-i-1} s^{n-i-1} + \frac{b'(s)}{d'(s)}$$

with degree $d'(s) = i$, degree $(b'(s)) < i$. Let Σ' be a system of dimension $\leqslant i$ with transfer function equal to $d'(s)^{-1}b'(s)$ and let D be the differential operator $e_0 + e_1 \frac{d}{dt} + \ldots + e_{n-i-1} \frac{d^{n-i-1}}{dt^{n-i-1}}$. The Laplace transform of the relation

$$y(t) = f(\Sigma')u(t) + Du(t)$$

for smooth input functions $u(t)$, is

$$\hat{y}(s) = T(s)\hat{u}(s)\,.$$

Because the Laplace transform is injective (on smooth functions) it follows that

$$\lim_{z\to\infty} f(\Sigma_z) = f(\Sigma') + D\,.$$

***8.13 Remarks on compactification, desingularization, symmetry breaking, etc.** There are more input, more output versions of theorems 8.3 and 8.12. To prove them it is more convenient to use another technique which is based on a continuity property of the inverse Laplace transform for certain sequences of functions. (The inverse Laplace transform is certainly not continuous in general; also it is perfectly possible to have a sequence of systems Σ_z such that their transfer functions $T_z(s)$ converge for $z \to \infty$, but such that the $f(\Sigma_z)$ do not converge, e.g. $T_z(s) = z(z-s)^{-1}$).

Let Σ be a co and cr system of dimension n with one input and one output. Let $T(s)$

$$T(s) = \frac{b_{n-1}s^{n-1} + \ldots + b_1 s + b_0}{s^n + d_{n-1}s^{n-1} + \ldots + d_1 s + d_0} = \frac{b(s)}{d(s)}$$

be the transfer function of Σ. Assign to $T(s)$ the point

$$(b_0 : \ldots : b_{n-1} : d_0 : \ldots : d_{n-1} : 1) \in \mathbb{P}^{2n}(\mathbb{R})\,,$$

real projective space of dimension 2n. This defines an embedding of $M_{1,n,1}^{co,cr}(\mathbb{R})$ into $\mathbb{P}^{2n}(\mathbb{R})$. The image is obviously dense so that $\mathbb{P}^{2n}(\mathbb{R})$ is a smooth compactification of $M_{1,n,1}^{co,cr}(\mathbb{R})$.

Let $\bar{M}_{1,n,1}(\mathbb{R})$ be the subspace of $\mathbb{P}^{2n}(\mathbb{R})$ consisting of those points $(x_0 : \ldots : x_{n-1} : y_0 : y_1 : \ldots : y_n) \in \mathbb{P}^{2n}(\mathbb{R})$ for which at least one $y_i, i = 0, \ldots, n$ is different from zero. For these points

$$\frac{x_0 + x_1 s + \ldots + x_{n-1}s^{n-1}}{y_0 + y_1 s + \ldots + y_n s^n}$$

has meaning and this rational function is then the transfer function of a generalized linear dynamical system:

$$\text{(8.14)} \qquad \begin{aligned} \dot{x} &= Fx + Gu \\ y &= Hx + Du \end{aligned}$$

where D is a differential operator. (The points in $\mathbb{P}^{2n}(\mathbb{R}) \setminus \overline{M}_{1,n,1}$ correspond to "systems" which tend to give infinite outputs for finite inputs; they are interpretable, however, in terms of correspondences $y(t) \mapsto u(t)$).

Further let $\hat{M}_{1,n,1}$, consist of those $(x_0 : \ldots : x_{n-1} : y_0 : \ldots : y_n)$ for which if $y_i = 0$ for $i > r$, then also $x_{i-1} = 0$ for $i \geqslant r$. For these points the D in (8.14) is zero and these points thus yield transfer functions of systems of dimension $\leqslant n$. (But many points in $\hat{M}_{1,n,1}$ have the same transfer functions). Assigning to a point in $\hat{M}_{1,n,1}$ the first $2n+1$ coefficients of

$$\frac{x_0 + x_1 s + \ldots + x_{n-1} s^{n-1}}{y_0 + y_1 s + \ldots + y_n s^n} = a_0 s^{-1} + a_1 s^{-2} + a_2 s^{-3} + \ldots$$

we find the following situation

$$\begin{array}{ccc} M^{co,cr}_{1,n,1} & \subset & \hat{M}_{1,n,1} \\ \downarrow H & & \downarrow \hat{H} \\ \mathbb{R}^{2n+1} & = & \mathbb{R}^{2n+1} \end{array} .$$

Here H is an embedding and its image is the subspace of all sequences $A = (a_0, \ldots, a_{2n})$ such that $\operatorname{rank} H_{n-1}(A) = \operatorname{rank} H_n(A) = n$. The image of $\hat{H}$ is the space of all sequences A such that $\operatorname{rank} H_n(A) = \operatorname{rank} H_{i-1}(A) = i$ for some $i \leqslant n$. This is a singular submanifold of $\mathbb{R}^{2n+1}$ and $\hat{H}$ is a resolution of singularities.

The points of $(\hat{M}_{1,n,1} \setminus M^{co,cr}_{1,n,1})$ correspond to transfer functions of lower dimensional co and cr systems. If a sequence $x_z \in M^{co,cr}_{1,n,1}$ converges to such a point, the internal symmetry group $GL_n(\mathbb{R})$ of x_z suddenly contracts to some $GL_m(\mathbb{R}) \subset GL_n(\mathbb{R})$ with $m < n$.

References

[1] *Casti, J. L.*, Dynamical Systems and their Applications: Linear Theory, Acad. Pr., 1977.

[2] *Denham, M. J.*, Canonical Forms for the Identification of Multivariable Linear Systems, IEEE Trans. Automatic Control 19, 6 (1974), 646–656.

[3] *Hazewinkel, M., Kalman, R. E.*, Moduli and Canonical Forms for Linear Dynamical Systems, Report 7504, Econometric Institute, Erasmus Univ. Rotterdam, 1975.

[4] *Hazewinkel, M., Kalman, R. E.*, On Invariants, Canonical Forms and Moduli for Linear, Constant, Finite Dimensional, Dynamical Systems. In: Proc. CNR-CISM Symp. Algebraic System Theory (Udine 1975). Lect. Notes Economics and Math Systems **131** (1976), Springer, 48–60.

[5] *Hazewinkel, M.*, Moduli and Canonical Forms for Linear Dynamical Systems II: The Topological Case, J. Math. System Theory **10** (1977), 363–385.

[6] *Hazewinkel, M.*, Moduli and Canonical Forms of Linear Synamical Systems III: The Algebraic-Geometric Case. In: C. Martin, R. Hermann (eds), The 1976 AMES Research Centre (NASA) Conf. on Geometric Control, Math. Sci. Press, 1977, 291–360.

[7] *Hazewinkel, M.*, Degenerating Families of Linear Dynamical Systems I: Proc. 1977 IEEE Conf. on Decision and Control (New Orleans), 258–264.

[8] *Hazewinkel, M.*, Invariants, Moduli and Canonical Forms for Linear Time-varying Dynamical Systems, Ricerche di Automatica **9** (1979), to appear.

[9] *Kalman, R. E.*, Lectures on Controllability and Observability. In: G. Evangilisti (ed), Controllability and Observability (CIME, 1968), Edizione Cremonese, 1969, 1–151.

[10] *Kalman, R. E., Falb, P. L., Arbib, M. A.*, Topics in Mathematical Systems Theory, McGraw-Hill, 1969.

[11] *Kalman, R. E.*, Kronecker Invariants and Feedback. In: L. Weiss (ed), Ordinary Differential Equations (1971, NRL-MRC conference), Acad. Press, 1972, 459–471.

[12] *Kar-Keung, Young, D., Kokotovic, P. V., Utkin, V. I.*, A Singular Perturbation Analysis of high-gain Feedback Systems. IEEE Trans. Automatic Control **22**, 6 (1977), 931–938.

[13] *Martin, C., Hermann, R.*, Applications of Algebraic Geometry to Systems Theory: The McMillan degree and Kronecker indices of transfer functions as Topological and Holomorphic System Invariants, SIAM J. Control and Opt. **16** (1978), 743–755.

[14] *Rosenbrock, H. H.*, State space and Multivariable Theory, Nelson, 1970.

[15] *Willems, J. C., Glover, K.*, Parametrizations of Linear Dynamical Systems: Canonical Forms and Identifiability, IEEE Trans. Automatic Control **19**, 6 (1974), 640–646.

[16] *Wilkinson, J. H., Golub, G. H.*, Ill-conditioned Eigensystems and Computation of the Jordan Canonical Form, SIAM Review **18**, 4 (1976), 578–619.

[17] *Wolowich, W. A.*, Linear Multivariable Systems, Springer, 1974.

[18] *Wonham, W. Murray*, Linear Multivariable Control. A geometric Approach, Lect. Notes Economics and Math. Systems **101**, Springer, 1974.

Subject index